Lecture Notes in Biomathematics

Managing Editor: S. Levin

52

Population Biology

Proceedings of the International Conference
held at the University of Alberta,
Edmonton, Canada, June 22 – 30, 1982

Edited by H. I. Freedman and C. Strobeck

Springer-Verlag
Berlin Heidelberg New York Tokyo 1983

Editors

Herbert I. Freedman
Department of Mathematics, University of Alberta
Edmonton, Alberta T6G 2G1 Canada

Curtis Strobeck
Department of Genetics, University of Alberta
Edmonton, Alberta T6G 2E1 Canada

ISBN 978-3-540-12677-5   ISBN 978-3-642-87893-0 (eBook)
DOI 10.1007/978-3-642-87893-0

PREFACE

This volume contains the Proceedings of the International Conference in Population Biology held at The University of Alberta, Edmonton, Canada from June 22 to June 30, 1982. The Conference was sponsored by The University of Alberta and The Canadian Applied Mathematics Society, and overlapped with the summer meeting of CAMS.

The main objectives of this Conference were: to bring mathematicians and biologists together so that they may interact for their mutual benefit; to bring those researchers interested in modelling in ecology and those interested in modelling in genetics together; to bring in keynote speakers in the delineated areas; to have sessions of contributed papers; and to present the opportunity for researchers to conduct workshops. With the exception of the last one, the objectives were carried out.

In order to lend some focus to the Conference, the following themes were adopted: models of species growth, predator-prey, competition, mutualism, food webs, dispersion, age structure, stability, evolution of ecological parameters, evolution of behaviour, life history strategies, group and social selection, and evolution of genetic systems. There were speakers (invited and/or contributed papers) in each of these areas.

Talks were given on Tuesday, June 22 to Friday, June 25 and on Monday, June 28 to Wednesday, June 30. On each day there were several talks by the principal speakers as well as contributed sessions. Altogether, there were ninety-one papers given, of which twelve were by the principal speakers. There were one hundred and twenty-three registered participants from twelve different countries.

There were a number of social activities, including a barbecue, two wine and cheese affairs, a day trip to Jasper and a family picnic at Fort Edmonton.

Financial support for the Conference was generous, and we would like to gratefully acknowledge the Natural Sciences and Engineering Research Council of Canada, the President's Fund, the Conference Fund, and the Mathematics Department of The University of Alberta for supporting the Conference.

There were many whose time and effort helped make the Conference a success. In particular, we wish to thank the other members of the steering committee, John Addicott, Peter Antonelli, Geoff Butler, and Ken Morgan, for their moral, financial and physical support. Last, but not least, we are eternally indebted to our secretary, Vivian Spak, whose diligence, organizational ability, and perseverance made the Conference and these Proceedings possible.

H.I. Freedman

C. Strobeck

May, 1983

LIST OF PARTICIPANTS

INTERNATIONAL CONFERENCE ON POPULATION BIOLOGY

ADACHI, N., Department of Applied Mathematics and Physics, Faculty of Engineering,
     Kyoto University, Kyoto 606, Japan.

ADDICOTT, John, Department of Zoology, University of Alberta, Edmonton, Alberta,
     T6G 2E1, Canada.

AGUR, Zvia, Service de Chimie Physique II, Université Libre de Bruxelles, Belgium.
     (Permanent address: Department of Applied Mathematics, The Weizmann Institute
     of Science, Rehovot 76100, Israel.)

BAGGS, Ivan, Department of Mathematics, University of Alberta, Edmonton, Alberta,
     T6G 2E1, Canada.

BALTZIS, Basil C., Department of Chemical Engineering and Materials Science,
     151 Amundson Hall, University of Minnesota, Minneapolis, Minnesota   55355,
     U.S.A.

BARCLAY, Hugh, Pacific Forest Research Centre, 506 W. Burnside Road, Victoria,
     British Columbia, V8Z 1M5, Canada.

BECK, Karen, Department of Mathematics, University of Utah, Salt Lake City, Utah,
     84112, U.S.A. (Permanent address: Dept. of Math., U. of Texas at Arlington.)

BHARGAVA, S.C., St. Stephen's College, University of Delhi, Delhi, 110 007, India.

BLYTHE, S.P., Department of Applied Physics, University of Strathclyde, Glasgow,
     Scotland.

BOJADZIEV, George, Department of Mathematics, Simon Fraser University, Burnaby,
     British Columbia, V5A 1S6, Canada.

BOTSFORD, Louis W., Department of Wildlife and Fisheries Biology, University of
     California at Davis, Davis, California   95616, U.S.A.

BOYLE, Phelin P., Faculty of Arts, Accounting Group, University of Waterloo,
     Waterloo, Ontario   N2L 3G1, Canada.

BRANER, Moshe, Ecology and Systematics, Cornell University, Ithaca, New York
     14850, U.S.A.

BUTLER, Geoff, Department of Mathematics, University of Alberta, Edmonton, Alberta
     T6G 2E1, Canada.

CABAY, S. Computing Science Department, University of Alberta, Edmonton, Alberta
     T6G 2E1, Canada.

CHARLES, Tony, 207 - 2115 Cypress Street, Vancouver, British Columbia, Canada.

CHELIAK, Bill, PNKI, Chalk River, Ontario   K0J 1J0, Canada.

CHESSON, Jean, Battelle Laboratories, 505 King Avenue, Columbus, Ohio   43212,
     U.S.A.

CHESSON, Peter L., Department of Zoology, Ohio State University, 1735 Neil Avenue,
     Columbus, Ohio   43210, U.S.A.

CLARK, Charles E., Department of Mathematics, University of Tennessee, Knoxville, Tennessee 37996-1300, U.S.A.

CLARK, Colin W., Department of Mathematics, University of British Columbia, Vancouver, British Columbia V6T 1Y4, U.S.A.

COHEN, Dan, Department of Biological Sciences, Stanford University, Stanford, California 94305, U.S.A.

COSTE, Jean, Laboratoire de Physique de la Matière Condensée, 06034 Nice Cedex, France.

CRESSMAN, Ross, Department of Mathematics and Statistics, University of Guelph, Guelph, Ontario.

CUSHING, Jim, Mathematics Department and Program in Applied Mathematics, Building #89, University of Arizona, Tucson, Arizona 85721, U.S.A.

DALE, M., Department of Botany, University of Alberta, Edmonton, Alberta T6G 2E1, Canada.

DASH, A.T., University of Guelph, Guelph, Ontario.

DE LA TORRE, Deborah, c/o M. Witten, Department of Mathematics, University of California, Santa Barbara, California 93106, U.S.A.

DERRICK, William R., Department of Mathematical Sciences, University of Montana, Missoula, Montana 59812, U.S.A.

DOEDEL, Eusebius, Department of Computer Science, Concordia University, 1455 de Maisonneuve Blvd. West, Montreal, Quebec H3G 1M8, Canada.

DUTT, Ranabir, Department of Physics, Visva-Bharati University, Santiniketan 731235, West Bengal, India.

EVANS, Geoffrey T., Northwest Atlantic Fisheries Centre, Box 5667, St. John's, Newfoundland Alc 5X1, Canada.

FINDLAY, C.S., Department of Biology, Queen's University, Kingston, Ontario K7L 3N6.

FREEDMAN, Herbert I., Department of Mathematics, University of Alberta, Edmonton, Alberta T6G 2E1, Canada.

GALPERIN, E.A., Department de Mathematique, Univ. de Quebec a Montreal, C.P. 8888, Succ "A", Montreal, Quebec H3C 3P8.

GARD, Thomas C., Department of Mathematics, University of Georgia, Athens, Georgia 30602, U.S.A.

GERAMITA, Joan M., Department of Mathematics, Queen's University, Kingston, Ontario K76 3N6, Canada.

GHISA, Dorin, Department de Physique - Mathematiques, Universite de Moncton, Moncton, New Brunswick E1A 3E9, Canada.

GHOSH, P.K., Department of Physics, Visva-Bharati University, Santiniketan 731235, West Bengal, India.

GILLIAM, James F., Kellogg Biological Station, Michigan State University, Hickory Corners, Michigan 49060, U.S.A. (Permanent address: Dept. of Biological Sciences, SUNY-Albany, Albany, New York 12222, U.S.A.)

GLASSER, John W., Department of Zoology, University of Georgia, Athens, Georgia 30602, U.S.A.

GOH, Bean San, Department of Mathematics, University of Western Australia, Nedlands, W.A., Australia.

GOPALSAMY, K., School of Mathematics, Flinders University, Bedford Park, S.A., 5042, Australia.

GOULDING, Brian, Genetics Department, University of Alberta, Edmonton, Alberta T6G 2E1, Canada.

GRANERO-PORATI, Maria Ilde, Institute of Physics, Section of Biophysics, GNCB-CNR, University of Parma, 43100 Parma, Italy.

GREEN, Richard F., Department of Mathematical Sciences, University of Minnesota, Duluth, Minnesota 55812, U.S.A.

GREGORIUS, Hans-Rolf, Lehrstuhl fur Forstgenetik, Univ. Göttingen, Busgenweg 2, 34 Göttingen, West Germany.

GURNEY, W.S.C., Department of Applied Physics, University of Strathclyde, Glasgow, Scotland.

HAEFNER, James W., Institute of Marine Sciences, University of North Carolina, 3407 Arendall Street, Morehead City, N.C. 28557, U.S.A.

HANSON, Floyd, Department of Mathematics, Statistics and Computer Sciences, University of Illinois at Chicago Circle, Box 4348, Chicago, Illinois 60680, U.S.A.

HARMSEN, Rudolf, Biology Department, Queen's University, Kingston, Ontario K7L 3N6, Canada.

HARRISON, Gary W., Department of Mathematics, University of Georgia, Athens, Georgia 30602, U.S.A.

HASSELL, Michael H., Department of Zoology and Applied Entomology, Imperial College, London, A.W. 7, England.

HASTINGS, Harold M., Department of Mathematics, Ecosystems Modelling Group, Hofstra University, Hempstead, New York 11550, U.S.A.

HESTBECK, Jay, Department of Wildlife and Fisheries Biology, University of California at Davis, Davis, California 95616, U.S.A.

HILBORN, Ray, Institute of Animal Resource Ecology, University of British Columbia, 2204 Main Mall, Hut B-8, Vancouver, British Columbia V6T 1W5, Canada.

HINES, W.G., Department of Mathematics and Statistics, University of Guelph, Guelph, Ontario N1G 2W1, Canada.

HOCHBERG, Kenneth J., Department of Mathematics and Statistics, Case Western Reserve University, Cleveland, Ohio 44106, U.S.A.

HOGAN, John, Department of Linguistics, University of Alberta, Edmonton, Alberta T6G 2E1, Canada.

HOLMES, John, Department of Zoology, University of Alberta, Edmonton, Alberta T6G 2E1, Canada.

HOLT, Robert D., University of Kansas, Museum of Natural History, Lawrence, Kansas 66045, U.S.A.

HUNT, Fern, Laboratory of Theoretical Biology, Building 10, Room 4B56, National Institute of Health, Bethseda, Maryland 20205, U.S.A. (Permanent address: Department of Mathematics, Howard University, Washington, D.C.)

ISAC, G. Départment de Mathématiques, Collège Militaire Royal, St. Jean, Québec J0J 1R0, Canada.

IWASA, Yoh, Department of Biological Sciences, Stanford University, Stanford, California 94305, U.S.A.

KAVUNCU, Orhan, Department of Genetics, University of Alberta, Edmonton, Alberta T6G 2E1, Canada.

KEYFITZ, Nathan, Department of Sociology, 300 Administration Building, 190 North Oval Mall, Columbus, Ohio 43210, U.S.A. (Address after Sept. 1983: Rosenstadt Professor, Department of Preventive Medicine and Biostatistics, Faculty of Medicine, University of Toronto, Toronto, Ontario M5S 1A8, Canada.)

KITT, Ron, Imperial Oil, Edmonton, Alberta, Canada.

KLONTZ, Calvin, #1, 10725 - 79 Avenue, Edmonton, Alberta, Canada.

KNOP, Larry E., Department of Mathematics, Hamilton College, Clinton, New York 13323, U.S.A.

KRISHNAN, P., Department of Sociology, University of Alberta, Edmonton, Alberta T6G 2E1, Canada.

LAL, Mohan, Department of Mathematics and Statistics, Memorial University, St. John's, Newfoundland, Canada.

LALLI, B.S., Department of Mathematics, University of Saskatchewan, Saskatoon, Saskatchewan.

LANGLAIS, Michel R., Department of Mathematics, Purdue University, West Lafayett, Indiana 47907, U.S.A. (Permanent address: Mathématiques et Informatique, Universite de Bordeaux L, 33405 Talence Cedex, France.)

LEEMING, David J., Department of Mathematics, University of Connecticut, Storrs, Connecticut, U.S.A. (Permanent address: Department of Mathematics, University of Victoria, Victoria, British Columbia, Canada.)

LEON, J.A., Gray Herbarium, Harvard University, 22 Divinity Avenue, Cambridge, Mass. 02138, U.S.A. (Permanent address: Instituto de Zoologia Tropical, Facultad de Ciencia, Central University of Venezuela, Apartado 47058, Caracas 1041-A, Venezuela.)

LEUNG, Anthony, Department of Mathematical Science, University of Cincinnati, Cincinnati, Ohio 45221, U.S.A.

LEVIN, Simon, Department of Ecology and Systematics, Langmuir Lab., Cornell University, Ithaca, New York 14850, U.S.A.

LEVINE, Dan, Mathematics Department, University of Houston, Central Campus, Houston, Texas 77004, U.S.A.

LUDWIG, Donald, Institute of Animal Resource Ecology, 2075 Wesbrook Mall, Univ. of British Columbia, Vancouver, British Columbia V6T 1W5, Canada.

MADDISON, David, Department of Entomology, University of Alberta, Edmonton, Alberta T6G 2E3, Canada.

McKELVEY, Robert, Department of Mathematics, University of Montana, Missoula, Montana 59801, U.S.A.

MELESHKO, Ron, Mathematics Department, University of Alberta, Edmonton, Alberta, T6G 2E1.

MERRILL, Steve, Department of Mathematics, Statistics and Computing Science, Marquette University, William Wehr Physics Building, Milwaukee, Wisconsin 53233, U.S.A.

MOODY, Michael, Mathematics Department, Washington State University, Pullman, Washington 99164, U.S.A.

MORGAN, Ken, Department of Genetics, University of Alberta, Edmonton, Alberta T6G 2E1, Canada.

NUNNEY, Len, Department of Biology, University of California at Riverside, Riverside, California 92521, U.S.A.

PORATI, Alfredo, Institute of Physics, Section of Biophysics, University of Parma, 43100 Parma, Italy.

POUNDER, J.R., Department of Mathematics, University of Alberta, Edmonton, Alberta T6G 2E1, Canada.

PUGLIESE, Andrea, Department of Ecology and Evolution, State University of New York, Stony Brook, New York 11794, U.S.A.

RAI, B., Department of Mathematics, University of Allahabad, Allahabad, India.

RAO, V.S.H., Department of Mathematics, University of Alberta, Edmonton, Alberta T6G 2E1, Canada. (Permanent address: Osmania University, Hyderabad 500007, India.)

RASMUSSEN, J.B., Department of Biology, University of Calgary, Calgary, Alberta, T2N 1N4, Canada.

REED, William J., Department of Mathematics, University of Victoria, P.O. Box 1700, Victoria, British Columbia V8W 2Y2, Canada.

ROSE, M.L., Department of Biology, Dalhousie University, Halifax, Nova Scotia B3H 4J1, Canada.

ROUGHGARDEN, Jonathan, Department of Biological Sciences, Stanford University, Stanford, California 94305, U.S.A.

ROUTLEDGE, Rick, Department of Mathematics, Simon Fraser University, Burnaby, British Columbia V5A 1S6, Canada.

SALEEM, Mohammed, Department of Mathematics, Building 89, University of Arizona, Tucson, Arizona 85721, U.S.A.

SARASWATI, Dipak K., Athabasca University, 14515 - 122 Avenue, Edmonton, Alberta T5L 2W4, Canada.

SEYMUR, Brian, Department of Mathematics, University of British Columbia, Vancouver, British Columbia, Canada.

SHONKWILER, Ron, School of Mathematics, Georgia Institute of Technology, Atlanta,
    Georgia 20332, U.S.A.

SILCOX, Susan, Department of Biology, Carleton University, Ottawa, Ontario, Canada.

SILVERT, William, Marine Ecology Lab., Bedford Institute of Oceanography, P.O. Box
    1006, Dartmouth, Nova Scotia, B2Y 4A2, Canada.

SLATKIN, Montgomery, Department of Zoology, NJ-15, University of Washington,
    Seattle, Washington 98195, U.S.A.

STROBECK, Curtis, Department of Genetics, University of Alberta, Edmonton, Alberta
    T6G 2E1, Canada.

TAKEUCHI, Yasuhiro, Department of Applied Mathematics, Faculty of Engineering,
    Shizuoka University, Hamamatsu 432, Japan.

TAYLOR, P.D., Department of Mathematics, Queen's University, Kingston, Ontario
    K7L 3N6, Canada.

TEMPLETON, Alan, Department of Biology, Washington University, St. Louis, Missouri
    63130, U.S.A.

THIERRIN, G., Department of Mathematics, University of Western Ontario, London,
    Ontario M6H 5B7, Canada.

TURELLI, Michael, Department of Genetics, University of California at Davis, Davis,
    California, U.S.A.

VAN DEN DRIESSCHE, Pauline, Department of Mathematics, University of Victoria,
    Victoria, British Columbia V8W 2Y2, Canada.

VANCE, R.R., Department of Biology, University of California at Los Angeles,
    Los Angeles, California 90024, U.S.A.

VINCENT, Thomas L., Aerospace and Mechanical Engineering, University of Arizona,
    Tucson, Arizona 85721, U.S.A.

VOORHEES, Burt, Department of Mathematics, University of Alberta, Edmonton, Alberta
    T6G 2E1, Canada.

WALTER, Gilbert, Department of Mathematics, University of Wisconsin at Milwaukee,
    Milwaukee, Wisconsin 53201, U.S.A.

WALTMAN, Paul, Department of Mathematics, Emory University, Atlanta, Georgia, U.S.A.

WAN, Fred, Institute of Applied Mathematics and Statistics, University of British
    Columbia, Vancouver, British Columbia V6T 1Y4, Canada.

WHITTAKER, James V., Department of Mathematics, University of British Columbia,
    Vancouver, British Columbia V6T 1W5, Canada.

WILCOX, David, Biology Department, Eastern College, St. Davids, Pennsylvania 19087,
    U.S.A.

WOLKOWICZ, Gail, Department of Mathematics, University of Alberta, Edmonton, Alberta
    T6G 2E1, Canada.

WOLLKIND, David J., Department of Pure and Applied Mathematics, Washington State
    University, Pullman, Washington 99164, U.S.A.

WORMS, Jean, Marine Biological Research Centre, Université de Moncton, Moncton, New Brunswick E1A 3E9, Canada.

WÖRZ-BUSEKROS, Angelika, Universitat Tubingen, Institut fur Biomathematik, Lehrstuhl fur Biomathematik, D 7400 Tubingen 1, Auf der Morgenstelle 28, West Germany.

LIST OF CONTRIBUTORS TO THE

NATIONAL CONFERENCE ON POPULATION BIOLOGY - PROCEEDINGS

ADACHI, N., Department of Applied Mathematics and Physics, Faculty of Engineering, Kyoto University, Kyoto 606, Japan.

ADDICOTT, John, Department of Zoology, University of Alberta, Edmonton, Alberta T6G 2E1, Canada.

AGUR, Zvia, Sevice de Chimie Physique II, Université Libre de Bruxelles, Bruxelles, Belgium. (Permanent address: Department of Applied Mathematics, The Weizmann Institute of Science, Rehovot 76100, Israel.)

BARCLAY, Hugh, Pacific Forest Research Centre, 506 W. Burnside Road, Victoria, British Columbia V8A 1M5, Canada.

BECK, Karen, Department of Mathematics, University of Utah, Salt Lake City, Utah 84112, U.S.A. (Permanent address: Math. Dept., U. of·Texas at Arlington.)

BHARGAVA, S.C., St. Stephen's College, University of Delhi, Delhi 110007, India.

BLYTHE, S.P., Department of Applied Physics, University of Strathclyde, Glasgow, Scotland.

BOJADZIEV, George, Department of Mathematics, Simon Fraser University, Burnaby, British Columbia V5A 1S6, Canada.

BOTSFORD, Louis W., Department of Wildlife and Fisheries Biology, University of California at Davis, Davis, California 95616, U.S.A.

BROWN, Joel S., Aerospace and Mechanical Engineering, University of Arizona, Tucson, Arizona 85721, U.S.A.

BUTLER, Geoff, Department of Mathematics, University of Alberta, Edmonton, Alberta T6G 2E1, Canada.

CHESSON, Peter L., Department of Zoology, Ohio State University, 1735 Neil Avenue, Columbus, Ohio 43210, U.S.A.

CLARK, Charles E., Department of Mathematics, University of Tennessee, Knoxville, Tennessee 37996-1300, U.S.A.

CLARK, Colin W., Department of Mathematics, University of British Columbia, Vancouver, British Columbia V6T 1Y4, Canada.

COOKE, Fred, Department of Biology, Queen's University, Kingston, Ontario K7L 3N6, Canada.

COSTE, Jean, Laboratoire de Physique de la Matière Condensée, 06034 Nice Cedex, France.

CUSHING, Jim, Mathematics Department and Program in Applied Mathematics, Building #89, University of Arizona, Tucson, Arizona 85721, U.S.A.

DUTT, Ranabir, Department of Physics, Visva-Bharati University, Santiniketan 731255, West Bengal, India.

FINDLAY, C.S., Department of Biology, Queen's University, Kingston, Ontario K7L 3N6, Canada.

FREEDMAN, Herbert I., Department of Mathematics, University of Alberta, Edmonton, Alberta T6G 2E1, Canada.

GARD, Thomas C., Department of Mathematics, University of Georgia, Athens, Georgia 30602, U.S.A.

GHOSH, P.K., Department of Physics, Visva-Bharati University, Santiniketan 731235, West Bengal, India.

GINZBURG, Lev R., Department of Ecology and Evolution, State University of New York, Stony Brook, New York 11794, U.S.A.

GLASSER, John W., Department of Zoology, University of Georgia, Athens, Georgia 30602, U.S.A.

GRANERO-PORATI, Maria Ilde, Institute of Physics, Section of Biophysics, University of Parma, 43100 Parma, Italy.

GREGORIUS, Hans-Rolf, Lehrstuhl fur Forstgenetik, Univ. Göttingen, Busgenweg 2, 34 Göttingen, West Germany.

GURNEY, W.S.C., Department of Applied Physics, University of Strathclyde, Glasgow, Scotland.

HANSON, Floyd, Department of Mathematics, Statistics and Computer Sciences, Univ. of Illinois at Chicago Circle, Box 4348, Chicago, Illinois 60680, U.S.A.

HASSELL, Michael H., Department of Zoology and Applied Entomology, Imperial College, London, S.W. 7, England.

HASTINGS, Harold M., Department of Mathematics, Hofstra University, Hempstead, New York 11550, U.S.A.

HESTBECK, Jay, Department of Wildlife and Fisheries Biology, University of California at Davis, Davis, California 95616, U.S.A.

HOCHBERG, Kenneth J., Department of Mathematics and Statistics, Case Western Reserve University, Cleveland, Ohio 44106, U.S.A.

HOLT, Robert D., University of Kansas, Museum of Natural History, Lawrence, Kansas 66045, U.S.A.

HUNT, Fern, Lab. of Theoretical Biology, Building 10, Room 4B56, National Institute of Health, Bethseda, Maryland 20205, U.S.A. (Permanent address: Dept. of Mathematics, Howard University, Washington, D.C., U.S.A.)

KEENER, J.P., Department of Mathematics, University of Utah, Salt Lake City, Utah 84112, U.S.A.

KEYFITZ, Nathan, Department of Sociology, 300 Administration Building, 190 North Oval Mall, Columbus, Ohio 43210, U.S.A. (Address after September 1983: Rosenstadt Professor, Department of Preventive Medicine and Biostatistics, Faculty of Medicine, University of Toronto, Toronto, Ontario M5S 1A8, Canada)

KRISHNAN, P., Department of Sociology, University of Alberta, Edmonton, Alberta, T6G 2E1, Canada.

KRON-MORELLI, R., Institute of Physics, Section of Biophysics, University of Parma, 43100 Parma, Italy.

LALU, N.M., Department of Sociology, University of Alberta, Edmonton, Alberta T6G 2E1, Canada.

LANGLAIS, Michel R., Department of Mathematics, Purdue University, West Lafayett, Indianna 47907, U.S.A. (Permanent address: Mathematiques et Informatique, Universite de Bordeaux, 33405 Talence Cedex, France.)

LEON, J.A., Gray Herbarium, Harvard University, 22 Divinity Avenue, Cambridge, Mass. 02138, U.S.A. (Permanent address: Instituto de Zoologia Tropical, Facultad de Ciencias, Central University of Venezuela, Apartado 47058, Caracas 1041-A, Venezuela.)

LEUNG, Anthony, Department of Mathematical Science, University of Cincinnati, Cincinnati, Ohio 45221, U.S.A.

LEVIN, Simon, Department of Ecology and Systematics, Langmuir Lab., Cornell Univ., Ithaca, New York 14850, U.S.A.

LEVINE, Dan, Mathematics Department, University of Houston, Central Campus, Houston, Texas 77004, U.S.A.

MANGEL, Marc, Department of Mathematics, University of British Columbia, Vancouver, British Columbia V6T 1Y4, Canada.

MERRILL, Steve, Department of Mathematics, Statistics and Computing Science, Marquette University, William Wehr Physics Building, Milwaukee, Wisonsin 53233, U.S.A.

MOODY, Michael, Mathematics Department, Washington State University, Pullman, Washington 99164, U.S.A.

NISBET, R.M., Department of Applied Physics, University of Strathclyde, Glasgow, Scotland.

NUNNEY, Len, Department of Biology, University of California at Riverside, Riverside, California 92521, U.S.A.

PORATI, Alfredo, Institute of Physics, Section of Biophysics, University of Parma, 43100 Parma, Italy.

PUGLIESE, Andrea, Department of Ecology and Evolution, State University of New York, Stony Brook, New York 11794, U.S.A.

RAI, Bindhyachal, Department of Mathematics, University of Allahabad, Allahabad, India.

RAO, V.S.H., Department of Mathematics, University of Alberta, Edmonton, Alberta T6G 2E1. (Permanent address: Osmania University, Hyderabad - 500007, India.)

RASMUSSEN, J.B., Department of Biology, University of Calgary, Calgary, Alberta T2N 1N4, Canada.

RICCIARDI, P., Department of Mathematics, University of Utah, Salt Lake City, Utah 84112, U.S.A.

ROCKWELL, Robert F., Department of Biology, Queen's University, Kingston, Ontario K7L 3N6, Canada.

ROSE, M.L., Department of Biology, Dalhousie University, Halifax, Nova Scotia B3H 4J1, Canada.

SALEEM, Mohammed, Department of Mathematics, Building 89, University of Arizona, Tucson, ARizona 85721, U.S.A.

SHONKWILER, Ron, School of Mathematics, Georgia Institute of Technology, Atlanta, Georgia 20332, U.S.A.

SILVERT, William, Marine Ecology Laboratory, Bedford Institute of Oceanography, P.O. Box 1006, Dartmouth, Nova Scotia B2Y 4A2, Canada.

STROBECK, Curtis, Department of Genetics, University of Alberta, Edmonton, Alberta T6G 2E1, Canada.

TAKEUCHI, Yasuhiro, Department of Applied Mathematics, Faculty of Engineering, Shizuoka University, Hamamatsu 432, Japan.

TAYLOR, P.D., Department of Mathematics, Queen's University, Kingston, Ontario, K7L 3N6, Canada.

TEMPLETON, Alan, Department of Biology, Washington University, St. Louis, Missouri 63130, U.S.A.

THOMPSON, Maynard, School of Mathematics, Georgia Institue of Technology, Atlanta, Georgia 20332, U.S.A.

TIER, C., Department of Mathematics, Statistics and Computer Sciences, University of Illinois at Chicago Circle, Box 4348, Chicago, Illinois 60680, U.S.A.

VAN DEN DRIESSCHE, Pauline, Department of Mathematics, University of Victoria, Victoria, British Columbia V8W 2Y2, Canada.

VINCENT, Thomas L., Aerospace and Mechanical Engineering, University of Arizona, Tucson, Arizona 85721, U.S.A.

WALTER, Gilbert, Department of Mathematical Sciences, University of Wisconsin at Milwaukee, Milwaukee, Wisconsin 53201, U.S.A.

WALTMAN, Paul, Department of Mathematics, Emory University, Atlanta, Georgia, U.S.A.

WHITTAKER, James V., Department of Mathematics, University of British Columbia, Vancouver, British Columbia V6T 1W5, Canada.

WILLIAMS, G.C., Department of Mathematics, Queen's University, Kingston, Ontario K7L 3N6, Canada.

WORZ-BUSEKROS, Angelika, Universitat Tubingen, Institut fur Biomathematik, Lehrstuhl fur Biomathematik, D 7400 Tubingen 1, Auf der Morgenstell 28, West Germany.

TABLE OF CONTENTS

PART I:

POPULATION GENETICS

THE USE OF COEFFICIENTS OF IDENTITY
TO STUDY RANDOM DRIFT

Curtis Strobeck

The use of coefficients of identity in population genetics has a long history. Wright (1922) defined F as the genetic correlation between uniting gametes and used it to describe the degree of inbreeding. Later the degree of inbreeding was defined by the probability that the two genes of an individual were identical by descent (Berstein, 1930; Haldance and Moschinsky, 1939; Cotterman, 1940; Malecot, 1948). Malecot (1948) also considered the probability that the genes of two individuals are identical by descent, the coefficient of kinship (Jacquard, 1974). These concepts can be extended to consider the probabilities of the genes in a given sample of gametes are identical. Such probabilities are defined as coefficients of identity by descent by Gillois (1964), Jacquard (1974), and Chevalet, Gillois, and Nasser (1977); K coefficients (Cotterman, 1940); inbreeding coefficients (Serant, 1974, 1976): and descent measures (Cocherham and Weir, 1973). I will use the term "coefficients of identity" to describe the probabilities that sets of gametes are, or are not, identical, i.e., the same or different alleles. (They may or may not be identical by descent depending on the model assumed.)

Coefficients of identity are useful to analyze certain types of problems that arise in the theory of random drift. Their utility lies in the fact that recursion equations for their expected values over replicate populations can be easily derived using simple probabilistic arguments. These recursion equations allow the moments of the transient and stationary distributions of the allele frequencies to be easily calculated. Although, these same moments can be obtained using diffusion equations or path coefficients, these methods are more difficult to apply for complex problems.

In the following pages, several different uses of identity coefficients will be illustrated. These include, the asymptotic rate to homozygosity, the expected homozygosity, the expected squared linkage disequilibrium, and the variance of homozygosity.

1. ASYMPTOTIC RATE OF HOMOZYGOSITY

The asymptotic rate to homozygosity is the largest non-unit eigenvalue of Markov chain defining the random drift process with no mutation. For the Wright Fisher process with N diploid individuals, the asymptotic rate to homozygosity can be obtained using the identity coefficient

$$\phi = P(a_i \equiv a_j) \quad i \neq j$$

where a_i and a_j are two distinct gametes drawn randomly from the population ("$\equiv$" is read "identical to"). The recursion relationship for the expected value of ϕ is

$$\phi' = \frac{1}{2N} + (1 - \frac{1}{2N}) \phi \tag{1}$$

where ϕ' is the value of ϕ in the next generation. The equilibrium value is $\hat{\phi} = 1$ and the rate to homozygosity is $\lambda = 1 - (1/2N)$.

In a partial selfing population two coefficients of identity are required

$$\psi = P(a_{i1} \equiv a_{i2})$$

and

$$\phi = P(a_{i1} \equiv a_{j1}) \quad i \neq j$$

where a_{i1} and a_{i2} are the two gametes which united to form the ith individual. If N is the population size and S the partial selfing rate, the recursion equations for the expected values are

$$\psi' = S(\tfrac{1}{2} + \tfrac{1}{2}\psi) + (1-S)\phi$$

$$\phi' = \frac{1}{N} (\tfrac{1}{2} + \tfrac{1}{2}\psi) + (1 - \frac{1}{N})\phi \tag{2}$$

(Golding and Strobeck, 1980). The eigenvalues of the Jacobian are

$$\lambda_1 = 1 - \frac{1}{2N(1 - \tfrac{1}{2}S)} + 0(\frac{1}{N^2})$$

$$\lambda_2 = 1 - \tfrac{1}{2}S + 0(\frac{1}{N})$$

Therefore the asymptotic rate to homozygosity is given by λ_1 and the effective population size is

$$N_e = (1 - \tfrac{1}{2}S)N.$$

The last model to be considered in this section is for a neutral locus, A, closely associated with an inversion (Strobeck, 1983). Let C_1 and C_2 be two chromosomal arrangements differing by an inversion and maintained in the population by a heterokaryotic advantage. Three coefficients of identity are required

$$\phi_{11} = P(a_i^1 \equiv a_j^1)$$

$$\phi_{12} = P(a_i^1 \equiv a_j^2)$$

$$\phi_{22} = P(a_i^2 \equiv a_j^2)$$

where a_i^1 and a_i^2 are gametes containing the C_1 and C_2 type chromosomes, respectively. If $P = Q = \frac{1}{2}$ are the frequencies of the two types of chromosomes in the population, the eigenvalues for the recursion equation for the three identity coefficients are

$$\lambda_1 = 1 - \left[\frac{1 + 2N\tilde{r} - \sqrt{1 + 4N^2\tilde{r}^2}}{2N}\right]$$

$$\lambda_2 = 1 - \left[\frac{1 + 2N\tilde{r} + \sqrt{1 + 4N^2\tilde{r}^2}}{2N}\right] \tag{3}$$

$$\lambda_3 = 1 - \frac{1}{N} - \tilde{r}$$

where r is the recombination value between the loci and the inversion and $\tilde{r} = r/\bar{W}$ and $\bar{W}$ is the average fitness (Strobeck, 1983).

2. EXPECTED HOMOZYGOSITY

If there is a non-zero mutation rate, there will be a stationary distribution of allele frequencies. The moments of this distribution can be obtained using coefficients of identity. Both the K-allele and the infinite allele model for the Wright-Fisher process will be illustrated. For the K-allele model, the recursion equation for the expected homozygosity is

$$\phi' = (1-\mu)^2[\frac{1}{2N} + (1 - \frac{1}{2N})\phi] + \frac{2\mu}{K-1}(1-\mu)(1-\phi) + 0\,(\frac{1}{N^2})$$

$$\simeq \frac{1}{2N} + (1 - \frac{1}{2N} - 2\mu)\phi + \frac{2\mu}{K-1}(1-\phi) \tag{4}$$

where μ is the mutation rate of each allele to any of the K-1 possible alleles. At equilibrium

$$\hat{\phi} = \frac{1 + \theta/K-1}{1 + K\theta/K-1}$$

where $\theta = 4N\mu$ (Ewens, W.J., 1979). For the infinite allele model (allowing $K \to \infty$)

$$\hat{\phi} = \frac{1}{1 + \theta}$$

In a partial selfing population, the equilibrium values for the expected value of the two identity coefficients (assuming the infinite allele model) are

$$\hat{\psi} = \frac{1 + 2N\mu S}{1 + 4N_e\mu}$$

$$\hat{\phi} = \frac{1}{1 + 4N_e\mu} \tag{5}$$

where
$$N_e = (1 - \tfrac{1}{2}S)N.$$

Thus the expected amount of variation in a completely selfing population is equal to that in a random mating population of half the population size (Golding and Strobeck, 1980).

For a neutral locus associated with an inversion, the expected values of the three identity at equilibrium with $P = Q = \tfrac{1}{2}$ (assuming the infinite allele model) are

$$\hat{\phi}_{11} = \hat{\phi}_{22} = \frac{2\theta + R}{2\theta + R + \theta^2 + \theta R}$$

$$\phi_{12} = \frac{R}{2\theta + R + \theta^2 + \theta R}$$

$$(6)$$

where $\theta = 4N\mu$ and $R = 4N\tilde{r}$ (Strobeck, 1983).

For the two locus models, five coefficients of identity are required

$$\phi_A = P(a_i \equiv a_j)$$

$$\phi_B = P(b_i \equiv b_j)$$

$$\phi_{AB} = P(a_i \equiv a_j \text{ and } b_i \equiv b_j)$$

$$\Gamma_{AB} = P(a_i \equiv a_j \text{ and } b_i \equiv b_k)$$

$$\Delta_{AB} = P(a_i \equiv a_k \text{ and } b_j \equiv b_l)$$

where $a_i b_i$ is an arbitrary gamete. The expected value of these coefficients at equilibrium is given in Strobeck and Morgan (1978).

In a partial selfing population, the two locus models require 10 coefficients of identity (Golding and Strobeck, 1980). These coefficients are given in Table 1. The equilibrium values for the expected value of these identity coefficients are given in Golding and Strobeck (1980).

3. EXPECTED SQUARED LINKAGE DISEQUILIBRIUM

The expected squared linkage disequilibrium can be obtained directly from the expected values of the coefficients of identity. Let

$$D_{ij} = f_{ij} - p_i q_j$$

where p_i is the frequency of the ith allele at the A locus, A_i; q_j is the frequency of the ith allele at the B locus, B_j; and f_{ij} is the frequency of the gamete $A_i B_j$.

The

Table 1 Definition of the Sixteen Coefficients of Identity Required for the Two Locus Models in a Partial Selfing Population

$$\Psi_{(A/A)} = P(a_{i1} \equiv a_{i2}) \qquad\qquad \Psi_{(B/B)} = P(b_{i1} \equiv b_{i2})$$

$$\Phi_{(A)(A)} = P(a_{i1} \equiv a_{j1}) \qquad\qquad \Phi_{(B)(B)} = P(b_{i1} \equiv b_{j1})$$

$$\Psi_{(AB/AB)} = P(a_{i1} \equiv a_{i2} \text{ and } b_{i1} \equiv b_{i2})$$

$$\Phi_{(AB)(AB)} = P(a_{i1} \equiv a_{j1} \text{ and } b_{i1} \equiv b_{j1})$$

$$\Phi_{(AB)(A/B)} = P(a_{i1} \equiv a_{j1} \text{ and } b_{i1} \equiv b_{j2})$$

$$\Phi_{(AB/B)(A)} = P(a_{i1} \equiv a_{j1} \text{ and } b_{i1} \equiv b_{i2})$$

$$\Phi_{(AB/A)(B)} = P(a_{i1} \equiv a_{i2} \text{ and } b_{i1} \equiv b_{j1})$$

$$\Phi_{(A/A)(B/B)} = P(a_{i1} \equiv a_{i2} \text{ and } b_{j1} \equiv b_{j2})$$

$$\Phi_{(A/B)(A/B)} = P(a_{i1} \equiv a_{j2} \text{ and } b_{i1} \equiv b_{j2})$$

$$\Gamma_{(AB)(A)(B)} = P(a_{i1} \equiv a_{j1} \text{ and } b_{i1} \equiv b_{k1})$$

$$\Gamma_{(B/B)(A)(A)} = P(a_{j1} \equiv a_{k1} \text{ and } b_{i1} \equiv b_{i2})$$

$$\Gamma_{(A/A)(B)(B)} = P(a_{i1} \equiv a_{i2} \text{ and } b_{j1} \equiv b_{k1})$$

$$\Gamma_{(A/B)(A)(B)} = P(a_{i1} \equiv a_{j1} \text{ and } b_{i2} \equiv b_{k1})$$

$$\Delta_{(A)(B)(A)(B)} = P(a_{i1} \equiv a_{k1} \text{ and } b_{j1} \equiv b_{l1})$$

$$E(\Sigma D_{ij}^2) = E(\Sigma [f_{ij} - p_i q_j]^2)$$

$$= E(\Sigma f_{ij}^2) - 2E(\Sigma p_i q_j f_{ij}) + E(\Sigma p_i^2 q_j^2)$$

$$= \phi_{AB} - 2\Gamma_{AB} + \Delta_{AB}$$

and at equilibrium

$$E(\Sigma D_{ij}^2) = \frac{\theta^2 (1+\theta)(2\theta + 2R + 5)}{(1+\theta)^2 [4\theta^3 + 12\theta^2 R + 8\theta^2 + 20\theta^2 + 38\theta R + 8R^2 + 27\theta + 26R + 9]} \qquad (7)$$

where $\theta = 4N\mu$ and $R = Nr$ (Strobeck and Morgan, 1978).

In a partial selfing population the $E(\Sigma D_{ij}^2)$ is given by (4) with $N_e = (1 - \frac{1}{2}S)N$ and $r_e = (1-S)r/(1 - \frac{1}{2}S)$ replacing N and r respectively (Golding and Strobeck, 1980). Thus the effect of partial selfing is to reduce the population size and to reduce the effective recombination rate.

For a neutral locus associated with an inversion, let $x_i (y_i)$ be the frequence of A_i on the $C_1 (C_2)$ type chromosone. Then, if $P = Q = \frac{1}{2}$

$$D_i = \tfrac{1}{4}(x_i - y_i)$$

and

$$E(\Sigma D_i^2) = \frac{1}{16}\left[E(\Sigma x_i^2) - 2E(\Sigma x_i y_i) + E(\Sigma y_i^2)^1\right]$$

$$= \frac{1}{16}\left[\phi_{11} - 2\phi_{12} + \phi_{22}\right]$$

At equilibrium, from (6)

$$E(\Sigma D_i^2) = \frac{N}{2\theta+R+\theta^2+\theta R}$$

(Strobeck, 1983).

4. VARIANCE OF HOMOZYGOSITY

Identity coefficients not only involve the probability that gametes are identical, but can be extended to three, four, and more gametes and include the possibility of non-identity. All possible coefficients of identity at a single locus with four or less gametes are given in Table 2. These are only four independent coefficients since

$$\phi_{11} + \phi_{1/1} = 1$$

$$\Gamma_{111} + \Gamma_{11/1} + \Gamma_{1/1/1} = 1$$

$$\Delta_{1111} + \Delta_{111/1} + \Delta_{11/1/1} + \Delta_{11/11} + \Delta_{1/1/1/1} = 1$$

$$\Gamma_{111} + 1/3\ \Gamma_{11/1} = \phi_{11}$$

$$\Delta_{1111} + 1/4\ \Delta_{111/1} = \Gamma_{111}$$

$$\Delta_{1111} + 1/2\ \Delta_{111/1} + 1/3\ \Delta_{11/11} + 1/6\ \Delta_{11/1/1} = \phi_{11}$$

The recursion equations for the expected value of the four independent identity coefficients are

Table 2 Definition of the Coefficients of Identity with Four or Less Gametes
Required for a Single Locus Model

$$\phi_{11} = P(a_1 \equiv a_2) \qquad \phi_{1/1} = P(a_1 \not\equiv a_2)$$

$$\Gamma_{111} = P(a_1 \equiv a_2 \equiv a_3) \qquad \Gamma_{11/1} = P(a_1 \equiv a_2 \not\equiv a_3) \qquad \Gamma_{1/1/1} = P(a_1 \not\equiv a_2 \not\equiv a_3)$$

$$\Delta_{1111} = P(a_1 \equiv a_2 \equiv a_3 \equiv a_4) \qquad \Delta_{111/1} = P(a_1 \equiv a_2 \equiv a_3 \not\equiv a_4) \qquad \Delta_{11/1} = P(a_1 \equiv a_2 \not\equiv a_3 \equiv a_4)$$

$$\Delta_{11/1/1} = P(a_1 \equiv a_2 \not\equiv a_3 \not\equiv a_4) \qquad \Delta_{1/1/1/1} = P(a_1 \not\equiv a_2 \not\equiv a_3 \not\equiv a_4)$$

$$\phi'_{11} = \frac{1}{2N} + (1 - \frac{1}{2N} - 2\mu)\phi_{11}$$

$$\Gamma'_{111} = \frac{3}{2N}\phi_{11} + (1 - \frac{3}{2N} - 3\mu)\Gamma_{111}$$

$$\Delta'_{1111} = \frac{6}{2N}\Gamma_{111} + (1 - \frac{6}{2N} - 4\mu)\Delta_{1111} \qquad (8)$$

$$\Delta'_{11/11} = \frac{2}{2N}\Gamma_{11/1} + (1 - \frac{6}{2N} - 4\mu)\Delta_{11/11}$$

$$= \frac{6}{2N}\phi_{11} - \frac{6}{2N}\Gamma_{111} + (1 - \frac{6}{2N} - 4\mu)\Delta_{11/11}$$

At equilibrium

$$\hat{\phi}_{11} = \frac{1}{1+\phi}$$

$$\hat{\Gamma}_{111} = \frac{2}{(2+\theta)(1+\theta)}$$

$$\hat{\Delta}_{1111} = \frac{6}{(3+\theta)(2+\theta)(1+\theta)}$$

$$\hat{\Delta}_{11/11} = \frac{3\theta}{(3+\theta)(2+\theta)(1+\theta)}$$

The variance of homozygosity can be obtained directly from these values since

$$Var_1 = E(\Sigma p_i^2 \cdot \Sigma p_i^2) - E^2(\Sigma p_i^2)$$

$$= E(\Sigma p_i^4) + E(\underset{i/5j}{\Sigma} \Sigma p_i p_j) - E^2(\Sigma p_i^2)$$

$$= \Delta_{1111} + \frac{1}{3}\Delta_{11/11} - \phi_{11}^2$$

Therefore at equilibrium

$$\text{Var}_1 = \frac{2\theta}{(3+\theta)(2+\theta)(1+\theta)^2} \qquad (9)$$

(Stewart, 1976).

This method can be extended to the calculation of the variance of homozygosity for a single locus in a subdivided population. This requires a minimum of 17 identity coefficients (Table 3). If there are n subpopulations the variance of homozygosity in the subpopulation is

$$\text{Var}_i = \Delta_{iiii} + \frac{1}{3}\Delta_{ii/ii} - \phi_{ii}^2 \qquad (10)$$

If the total population is sampled at random then the expected homozygosity at equilibrium is

$$\frac{1}{n}\phi_{ii} + \frac{n-1}{n}\phi_{ij} \qquad (11)$$

and the variance of homozygosity is

$$\begin{aligned}
\text{Var} = \frac{1}{n^3}\Big[&\Delta_{iiii} + 4(n-1)\Delta_{iiij} + 3(n-1)\Delta_{iijj} \\
&+ 6(n-1)(n-2)\Delta_{iijk} + (n-1)(n-2)(n-3)\Delta_{ijk} \\
&+ \frac{1}{3}\Delta_{ii/ii} + \frac{4}{3}\Delta_{ii/ij} + 2(n-1)\Delta_{ij/ij} \\
&+ (n-1)\Delta_{ii/jj} + 2(n-1)(n-2)\Delta_{ii/jk} \\
&+ 4(n-1)(n-2)\Delta_{ij/ik} + (n-1)(n-2)(n-3)\Delta_{ij/k} \Big] \\
&- (\frac{1}{n}\phi_{11} + \frac{n-1}{n}\phi_{ii})^2
\end{aligned} \qquad (12)$$

The recursion equations for the expected value of these identity coefficients (Golding and Strobeck, 1983a) can be solved numerically on the computer and therefore the variances of homozygosity in the ith subpopulation and in total population can be obtained.

Since most natural populations are subdivided to some extent, it is necessary to understand how this subdivision affects the sampling theory of neutral alleles (Ewens, 1979) therefore the variance of homozygosity in the ith subpopulation (10) and the variance of homozygosity in the total population (12) are compared to the variance of homozygosity in a non-subdivided population (9) with the same expected homozygosity, i.e., with

Table 3   Definition of the Coefficients of Identity Required for a Single Locus in
a Subdivided Population

$$\phi_{ii} = \text{Prob}(a_{i1} \equiv a_{i2}) \qquad \Phi_{ij} = \text{Prob}(a_{i1} \equiv a_{j2})$$

$$\Gamma_{iij} = \text{Prob}(a_{i1} \equiv a_{i2} \equiv a_{j3}) \qquad \Gamma_{ijk} = \text{Prob}(a_{i1} \equiv a_{j2} \equiv a_{k3})$$

$$\Delta_{iiij} = \text{Prob}(a_{i1} \equiv a_{i2} \equiv a_{i3} \equiv a_{j4}) \qquad \Delta_{iijj} = \text{Prob}(a_{i1} \equiv a_{i2} \equiv a_{j3} \equiv a_{j4})$$

$$\Delta_{ijk\ell} = \text{Prob}(a_{i1} \equiv a_{i2} \equiv a_{k3} \equiv a_{4}) \qquad \Delta_{ii/ii} = \text{Prob}(a_{i1} \equiv a_{i2} \not\equiv a_{i3} \equiv a_{i4})$$

$$\Delta_{ij/ij} = \text{Prob}(a_{i1} \equiv a_{j2} \not\equiv a_{i3} \equiv a_{j4}) \qquad \Delta_{ii/jj} = \text{Prob}(a_{i1} \equiv a_{i2} \not\equiv a_{j3} \equiv a_{j4})$$

$$\Delta_{ii/jk} = \text{Prob}(a_{i1} \equiv a_{i2} \not\equiv a_{j3} \equiv a_{k4}) \qquad \Delta_{ij/k\ell} = \text{Prob}(a_{i1} \equiv a_{j2} \not\equiv a_{k3} \equiv a_{\ell4})$$

$$\Gamma_{iii} = \text{Prob}(a_{i1} \equiv a_{i2} \equiv a_{i3})$$

$$\Delta_{iiii} = \text{Prob}(a_{i1} \equiv a_{i2} \equiv a_{i3} \equiv a_{i4})$$

$$\Delta_{iijk} = \text{Prob}(a_{i1} \equiv a_{i2} \equiv a_{j3} \equiv a_{k4})$$

$$\Delta_{ii/ij} = \text{Prob}(a_{i1} \equiv a_{i2} \not\equiv a_{i3} \equiv a_{j4})$$

$$\Delta_{ij/ik} = \text{Prob}(a_{i1} \equiv a_{j2} \not\equiv a_{i3} \equiv a_{k4})$$

$$\text{where} \quad i \neq j \neq k \neq \ell$$

$$\theta = \frac{1}{\phi_{ii}} - 1$$

and the   ith   subpopulation and

$$\theta = \frac{n}{\phi_{ii} + (n-1)\phi_{ij}} - 1$$

for the total population. The variance of homozygosity in the ith subpopulation
is approximately the same as in a nonsubdivided population. However, if the
migration rates between the subpopulation is small, the variance of homozygosity

in the total population is much smaller than in a nonsubdivided population (Figure 1). This implies that Ewens sampling theory does not apply to a subdivided population if the migration rates between the subpopulations are small (Kingman, 1977) (see also Slatkin, 1982).

The methodlogy can also be used to look at the variance of the linkage disequilibrium squared. This requires a minimum of 50 coefficients of identity (Golding and Strobeck, 1983b).

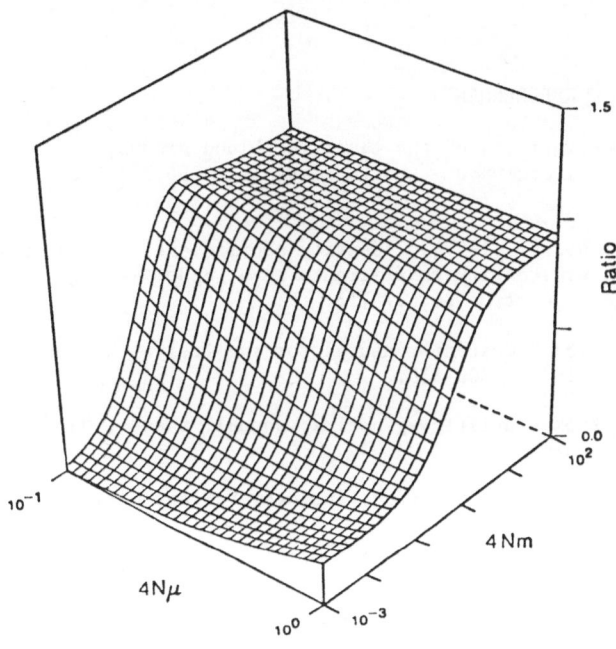

Figure 1   Ratio of the actual variance of homozygosity sampling at random from four subpopulations to the expected variance for a single population with the same variability.

5. SAMPLING THEORY

The coefficients of identity used to calculate the variance of homozygosity (Table 2) are all possible samples of gametes with sample sizes two, three, and four. Also, the recursion equations are identical to those of Karlin and McGregor (1972) for these samples. Thus there is an equivalence between sampling theory and the use of coefficients of identity.

6. SUMMARY

Some of the uses of identity coefficients to study random drift have been illustrated by specific examples. The ease with which the recursion equations for the expected values of these coefficients can be written down make this methodology applicable to many situations which cannot be easily handled by studying the moments of diffusion equations.

REFERENCES

Berstein, F. (1930): Further investigations on the theory of blood groups, *Zeitschrift fuer induktive - abstamungs - und Verebungslehre* 56:233-273 (German).

Chevalet, C., M. Gillois, and R.F. Nasser (1977): Identity coefficients in finite populations. I. Evolution of identity coefficients in a random mating diploid dioecious population, *Genetics* 86:697-713.

Cockerburn, C.C., and B.S. Weir (1973): Descent measures for two loci with some applications, *Theoret. Pop. Biol.* 4:300-330.

Cotterman, C.W. (1940): A calculus of statistico-genetics, Unpubl. thesis, Ohio State University, Columbus, Ohio.

Ewens, W.J. (1979): *Mathematical Population Genetics*, Springer-Verlag, N.Y.

Gillois, M. (1964): La relation d'identité génétique, Thesis, Faculte des Science, Paris.

Golding, G.B., and C. Strobeck (1980): Linkage disequilibrium in a finite population which is partiall selfing, *Genetics* 94:777-789.

_____ (1983a): Variance and covariance of homozygosity in a structured population, *Genetics* (in press).

_____ (1983b): Two-locus, fourth order gene frequency moments: Implications for the variance of squared linkage disequilibrium and the variance of homozygosity, *Theor. Pop. Biol.* (in press).

Haldane, J.B.S., and P. Moshinsky (1939): Inbreeding in Mendelian populations with special reference to human cousin marriage, *Ann. Eugenics* 9:321-340.

Jacquard, A. (1974): *The Genetic Structure of Populations*, Springer-Verlag, N.Y.

Kingman, J.F.C. (1977): The populations structure associated with Ewens sampling formula, *Theor. Pop. Biol.* 11:274-283.

Malecot, G. (1948): *Les Mathématiques de l'Hérédite*, Masson et Cie, Paris.

Serant, D. (1974): Linkage and inbreeding coefficients in a finite random mating populations, *Theoret. Pop. Biol.* 6:251-263.

  (1976): An application of kinship process to the gene frequencies. Linkage disequilibrium due to random drift in Mendelian genetics with reversible mutation and in molecular genetics, *Theoret. Pop. Biol.* 9:1-11.

Slatkin, M. (1982): Testing neutrality in subdivided population, *Genetics* 100:533-545.

Stewart, F.M. (1976): Variability in the amount of heterozygosity maintained by neutral mutations, *Theoret. Pop. Biol.* 9:188-201.

Strobeck, C. (1983): Expected linkage disequilibrium for a neutral locus linked to a chromosomal arangement, *Genetics* 103:545-555.

Strobeck, C., and K. Morgan (1978): The effect of intragenic recombination on the number of alleles in a finite population, *Genetics* 88:829-844.

Wright, S. (1922): Coefficients of inbreeding and relationship, *Amer. Nat.* 56:330-338.

INSTABILITY AND REPULSIVITY OF THE FIXATION STATES
IN BIALLELIC·SELECTION MODELS

Hans-Rolf Gregorius

ABSTRACT

The difference between the concepts of instability and repulsivity is demonstrated, and a general method for analysing protectedness of alleles in biallelic selection models is introduced. The relationship to the spectral radius of the derivative is demonstrated.

1. THE PROBLEM

Consider a discrete-time dynamical system given by the iterates of a function in the finite dimensional Euclidean space, and assume that this function has a fixed point at which it is differentiable. Since the fundamental paper of Oskar Perron (1929) about stability of difference equations we know that the fixed point is asymptotically stable and repulsive, according to whether the eigenvalues of the derivative are all smaller or all greater than one in absolute value. If some of the eigenvalues are greater and some smaller than one in absolute value conditional stability occurs, where the latter is a particular form of instability.

A particular case of interest arises if the fixed point is located on the boundary of an invariant set, and if stability of the fixed point is to be analysed with respect to this invariant set only. In population genetics such invariant sets are the simplices of genotypic frequencies and the fixed points of interest are the states of genetic fixation. Now the above mentioned conditions for local asymptotic stability and repulsivity may be far too strong for analysing these stability properties within the invariant set. Different situations of conditional stability may imply that relative to the invariant set the fixed point is attractive in the one case and repulsive in the other.

Hence, in such a situation it is desirable to have a better fitted method of analysis. For deterministic biallelic selection models based on the state space of the three pertinent genotypic frequencies, an appropriate analysis may be developed by applying a particular concept of fitness which is of general significance (c.f. Gregorius 1983a).

2. A NEW METHOD

Let P_A, P_B and P_H be the frequencies of the two homozygotes and the heterozygote, respectively, and let p_A be the frequency of the allele specified by the A-homozygote. Assuming regular Mendelian segregation, the transition equations may always be written in the form

$$P_A' = [w_{A;A} \cdot P_A + (w_{A;H} + \tfrac{1}{2} w_{H;H}) \cdot \tfrac{1}{2} P_H]/\bar{w} ,$$

$$P_A' = \tfrac{1}{2} (w_A \cdot P_A + w_H \cdot \tfrac{1}{2} P_H)/\bar{w} .$$

Herein the $w_{X;Y}$ are the "fractional fitnesses" of the genotype Y with respect to the genotype X, which corresponds to the average number of successful gametes of Y that result from fertilization by X; w_X is the "gametic fitness" of the genotype X, i.e. its average number of successful gametes; and $\bar{w}$ is the population average zygotic fitness. For example, $w_A = 2 \cdot w_{A;A} + w_{H;A} + w_{B;A}$ and $\bar{w} = \tfrac{1}{2} (w_A P_A + w_H P_H + w_B P_B)$.

By assumption the fractional fitnesses are functions of the three genotypic frequencies only, and we assume that they are continuous in the vicinity of the fixation states. Furthermore, since some of the fractional fitnesses are not properly defined at the fixation states $p_A = 0$ and $p_A = 1$, we assume that they allow for a continuous extension to these states. Differentiability is not required!

The principle of the analysis consists in investigating the change of the multiplication rate, p_A'/p_A of the A-allele along a straight line given by $x = P_A/p_A$ as p_A tends to zero.

It can be shown easily that for given x

$$\lim_{p_A \to 0} p_A'/p_A =: r(x) = c \cdot x + d \cdot (1-x)$$

where $c = w_A/w_B$, $d = w_H/w_B$ and all fitnesses are evaluated at $p_A = 0$. Moreover, for given $x = P_A/p_A$, this value in the next generation i.e. P_A'/p_A' obeys

$$\lim_{P_A \to 0} P_A'/p_A' = x' = g(x) = \frac{a \cdot x + b \cdot (1-x)}{r(x)}$$

where $a = 2 \cdot w_{A;A}/w_B$, $b = w_{H;H}/w_B$, and again all fitnesses are evaluated at $p_A = 0$ (note that always $a \le c$ and $b \le \tfrac{1}{2}d$). This provides us with a difference equation, i.e. $x' = g(x)$, with a fixed point $\hat{x}$, $0 \le x \le 1$, which is globally attractive for x, $0 < x < 1$. It can also be shown that

$$\theta := r(\hat{x}) = K + \sqrt{L^2 + M}$$

where $K = \tfrac{1}{2} (a+d-b)$, $L = \tfrac{1}{2} (a+b-d)$ and $M = b \cdot (c-a)$ (cf. Gregorius 1983b). Clearly, θ *is the multiplication rate of the A-allele after many generations in the close vicinity of the fixation state* $p_A = 0$. Consequently, for $\theta < 1$ and $\theta > 1$ one expects the allelic frequency of A to decrease and increase, respectively, if p_A is close to 0, and a sufficient number of generations has passed.

More precisely, $p_A = 0$ is (locally) attractive if $\theta < 1$, and, consequently, the A-allele cannot become established.

In order to illustrate the case $\theta > 1$, consider for $0 \leq x_1 < x_2 \leq 1$ the set $S(x_1, x_2) := \{P \mid x_1 \leq P_A/p_A \leq x_2\}$, which shall be called a sector. Now, if $\theta > 1$ then there exists an open neighbourhood V of $p_A = 0$ and a sector $S(x_1, x_2)$ such that each trajectory starting in V enters the intersection $V \cap S(x_1, x_2)$ after a finite number of steps and remains there (with $p'_A > p_A$) until it leaves V. If $p_A = 0$ cannot be reached from outside V in a single step, this implies protectedness of the A-allele. The results of this analysis can be utilised to derive fairly general conditions for protectedness of biallelic polymorphisms, and they can be stated in a form that relies merely on the rankings between gametic and fractional fitnesses (cf. Gregorius 1983b).

3. THE SPECTRAL VALUE METHOD IN COMPARISON TO THE θ-METHOD

In addition to the previous requirements for the fractional fitnesses it will now be assumed that they are differentiable at $p_A = 0$. Since P_A and p_A are chosen to be the independent variables, the partial derivatives of P'_A and p'_A with respect to P_A and p_A have to be computed at the fixed point $p_A = 0$. Considering that $w_{A;H} = 0$ at $p_A = 0$ and using the quantities $\underline{a}$, $\underline{b}$, $\underline{c}$ and $\underline{d}$ defined previously it turns out that $\partial P'_A/\partial P_A = a - b$, $\partial P'_A/\partial p_A = b$, $\partial p'_A/\partial P_A = c - d$ and $\partial p'_A/\partial p_A = d$. Consequently, the eigenvalues of the derivative are

$$\lambda_{1,2} = \frac{1}{2}(a-b+d) \pm \sqrt{[1/2(a-b+d)]^2 + bc - ad} \; .$$

It is a straightforward task to show that the term under the root is identical for $\lambda_{1,2}$ and θ, and it is non-negative. Consequently, *the eigenvalues of the derivative are real, the maximum eigenvalue is non-negative and equal to θ and has therefore the interpretation previously derived for θ.*

This result is a characteristic of all deterministic biallelic selection models depending only on the three frequencies of the unordered genotypes under the assumption of separated generations and regular Mendelian segregation. Treating a particular biallelic selection model in subdivided populations with migration Nagylaki (1977, p. 134 f) arrived at a similar conclusion concerning the significance of the maximum eigenvalue. However, he had to assume that the derivative be non-negative and irreducible. This is not required for the present model, since $\partial P'_A/\partial P_A$ and $\partial p'_A/\partial P_A$ may be negative.

This work and its presentation at the International Conference on Population Biology was supported by a Heisenberg Fellowship.

REFERENCES

Gregorius, H.-R. (1983a): Fractional fitnesses in exclusively sexually reproducing populations, Manuscript.

_____(1983b): Allele protectedness in frequency dependent biallelic selection models with separated generations, Manuscript.

Nagylaki, T. (1977): *Selection in One-and-Two-Locus Systems*, Springer-Verlag, New York.

Perron, O. (1929): Über Stabilität und asymptotisches Verhalten der Lösungen eines Systems endlicher Differenzengleichungen, *Journal für Mathematik* 161:41-64.

MICROSCOPIC CLUSTERING OF POPULATION PROCESSES

Kenneth J. Hochberg*

1. INTRODUCTION

Mathematical descriptions of the variations in the distribution of populations due to such factors as spatial motion of the members, local interaction between neighboring or related individuals, and stochastic fluctuation effects can be formulated in terms of stochastic evolution equations, also called Ito-Langevin equations, and their associated stochastic processes. These processes may be measure-valued, in the sense that a realization of the process consists of an assignment, for each fixed time, of some mass distribution on the space of all possible states of the individuals in the population. Such measure-valued processes and their associated random measures have been shown to arise in the characterization of problems in such fields as epidemiology, ecology, chemical kinetics, population genetics, economics, and neutron and radiative transport. In this paper, we describe our study, performed jointly with D.A. Dawson, of the microscopic clustering of such population processes, with specific applications to problems in epidemiology and population genetics.

We begin by introducing three mathematical notions that can be utilized in determining the degree of clustering or patchiness in spatially distributed populations.

(i) At each fixed time t, the value of a measure-valued stochastic process is given by a random measure. This measure is said to be *singular* if there exists a random support for the process that has Lebesgue measure zero. In general, a set is said to be singular if it has zero Lebesgue measure.

(ii) The *Hausdorff-Besicovitch dimension of support* of a Borel set E is defined by

$$H\text{-dim}(E) = \sup\{\beta > 0: \liminf_{\delta \downarrow 0} \sum_{\mathcal{E}} \sum_i [d(E_i)]^\beta = \infty\},$$

where $d(E_i)$ is the diameter of the set E_i and

$$\mathcal{E} = \{\{E_i\}: E \subset \bigcup_i E_i, \quad d(E_i) < \delta \quad \text{for each} \quad i\},$$

that is, $\mathcal{E}$ is the set of all coverings of the set E by sets of diameter less than δ. A fundamental relationship between the Hausdorff-Besicovitch dimension of support and singularity of a set lies in the following result:

---

*Research supported in part by Public Health Service grant GM 28336.

THEOREM 1 *If* E *is a subset of* R^d, d-*dimensional Euclidean space, and the Hausdorff-Besicovitch dimension of* E *is less than* d, *then the set* E *is singular.*

(iii) A set B in R^d is called a *generalized random Cantor set* if B can be expressed as the intersection of an infinite decreasing sequence of sets $\{B_n: n = 0,1,2,...\}$, such that B_0 is a unit cube in R^d, and each B_n is the union of some number, say Λ_n, of disjoint subcubes of volume $(\Gamma_n)^{-d}$, where $\{\Gamma_n\}$ is an increasing sequence of non-negative integers. In other words, the distin-guishing feature of generalized random Cantor sets and, as we shall see shortly, of several population processes in biology and elsewhere, is that it consists of a limit of sets B_n which, as n increases, themselves are the unions of progres-sively increasing numbers of subsets, each of which is successively smaller in volume (see Figure 1). The terminology has been borrowed from the Cantor ternary set of classical mathematics, which is obtained (in one dimension) by first removing the middle third (1/3, 2/3) from the unit interval [0,1], then removing the middle thirds (1/9, 2/9) and (7/9, 8/9) of the remaining intervals, and so on. At each successive step, the number Λ_n of intervals increases ($\Lambda_n = 2^n$) while the length $(\Gamma_n)^{-1}$ of each interval decreases ($(\Gamma_n)^{-1} = 3^{-n}$). A "generalized random" Cantor set differs from the Cantor ternary set in that the sequences of numbers $\{\Lambda_n\}$ and $\{\Gamma_n\}$ are not necessarily deterministic but are permitted to be random, as are the locations of the surviving intervals in the successive B_n's.

The relationship between generalized random Cantor sets and Hausdorff-Besicovitch dimensions that we shall exploit is the following, proven in Dawson and Hochberg (1982, Lemma 3.1):

THEOREM 2 *If* B *is a generalized random Cantor set, then*

$$H\text{-dim}(B) \le \lim_{n \to \infty} \inf \frac{\log \Lambda_n}{\log \Gamma_n} .$$

For example, application of this theorem to the Cantor ternary set C yields

$$H\text{-dim}(C) \le \lim_{n \to \infty} \inf \frac{\log 2^n}{\log 3^n} = \frac{\log 2}{\log 3} ,$$

which, in fact, is known to be the exact Hausdorff-Besicovitch dimension of C.

Our method of application of these mathematical notions has been to find a population process -- or some derivative process related to it -- whose evolution-ary behavior follows the pattern of a generalized random Cantor set and to use Theorem 2 to find an upper bound for the Hausdorff-Besicovitch dimension of its support. If this value is less than the dimension d of the space on which the process is defined, then Theorem 1 asserts that the distribution describing the process is singular, i.e., it is highly concentrated or clustered on the state space.

In addition, motivated by the results in Theorems 1 and 2, we define a new quantity $I(B)$, called the *limiting clustering* (or *occupation*) *index* of the set B, by

$$I(B) = \lim_{n \to \infty} I_n(B)$$

where

$$I_n(B) = d \frac{\log \Gamma_n}{\log \Lambda_n} = \frac{\log \Gamma_n^d}{\log \Lambda_n} .$$

Clearly, $I(B) \geq 1$ for any set B. A set with clustering index near one is highly diffused over the space, whereas, in contrast, a finite point set has an infinite limiting occupation index.

In the following sections, we will apply these concepts to two measure-valued processes that arise from biologically relevant mathematical models.

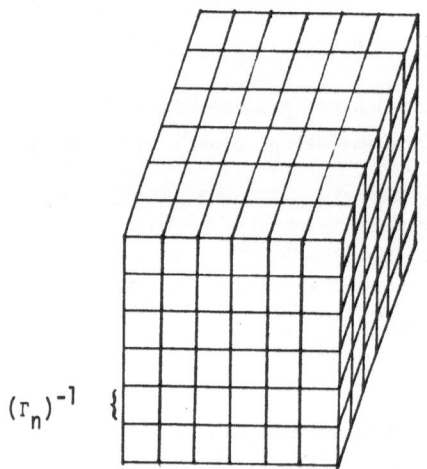

$(\Gamma_n)^{-1}$ {

Figure 1

2. THE FLEMING-VIOT MODEL OF POPULATION GENETICS

Fleming and Viot (1979) have introduced a mathematical model for the distribution of multi-dimensional genetic characteristics in large natural populations. In this model there is a continuum of states, and each coordinate represents a different observable numerical characteristic of the individuals in the population. In Dawson and Hochberg (1983) it is shown that the Fleming-Viot model arises as the weak limit of two different formulations of the continuous-time Ohta-Kimura ladder or stepwise-mutation model for selectively neutral allelic populations evolving under the genetic forces of random genetic drift, via multinomial sampling from the empirical distribution of allelic frequencies in the host population, and a symmetric mutation structure in which mutations always result in some pre-existing state, where the large population limit is taken so that the mutation rate remains constant while the incremental effect of each mutation is assumed to decrease at a rate that is inversely proportional to the square root of the population size. It should be noted that although the allelic state space for the Ohta-Kimura ladder model is restricted to the set of integers -- the model itself having originated to describe charge states obtained from the analysis of electrophoretic data -- the Fleming-Viot model has a continuous state space consisting of all of R^d and is more generally applicable to various observable traits that might characterize individuals in reproducing populations, such as height, weight, shade of eye or hair color, or length of wing span.

For the Fleming-Viot process, microscopic clustering of the allelic distribution follows from the observation that the two genetic forces that we described above tend to have opposing influences on the distribution: multinomial sampling tends to eliminate certain genetic traits or alleles from the population, should they fail to be sampled from the parent or host population, thus decreasing diversity, and mutation tends to reassert alleles that would otherwise disappear from the population, thus preserving variability in the population and diffusing the distribution.

Essentially, one discovers generalized Cantor sets in this process as follows. The Fleming-Viot process is a measure-valued process; i.e., for any given time, the genetic distribution is described by a random measure on R^d. A family of moment measures exists for such stochastic processes, as is shown in Dawson and Hochberg (1982, section 6). Here, the moment measures determine the evolution of an infinite system of interacting particles $\{Z_j(\omega): j = 1,2,...\}$ which is related to the Fleming-Viot random measure $X(A,\omega)$ via the relationship

$$X(A,\omega) = \lim_{N \to \infty} N^{-1} \sum_{j=1}^{N} I_A(Z_j(\omega)),$$

where $I_A(x)$ is one if x is an element of the set A and is zero otherwise. The random motion of this infinite particle system is described by the motion of its k-particle subsystems, as follows: each particle performs an independent Brownian motion on R^d, and, at a constant rate, one particle disappears and another splits into two particles, each of which continues to perform an independent Brownian motion on the space.

If we look at the infinite system at time t_0 and trace back the genealogy of these particles to a time $t_0 - T_n^\infty$ at which the infinitely-many particles had exactly n common ancestors and now look at only these n "surviving family trees," we find that they form n random clusters (see Figure 2). Moreover, within each cluster the radius is bounded, with the bound determined from the scaling property of Brownian motion diffusion. Now, as n increases, we obtain increasing numbers of clusters, each with progressively smaller radii, so the clusters form a generalized Cantor set. We can now proceed as detailed earlier and conclude that the Fleming-Viot random measure is singular in three or more dimensions, indicating a highly clustered distribution and, thus, a high correlation between the genetic traits.

time:

$t_0 - T_n^\infty$

number of ancestors:

n

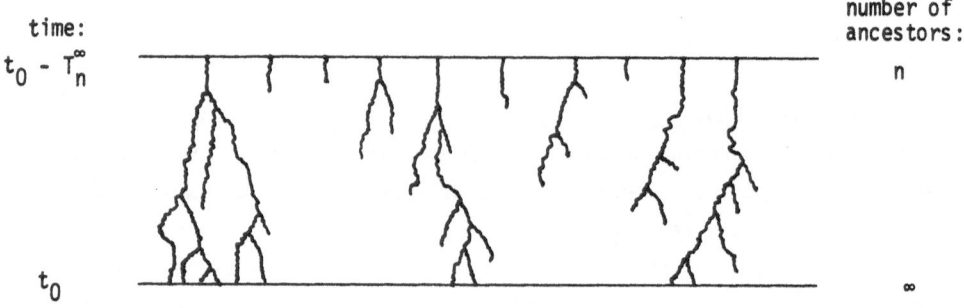

t_0

∞

Figure 2

3. THE STOCHASTIC MEASURE DIFFUSION PROCESS

Branching diffusion processes arise prominently as models for the spread of communicable diseases and epidemics, as well as in other areas. These processes consist of individuals or particles that move in space according to some deterministic diffusion and independently, at random times, each particle either splits into k ($k = 2, 3, ...$) particles with probability $p_k \geq 0$ or disappears with probability $p_0 = 1 - \sum_{k=2}^{\infty} p_k \geq 0$. The process is called "critical" if the expected number of particles remains constant, as occurs, for example, in a binary branching situation with $p_0 = p_2 = 1/2$.

We now consider a critical branching diffusion process where the particles move in R^d according to a symmetric stable diffusion of index α, where $0 < \alpha \leq 2$ (e.g., $\alpha = 2$ corresponds to Brownian motion diffusion), and, independently, after exponentially distributed holding times, either branch or die. The "high density" limit of this process is called the "stochastic measure diffusion process" and is obtained by considering a succession of such branching diffusions with more and more particles of smaller and smaller individual mass in such a way that the expected total mass remains constant.

Again, for this measure-valued process, the two evolutionary mechanisms -- branching and diffusion -- tend to act in opposition, with branching leading on occasion to disappearance of individuals from the population, while diffusion leads to a spread of the process to new members. Here, again, a generalized Cantor set can be found by tracing the past history or genealogy of surviving members of the population over various fixed time periods. One obtains a picture that resembles Figure 2, except that disappearances and splits in the trees need not coincide.

Another approach is to approximate the stochastic measure diffusion process, for each fixed t > 0, by alternating the branching and diffusion mechanisms over successive time intervals of length t/m. This method is carried out and justified in Dawson and Hochberg (1979) and leads to a succession of distributions that are summarized in Figure 3.

Specifically, by observing alternate circles in Figure 3, one obtains a series of distributions consisting of greater numbers of sets that are progressively smaller in volume. In this case, analysis of the generalized Cantor set and the Hausdorff-Besicovitch dimension of support leads to the conclusion that the stochastic measure diffusion process is singular if $d \geq \alpha$. In other words, any disease with spatial motion given by a symmetric stable diffusion in two or three dimensions with critical branching at random times will tend to concentrate on a small subset of the population as part of its natural course. Knowledge of this proclivity to cluster might therefore be considered in determining the manner in which limited resources, such as vaccine supply, medical personnel, and finances, should be utilized to challenge the contagion.

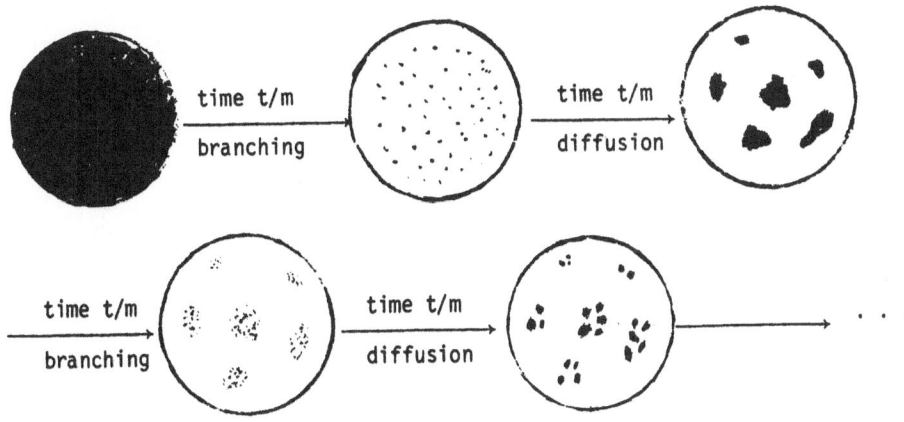

Figure 3

REFERENCES

Dawson, D.A., and K.J. Hochberg (1979): The carrying dimension of a stochastic measure diffusion, *Ann. Probability* 7:693-703.

_____(1982): Wandering random measures in the Fleming-Viot model, *Ann. Probability* 10:554-580.

_____(1983): Qualitative behavior of a selectively neutral allelic model, *Theor. Pop. Biol.* 23:1-18.

Fleming, W.H., and M. Viot (1979): Some measure-valued Markov processes in population genetics theory, *Indiana Univ. Math. J.* 28:817-843.

MODELS FOR PERIPHERAL POPULATIONS:
THE ROLE OF IMMIGRATION

Robert D. Holt

The movement of organisms over space has manifold consequences for both the ecology (Levin, 1976; McMurtrie, 1978) and genetics (Endler, 1977; Karlin, 1982) of populations. In this paper I examine how the rate of immigration influences the size, stability and genetic composition of a peripheral population. I contrast two classes of discrete-generation population models. (Comparable continuous-time models have been discussed elsewhere (Holt, 1983).) In the first, immigrants are genetically identical to residents, whereas in the second, immigrants differ at a haploid locus and are less fit in the local environment. For both we can ask how the rate of immigration affects equilibrial population density, N^*, local stability, and the pattern of fluctuations around N^* in unstable populations. For the genetic model we must also determine conditions for the persistence of a polymorphism. The study of the maintenance of pockets of local adaptation in the face of gene flow is a classical problem in population genetic theory (Haldane, 1930; Nagylaki, 1977). Here I place this problem into an ecological context. If selection coefficients are functions of population density, and immigration can change population size, I show that we cannot understand the ecological consequences of immigration without also understanding the genetic consequences.

Peripheral populations may exist at a species' geographical border or in marginal habitats within a species' range. Such populations are important in the study of factors limiting species' distributions (Krebs, 1978), and as crucibles for speciation (Mayr, 1963). I assume that non-peripheral populations collectively comprise a "bath" (Levin, 1976) from which immigrants are drawn. If relative to the bath a peripheral population is small in size, its population dynamics and genetic composition may be strongly perturbed by immigration, yet it may exert negligible reciprocal effects on the bath.

The first, purely ecological model is for a peripheral population with non-overlapping generations in a constant environment. If there is no immigration, the dynamics are described by the recursion $N_{t+1} = G(N_t)$. I assume that $G(0) = 0$, that there exist a unique positive equilibrium $N^* = G(N^*) \equiv K$, (the carrying capacity) and at most a single critical point N_c, $0 < N_c < K$, and that $dG/dN \leq 1$ for $N > N_c$. Geometrically, the growth function $G(N)$ may either rise monotonically with N or have a "hump". Each generation $\hat{I}$ immigrants enter the population. If we census immediately after immigration, the population model is $N_{t+1} = G(N_t) + \hat{I} \equiv F(N_t)$. Prout (1980) has cautioned that a full understanding of the implications of discrete-generation models requires a careful specification of the census stage being used. By censusing after rather than just before immigration, we highlight the similarity between the effects of immigration in discrete-time and continuous-time models (Holt, 1983). The stability properties of discrete-time growth models have received considerable attention in the literature of mathematical ecology (May and Oster, 1976; Guckenheimer et. al., 1977). The map $F(N)$ is locally stable with monotonic or oscillatory convergence to N^* if $|\lambda| < 1$, where $\lambda \equiv dF/dN|_*$. If unstable, the population exhibits stable limit cycles or 'chaotic' behaviour around N^*. How does population size and stability vary with $\hat{I}$? Since $dN^*/d\hat{I} = (1 - dG/dN|_*)^{-1} = (1 - \lambda)^{-1} > 0$, an increase in the rate of immigration always increases equilibrial density. The effect of immigration on population stability depends upon the shape of $G(N)$. Since $d\lambda/d\hat{I} = (d^2G/dN^2)|_*(dN^*/d\hat{I})$, immigration may destabilize populations with growth curves convex at N^*, and, conversely, stabilize populations with concave growth curves. As examples of these disparate effects, consider two familiar discrete-time versions of logistic growth: the linear-logistic, $N_{t+1} = N_t(1 + r(1-N_t/K)) + \hat{I}$, and the exponential-logistic, $N_{t+1} = N_t \exp(r(1-N_t/K)) + \hat{I}$. The dynamics of some natural populations are reasonably described by the latter model (Hassell, 1978). At $\hat{I} = 0$ the two models have similar stability properties (May and Oster, 1976). Figure 1 compares their stability domains as a function of $\hat{I}$. The infeasible region in Figure 1B exists because $G(N) < 0$ if $N > K(1+r)/r$ - a biological absurdity. A simple modification of the linear-logistic is to let $N_{t+1} = \hat{I}$ wherever $G(N) < 0$. With this change, the population persists stably at $N^* = \hat{I}$ when $\hat{I}/K > (1+r)/r$. At lower $\hat{I}$, $\lambda = 1 - r(1+4\hat{I}/Kr)^{\frac{1}{2}}$, and the population is locally unstable if $(4-r^2)/4r < \hat{I}/K < (1+r)/r$. The influence of immigration upon population stability is hence model-dependent. In the modified linear-logistic, immigration may destabilize a peripheral population, although population stability re-emerges at very high rates of immigration. By contrast, in the exponential-logistic a small influx of immigrants may suffice to produce point or cyclic stability. Little is known about the shapes of growth curves and the magnitudes of growth parameters characterizing natural peripheral populations. $\hat{I}$ could exceed K if a peripheral population is at one end of a steep spatial gradient in density. Whether r is high or low should depend on the mechanisms responsible for low density. If K is low because of high rates of density-independent mortality, so too should r be low, and the peripheral population should be dynamically stable in a stable environment. Conversely, r may be high even though K is low. For example, in MacArthur's resource-consumer model, if resources equilibrate rapidly the consumer follows a logistic-like growth model (Schaffer, 1981). The expression for K is proportional to resource productivity, and K may be low even though r, which depends upon the maximal standing crop of the resource, is high.

Vandermeer (1982) remarks that a highly chaotic population will be rare much of the time - after a population crash, several generations must pass before numbers suffice to overshoot K once again. A constant flow of immigrants reduces the time

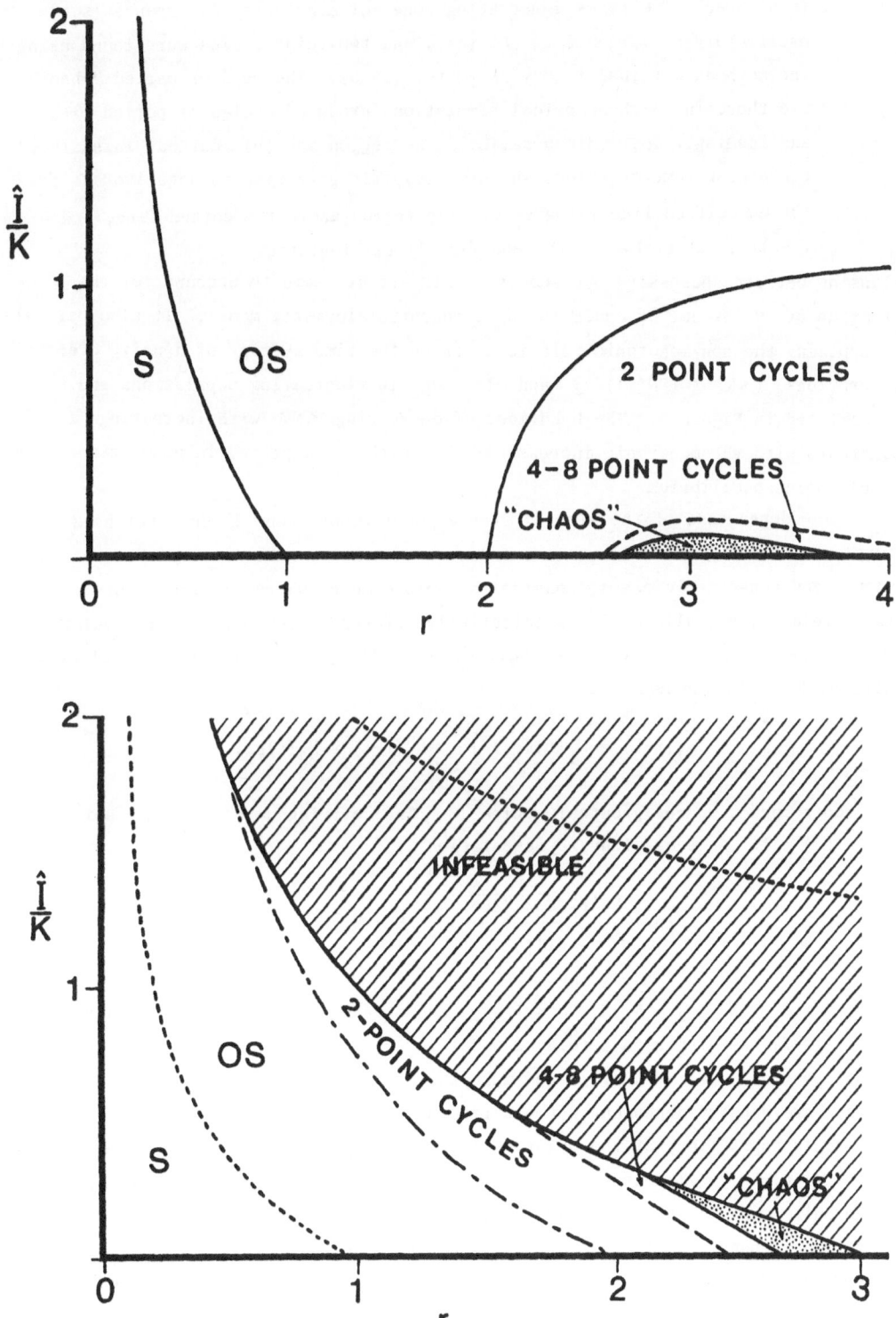

Figure 1 Stability regions for the exponential-logistic (a) and linear-logistic (b)

Figure 1 (continued) The lines demarcating zones of monotonic convergence (s),
oscillatory convergence to N* (os), and two-point cycles were found using
the methods outlined in May and Oster (1976). The regions marked "chaos"
are those in which numerical simulations produced cycles of period > 8
and seemingly aperiodic behavior. The region in (b) marked "infeasible"
contains parameter values which necessarily give rise to negative N for
the unmodified linear-logistic. The region above the dotted line, $\hat{I}/K$ =
$(1+r)/r$, is stable in the modified linear-logistic.

elapsing between successive overshoots. This effect seems to account for the
creation of cycles out of chaos in the exponential-logistic model. It also partially explains the non-monotonic relation between the time-average of density over τ
generations ($<N> \underline{\triangle} (\Sigma N(i))/\tau$) and $\hat{I}$ found in fluctuating populations and
illustrated in Figure 2. The behaviour of decreasing $<N>$ with increasing $\hat{I}$
contrasts with the monotonic increase of N* with $\hat{I}$ expected in populations at a
stable point equilibrium.

The dependence of the size of stable populations upon $\hat{I}$ does not hold if
immigrants differ genetically from residents. To sharpen the contrast between the
genetic and non-genetic cases, consider a haploid model where all immigrants contain allele 1, but allele 2 is selectively favored in the peripheral population.
Populations are censused just afer immigration. If g_i is the absolute selective
value of type i, the model is

$$N_1(t+1) = N_1(t)g_1 + \hat{I}, \quad N_2(t+1) = N_2(t)g_2. \tag{1}$$

$N_i g_i$ corresponds to the growth function G. The carrying capacities are defined
by $g_i(K_i) = 1$. An equivalent representation is

$$\Delta p = \frac{pq(g_1-g_2)+q\hat{I}/N}{\bar{g}+\hat{I}/N}, \quad \Delta N = N(\bar{g}-1) + \hat{I}$$

where $N = N_1 + N_2$, $p = N_1/N$, $q = 1 - p$, and $\bar{g} = pg_1 + qg_2$ (the mean fitness).
Comins (1977) has discussed a related model for how immigration hampers the evolution of pesticide resistance. I now make the important simplifying assumption that
the g_i are density-dependent but *not* frequency-dependent, so $g_i = g_i(N)$. In a
polymorphic population, at the point equilibrium $N_2^* = N_2^* g_2(N^*)$. Hence $(N^*,p^*) =$
$(K_2, \hat{I}/K_2(1-g_1(K_2)))$. Two necessary conditions for the existence of a stable point
polymorphism are 1) $K_1 < K_2$, and 2) $\hat{I} < K_2(1-g_1(K_2))$. In a stable, polymorphic
population a change in $\hat{I}$ does not perturb total population size *at all*. Instead,
a change in immigration rate is absorbed by a shift in gene frequency. This simple
result readily generalizes to multiple alleles if the resident allele has the
highest K and is not part of the immigrant pool.

As with the purely ecological model, the relation between $\hat{I}$ and local

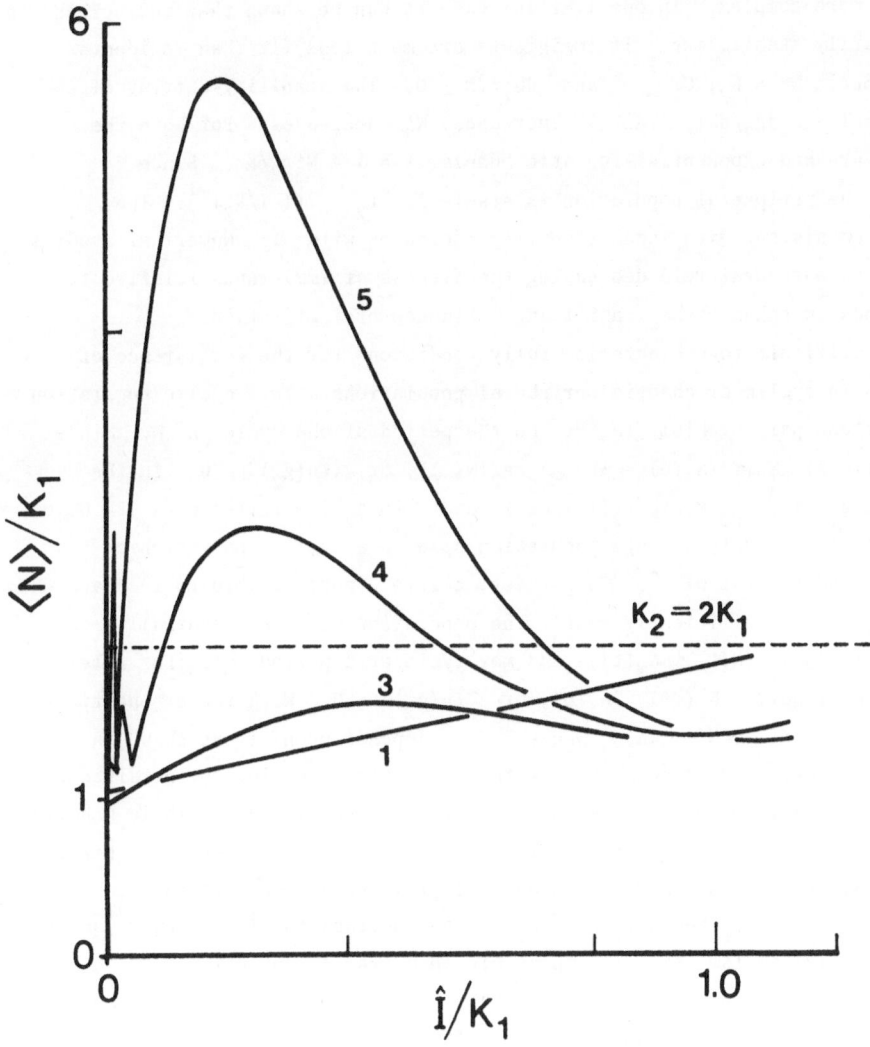

Figure 2   The time-average of density for the exponential-logistic. The solid lines show how $\langle N \rangle$ depends upon r_1 (the number beside each line) and $\hat{I}$ in populations monomorphic for type 1. From Figure 1, if $.1 < \hat{I}/K_1 < 1$ most values of $r > 2$ lead to stable two-point cycles, for which $\langle N \rangle \simeq \hat{I} +$ $G(\hat{I})/2$; the approximation is excellent for $r = 3$ if $\hat{I}/K_1 > .3$, and for $r = 4$ and 5 if $\hat{I}/K_1 > .05$. At low $\hat{I}/K_1$ and $r_1 = 4$ or 5, $\langle N_1 \rangle$ exhibits a complex pattern of multiple peaks as $\hat{I}/K_1$ increases; this is marked by a "blip" in the curve for $r = 4$ near the ordinate. Polymorphism requires $K_2 > \langle N_1 \rangle$. For example, if $K_2 = 2K_1$ the position of the broken line relative to $\langle N_1 \rangle$ determines the persistence of allele 2; at $r = 4$ or 5 polymorphism may be less likely at intermediate levels of $\hat{I}$ than at either low or high levels.

stability is more complex. In one limiting case it can be shown that an increase in $\hat{I}$ is necessarily stabilizing. If immigrants are much less fit than residents $(K_1 \ll K_2)$, near $N^* = K_2$, $G_1 \simeq 0$ and $dG_1/dN \simeq 0$. The stability-setting eigenvalue is $\lambda = 1 + N_2^* dg_2/dN|_*$. As $\hat{I}$ increases, N_2^* decreases. For both the modified linear- and exponential-logistic models, $\lambda = 1 - N_2^* r_2/K_2$. Since $N_2^* \simeq K_2 - \hat{I}$, the peripheral population is stable if $r_2 < 2(1-\hat{I}/K_2)^{-1}$. The maximum r_2 consistent with local stability increases with $\hat{I}$. Numerical studies suggest that as a general rule decreasing the fitness of immigrants relative to residents tends to enhance the stabilizing influence of immigration.

It is difficult to characterize fully conditions for the persistence of polymorphisms in cyclic or chaotic peripheral populations. In a cyclic population with a persistent polymorphism, let τ be the period of one cycle in N. Since allele 2 persists, $N_2(\tau)/N_2(0) = 1 = \prod_{i=0}^{\tau} g_2(N(i))$, or $\langle \ln(g_2) \rangle = 0$. In the exponential-logistic, $g_2 = \exp(r_2(1-N/K_2))$, and $\langle \ln(g_2) \rangle = r_2(1-\langle N \rangle/K_2) = 0$, or $\langle N \rangle = K_2$. For this model, average population size in a cyclic, polymorphic population is independent of $\hat{I}$. To persist, allele 2 must be able to increase when rare. If $N_2(0)$ is sufficiently small, the population will be essentially monomorphic for allele 1 $(N(t) \simeq N_1(t))$ and may cycle at a period τ'. For allele 2 to increase, we require $N_2(\tau') > N_2(0)$, or $\langle \ln(g_2) \rangle > 0$. With the exponential-logistic, this is $K_2 > \langle N \rangle \simeq \langle N_1 \rangle$ where $\langle N_1 \rangle$ depends upon $\hat{I}$ as shown in Figure 2. Numerical studies suggest that the heuristic condition for persistence, $K_2 > \langle N_1 \rangle$, is valid in populations with long cycles or seemingly aperiodic behavior. This condition, in conjunction with the figure, leads to two conclusions. First, for $\hat{I}/K \lesssim .75$, as r_1 increases it becomes progressively more difficult to maintain a polymorphism. Second, at high r_1 an increase in $\hat{I}$ may *increase* the likelihood of polymorphism; this is impossible in a stable population.

The ecological significance of immigration into a peripheral population thus depends upon the existence of local genetic differentiation. In turn, the maintenance of pockets of local adaptation reflects local ecological processes. In particular, overcompensatory, density-dependent growth processes may make it difficult for local adaptations to persist. For a particular model we have seen that dynamic instability through its effect on $\langle N \rangle$ may magnify the swamping effect of gene flow. Although this phenomenon does not occur in all discrete-time population models, it does occur in many commonly used models. Moreover, a cyclic or chaotic population repeatedly forces itself through bottlenecks of low density. A rate of immigration that is small in absolute numbers may actually be large relative to the number of residents then present. Locally adapted alleles may be low by drift during each bottleneck; this stochastic effect compounds the deterministic effect of instability. If local adaptation is requisite for speciation in peripheral isolates, speciation may be less likely in dynamically unstable (high r) than in stable (low r) peripheral populations. One key to a better understanding of patterns of speciation may thus be provided by the study of the ecological factors responsible for the existence of species borders and the shapes of growth curves in peripheral populations.

Acknowledgements

I thank Robert May for very useful comments. I would also like to thank the Conference participants for stimulating conversations, and in particular Dan Cohen, Yoh Iwasa, and Zvia Agur for helpful suggestions about modelling peripheral populations. The research was supported by Grant #3593, General Research Fund, University of Kansas.

REFERENCES

Comins, H.N. (1977): The development of insecticide resistance in the presence of migration, *J. Theor. Biol.* 64:177-197.

Endler, J.A. (1977): *Geographic Variation, Speciation and Clines*, Princeton: Princeton University Press.

Guckenheimer, J., G. Oster, and A. Ipaktchi (1977): The dynamics of density-dependent population models, *J. Math. Biol.* 4:101-147.

Haldane, J.B.S. (1930): A mathematical theory of natural and artificial selection, VI. Isolation, *Proc. Cambridge Philos. Soc.* 26:220-230.

Hassell, M.P. (1978): *The Dynamics of Arthropod Predator-Prey Systems*, Princeton: Princeton University Press.

Holt, R.D. (1983): Immigration and the dynamics of peripheral populations, Pp. 680-694, in A. Rhodin and K. Miyata (eds.) *Advances in Herpetology and Evolutionary Biology*, Cambridge, Mass: Harvard University, Museum of Comparative Zoology Publication Series, 713 pp.

Karlin, S. (1982): Classifications of selection-migration structures and conditions for a protected polymorphism, *Evol. Biol.* 14:61-204.

Krebs, C.J. (1978): *Ecology: The Experimental Analysis of Distribution and Abundance*, Second Edition, New York: Harper and Row.

Levin, S. (1976): Population dynamic models in heterogeneous environments, *Ann. Rev. Ecol. Syst.* 7:287-310.

May, R.M., and G.F. Oster (1976): Bifurcation and dynamic complexity in simple ecological models, *Amer. Natur.* 110:573-599.

Mayr, E. (1963): *Animal Species and Evolution*, Cambridge, Mass.: Harvard Univ. Press.

McMurtrie, R. (1978): Persistence and stability of single-species and prey-predator systems in spatially heterogeneous environments, *Math. Biosci.* 39:11-51.

Nagylaki, T. (1977): *Selection in One- and Two-Locus Systems*, Berlin: Springer-Verlag.

Prout, T. (1980): Some relationships between density-independent selection and density-dependent population growth, *Evol. Biol.* 13:1-68.

Schaffer, W.M. (1981): Ecological abstraction: The consequences of reduced dimensionality in ecological models, *Ecol. Monogr.* 51:383-401.

Vandermeer, J. (1982): To be rare is to be chaotic, *Ecology* 63:1167-1169.

A MATHEMATICAL ANALYSIS OF THE CHITTY HYPOTHESIS

Fern Hunt

1. INTRODUCTION

Chitty's hypothesis is an attempt to explain the widely observed cyclic fluctuations of vole (microtus) populations in terms of natural selection. If this is indeed a necessary condition then such cycling is an example of evolution on an ecological time scale where no steady state of local or global maximum fitness is reached.

The hypothesis supposes that fecundity and survival are controlled by a single autosomal locus with two alleles which we denote by A and a. Therefore the vole population consists of three genotype subpopulations AA, Aa and aa. The cycle is driven by density dependent changes in genotype birth and death rates that can be summarized as follows: at low densities AA and Aa have relatively high birth rates and higher survival rates than those experienced at high densities, while aa has the lowest birth rate and a relatively constant low death rate. Thus the total population rises quickly. At high densities the homozygote aa has a survival advantage while Aa and AA die out. As the population shifts to predominantly aa, there is a decline because aa has a relatively low birth rate. Once the population reaches a level low enough for AA and Aa to be at advantage, the increase phase of the cycle begins again.

A set of equations incorporating these hypotheses is presented below and we state sufficient conditions for the existence of periodic solutions. These solutions arise by Hopf bifurcation in a parameter determined by the birth and death rates of AA. Numerical computations suggest that this is a subcritical bifurcation so that small amplitude solutions are unstable. Further investigation of this conjecture is in progress. Using E. Doedel's program AUTO, global bifurcation information over a limited range of another parameter controlling the period has been obtained. Preliminary results of these computations and pictures of some numerical solutions are found in Section 3. The bifurcation results suggest that the model has stable periodic solutions for some range of parameters but that the periods have a minimum length.

Clearly this or any other mathematical model cannot prove Chitty's hypothesis to be valid. The existence of stable periodic solutions does imply that it is at least consistent and in fact plausible if parameter ranges and periods correspond to those that are actually observed (or more likely, estimated). Consequently more study and analysis will be needed to determine these. If in fact this is an explanation, what is the adaptive value of these oscillations? The numerical solutions show that the decline phase is much longer than the increase phase. Thus most of the time the population consists mainly of aa's. The increase in the heterozygote (which is brough about by an explosive increase in AA) allow the aa's to persist at far higher mean populations sizes than the size they would have if the population were fixed at a. If AA and Aa increase so rapidly that a stable polymorphism could not be reached, cycling could be the compromise.

2.   We consider a continuously breeding population composed of equal numbers of males and females. It suffices then to keep track of the number of females of each type. Thus let

$$G_1(t) = \text{number of AA females at time } t$$

$$G_2(t) = \text{number of Aa females at time } t$$

$$G_3(t) = \text{number of aa females at time } t$$

The corresponding per capita birth and death rates for each genotype are b_1, b_2, and b_3 and d_1, d_2, and d_3 respectivley. These rates are independent of age but dependent on total population size $N(t)$. Specifically,

$$b_i = b + \varepsilon\beta_i(1-\theta_3 N)$$
$$d_i = b + \varepsilon\delta_i(\theta_1 N - \theta_2) \qquad i = 1,2,3$$

Where b and θ_k, $k = 1,2,3$ are positive constants and $0 < \varepsilon \ll 1$. Let

$$p(t) = \frac{G_1(t) + G_2(t)/2}{N(t)}$$

denote the frequency of allele A in the adult gamete pool at time t. G_i satisfies the following equations.

$$\frac{dG_1}{dt} = (b_1 G_1 + b_2 G_2/2)p - d_1 G_1$$

$$\frac{dG_2}{dt} = b_1 G_1 q + b_3 G_3 p + \frac{b_2 G_2}{2} - d_2 G_2 \qquad (1)$$

$$\frac{dG_3}{dt} = (b_3 G_3 + b_2 G_2/2)q - d_3 G_3$$

where $q = 1 - p$. Since $\varepsilon \ll 1$, the solutions of (1) approach a quasi-Hardy-Weinberg equilibrium. Thus the G_i can be determined up to a term of order ε by p_0 and N_0 where

$$p = p(t,\varepsilon) = p_0(\varepsilon t) + 0(\varepsilon) \quad \text{and} \quad N = N(t,\varepsilon) = N_0(\varepsilon t) + 0(\varepsilon)$$

Equations for p_o and N_o are derived and analyzed in Hunt (1982). Thus (1) is reduced to the analysis of a planar system. Let

$$r = \frac{\beta_1 + \beta_3 - 2\beta_2}{2} + \mu\theta_2(\delta_1 - 2\delta_2 + \delta_3)$$

$$s = \frac{\beta_2 - \beta_3}{2} + \mu\theta_2(\delta_2 - \delta_3)$$

$$R = \beta_1 + \beta_3 - 2\beta_2 + \mu\theta_2(\delta_1 - 2\delta_2 + \delta_3)$$

$$S = \beta_2 - \beta_3 + \mu\theta_2(\delta_2 - \delta_3)$$

where $\mu = \theta_1/\theta_2\theta_3$ and for convenience we set $\delta_3 = 0$. The following conditions are assumed.

(i) $0 < \mu < 1$

(ii) $s > 0$, $r + s > 0$

(iii) $\beta_3 > 0$, $R + 2S + \beta_3 > 0$ and either $R > 0$, $S > 0$ or $R < 0$.

(iv) $\beta_1 < 0$, $\beta_2 > \beta_3$ and $\dfrac{\beta_2}{2|\beta_1|} + \dfrac{\delta_2}{\delta_1} < \dfrac{1}{2}$

Under (i)-(iv) the following theorem holds.

THEOREM *Given* $\beta_2\beta_3$, δ_2 *fixed, and sufficiently large* $|\beta_1|$ *and* δ_1 *such that*

(a) $\beta_2/|\beta_1| < 0.1$ *and* (b) $\beta_3/|\beta_1|$, δ_2/δ_1

are sufficiently small, there is for each $0 < \mu \le \mu_o$ *a family of periodic solutions of the planar system parameterized by* $\lambda = \theta_2\delta_1/|\beta_1|$ *with* $1/\mu < \lambda < \lambda_c$. *These solutions arise by Hopf bifurcation in the parameter* λ *and* λ_c *is its critical value.*

A proof of this theorem, other details of the analysis and a discussion of the condition on the parameters can be found in Hunt (1982).

3. Figures 1-3 show numerical solutions of (1) for the indicated parameter values. Note that the period decreases as δ_2 increases, keeping other parameters fixed. This is confirmed by the stability calculation carried out by E. Doedel. The results, obtained through the use of AUTO are illustrated in the bifurcation diagram in Figure 4.

Proceeding along the periodic branch as indicated by the arrow, we find that the period increases monotonically. At the elbow of the curve the periodic solutions become stable and δ_2 decreases. Thus the periods of stable solutions are bounded below. Our calculations indicate that this bound is $0(1/\varepsilon)$.

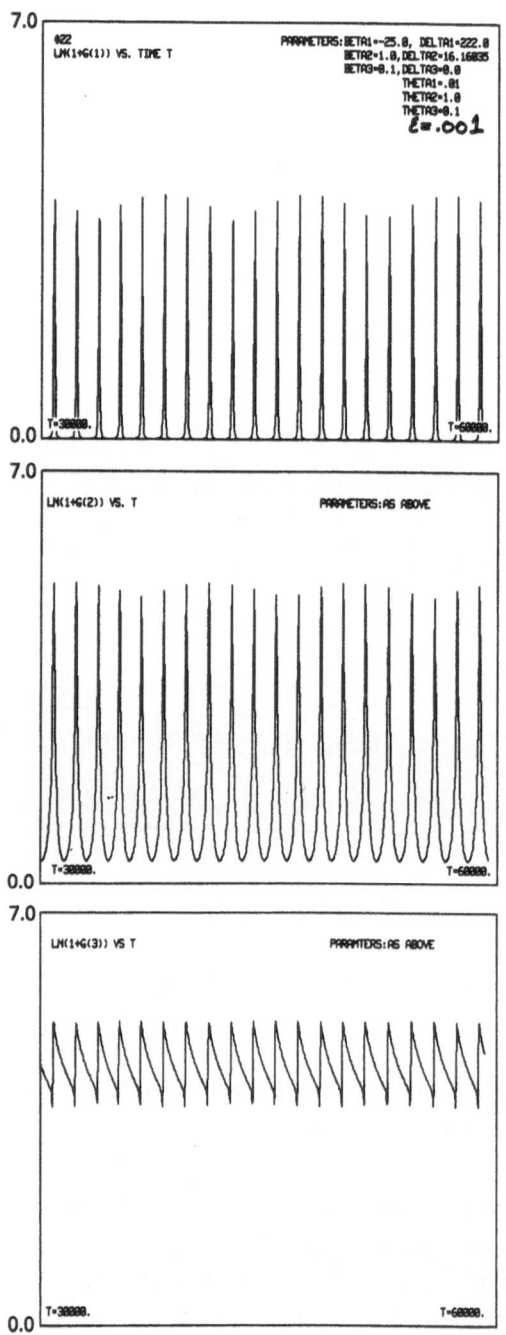

Figure 1

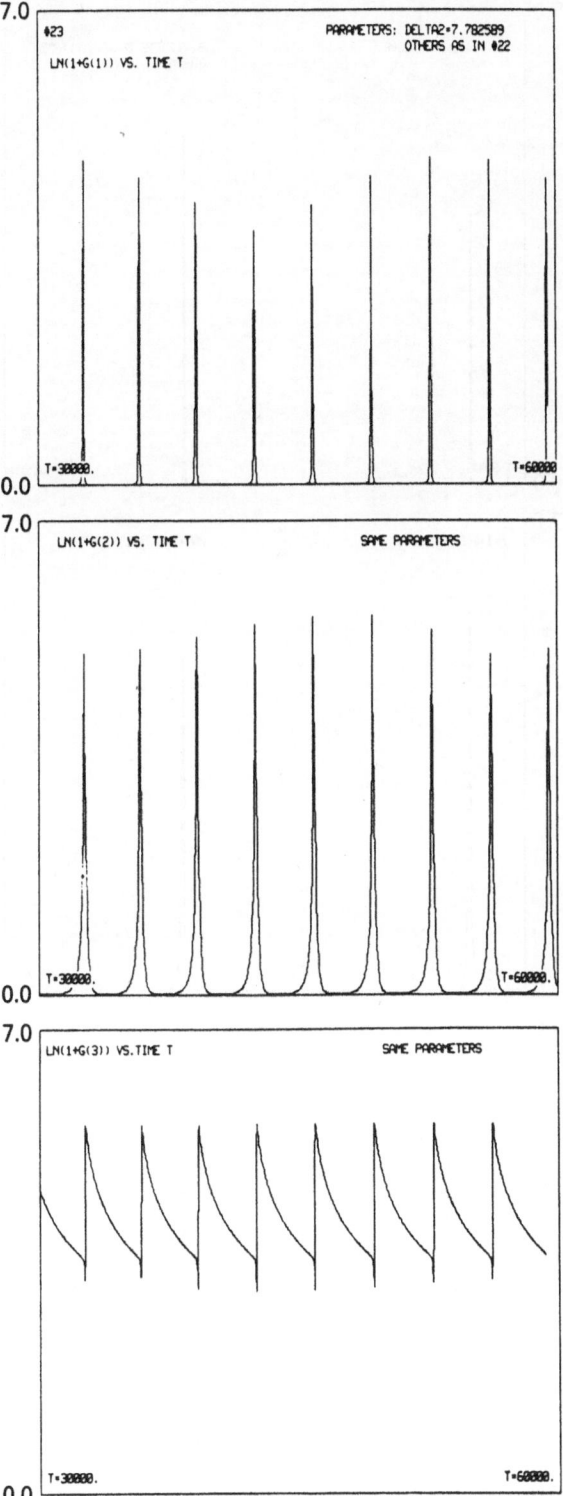

Figure 2

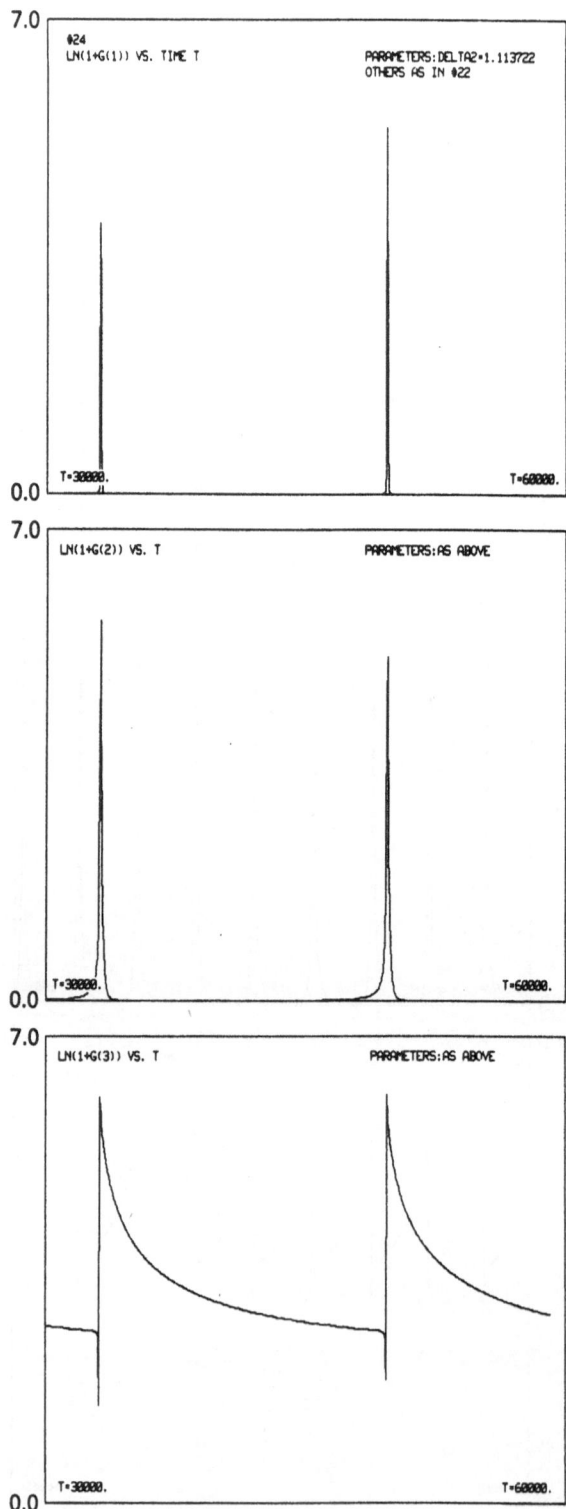

Figure 3

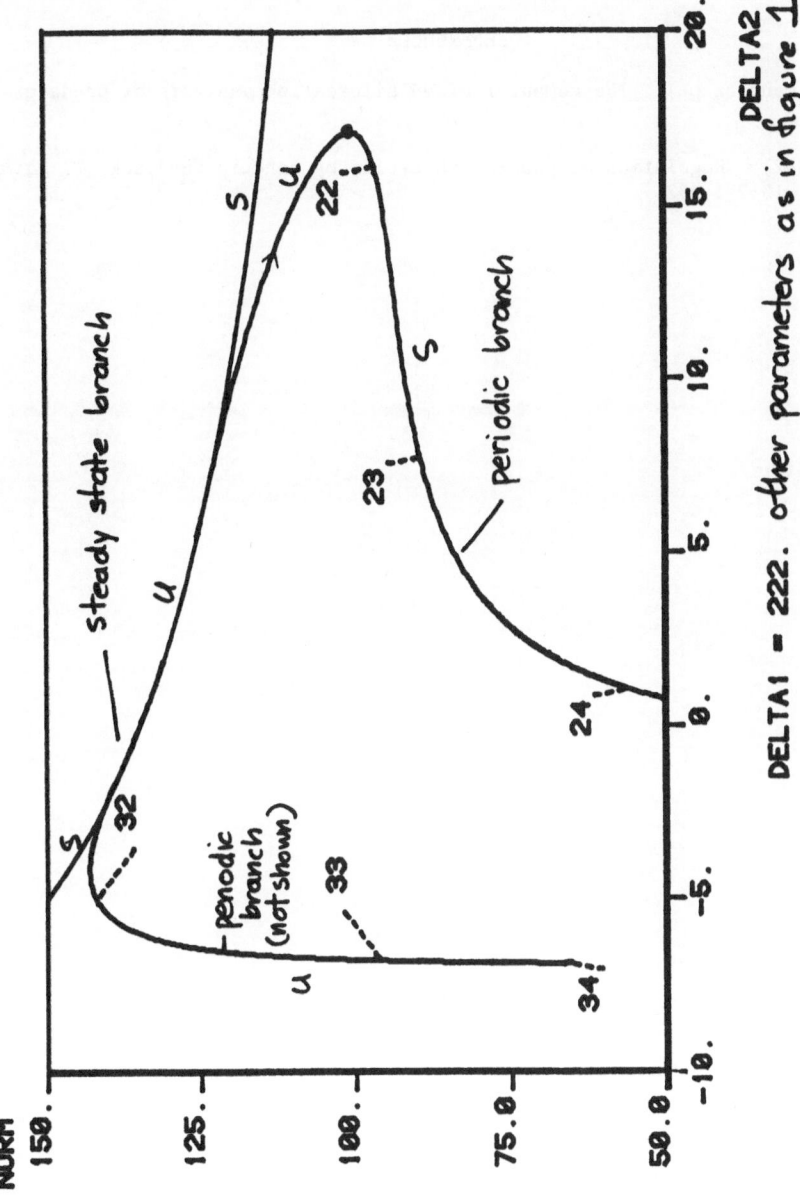

Figure 4

Acknowledgement

The author thanks E. Doedel for carrying out the calculation shown in Figure 4 as well as other stability calculations.

REFERENCES

Doedel, E. (manuscript): The computer-aided bifurcation analysis of predator-prey models.

Hunt, F. (1982): Regulation of population cycles by genetic feedback, *T. Math. Biol.* 13:271-282.

AN HAPLOID MODEL FOR GENOTYPE-
DEPENDENT POPULATION DIFFUSION

Michael E. Moody

1. INTRODUCTION

Recent work (Moody, 1981; Nagylaki and Moody, 1980) has extended the analysis of diploid migration models to the biologically more reasonable case of genotype-dependent migration. We here consider a similar model applied to haploid populations. For a diallelic, neutral, deterministic, discrete-space, discrete-time haploid model of selective migration, a weak-migration diffusion approximation is developed and analyzed. It is shown that alleles can be lost in the absence of natural selection, and that if the distributions of migration distance for the two types have different means, then all polymorphic equilibria are clines. A specific example is given.

2. FORMULATION

Suppose an haploid population occupies $N+1$ equally spaced colonies in a linear array. At a given locus there are alleles A_i, $i = 1,2,\dots$, and the population evolves in discrete time with nonoverlapping generations and asexual reproduction. The frequency of allele A_i in deme k ($k = 0,1,2,\dots,N$) and the relative size of deme k in generation t will respectively be denoted by $p_{i,k}(t)$ and $c_k(t)$. We further assume the absence of natural selection either through viability or fertility differences, and neglect mutation and random drift. Each generation, the zygotic distribution of relative deme size (the $c_k(t)$) will be determined by genotype-independent population regulation, and a fraction $\tilde{m}_{i;jk}$ of A_i adults in deme j migrate to deme k. For fixed i, the matrix $\tilde{M}_i = (\tilde{m}_{i;jk})$ is the forward migration matrix for genotype A_i. For this model, Moody (in preparation) has given the recursion relations satisfied by the $p_{i,k}$:

$$p_{i,k}(t+1) = \frac{\sum_{\ell} p_{i,\ell} c_\ell \tilde{m}_{j,\ell k}}{\sum_{j,\ell} p_{j,\ell} c_\ell \tilde{m}_{j,\ell k}} , \tag{1}$$

where the time argument has been suppressed on the right. In Moody (in preparation), the linearized version of (1) is analyzed as well as the associated dynamics of the genotypic numbers. The similarity of (1) to the analagous equation for diploids should be noted (Moody, 1981; Nagalaki and Moody, 1980).

We seek to replace (1) by an appropriate system of partial differential equations. To this end, suppose the number of colonies is large ($N \gg 1$), and the separation between colonies, ϵ, is small $(\epsilon \ll 1)$; rescale time so that δ^{-1} generations of the old time units corresponds to one generation on our new scale, where $\delta > 0$ is a small parameter. Suppose that $\delta \to 0$ as $\epsilon \to 0$ in such a way that ϵ^2/δ tends to a positive constant; also suppose that $N\epsilon = L$ (the habitat length) is fixed as $\epsilon \to 0$, $N \to \infty$. The rescaled position and time variables are defined in the obvious way,

$$x = \epsilon k, \quad y = \epsilon \ell, \quad \tau = \delta t, \tag{2a}$$

and the local population density $\rho(x,\tau)$ is taken to be proportional to $c_k(t)$

$$c_k(t) = \epsilon \rho(x,\tau). \tag{2b}$$

In our new variables the gene frequencies read $u_i(x,\tau) = p_{i,k}(t)$.

Henceforth, we will restrict consideration to nearest-neighbor migration and will follow the notation in Nagylaki and Moody (1980). Define

$$a_{i,k} = \tilde{m}_{i;k,k+1}, \qquad 0 \le k \le N - 1, \tag{3a}$$

$$b_{i,k} = \tilde{m}_{i;k,k-1}, \qquad 1 \le k \le N, \tag{3b}$$

and make the diffusion assumptions as $\epsilon \to 0$:

$$\epsilon \delta^{-1}[a_{i,k} - b_{i,k}] = M_i(x,\tau) + o(\epsilon), \qquad 0 \le k \le N, \tag{3c}$$

$$\epsilon^2 \delta^{-1}[a_{i,k} + b_{i,k}] = V_i(x,\tau) + o(\epsilon^2), \qquad 0 \le k \le N. \tag{3d}$$

As in the derivation for the diploid case, we choose $a_{i,N}$ and $b_{i,0}$ so that (3c) and (3d) hold with smooth drift and diffusion coefficients, $M_i(x,\tau)$ and $V_i(x,\tau)$ for genotype A_i. In the limit we recover the appropriate system of partial differential equations for the $u_i(x,\tau)$:

$$\rho(u_i)_\tau = u_i \sum_j (J_j)_x - (J_i)_x \qquad 0 < x < L, \quad \tau > 0, \tag{4a}$$

where

$$J_i(x,\tau) = \rho(x,\tau)M_i(x,\tau)u_i(x,\tau) - \frac{1}{2}[\rho(x,\tau)V_i(x,\tau)u_i(x,\tau)]_x \tag{4b}$$

represents the flux of A_i genes. Requiring that $u_\tau(x,\tau)$ be finite as $x \to 0^+, L^-$ leads to the appropriate boundary conditions

$$J_i - u_i \sum_j J_j = 0, \qquad x = 0,L, \quad \tau \ge 0. \tag{4c}$$

Due to the complexity of (4), we will henceforth restrict our attention to the case of two alleles, uniform (regulated) population density, and homogeneous drift and diffusion coefficients. With these assumptions (4) reduces to

$$u_\tau = -f_M(u)u_x + \frac{1}{2} f_V(u)u_{xx}, \qquad 0 < x < L, \qquad \tau > 0 \qquad (5a)$$

$$f_V(u)u_x = 2(M_1-M_2)u(1-u), \qquad x = 0,L, \qquad \tau \geq 0, \qquad (5b)$$

where $u = u_1$ and

$$f_M(u) = (M_2-M_1)u + M_1, \qquad (5c)$$

$$f_V(u) = (V_2-V_1)u + V_1; \qquad (5d)$$

M_1, M_2, V_1 and V_2 are independent of x and τ.

3. ANALYSIS

We can easily show that the solutions to (5) are always biologically acceptable (Nagylaki and Moody, 1980): $0 \leq u(x,0) \leq 1$ implies $0 \leq u(x,\tau) \leq 1$ for $\tau > 0$, $0 \leq x \leq L$.

3.1 *Equal mean displacements* $(M_1 = M_2)$. If the mean displacements are genotype-independent, then (5) yields immediately that the only equilibrium (i.e. time-independent) solutions are constants: $\hat{u}(x) = u_0$ for some $0 \leq u_0 \leq 1$. By an application of the maximum principle for linear parabolic systems, as outlined in Nagylaki and Moody (1980), we may show that $u(x,\tau)$ is bounded for all $\tau > 0$ between its initial minimum and maximum values: $\inf_x u(x,0) \leq u(x,\tau) \leq \sup_x u(x,0)$.

Thus all constant solutions are *a fortiori* stable equilibria. Furthermore, the function

$$\Gamma(\tau) = \int_0^L [u_x(x,\tau)]^2 dx$$

is a Lyapunov functional for (5), $(\Gamma(\tau) \geq 0, \Gamma'(\tau) \leq 0)$ we conclude that $u(x,\tau) \to \hat{u}_0$, as $\tau \to \infty$, where $\hat{u}_0$ is constant.

3.2 *Protection of rare alleles*. If we linearize (5) near $u = 0$ we obtain the boundary value problem

$$u_\tau = \frac{1}{2} V_1 u_{xx} - M_1 u_x \qquad 0 < x < L, \qquad \tau > 0, \qquad (6a)$$

$$V_1 u_x = 2(M_1-M_2)u \qquad x = 0,L, \qquad \tau \geq 0. \qquad (6b)$$

Solving the eigenvalue problem associated with (6) allows us to conclude that, if $M_1 = 0$, then $u = 0$ is unstable, whereas if $M_1 \neq 0$, then $u = 0$ is asymptotically stable only if $|M_1| < |M_2|$ and $M_1 M_2 > 0$. Thus allele A_1 will be eliminated when initially sufficiently rare if the mean displacement of A_1 is in the same direction as A_2 but of smaller absolute value. It will otherwise be maintained in the population. Interchanging 1 and 2 in the above inequalities yields, by symmetry, the appropriate protection conditions for allele A_2. Also, if the mean displacements are unequal but of the same sign, then both fixation states cannot be simultaneously stable. Thus genetic variability can be completely lost in the absence of natural selection.

By the results of the preceding section, if $M_1 = M_2$ then both alleles will be protected.

3.3 *Equilibria*. We suppose $M_1 \neq M_2$; otherwise the results of Section 1 apply. We are interested in the time-independent analog of (5): Put $u(x,\infty) = \hat{u}(x)$. Then (5) yields

$$f_V(\hat{u})\hat{u}''(x) = 2f_M(\hat{u})\hat{u}'(x) \qquad 0 < x < L, \tag{7a}$$

$$f_V(\hat{u})\hat{u}'(x) = 2h_M(\hat{u}) \qquad x = 0, L, \tag{7b}$$

where

$$h_M(\hat{u}) = (M_1 - M_2)\hat{u}(1-\hat{u}). \tag{7c}$$

It can be readily shown that all nontrivial solutions to (7) must be monotone: $\hat{u}'(x) \neq 0$, $0 \leq x \leq L$. As per the method in Nagylaki and Moody (1980), we obtain any polymorphic solution to (7) implicitly as

$$x = \int_\alpha^{\hat{u}} \frac{du}{U(u)} , \tag{8a}$$

where

$$U(\hat{u}) = 2[h_M(\alpha)f_V^{-1}(\alpha) + \int_\alpha^{\hat{u}} f_M(u)f_V^{-1}(u)du] \tag{8b}$$

and where the unknown constants $\alpha \equiv \hat{u}(0)$, $\beta \equiv \hat{u}(L)$ are determined by (7b) at $\hat{u} = \beta$ and by L:

$$f_V(\beta)U(\beta) = 2h_M(\beta) \tag{8c}$$

$$\hat{u}(L) = \beta. \tag{8d}$$

In (8c), (8d), we substitute for u and U from (8a) and (8b), respectively.

As the ratio of two linear functions, the integral in (8b) is easily reduced to quadrature.

The biologically interesting aspect of the above result is that all polymorphic equilibria, should they exist, must be monotone, i.e., a cline. Based on the stability criteria for the fixation equilibria, we are led to conjecture that a sufficient condition for an equilibrium cline to exist is $sgn(M_1 M_2) < 0$. We record by way of example the following special case: $M_1 = M = -M_2$, $V_1 = V_2$. With these parameters the solution is given by

$$\hat{u}(x) = \frac{\mu^- e^{\eta x} - \nu \mu^+}{e^{\eta x} - \nu} \tag{9a}$$

where

$$\mu^{\pm} = \frac{1}{2} \{1 \pm [1 + 4\alpha(1-\alpha)^{\frac{1}{2}}\} \tag{9b}$$

$$\eta = -\frac{2M}{V} [1 + 4\alpha(1-\alpha)]^{\frac{1}{2}} \tag{9c}$$

$$\nu = \frac{2\alpha - 1 + [1 + 4\alpha(1-\alpha)]^{\frac{1}{2}}}{2\alpha - 1 - [1 + 4\alpha(1-\alpha)]^{\frac{1}{2}}} \tag{9d}$$

and α, which equals $1-\beta$, is given by the solution to

$$\alpha = \frac{\mu^+ e^{\eta L} - \nu \mu^-}{e^{\eta L} - \nu} \tag{9e}$$

which lies in the interval $[0, \frac{1}{2})$. Since $\alpha + \beta = 1$, we observe that there exists an x for which $\hat{u}(x) = 1/2$, and hence $u'' = 0$ at this point; thus the cline has an inflection point, in contrast to the examples studied in Nagylaki and Moody (1980) for diploids.

Hence geographic variability can be maintained in the absence of the effects of natural selection.

4. SUMMARY

The preceding analysis complements the work of Moody (1981) and Nagylaki and Moody (1980) on selective migration for diploid populations. Qualitatively similar results obtain in the simpler haploid genetic situation: alleles can be eliminated from the population in the absence of natural selection, and all polymorphic equilibria, when extant, are clines. By example it was shown that an equilibrium cline could possess an inflection point, which was unobserved in the diploid models (Nagylaki and Moody, 1980).

Hence, the joint action of genotype-dependent (selective) migration and genotype-independent population regulation can mimic the effects of natural selection both to reduce and to maintain geographic heterogeneity.

REFERENCES

Moody, M.E. (1981): Polymorphism with selection and genotype-dependent migration, *J. Math. Biol.* 11:245-267.

_____ (in preparation): A model for genotype-dependent migration in haploid, subdivided populations.

Nagylaki, T., and M.E. Moody (1980): Diffusion model for genotype-migration, *Proc. Nat. Acad. Sci. USA.* 77:4842-4846.

FURTHER MODELS OF SELECTION WITH ANTAGONISTIC PLEIOTROPY

Michael R. Rose

1. INTRODUCTION

Evidence supporting Wright's (1980) view that pleiotropy is of importance in evolution has begun accumulating (e.g., Simmons et al., 1980). As has been pointed out on many occasions, if such pleiotropy has antagonistic effects, such that the enhancement of one fitness-component is associated with deleterious effects on another, then it may lead to the maintenance of allelic polymorphism at loci affecting fitness-components (Caspari, 1950). This polymorphism may in turn explain the apparent maintenance of significant additive genetic variances for fitness-components without significant additive genetic variance in fitness itself (Falconer, 1977).

I attempted to assess the potential scope for the maintenance of genetic variability by means of antagonistic pleiotropy using a variety of simple population genetic models (Rose, 1982). These were models of monoecious populations of infinite size with discrete generations and two fitness-components. Fitnesses were assumed to equal the sum or the product of these two fitness-components. The two major results were that (i) antagonistic pleiotropy could indeed readily maintain genetic variability, and (ii) such maintenance was fostered by recessive deleterious gene action. These conclusions held for (a) summed or multiplied fitness-component pairs, (b) two or three alleles at a single locus, and (c) two diallelic loci in the absence of epistasis affecting individual fitness components.

In view of this pattern of robustness, it is tempting to push the models further. In the present article, the previous analysis is generalized to include (a) an arbitrary number of fitness-components, (b) both summed and multiplied components simultaneously determining fitness, (c) arbitrary weak epistasis patterns in two-locus models, and (d) populations with two sexes.

2. MULTIPLE FITNESS-COMPONENTS

2.1 *Two Fitness-Components*

In order to provide some orientation, the simplest model of Rose (1982) will be reviewed. All subsequent models will be variations of it. As in all these models, it is assumed that the population in question is infinite in size, randomly mating, and diploid, with discrete generations.

Let there be two alleles at a locus, one designated A_1 and one A_2. Fitness of genotype A_iA_j is designated $W(A_iA_j)$, with fitness-components $W_1(A_iA_j)$ and $W_2(A_iA_j)$. Let V and f be positive scalar variables of magnitude about unity. Let ε and δ be strictly positive scalar variables much less than one in magnitude. Let h_1 and h_2 be non-negative scalars. The following table gives the fitness-components and fitnesses for additive and multiplicative fitness-components:

	A_1A_1	A_1A_2	A_2A_2
W_1	$V + h_1\varepsilon$	V	$V - \varepsilon$
W_2	$f - \delta$	f	$f + h_2\delta$
$W_1 + W_2 = W$	$V + f + h_1\varepsilon - \delta$	$V + f$	$V + f + h_2\delta - \varepsilon$
$W_1W_2 = W$	$Vf + h_1 f\varepsilon$	Vf	$Vf + h_2 V\delta$
	$-V\delta - h_1\varepsilon\delta$		$-f\varepsilon - h_2\varepsilon\delta$

When $W = W_1 + W_2$, protected polymorphism is achieved when $\varepsilon > h_2\delta$ and $\delta > h_1\varepsilon$. When $W = W_1W_2$, protected polymorphism is achieved when $V\delta > h_1\varepsilon(f-\delta)$ and $f\varepsilon > h_2\delta(V-\varepsilon)$. Two things should be noted. Firstly, it is easy to meet these parametric conditions. Secondly, for $h_i = 0$, they are always met, while for $h_i \approx 1$ they are difficult or impossible to meet. This analysis will now be generalized.

2.2 *Many Additive or Multiplicative Fitness-Components*

Let there be two alleles, as before. Let the number of fitness-components be n, with parameters f_i, ε_i, and h_j corresponding to the f, ε, and h_i, above, in magnitude and sign. The fitness-components are now as follows:

	A_1A_1	A_1A_2	A_2A_2
W_1	$f_1 + h_1\varepsilon_1$	f_1	$f_1 - \varepsilon_1$
W_2	$f_2 + h_2\varepsilon_2$	f_2	$f_2 - \varepsilon_2$
$\vdots$	$\vdots$	$\vdots$	$\vdots$
W_k	$f_k + h_k\varepsilon_k$	f_k	$f_k - \varepsilon_k$
W_{k+1}	$f_{k+1} - \varepsilon_{k+1}$	f_{k+1}	$f_{k+1} + h_{k+1}\varepsilon_{k+1}$
$\vdots$	$\vdots$	$\vdots$	$\vdots$
W_n	$f_n - \varepsilon_n$	f_n	$f_n + h_n\varepsilon_n$

If $W(A_iA_j) = \sum_{m=1}^{n} W_m(A_iA_j)$, and $\sum_{m=1}^{n} f_m = F_a$, then $W(A_1A_2) = F_a$,

$W(A_1A_1) = F_a + \sum_{i=1}^{k} h_i\varepsilon_i - \sum_{i=k+1}^{n} \varepsilon_i$, and $W(A_2A_2) = F_a - \sum_{i=1}^{k} \varepsilon_i + \sum_{i=k+1}^{n} h_i\varepsilon_i$.

As before, if the $h_i = 0$, then

$$W(A_1A_2) - W(A_jA_j) = -\sum \varepsilon_i > 0$$

with the appropriate summation limits. This guarantees protected polymorphism. If all $h_i = 1$, $W(A_1A_1) + W(A_2A_2) = 2W(A_1A_2)$, which implies that either there exists a j such that $W(A_jA_j) > W(A_1A_2)$ or $W(A_1A_2) = W(A_jA_j)$, $j = 1,2$. In either case, protected polymorphism cannot be achieved. If all $h_i > 1$, there exists a j such that $W(A_jA_j) > W(A_1A_2)$. As was found before, recessive deleterious gene action fosters protected polymorphism.

If $W(A_iA_j) = \prod_{m=1}^{n} W_m(A_iA_j)$, and $\prod_{m=1}^{n} f_m = F_p$, then $W(A_1A_2) = F_p$,

$$W(A_1A_1) = F_p + F_p \sum_{i=1}^{k} \frac{\varepsilon_i h_i}{f_i} - F_p \sum_{i=k+1} \varepsilon_i$$

$$+ F_p \sum_{i=1}^{k} \frac{h_i \varepsilon_i}{f_i} \left\{ \sum_{\substack{j=1 \\ j \neq i}}^{k} \frac{h_j \varepsilon_j}{f_j} - \sum_{j=k+1}^{n} \varepsilon_j/f_j \right\} - F_p \sum_{i=k+1}^{k} \left[\frac{\varepsilon_i}{f_i} \left\{ \sum_{j=1}^{k} \frac{h_j \varepsilon_j}{f_j} - \sum_{\substack{j=k+1 \\ j \neq i}}^{n} \frac{\varepsilon_j}{f_j} \right\} \right]$$

$$+ \ldots = F_p \left\{ 1 + \sum_{i=1}^{k} \frac{\varepsilon_i h_i}{f_i} - \sum_{i=k+1}^{n} \frac{\varepsilon_i}{f_i} \right\} + 0(\varepsilon^2)$$

where $\varepsilon = \sum_{i=1}^{n} |\varepsilon_i|$. The equation for $W(A_2A_2)$ is symmetric, and need not be given. If all $h_i = 0$, we have $W(A_1A_2) = F_p \left\{ 1 - \sum_{i=k+1}^{n} \frac{\varepsilon_i}{f_i} \right\} + 0(\varepsilon^2)$,

with a similar expression for $W(A_2A_2)$. With this result, if $\varepsilon \ll 1$, then $W(A_1A_2) > W(A_jA_j)$, $j = 1,2$. This result is not as clean as the result for additive fitness-components or that for two multiplicative fitness-components, but it still shows that recessive deleterious gene action tends to foster protected polymorphism. As was the case for additive fitness-components, if all $h_i = 1$, then $2W(A_1A_2) = W(A_1A_1) + W(A_2A_2)$, precluding protected polymorphism.

2.3 *Combined Multiplicative-Additive Model*

Let f_{ij}, h_{ij}, and ε_{ij} be similar to f, h_i and ε, as before. Say there is the gene effects pattern shown on the following page.

If $W(\cdot) = \sum_{i=1}^{m} \prod_{j=1}^{\ell_i} W_{ij}(\cdot)$, $F_i = \prod_{j=1}^{\ell_i} f_{ij}$, and $F_{pa} = \sum_{i=1}^{m} (\prod_{j=1}^{\ell_i} f_{ij})$, then

$W(A_1A_2) = F_{pa}$ and

$$W(A_1A_1) = \sum_{i=1}^{m} \left\{ F_i \left[1 + \sum_{j=1}^{k_i} \frac{h_{ij}\varepsilon_{ij}}{f_{ij}} - \sum_{j=k_i+1}^{\ell_i} \frac{\varepsilon_{ij}}{f_{ij}} + 0(\varepsilon_i^2) \right] \right\},$$

where $\varepsilon_i = \sum_j \varepsilon_{ij}$.

	A_1A_1	A_1A_2	A_2A_2
W_{11}	$f_{11} + h_{11}\varepsilon_{11}$	f_{11}	$f_{11} - \varepsilon_{11}$
$\vdots$	$\vdots$	$\vdots$	$\vdots$
W_{1k_1}	$f_{1k_1} + h_{1k_1}\varepsilon_{1k_1}$	f_{1k_1}	$f_{1k_1} - \varepsilon_{1k_1}$
$W_{1(k_1+1)}$	$f_{1(k_1+1)} - \varepsilon_{1(k_1+1)}$	$f_{1(k_1+1)}$	$f_{1(k_1+1)} + h_{1(k_1+1)}\varepsilon_{1(k_1+1)}$
$\vdots$	$\vdots$	$\vdots$	$\vdots$
$W_{1\ell_1}$	$f_{1\ell_1} - \varepsilon_{1\ell_1}$	$f_{1\ell_1}$	$f_{1\ell_1} + h_{1\ell_1}\varepsilon_{1\ell_1}$
$\vdots$	$\vdots$	$\vdots$	$\vdots$
$W_{m\ell_m}$	$f_{m\ell_m} - \varepsilon_{m\ell_m}$	$f_{m\ell_m}$	$f_{m\ell_m} + h_{m\ell_m}\varepsilon_{m\ell_m}$

Thus

$$W(A_1A_1) - W(A_1A_2) = \sum_{i=1}^{m} \left\{ F_i \left[\sum_{j=1}^{k_i} \frac{h_{ij}\varepsilon_{ij}}{f_{ij}} \right] \right\} - \sum_{i=1}^{m} \left\{ F_i \sum_{j=k_i+1}^{\ell_i} \frac{\varepsilon_{ij}}{f_{ij}} \right\} + 0(\varepsilon^2),$$

and likewise for $W(A_2A_2)$. As before, as the $h_{ij} \to 0$, for small ε, heterozygote superiority is rendered more likely, and conversely.

3. ARBITRARY WEAK EPISTASIS

In Rose (1982), two diallelic loci affecting two fitness-components were treated theoretically under the assumption of no epistasis in effects on single fitness-components, although epistasis arose for fitness itself in the multiplicative case. Here I allow epistasis in fitness-component effects. As before, let there be two loci, A and B, each with two alleles, A and a, B and b respectively. The within-locus fitness-component effects are as follows:

	W_1 Effect	W_2 Effect
AA	$h_{11}\varepsilon_1$	$-\delta_1$
Aa	0	0
aa	$-\varepsilon_1$	$h_{12}\delta_1$
BB	$h_{21}\varepsilon_2$	$-\delta_2$
Bb	0	0
bb	$-\varepsilon_2$	$h_{22}\delta_2$

Here the ε_i, δ_j, and $h_{k\ell}$ parameters are generalizations of the parameters from the one-locus two-component model.

Given arbitrary weak epistasis and no position effects, the fitnesses of w_{ij} genotypes composed of gametes i and j, where AB, Ab, aB, and ab are gametes 1, 2, 3, and 4 respectively, are:

$$w_{11} = C + a_1 + a_2 + e_{11}, \quad w_{13} = C + a_2 + e_{13}, \quad w_{33} = C + a_2 + a_3 + e_{33},$$

$$w_{12} = C + a_1 + e_{12}, \quad w_{23} = w_{14} = C, \quad w_{34} = C + a_3 + e_{34}, \quad w_{22} = C + a_1 + a_4 + e_{22},$$

$$w_{24} = C + a_4 + e_{24}, \quad w_{44} = C + a_3 + a_4 + e_{44}, \quad \text{where the } e_{ij} \text{ give the epistatic}$$

effects, and the a_i depend on whether the fitness-components are multiplicative or additive. When they are multiplicative:

$$a_1 = fh_{11}\varepsilon_1 - V\delta_1 - h_{11}\varepsilon_1\delta_1, \qquad a_2 = fh_{21}\varepsilon_2 - V\delta_2 - h_{21}\varepsilon_2\delta_2,$$

$$a_3 = Vh_{12}\delta_1 - f\varepsilon_1 - h_{12}\varepsilon_1\delta_1, \text{ and } a_4 = Vh_{22}\delta_2 - f\varepsilon_2 - h_{22}\varepsilon_2\delta_2.$$

When they are additive: $a_1 = h_{11}\varepsilon_1 - \delta_1$, $a_2 = h_{21}\varepsilon_2 - \delta_2$, $a_3 = h_{12}\delta_1 - \varepsilon_1$, and $a_4 = h_{22}\delta_2 - \varepsilon_2$.

The gamete monomorphism equilibria are unstable if there is always an i such that, $w_{ij} > w_{jj}$ for $j = 1,2,3,4$ (Bodmer and Felsenstein, 1967). This will always be true, in both multiplicative and additive cases, when $0 > a_i + 0(e)$, $i = 1,2,3$, and 4.

The two-gamete equilibria depend on more complex conditions. If $u = (w_{34}-w_{44})/(2w_{34}-w_{33}-w_{44})$, $v = 1 - u$, $w_1^* = uw_{13} + vw_{14}$, $w_2^* = uw_{14} + vw_{24}$, and $w = uw_{33} + vw_{34}$, then a sufficient condition for the evolutionary instability of a population polymorphic for aB and ab is $w_i^* > w$, $i = 1,2$ (Bodmer and Felsenstein, 1967). Similar conditions apply for all other two-gamete boundary equilibria. Taking the aB/ab polymorphic case explicitly,

$$u = \frac{a_4}{a_2+a_4} + 0(e), \quad v = \frac{a_2}{a_2+a_4} + 0(e), \quad w_1^* = C + \frac{a_2 a_4}{a_2+a_4} + 0(e), \quad w_2^* = C + \frac{a_2^2 a_4}{a_2+a_4} + 0(e),$$

and $w = C + a_3 + \dfrac{a_2 a_4}{a_2+a_4} + 0(e)$.

Thus instability arises when $0 > a_3 + 0(e)$, and similarly for all other two-gamete equilibria. Thus the general sufficient condition for protected polymorphism, with all boundary equilibria vulnerable to perturbations establishing four-gamete polymorphism, is simply:

$$0 > a_i + 0(e), \quad i = 1, \ldots 4.$$

In the additive case, this condition is equivalent to: $\delta_1 > h_{11}\varepsilon_1 + 0(e)$,

$\delta_2 > h_{21}\epsilon_2 + 0(e)$, $\epsilon_1 > h_{12}\delta_1 + 0(e)$, $\epsilon_2 > h_{22}\delta_2 + 0(e)$. In the multiplicative

case, it suffices to have: $V\delta_1 > h_{11}(f-\delta_1)\epsilon_1 + 0(e)$, $V\delta_2 > h_{21}(f-\delta_2)\epsilon_2 + 0(e)$,

$f\epsilon_1 > h_{12}(V-\epsilon_1)\delta_1 + 0(e)$, $f\epsilon_2 > h_{22}(V-\epsilon_2)\delta_2 + 0(e)$. In either case, it is clear

that $h_{ij} = 0$ leads to protected polymorphism for sufficiently small $e_{k\ell}$. If the

$h_{ij} \cong 1$, the evolutionary outcome depends on the $e_{k\ell}$.

4. TWO SEXES

The results of Kidwell et al. (1977), also discussed in Ewens (1979, pp. 36-40), can be re-formulated to show their affinity with the present results. Let w_{ij} give the fitness of *males* of genotype A_iA_j and v_{ij} give the fitness of females of genotype A_iA_j, with two alleles at the A locus. Take $w_{11} = 1$, $w_{12} = 1 - h_1\epsilon_1$, $w_{22} = 1 - \epsilon_1$, $v_{11} = 1 - \epsilon_2$, $v_{12} = 1 - h_2\epsilon_2$, and $v_{22} = 1$, with the h_i and ϵ_j as before. Protected polymorphism now requires $(1-h_1\epsilon_1/1-\epsilon_1) + 1 - h_2\epsilon_2 > 2$ and $1 - h_1\epsilon_1 + (1-h_2\epsilon_2/1-\epsilon_2) > 2$. Rearranged, this condition becomes

$$\epsilon_1 - h_1\epsilon_1 - h_2\epsilon_2(1-\epsilon_1) > 0$$

and

$$\epsilon_2 - h_2\epsilon_2 - h_1\epsilon_1(1-\epsilon_2) > 0.$$

As before, if the $h_i = 0$, protected polymorphism is assured.

5. CONCLUSION

After further generalization, it remains the case that protected polymorphism is readily achievable if there is antagonistic pleiotropy in gene effects on fitness-components. In particular, such polymorphism is fostered if dominance patterns lead to recessive deleterious effects on individual fitness-components, as found before (Rose, 1982).

REFERENCES

Bodmer, W.F., and J. Felsenstein (1967): Linkage and selection: Theoretical analysis of the deterministic two locus random mating model, *Genetics* 57:237-265.

Caspari, E. (1950): On the selective value of the alleles Rt and rt in *Ephestia Kühniella*, *Amer. Nat.* 84:367-380.

Ewens, W.J. (1979): *Mathematical Population Genetics*, Springer-Verlag, New York.

Falconer, D.S. (1977): Why are mice the size they are? In *International Conference on Quantitative Genetics*, eds. E. Pollak, O. Kempthorne, and E.J. Bailey, Iowa State University Press, Ames, Iowa.

Kidwell, J.F., M.T. Clegg, F.M. Stewart, and T. Prout (1977): Regions of stable equilibria for models of differential selection in the two sexes under random mating, *Genetics* 85:171-183.

Rose, M.R. (1982): Antagonistic pleiotropy, dominance, and genetic variation, *Heredity* 48:63-78.

Simmons, M.J., C.R. Preston, and W.R. Engels (1980): Pleiotropic effects on fitness of mutations affecting viability in *Drosophila melanogaster*, *Genetics* 94: 467-475.

Wright, S. (1980): Genic and organismic selection, *Evolution* 34:825-843.

ALGEBRAIC METHODS IN GENETICS

Angelika Wörz-Busekros

1. INTRODUCTION

Algebras arise in population genetics in a quite natural way. Let us consider an infinitely large population of diploid (or more generally 2r-ploid individuals which differ genetically in one or several autosomal loci. Let $a_1, \ldots, a_n$ be the genetically distinct gametes produced by this population. Then the zygotes can be represented by all pairs (a_i, a_j), i, j = 1,...,n, whereby we formally distinguish between (a_i, a_j) and (a_j, a_i) for $i \neq j$, i, j = 1,...,n. Let us assume that a zygote (a_i, a_j) produces a number $\gamma_{ijk} \geq 0$ of gametes a_k which survive the next generation, if we have separated generations, or a number $\gamma_{ijk} \cdot \Delta t + o(\Delta t)$ of gemetes a_k during a time interval of length Δt, if we have continuously overlapping generations, respectively. Let the gametic frequencies in the t-th generation or at time t be given by the vector $(\alpha_1(t), \ldots, \alpha_n(t))$ with

$$0 \leq \alpha_i(t) \leq 1, \quad i = 1,\ldots,n, \quad \sum_{i=1}^{n} \alpha_i(t) = 1. \qquad (1)$$

In the case of separated generations the gametic frequencies in the (t+1)-th generation are obtained from those in the t-th generations by

$$\alpha_k(t+1) = \frac{1}{\mu} \sum_{i,j=1}^{n} \alpha_i(t)\gamma_{ijk}\alpha_j(t), \quad k = 1,\ldots,n, \qquad (2)$$

where the normalizing term μ depending on $(\alpha_1(t), \ldots, \alpha_n(t))$ is given by

$$\mu = \sum_{p,q,r=1}^{n} \alpha_p(t)\gamma_{pqr}\alpha_q(t) . \qquad (3)$$

In the case of continuously overlapping generations we assume that the diploid (2r-ploid) population is kept in Hardy-Weinberg equilibrium at every time t. Hence it is sufficient to consider the gametic frequencies. For the frequency $\alpha_k(t+\Delta t)$ of a_k at time $t + \Delta t$ we have

$$\alpha_k(t+\Delta t) = \alpha_k(t) + \Delta t \cdot \sum_{i,j=1}^{n} \alpha_i(t)\gamma_{ijk}\alpha_j(t) - \Delta t \cdot \mu\alpha_k(t) + o(\Delta t), \quad k = 1,\ldots,n.$$

In the limit $\Delta t \to 0$ we obtain the following system of differential equations

$$\dot{\alpha}_k(t) = \sum_{i,j=1}^{n} \alpha_i(t)\gamma_{ijk}\alpha_j(t) - \mu\alpha_k(t), \quad k = 1,\ldots,n, \qquad (4)$$

where μ is given by (2).

Let us now consider the gametes $a_1,\ldots,a_n$ as abstract elements which are free over the field **R** of real numbers. They span an n-dimensional vector space

$$V: = \left\{ \sum_{i=1}^{n} \alpha_i a_i \Big| \alpha_i \in \mathbf{R}, \quad i = 1,\ldots,n \right\}. \tag{5}$$

With the assistance of the zygotic parameters γ_{ijk}, $i,j,k = 1,\ldots,n$, we introduce in V a multiplication by

$$a_i a_j: = \sum_{k=1}^{n} \gamma_{ijk} a_k, \quad i, j = 1,\ldots,n, \tag{6}$$

and its bilinear extension onto $V \times V \to V$. Thereby V becomes an algebra G over **R** which is called the *gametic* algebra of the population, cf. Etherington (1939) who has introduced this notation for populations of genotypes with equal fertilities and separated generations. The algebra G is commutative if and only if the zygotic parameters γ_{ijk} satisfy

$$\gamma_{ijk} = \gamma_{jik} \quad \text{for all} \quad i,j,k = 1,\ldots,n,$$

and G is associative if and only if

$$\sum_{k=1}^{n} \gamma_{ijk} \gamma_{k\ell m} = \sum_{k=1}^{n} \gamma_{ikm} \gamma_{j\ell k} \quad \text{for all} \quad i,j,\ell,m = 1,\ldots,n.$$

In terms of the algebra multiplication in G the systems of difference equations (2) and of differential equations (4) can be written as

$$x(t+1) = \frac{1}{\mu(x(t))} x(t)^2 \tag{7}$$

and

$$\dot{x}(t) = x(t)^2 - \mu(x(t))x(t), \tag{8}$$

respectively, where

$$x(t) = \sum_{i=1}^{n} \alpha_i(t) a_i \tag{9}$$

and $\mu(x(t))$ equals the sum of the coefficients of $x(t)^2$, i.e. $\mu(x(t))$ is given by the right hand side of (3).

There is a one-to-one correspondence between the stationary solutions x* of

(7) or (8) with $\mu(x^*) \neq 0$ and the idempotents e of G with $\mu(e) \neq 0$.

In the classical selection model of Fisher, Wright and Haldane (e.g. Hadeler, 1973) it is assumed that the zygote (a_i, a_j) has fertility $f_{ij} \geq 0$ with $f_{ij} = f_{ji}$ and segregates gametes a_i and a_j in equal proportions, $i,j = 1,\ldots,n$. Hence the multiplication in the gametic algebra G reads

$$a_i a_j = \frac{1}{2} f_{ij}(a_i + a_j), \quad i,j = 1,\ldots,n,$$

and the multiplication constants are given by

$$\gamma_{iik} = \begin{cases} f_{ii} & \text{if } k = i \\ 0 & \text{otherwise} \end{cases}$$

and

$$\gamma_{ijk} = \begin{cases} \frac{1}{2} f_{ij} & \text{if } k = i \text{ or } k = j \\ 0 & \text{otherwise} \end{cases} \quad \text{for } i \neq j$$

The normalizing term

$$\mu(x(t)) = \sum_{p,q=1}^{n} \alpha_p(t) f_{pq} \alpha_q(t)$$

can be interpreted as mean fitness of the population. It is well known that $\mu(x)$ serves as a Ljapunov function for (7) as well as for (8). Hence the ω-limit set of every trajectory is contained in a continuum of equilibrium states. Beyond this we have convergence for $n = 2$ (trivial), and for $n = 3$ (cf. an der Heiden, 1975), and for arbitrary n in case (8), if the ω-limit set of every trajectory contains a regular point, (cf. Aulbach and Hadeler, manuscript).

Furthermore there exist other genetic situations where again $\mu(x(t))$ is a Ljapunov function, e.g. two loci with additive selection parameters, (cf. Ewens, 1969). But it is an open problem to determine all commutative algebras G with nonnegative multiplication constants with respect to a certain basis in which $\mu(x)$ is nondecreasing along solutions of (7) or (8). On the other hand there exist algebras, so called genetic algebras, where we can prove convergence under some additional conditions without the assistance of a Ljapunov function, (cf. Theorem 3 in Section 2).

In the absence of selection, i.e..

$$\sum_{k=1}^{n} \gamma_{ijk} = c \quad \text{for all } i,j = 1,\ldots,n,$$

we can assume without loss of generality that $c = 1$, i.e.

$$\sum_{k=1}^{n} \gamma_{ijk} = 1 \quad \text{for all} \quad i,j = 1,\ldots,n. \tag{10}$$

Hence

$$\mu(x) = 1 \quad \text{for all} \quad x \in H,$$

where

$$H: = \left\{ x = \sum_{i=1}^{n} \alpha_i a_i \in V \;\middle|\; \sum_{i=1}^{n} \alpha_i = 1 \right\}. \tag{11}$$

Then systems (7) and (8) restricted to H reduce to

$$x(t+1) = x(t)^2 \tag{12}$$

and

$$\dot{x}(t) = x(t)^2 - x(t), \tag{13}$$

respectively, which are difference and differential equations of Riccati type, (cf. Heuch, 1973).

Gametic algebras describing the inheritance of populations without selection e.g. are commutative but not associate. Futhermore they do no belong to one of the well known classes of algebras such as Lie-, Jordan- or alternative algebras. But many algebras associated with specific genetic traits (e.g. coupled or partially coupled loci, polyploidy etc.) have special algebraic properties which we shall discuss in the following section.

2. GENETIC ALGEBRAS

In the following let $\mathbb{K}$ be a field, $M_n(\mathbb{K})$ the algebra of all $n \times n$ matrices over $\mathbb{K}$, and $\mathbb{K}\{X_1,\ldots,X_r\}$ the algebra of all polynomials in r associative, noncommuting indeterminants $X_1,\ldots,X_r$ over $\mathbb{K}$, in which two monomials are multiplied by writing them side by side. Furthermore, let A be an n-dimensional algebra over $\mathbb{K}$ which need not to be associative or commutative. Let $T(A)$ denote the transformation algebra of A which is generated by all left and right transformations $L_x, R_x: A \to A$, $x \in A$, defined by

$$L_x: a \mapsto xa, \quad R_x: a \mapsto ax, \quad a \in A.$$

Hence every $T \in T(A)$ can be written as

$$T = f(L_{x_1}, R_{x_2}, \ldots, R_{x_r}) \quad \text{for} \quad f \in \mathbb{K}\{X_1, \ldots, X_r\} \quad \text{and} \quad x_1, \ldots, x_r \in A.$$

For further standard definitions concerning nonassociative algebras we refer to the book of Schafer (1966). The following definition has been given by Etherington (1939):

DEFINITION. An algebra A is called *baric*, if it admits a nontrivial algebra homomorphism $\omega: A \to \mathbb{K}$. Then ω is called a *weight (homomorphism)*.

Every gametic algebra G whose zygotic parameters γ_{ijk} satisfy (10) is a baric algebra with weight homomorphism $\omega: G \to \mathbb{R}$ defined by

$$\omega\left(\sum_{i=1}^{n} \alpha_i a_i\right) := \sum_{i=1}^{n} \alpha_i.$$

The concept of genetic algebras (intermediate between the classes of train and special train algebras, cf. Etherington, 1939) has been introduced by Schafer (1949) and was characterized subsequently in various equilvent ways. We begin with Gonshor's characterization (cf. Gonshor, 1971), and summarize the main characterizations in Theorem 1 after we have introduced McCoy's generalization of commutative matrices (cf. McCoy, 1936), which we call semicommutative matrices.

DEFINITION. An algebra A is called *(Gonshor) genetic*, if it is commutative and it possesses a basic $c_1, \ldots, c_n$ over an algebraic extension field $\mathbb{L}$ of $\mathbb{K}$ such that the multiplication constants λ_{ijk}, $i,j,k = 1, \ldots, n$, with respect to this basis satisfy

$$\lambda_{111} = 1$$
$$\lambda_{1jk} = 0 \quad \text{if} \quad k < j$$
$$\lambda_{ijk} = 0 \quad \text{if} \quad k \leq \max\{i,j\}.$$

Then $c_1, \ldots, c_n$ is called a *canonical* basis of A.

Although a canonical basis of a genetic algebra is not uniquely determined, the multiplication constants $\lambda_{1ii} =: \lambda_i$, $i = 1, \ldots, n$, are invariants of A, they are called the *(train) roots* of A. It follows immediately that every genetic algebra A is baric with $\omega: A \to \mathbb{K}$ defined by

$$\omega\left(\sum_{i=1}^{n} \beta_i c_i\right) = \beta_1,$$

and ω is the unique weight homomorphism of A.

DEFINITION. Matrices $A_1, \ldots, A_r \in M_n(\mathbb{K})$ are called *semicommutative*, if all matrices

$$f(A_1, \ldots, A_r)(A_i A_j - A_j A_i), \quad i,j = 1, \ldots, n, \quad f \in \mathbb{K}\{X_1, \ldots, X_r\}$$

are nilpotent.

THEOREM 1. *Let* A *be an* n-*dimensional commutative algebra over* $\mathbb{K}$ *with weight homomorphism* ω. *Let* $\Gamma_1, \ldots, \Gamma_n$ *be the matrices corresponding to the left transformations induced by basis elements of weight* 1. *Then the following properties are equivalent:*

(a) A *is (Gonshor) genetic.*

(b) *The characteristic equation of every transformation* $T \in T(A)$, $T = f(L_{x_1}, \ldots L_{x_r})$ *insofar as it depends on* $x_1, \ldots, x_r$ *only depends on their weights* $\omega(x_1), \ldots, \omega(x_r)$ *(i.e.* A *is genetic in the sense of Schafer).*

(c) *The Lie-algebra* $T(A)^-$ *associated with* $T(A)$ *is solvable and every element of* $N := \ker \omega$ *is nilpotent.*

(d) *The matrices* $\Gamma_1, \ldots, \Gamma_n$ *can be simultaneously transformed into lower triangular matrices with the same diagonal.*

(e) *The matrices* $\Gamma_1, \ldots, \Gamma_n$ *are semicommutative and their differences* $\Gamma_i - \Gamma_j$, $i,j = 1, \ldots, n$, *are nilpotent.*

Since we cannot present the proofs of these results in the frame of this article, we restrict ourselves to mention the original papers: The equivalence of (a) and (b) has been proved by Gonshor (1971), (a) $\Longleftrightarrow$ (c) is shown by Holgate (1972), (a) $\Longleftrightarrow$ (d) is contained in a more general result on noncommutative algebras (cf. Wörz-Busekros, 1981), and finally the equivalence of (d) and (e) essentially is due to McCoy (1936), and Goldhaber and Whaples (1953). In view of the fact that a commutative algebra A is associative if and only if the matrices $\Gamma_1, \ldots, \Gamma_n$ are pairwise commutative and the fact that semicommutativity generalizes commutativity, a genetic algebra is a generalized associative one.

In the following we shall investigate the qualitative behaviour of solutions of (12) and (13) in a genetic algebra A over $\mathbb{R}$ or $\mathbb{C}$. Gonshor (1960) has shown that a genetic algebra with roots different from $\frac{1}{2}$ possesses exactly one idempotent. If in addition all roots apart from $\lambda_1 = 1$ are smaller than $\frac{1}{2}$ in modulus, then every solution of the difference equation (12) converges to this idempotent. Heuch (1973) mentioned that this result remains true for the differential equation (13) if the real parts of all roots different from $\lambda_1 = 1$ are smaller than $\frac{1}{2}$. Since many genetic algebras arising in concrete genetic situations have roots equal to $\frac{1}{2}$, it was desirable to generalize these results.

THEOREM 2. *Let* A *be a genetic algebra over* **R** *or* **C** *with a canonical basis* $c_1, \ldots, c_n$ *and roots* $\lambda_1 = 1$, $\lambda_2, \ldots, \lambda_n$. *Assume that exactly* r *roots equal* $\frac{1}{2}$, *namely* $\lambda_{t_1}, \ldots, \lambda_{t_r}$, $2 \le t_1 < \ldots < t_r \le n$, *and that* $c_i c_j$ *has no component with respect to* c_{t_ρ} *for* i, j = 1, \ldots, t_ρ - 1, $\rho = 1, \ldots, r$. *Then* A *possesses an* r-*parametric family* E *of idempotents.*

THEOREM 3. *Let* A *be a genetic algebra over* **R** *or* **C**. *If the assumptions of Theorem 2 are fulfilled and in addition*

$$|\lambda_i| < \tfrac{1}{2} \text{ for all } i \in \{2, \ldots, n\} - \{t_1, \ldots, t_r\}$$

or

$$|\lambda_i| \le \tfrac{1}{2} \text{ but } \lambda_i \ne \tfrac{1}{2} \text{ for all } i \in \{2, \ldots, \} - \{t_1, \ldots, t_r\},$$

then every solution of (12) *or* (13), *respectively, with* $\omega(x(0)) = 1$ *converges to an idempotent in* E.

Let us mention first that the assumptions on the products $c_i c_j$, i, j = $1, \ldots, t_\rho - 1$, $\rho = 1, \ldots, r$, are independent of the special choice of a canonical basis. The proofs of Theorem 2 and the first part of Theorem 3 are given in Wörz-Busekros (1980), Sections 3A and 3C, respectively. The second part of Theorem 3 can be proved similarly, namely represent (13) like (12) in coordinates with respect to a canonical basis. Then the above result follows by induction on k, $1 \le k \le n$, using the fact that $|\lambda_i| \le \frac{1}{2}$ but $\lambda_i \ne \frac{1}{2}$ implies $\text{Re}(2\lambda_i - 1) < 0$. In the case of a genetic algebra which can be "interpreted genetically" (for this notation, cf. Heuch, 1978) the latter assumption of Theorem 3 is not so very restrictive since we have the following result, (cf. Wörz-Busekros, 1980, Section 4B):

THEOREM 4. *Let* A *be a genetic algebra over* **R** *or* **C** *with roots* $\lambda_1 = 1$, $\lambda_2, \ldots, \lambda_n$. *If* A *possesses a basis with multiplication constants* $\gamma_{ijk} \ge 0$, i, j, k = 1, \ldots, n, *satisfying* (10), *then*

$$|\lambda_i| \le \tfrac{1}{2} \text{ for } i = 2, \ldots, n.$$

For the numerous applications of Theorem 3 to various genetic situations we refer to Wörz-Busekros (1980), Section 7.

REFERENCES

Aulbach, B., and K.P. Hadeler (manuscript): Convergence to equilibrium in the classical model of population genetics.

Etherington, I.M.H. (1939): Genetic algebras, *Proc. Roy. Soc. Edinburgh* 59:242-258.

Ewens, W.J. (1969): Mean fitness increases when fitnesses are additive, *Nature* 221: 1076.

Gonshor, H. (1960): Special train algebras arising in genetics, *Proc. Edinburgh Math. Soc.* (2), 12:41-53.

_____ (1971): Contributions to genetic algebras, *Proc. Edinburgh Math. Soc.* (2), 17:289-298.

Goldhaber, J.K., and G. Whaples (1953): On some matrix theorems of Frobenius and McCoy, *Can. J. Math.* 5:332-335.

Hadler, K.P. (1973): Selektionsmodelle in der Populationsgenetik, *Methoden und Verfahren der Mathematischen Physik (BI-Verlag)* 9:136-160.

an der Heiden, U. (1975): On manifolds of equilibria in the selection model for multiple alleles, *J. Math. Biol.* 1:321-330.

Heuch, I. (1973): Genetic algebras and continuous time models, *Theor. Pop. Biol.* 4:133-144.

_____ (1978): Genetic algebras considered as elements in a vector space, *SIAM J. Appl. Math.* 35:695-703.

Holgate, P. (1972): Characterizations of genetic algebras, *J. London Math. Soc.* (2) 6:169-174.

McCoy, N.H. (1936): On the characteristic roots of matric polynomials, *Bull. Amer. Math. Soc.* 42:592-600.

Schafer, R.D. (1949): Structure of genetic algebras, *Amer. J. Math.* 71:121-135.

_____ (1966): *An Introduction to Nonassociative Algebras,* Acad. Press, New York.

Wörz-Busekros, A. (1980): *Algebras in Genetics,* Lecture Notes in Biomathematics, No. 36, Springer-Verlag, Berlin.

_____ (1981): Relationship between genetic algebras and semi-commutative matrices, *Lin. Alg. Appl.* 39:111-123.

PART II:

LIFE HISTORY STRATEGIES

THE EVOLUTION OF LIFE HISTORIES UNDER
PLEIOTROPIC CONSTRAINTS AND K-SELECTION

Alan R. Templeton*

1. INTRODUCTION

Caswell (1978) formulated a generalized sensitivity measure of population growth rate to changes in life history parameters. His measure is based upon certain inner products involving v, the reproductive value vector (the dominant row eigenvector of the Leslie matrix), and s, the stable age distribution vector (the dominant column eigenvector). Let L be a Leslie matrix and dL be a matrix whose elements are differential perturbations of the elements of L. Defining the inner products

$$(v,s) = v \cdot s$$
$$((v,s)) = v \cdot L \cdot s,$$

Then Caswell's sensitivity measure is

$$d\lambda = ((v,s))/(v,s) \tag{1.1}$$

where $d\lambda$ is the differential perturbation of λ, the maximum real eigenvalue of L. Caswell then used this criterion to investigate the impact on λ of changes in a single life history parameter.

Templeton (1980) showed how Caswell's criterion (1.1) could be used to study life history evolution when several different life history parameters are changed simultaneously, thereby allowing an investigation of the importance of pleiotropy. Templeton also showed how Caswell's criterion could be generalized to cases in which the stable age distribution assumption does not hold, a situation applicable to many "r-selected" species. However, both Caswell's work (1978) and Templeton's (1980) assume density independent population growth. It is the purpose of this paper to extend Templeton's (1980) results to populations that are at a stable population size and structure due to the operation of density-dependent life history parameters.

2. DENSITY DEPENDENT LESLIE MATRICES

I assume that the population growth is governed by a Leslie matrix which under conditions of low density (density independent growth) has the form

---

* This work was supported by National Institutes of Health Grant R01 GM27021.

$$
L = \begin{bmatrix}
P_0 m_1 & P_1 m_2 & \cdots\cdot & P_r m_{r+1} \\
P_0 & 0 & \cdots\cdot & 0 \\
0 & P_1 & \cdots\cdot & 0 \\
\vdots & \vdots & & \vdots \\
0 & 0 & P_{r-1} & 0
\end{bmatrix} ,
$$

Where p_i is the probability of survival from age i to $i+1$ and m is the age specific fecundity under the conventions that reproduction occurs at the beginning of each age interval, the pivotal age is at the beginning of the age interval, and the census is immediately after reproduction.

Density dependence can be introduced into this model by either directly making the p's and m's density dependent or by multiplying the density independent p's and m's by density adjusting parameters. I take the latter course in this paper as it allows the life history responses to be partitioned into density independent and density dependent components.

The effect of density upon survivorship is most conveniently modeled by constructing a matrix, $Q(t)$, whose diagonal elements are $q_i(t)$ with 0 everywhere else. The realized survivorship of a population characterized by the age vector $N(t)$ at time t is given by $q_i(t)p_i$ for all i. Hence, $q_i(t)$ measures the change in survivorship at age i induced by the age vector $N(t)$. If fecundities are not density dependent, then the age vector of the population at time $t+1$ is given by $N(t+1) = LQ(t)N(t)$. I now assume that the q's are such that a stable equilibrium exists and that the population is at this stable equilibrium (see Charlesworth, 1980, for a discussion of the stability properties associated with density-dependent Leslie matrices). This implies that $LQ(eq)$ has a dominant real eigenvalue of 1 when the q's are at their equilibrium value. For example, suppose viability is depressed in an age-independent fashion by a fraction $1 - N/K$ from its density independent values where N is the total population size and K is a parameter (Emlen, 1973). Then $LQ(eq)$ will have a maximum real eigenvalue of 1 if $q(eq) = \lambda^{-1} = 1 - N(eq)/K$ where λ is the dominant real eigenvalue of L (Emlen, 1973). Alternatively, if one assumes each increment in total population size decreases all viabilities by a factor $\exp(-b)$, then $q(eq) = \exp[-bN(eq)]$ where $N(eq) = \ln(\lambda)/b$ (Charlesworth, 1980). These examples illustrate the fact that the equilibrium population density tends to increase with increasing values of λ.

Density dependent fecundity affects could also exist, and they are modeled by replacing the m's in the first row of L by $mf(t)$ where $f(t)$ measures the impact of the age vector $N(t)$ upon fecundity at age i. This means that L must be replaced by $L(t)$ in describing population growth. Hence, the model for population growth is $N(t+1) = L(t)Q(t)N(t)$. Assuming a stable equilibrium exists and the population is at that equilibrium, the dominant eigenvalue of $L(eq)Q(eq)$ is 1 and the stable age distribution at equilibrium is proportional to the column vector whose elements are:

$$\ell_i(eq) = \prod_{j=0}^{i-1} q_j(eq)p_j$$

and the reproductive value of age i at equilibrium is given by:

$$v_i(eq) = \left[\sum_{j=i+1}^{r+1} \ell_j(eq)m_j f_j(eq) \right] / \ell_i(eq).$$

3. PLEIOTROPIC CHANGES IN LIFE HISTORY PARAMETERS

Under pleiotropy, changes in one life history parameter automatically induce changes in other parameters. In this section, I assume the most general case possible; that is, changes in one life history parameter induce alternations in all others, both density independent (p's and m's) and density dependent (q's and f's). Pleiotropy is quantified in a manner analagous to that given in Templeton (1980).

The first step in quantifying pleiotropy is to choose some reference life history parameter affecting age g, say y_g. The parameter y_g is most conveniently choosen as one of life history parameters most strongly affected by the alteration whose evolutionary impact is to be studied. The only additional constraint upon choosing y_g is that it is increased in a positive fashion by the alteration under consideration. Once y_g is given, the pattern of pleiotropic change can be measured by partial derivatives as follows:

$$\partial y_g / \partial y_g = 1$$
$$u_{g,i} = \partial \ln q_i(eq) / \partial y_g$$
$$t_{g,i} = \partial \ln p_i / \partial y_g \qquad (3.1)$$
$$w_{g,i} = \partial m_i / \partial y_g$$
$$z_{g,i} = \partial f_i(eq) / \partial y_g.$$

Now consider a perturbation from the equilibrium state such that $L(eq)Q(eq)$ is changed to $L(eq)Q(eq) + d[L(eq)Q(eq)]$. Although the eigenvalue of $L(eq)Q(eq)$ is by definition 1, the eigenvalue of this perturbed matrix is not necessarily 1, so we can still meaningfully speak of the perturbation induced in λ by this life history alteration. However, also note that throughout we are dealing only with small perturbations from the equilibrium condition, and this restriction to small perturbations is also true for the results given in Caswell (1978) and Templeton

(1980). Finally, note that any changes induced in the dominant real eigenvalue of L(eq)Q(eq) will cause a departure from the equilibrium state. As mentioned before, any increase in λ will tend to induce an increase in the carrying capacity. Consequently, changes in λ will induce a period of population growth until a new population size is stabilized. Nagylaki (1979) has derived a density-dependent extension of Fisher's fundamental theorem of natural selection which shows that the operation of weak selection (as is assumed here) brings about an increase in population size. Therefore, changes in λ still represent an excellent life history criterion under K-selection, although the biological manifestation of λ is quite different in this case (increased carrying capacity) from the density independent interpretation (increased growth rate).

An alteration in life history will be regarded as being favored in an evolutionary sense if it increases the value of the eigenvalue of the matrix L(eq)Q(eq). Thus, an analogue of Caswell's life history criterion, equation (1.1), can be used here; namely, the alteration d[L(eq)Q(eq)] will be favored whenever

$$d\lambda = ((v,s))/(v,s) > 0 \qquad (3.2)$$

where v and s are the reproductive values and stable age distributions associated with the original equilibrium population described by L(eq)Q(eq) and where the inner product ((,)) is defined with respect to d[L(eq)Q(eq)]. Using equations (3.1) and the convention that the perturbation induced in the index life history parameter, y, is positive, inequality (3.2) can be written in terms of a simple derivative (see Templeton, 1980) as

$$d\lambda/dy_g > 0. \qquad (3.3)$$

The left-hand side of inequality (3.3) can be shown to be, using identical techniques to those given in Templeton (1980):

$$\frac{d\lambda}{dy_g} = \sum_{i=0}^{r} [a_{g,i}\ell_i(eq)v_i(eq) + c_{g,i+1}\ell_{i+1}(eq)]/(v,s) \qquad (3.4)$$

where

$$a_{g,i} = t_{g,i} + u_{g,i}$$

$$c_{g,i} = f_i(eq)w_{g,i} + m_i z_{g,i}$$

$$(v,s) = \sum_{j=1}^{r+1} j\ell_j(eq)m_j f_j(eq).$$

The change in life history described by the t's, u's, w's and z's will be favored if equation (3.4) is positive. Hence, this equation is my primary life history criterion under K-selection.

4. IMPLICATIONS AND SPECIAL CASES

Equation (3.4) can be subdivided into two major components: a component reflecting how changes in fecundity alter λ and a component reflecting how changes in survivorship alter λ. It is commonplace to predict that K-selection tends to favor survivorship over fecundity, and from equation (3.4), it can be seen that survivorship will be favored over fecundity whenever

$$\sum_{i=0}^{r} a_{g,i} \ell_i(eq) v_i(eq) > - \sum_{i=0}^{r} c_{g,i+1} \ell_{i+1}(eq). \tag{4.1}$$

Obviously, (4.1) is most likely to be satisfied when both the a's and the c's tend to be positive, but as this corresponds to a simultaneous increase in both fecundity and survivorship, this prediction is straightforward and uninteresting. The more interesting cases occur when some sort of trade-off is encountered; for example, if positive a's tend to be associated with negative c's and vice versa. In this trade-off situation, inequality (4.1) shows that survivorship tends to be favored at the expense of fecundity if the reproductive values are high for many age categories, particularly for the reproductively mature age categories. This in turn implies that survivorship is high in the adult stage and reproduction tends to be delayed. This is, of course, a classic "K-selected" phenotype. Hence, starting from a "K-selected" phenotype, inequality (4.1) implies that selection would tend to reinforce it, or at the very least, select against perturbations favoring increased fecundity at the expense of survivorship. In this regard, the model developed here tends to support the classic K-selection predictions.

However, there is nothing in inequality (4.1) to indicate that a K-selected phenotype will evolve from a starting population with low survivorship and high early fecundities. Under these initial conditions, the weighting of the a's in (4.1) tends to be small, but the values of the c's tend to be high for the lower age categories (which have the largest ℓ's). Under these initial conditions, it would be difficult for a trade-off favoring survivorship over fecundity, particularly at the expense of early fecundity, to satisfy (4.1). Hence, the initial conditions and the exact nature of the pleiotropic patterns (the magnitudes and signs of the a's and c's) play a critical role in determining what type of life history is favored by selection. This blurs the distinction between the r vs. K-selection dichotomy.

This dichotomy is further obscured by the fact that the density-independent and density-dependent perturbations contribute to (3.4) or (4.1) in virtually the same fashion. Thus, under this model of K-selection, it makes little difference if one achieves greater survivorship (or fecundity) by increasing one's inherent, density-independent survivorship or by decreasing one's viability sensitivity to density. Thus, many perturbations that would be favored under "r-selection" could also be favored under K-selection. For example, suppose there is no variation for the density dependent parameters at all (i.e., the u's and z's are 0 in (3.4) and (4.1)). Then, all selection is directed towards the p's and m's in such a way as to maximize λ. This, of course, is completely equivalent to the classic view of r-selection. Moreover, Templeton (1980) has pointed out that the ecological situations usually associated with r-selection direct selection not only upon λ, but upon the other eigenvalues of the Leslie matrix as well. Hence, if no or little variation for density sensitivity exists in the initial population, K-selection results in more direct selection on "r" (ℓn λ) than "r-selection".

Of course, if variability exists for the density sensitive parameters, they will also be selected in this model whereas they could not be selected under density-independent growth. For example, consider the special case in which variation exists only for the density-sensitive parameters; that is, all t's and w's are 0 in (3.4) or (4.1). Let us further assume that only survivorship is density dependent and that the Q matrix has diagonal elements of value $\exp(-b_i N)$. Under these assumptions, inequality (4.1) reduces to

$$- \sum_{i=0}^{r} k_{g,i} \ell_i (eq) v_i (eq) > 0 \qquad (4.2)$$

where

$$k_{g,i} = \partial b_i / \partial b_g.$$

Obviously, (4.2) cannot be satisfied unless negative k's predominate over positive k's as weighted by the stable age distribution and reproductive values. Negative k's mean decreased b's; that is, less viability sensitivity to density.

In general, it is obvious from (4.1) that if little or no variation exists in the density independent parameters, K-selection will tend to favor those phenotypes with positive u's and z's; namely, those phenotypes least adversely affected by density. These results when coupled with the results given in the previous paragraph indicate that K-selection can favor phenotypes that would not be favored under r-selection, but phenotypes that are favored under r-selection could also be favored under K-selection as long as they had no or little effect on density sensitivity. This further obscures the distinction between r and K-selection.

Of more general interest is the case in which variation for both the density independent and density dependent parameters simultaneously exists in a trade-off situation; that is, an increase in the inherent, density-independent life history parameters tends to be associated with an increased and deleterious sensitivity to density, and vice versa. Because both density independent and density dependent alterations contribute to (3.4) or (4.1) in equivalent fashions, no general predictions can be made about which aspect of this trade-off will be favored by selection. Rather the outcome will vary from case to case depending upon the precise pattern of pleiotropy and the magnitude of the alterations. However, these equations do show that given certain patterns of pleiotropy, K-selection can actually favor those phenotypes showing an increased deleterious response to population density.

In summary, these results demonstrate that K-selected and r-selected regimes are not equivalent in an evolutionary sense, but neither are they dichotomies favoring mutually exclusive or opposing phenotypic traits. Not only are the distinctions between r and K selection blurred in light of this model, but they cannot even legitimately be thought of as opposite ends of a continuum. Differeneces do exist, but it is impossible to specify a single phenotype that is "optimal" under one and not the other. In both cases, the exact pattern of pleiotropy and the initial conditions play a critical role in determining what is "optimal".

5. LIMITATIONS OF THE MODEL

I have made two fundamental assumptions in constructing this model: 1) the initial population is at equilibrium, and 2) all perturbations are small in magnitude and selection is weak. These assumptions are by no means unique to this model; rather, they are virtually universal in life history theory (e.g., see Charelesworth, 1980). However, recent field work by Templeton and Johnston (1982) on natural populations of fruit flies raises serious questions about the biological validity of these assumptions. This work has shown that density-limited natural populations not only undergo extreme fluctuations in density, but in age structure as well. Moreover, large spatial fluctuations in age structure exist on a microgeographic scale. Hence, an individual fly can disperse in a single day (and some, in fact, do) from a population with a very young age structure to one having a very old age structure, and vice versa. The assumption of an equilibrium population is clearly untenable in light of these radical spatial and temporal heterogeneities.

Templeton and Johnston (1982) have also shown that the natural populations are polymorphic for alleles causing life history alterations that are extensively pleiotropic and that represent large -- not small -- deviations from each other. Hence, the validity of the small perturbation assumption is also seriously undermined. A non-equilibrium life history theory is therefore required that acts upon large phenotypic deviations. Such a theory does not exist at present, but it is sorely needed if life history theory is to be properly integrated with field studies.

REFERENCES

Caswell, H. (1978): A general formula for the sensitivity of population growth rate to changes in life history parameters, *Theor. Pop. Biol.* 14:215-230.

Charlesworth, B. (1980): *Evolution in Age-Structured Populations*, Cambridge University Press, Cambridge.

Emlen, J. (1973): *Ecology: An Evolutionary Approach*, Addison-Wesley, Reading, Mass.

Nagylaki, T. (1979): Dynamics of density- and frequency-dependent selection, *Proc. Natl. Acad. Sci. U.S.A.* 76:438-441.

Templeton, A.R. (1980): The evolution of life histories under pleiotropic constaints and r-selection, *Theor. Pop. Biol.* 18:279-289.

Templeton, A.R., and J.S. Johnston (1982): Life history evolution under pleiotropy and K-selection in a natural population of Drosophila mercatorum. In: *"Ecological Genetics and Evolution,"* J.S.F. Baker and W.T. Starmer (eds.), pp. 225-239. Academic Press, N.Y.

ON THE r - K TRADEOFF IN DENSITY DEPENDENT SELECTION

C.E. Clark

1. INTRODUCTION

Laboratory experiments designed to test the hypothesis of an r - K tradeoff have generally not been successful. Experiments by Luckinbill (1978,1979), Gill (1972), and Mueller and Ayala (1980) all failed to give firm evidence for the existence of an r - K tradeoff. The r - K tradeoff has also been criticized on theoretical grounds. Stearns (1977) pointed out that r and K are not directly comparable, r being a life history parameter and K a population parameter. In fact, K is a composite parameter whose value is determined by such factors as competitive ability, efficiency of resource utilization, predator avoidance, environmental quality, and the intrinsic growth rate r. The importance of the dependence of K on r was noted by Prout (1980) who pointed out that the traditional Verhulst-Pearl form of the logistic equation ((1), below) disguises the relationship between r and K.

When one speaks of an r - K tradeoff what is actually implied is a tradeoff between r and sensitivity to the effects of density, and K is perhaps not the appropriate parameter by which to measure the latter quantity. The purpose of this note is to reformulate the logistic equation in a manner which delineates the concepts of equilibrium value of the population, environmental quality, and sensitivity to density. In the traditional form of the logistic equation:

$$\left.\begin{aligned} \frac{dN}{dt} &= Nr(1 - N/K), \quad \text{(or, for future reference)}, \\ w(N) &= \frac{1}{N}\frac{dN}{dt} = 1 - N/K \end{aligned}\right\} \quad , \tag{1}$$

K serves as a measure of all three quantities. The equation

$$\left.\begin{aligned} \frac{dN}{dt} &= N(r - \frac{c}{B} N), \quad \text{or} \\ w(N) &= \frac{1}{N}\frac{dN}{dt} = r - \frac{c}{B} N \end{aligned}\right\} \quad (r \leq c) \tag{2}$$

was used by Clark and Hallam (1981) to explain anamolous behavior in a non-autonomous form of (1). In (2), r is the intrinsic growth rate, c is a parameter which measures sensitivity to density, and B is the saturation capacity of the environment, that is, the maximum number of organisms which the environment can maintain. Although this is an interpretation often given to K in (1), the value of B is not affected by the value of r, nor by most life history parameters which determine sensitivity to density, (body size and metabolic efficiency are exceptions).

The equilibrium value (or carrying capacity) in (2) is $K = \frac{rB}{c} \leq B$, an expression similar in interpretation to that derived by Roughgarden (1971) from principles of ecological energetics. Equation (2) is effectively equivalent to the familiar equation $dN/dt = N(r-aN)$. Equation (2) is used here because it allows for the assumption of a tradeoff between two life history parameters, r and c, which are measured in the same units (time^{-1}), and it provides a direct means of modeling variations in environmental quality through the parameter B. It is important to note that (2) is not equivalent to (1) in a situation in which genotypic variability affects r values.

2. EXPREIMENTAL EVIDENCE OF AN $r - c$ TRADEOFF

We will consider two experiments which failed to confirm an $r - K$ tradeoff and show that these experiments do, in fact, support the existence of a tradeoff between r and sensitivity to density if the latter quantity be measured by c instead of K. Luckinbill (1979) applied r-selection to a mixed population of *paramecium* strains to test the hypothesis that r-selection reduces K. In addition to the populations subjected to r-selection, there were two controls for the experiment. One group of populations underwent no selection and another group was subjected to random selection. The results are summarized below.

	random selection	no selection	r selection
r	1.02	1.31	2.67
K	1754	1686	2225

Instead of r-selection reducing K, the experiments showed a positive correlation between r and K. Since the parameter c $(= \frac{rB}{K})$ is proportional to r/K, it is possible to compare relative sizes of c. The results are shown below with B arbitrarily assigned a value of $B = 2000$.

	random selection	no selection	r selection
r	1.02	1.31	2.67
c	1.16	1.55	2.40

The table indicates that an increase in r is accompanied by an increase in c as predicted in the hypothesis of an r - c tradeoff.

Gill (1979) grew populations of two species of *paramecium* at three different temperatures and compared the r and K values of each species at each of the three temperatures. The results are shown below.

P - *aurelia* - 2	15°C	20°C	25°C
r	0.62	1.07	1.60
K	815	758	561

P - *aurelia* - 5	15°C	20°C	25°C
r	0.30	0.69	1.82
K	723	818	744

The parameters r and K are negatively correlated for P - *aurelia* - 2 but uncorrelated for P - *aurelia* - 5. Again we compare the c values by comparing the ratios r/K, (with $B = 1000$). The results in the table below again show a positive correlation between r and c.

P - *aurelia* - 2	15°C	20°C	25°C
r	0.62	1.07	1.60
c	0.76	1.41	2.85

P - *aurelia* - 5	15°C	20°C	25°C
r	0.30	0.69	1.82
c	0.41	0.84	2.45

Two other sets of experiments, one by Luckinbill (1978) with E. *coli* and one by Mueller and Ayala (1981) with *Drosophila* which failed to confirm an r - K tradeoff do not lend themselves to analyses as simple as those above. The experiments were based roughly on the premise that, in growth competition between r-selected and K-selected strains, the r-selected strain should perform better at low densities and the K-selected strain better at high densities. It will be shown in the next section that equation (2) at least has the flexibility to explain the outcomes of these experiments.

3. THEORETICAL CONSIDERATION OF THE r - c TRADEOFF

We consider a sexually reproducing population with fixed r and c values which are affected pleiotropically by allelic combinations at a single locus. A new mutant genotype has intrinsic growth rate $\hat{r}$ and sensitivity to density $\hat{c}$. We assume that the value of B is the same for both genotypes. Let $K = \frac{rB}{c}$ $(\hat{K} = \frac{\hat{r}B}{\hat{c}})$ denote the equilibrium values of a population consisting entirely of the original (mutant) genotype. (There is no loss of generality here in assuming dominance of the mutant allele.) We use two conclusions drawn from theoretical studies of density-dependent selection, (e.g. Roughgarden (1971), Anderson (1971), Charlesworth (1971), Smouse (1976)).

1) In a relatively stable environment with little or no density-independent mortality and in which the population is maintained near its equilibrium value, the new allele will become fixed in the population if $\tilde{K} > K$.

2) In a situation in which high density-independent mortality maintains the population at a level well below its equilibrium value, the new allele will become fixed if a genotype carrying that allele has a larger density-dependent per capita growth rate than a genotype carrying the original allele, i.e., if

$$\tilde{w}(N) = \tilde{r} - \frac{\tilde{c}}{B} N > w(N) = r - \frac{c}{B} N.$$

The assumption of an r - c tradeoff requires that $\tilde{r} > r$ if and only if $\tilde{c} > c$. We consider two strategies: an r-strategy in which $\tilde{r} = r + \Delta r$, $\tilde{c} = c + \Delta c$, (Δr, $\Delta c > 0$), and a c-strategy in which $\tilde{r} = r - \Delta r$, $\tilde{c} = c - \Delta c$, (Δr, $\Delta c > 0$).

We consider first the case of a constant environment, with low density-independent mortality, and with the population maintained near its equilibrium value, ($N \approx K$). These conditions place a high premium on those life history traits which decrease an organism's sensitivity to density, i.e., decrease c, Dobzhansky (1950), Pianka (1970). So the price paid by an r-strategist in terms of increased c is likely to be higher than the price paid by a c-strategist in terms of increased r. As noted above a new genotype in this situation will be successful if $\tilde{K} > K$. For an r-stategist, $\tilde{K} > K$ if $\Delta r > \frac{r}{c} \Delta c$; and for a c-strategist, $\tilde{K} > K$ if $\Delta r < \frac{r}{c} \Delta c$. If the equilibrium value, K, of the population is near the saturation capacity, B, of the environment, that if, $\frac{r}{c} \approx 1$, then, since the situation considered here favors efficient tradeoffs for c-strategists, the c-strategist is the more likely to be successful. This is K-selection in the traditional sense, that is, increased K and decreased r. However, if r is significantly less than c, or, equivalently, if K is well below B, then a different conclusion is possible. The r-strategist will be successful if $\Delta r > \frac{r}{c} \Delta c$, and, if $\frac{r}{c}$ is small, then this inequality may hold even with relatively inefficient tradeoffs, (Δc larger than Δr). So this situation could favor an r-strategist which would result in a simultaneous increase in both r and K, a result contrary to traditional theories of r - K selection.

We now consider the case of a population which is maintained at a level well below its equilibrium value by density-independent mortality, ($N < K$). In this case, the new genotype will be successful if $\tilde{w}(N) = \tilde{r} - \frac{\tilde{c}}{B} N > r - \frac{c}{B} N = w(N)$. For an r-strategist, $\tilde{w}(N) > w(N)$ if $\Delta r > \frac{N}{B} \Delta c$; and for a c-strategist, $\tilde{w}(N) > w(N)$ if $\Delta r < \frac{N}{B} \Delta c$. Since $N < K \leq B$, the r-strategist will be more likely to succeed, even with relatively inefficient tradeoffs. Note that $\frac{N}{B} < \frac{r}{c}$ (since $N < K = \frac{rB}{C}$), and recall that, for an r-strategist, $\tilde{K} > K$ if and only if $\Delta r > \frac{r}{c} \Delta c$. If $\frac{N}{B} \Delta c < \Delta r < \frac{r}{c} \Delta c$, then an r-strategy would succeed and would result in $\tilde{r} > r$, $\tilde{K} < K$, which is traditional r-selection. However, it is at least theoretically possible, especially if r/c is small, that an r-strategy could succeed and satisfy $\frac{N}{B} \Delta c < \frac{r}{c} \Delta c < \Delta r$, in which case the new genotype would have $\tilde{r} > r$ and $\tilde{K} > K$.

Laboratory experiments are normally conducted with r-selected species because of the requirement of a short generation time, and r-selected species are likely to have a low value of r/c, (high sensitivity to the effects of density). The preceding analysis shows that it is possible for such species to evolve both an increased r value and an increased K value, especially if maintained at high densities. This presents a possible explanation for the results of the experiments of Mueller and Ayala (1981) and Luckinbill (1978).

4. DENSITY-DEPENDENT SELECTION IN A RANDOM ENVIRONMENT

The purpose of this section is to present one example in which the applications of equations (1) and (2) in the same model give conflicting results. Heckel and Roughgarden (1980) used a linearized discrete form of (1) in the fitness equation for W(N) to show that a population at equilibrium in a fluctuating environment can be successfully invaded by a genotype with a lower r value, and, furthermore, the result does not require the assumption of pleiotropy; a tradeoff with K is not necessary. These results are also obtained in more general models by Turelli and Petry (1980) and Turelli (1981).

We refer to Heckel and Roughgarden (1980) for the assumptions and details of the procedure. Parameter values for the homozygote and the heterozygote are designated by use of the subscripts AA and Aa, respectively. The criterion for the successful invasion of a mutant allele a is:

$$E \log \frac{W_{AA}}{W_{Aa}} < 0,$$

where

$$W_{xy}(N_t, t) = 1 - r_{xy} + \frac{r_{xy}(K_{xy} + k_t)}{N_t}, \qquad (3)$$

where k_t is an ergodic random variable with zero expectation, and $E(F(t))$ denotes the expected value of $F(t)$. Equation (3) is obtained from the discrete form of (1) by linearization about the equilibrium. Using the discrete form of (2) instead of (1) and following the same procedure, one obtains:

$$\hat{W}_{xy} = 1 - r_{xy} + \frac{r_{xy}^2(B + k_t)}{c_{xy} N_t}.$$

It is assumed that the environmental fluctuations modeled by k_t are small, so if

$r_{Aa} < r_{AA}$, then $\tilde{W}_{Aa}(N_t,t) < \tilde{W}_{AA}(N_t,t)$ for every t. Hence $E \log \dfrac{\tilde{W}_{AA}}{\tilde{W}_{Aa}} > 0$, and a genotype with a lower r and no tradeoff, $(c_{Aa} = c_{AA})$, cannot become established in the population. This result will hold, in fact, for any generic form of W for which $\left.\dfrac{\partial W}{\partial r}\right|_{N=K} > 0$. It seems apparent that an increase in r should increase W at any density, including densities exceeding the carrying capacity, a property which does not hold for (1).

The introduction of stochasticity can still reverse the expected result for the deterministic case. Using the same approximation procedures as Heckel and Roughgarden, one obtains:

$$E \log \frac{\tilde{W}_{AA}}{\tilde{W}_{Aa}} = f(\beta) - g(\beta),$$

where

$$f(\beta) = \frac{r_{Aa}^2 r_{AA} \beta \sigma^2}{(2-r_{AA})B^2(r_{AA}-r_{Aa}r_{AA}+r_{Aa}^2\beta)^2} (r_{Aa}^2(1-r_{AA})\beta - r_{AA}^2(1-r_{Aa})),$$

$$g(\beta) = \log(1-r_{Aa}+\frac{r_{Aa}^2}{r_{AA}}\beta), \qquad \beta = \frac{c_{AA}}{c_{Aa}}, \qquad \sigma^2 = \text{Var } k_t.$$

If $r_{Aa} > r_{AA}$, then $f(\beta) - g(\beta) < 0$ if and only if $\beta > \beta^*$ where $\dfrac{r_{AA}}{r_{Aa}} < \beta^* < 1$. So a mutant allele will persist only if $\dfrac{c_{AA}}{c_{Aa}} > \dfrac{r_{AA}}{r_{Aa}}$, or if $K_{Aa} = \dfrac{r_{Aa}B}{c_{Aa}} > K_{AA} = \dfrac{r_{AA}B}{c_{AA}}$. This is the expected result for the deterministic case.

However, if $r_{Aa} < r_{AA}$, then $f(\beta) - g(\beta) < 0$ for $\beta > \beta^*$, where $1 < \beta^* < \dfrac{r_{AA}}{r_{Aa}}$. In this case, the new allele will persist if $\beta^* < \dfrac{c_{AA}}{c_{Aa}} < \dfrac{r_{AA}}{r_{Aa}}$, which implies that $K_{Aa} < K_{AA}$, a reversal of the deterministic result. Note, however, that since $\beta = \dfrac{c_{AA}}{c_{Aa}} > 1$, then $c_{Aa} < c_{AA}$, so the decrease in r must be accompanied by a decrease in c.

These results are similar to and were motivated by those of Heckel and Roughgarden (1980). The important difference is that in the present case, the evolution of a lower r is impossible without a tradeoff resulting in a lower c.

REFERENCES

Anderson, W.W. (1971): Genetic equilibrium and population growth under density regulated selection, *Amer. Nat.* 105:489-498.

Charlesworth, B. (1971): Selection in density-regulated populations, *Ecology* 52:469-474.

Dobzhansky, T. (1950): Evolution in the tropics, *Amer. Sci.* 38:209-221.

Gill, Douglas E. (1972): Intrinsic rates of increase, saturation densities, and competitive ability. I. An experiment with *paramecium*, *Amer. Nat.* 106:461-471.

Hallam, T.G., and C.E. Clark (1981): Non-autonomous logistic equations as models of populations in a deteriorating environment, *J. Theor. Biol.* 93:303-311.

Heckel, David G., and Jonathan Roughgarden (1980): A species near its equilibrium size in a fluctuating environment can evolve a lower intrinsic rate of increase, *Proc. Natl. Acad. Sci. USA* 77:7497-7500.

Luckinbill, L.S. (1978): r and K selection in experimental populations of *Escherichia coli-*, *Science* 202:1201-1203.

_____ (1979): Selection and the r-K continuum in experimental populations of protozoa, *Amer. Nat.* 113:427-437.

Mueller, L.D., and F.J. Ayala (1981): Trade-off between r-selection and K-selection in *Drosophila* populations, *Proc. Natl. Acad. Sci. USA* 78, No. 2: 1303-1305.

Pianka, E.R. (1970): On r- and K-selection, *Amer. Nat.* 104:592-597.

Prout, T. (1980): Some relationships between density-independent selection and density-dependent population growth, *Evol. Biol.* 13:1-68.

Roughgarden, J. (1971): Density dependent natural selection, *Ecology* 52:453-468.

Smouse, P.E. (1976): The implications of density-dependent population growth for frequency- and density-dependent selection, *Amer. Nat.* 110:849-860.

Stearns, Stephen C. (1977): The evolution of life history traits: A critique of the theory and a review of the data, *Ann. Rev. Ecol. Syst.* 8:145-171.

Turelli, Michael and Doug Petry (1980): Density dependent selection in a random environment: An evolutionary process that can maintain stable population dynamics, *Proc. Natl. Acad. Sci. USA.* 77:7501-7505.

Turelli, Michael (1981): Niche overlap and invasion of competitors in random environments. 1. Models without demographic stochasticity, *Theor. Pop. Biol.* 20:1-56.

A TEST OF IDEAS ABOUT THE EVOLUTION OF EFFICIENCIES
AND STRATEGIES OF RESOURCE USE

John W. Glasser

1. INTRODUCTION

In the ecological literature, the terms generalist and specialist refer to
organisms that use many and few resources, but these definitions specify neither the
number of alternative resources nor how available ones are used. Is an organism
that consumes a few resources indiscriminately more or less specialized than one
that consumes many selectively? The concept of environmental grain (MacArthur and
Levins 1964) provides a standard for quantifying behavior resulting in different
patterns of resource use. Generalists behave as though heterogeneous environments
were fine grained, using alternative resources in proportion to their frequencies,
possibly weighted by differences in conspicuousness, and specialists behave as
though heterogeneous environments were coarse grained. Although there are as many
ways of specializing as there are conceivable strategies for exploiting alternative
resources, doubtless some are not biologically realistic. Nonetheless, two extremes
may be distinguished, one facultative and the other obligate: Obligate specialists
order resources according to their intrinsic values, exhausting the most valuable
one before using the next most valuable resource, if any exist, and so on. In
contrast, the alternative resources of facultative specialists have values that
depend partly on extrinsic factors such as their abundances and the efficiencies
with which consumers use them.

Because resource acquisition determines fitness ultimately, individuals that
choose the most valuable resources available should comprise contemporary populations. The niche breadth trajectories of model populations with different
characteristics (Glasser 1982, figures 1 and 2) suggest that generalists and
facultative strategists with even efficiencies of resource use are adapted to
environments in which resources are more and less constantly scarce, respectively,
relative to consumer requirements. At the opposite extreme, obligate and facultative
specialists with uneven efficiencies of resource use apparently are adapted to
environments in which resources are more and less constantly abundant, respectively,
relative to consumer requirements. Accordingly, when resources are chronically
relatively abundant and scarce, obligate specialists and generalists, respectively,
should evolve. Facultative specialists with efficiencies whose evenness is related
directly to the mean relative resource abundance should evolve in variable environments. Because most environments vary and most populations track them imperfectly,
most consumers should be facultative (Glasser 1982).

Ultimately, these ideas must be evaluated empirically, in which endeavor
Vandermeer's (1972) grain-matrix methodology should prove useful, but it is possible

to subject them to a theoretical test. The objective of this paper is to report some results of this less ambitious task using a multiresource, multispecies generalization of Verhulst's (1838) logistic equation. In that model, the number of individuals of a species that the environment can sustain is determined as the quotient of the amount of some limiting resource, or its rate of supply, and the species' per capita requirements. In this generalization, the carrying capacity of the ith species, K_i, is a sum of T such quotients:

$$K_i = \sum_{j=1}^{T} A_j/\alpha_{ij},$$

where A_j is the abundance of the jth resource, or its rate of supply, and α_{ij} is the amount or rate at which resource j is required by each individual of species i. This per capita requirement may be written as a quotient of the individual ration, R_i, and efficiency with which the jth resource is used by the ith species, e_{ij}. That is,

$$\alpha_{ij} = R_i/e_{ij}.$$

2. THE MODEL

Substituting for the carrying capacity in Verhulst's equation and generalizing to many species,

$$\frac{dN_i}{dt} = r_i \sum_{j=1}^{T} n_{ij} \left(1 - \sum_{k=1}^{S} \frac{\alpha_{kj} n_{kj}}{A_j} \right),$$

where dN_i/dt is the population growth rate of species i, r_i is its biotic potential or intrinsic rate of natural increase, n_{ij} is the number of individuals of the ith species that is using the jth resource and S is the number of species in the guild of potential competitors. The effect of the kth species on the specific growth rate of the ith is:

$$\frac{\partial (dN_i/N_i dt)}{\partial N_k} = -r_i \sum_{j=1}^{T} \frac{f_{ij} \alpha_{kj} f_{kj}}{A_j},$$

where f_{ij} and f_{kj} are proportions of the populations of species i and k that are using the jth resource. That is,

$$f_{ij} = n_{ij}/N_i,$$

where $N_i = \sum_{j=1}^{T} n_{ij}$. If each element in the S by S matrix of these partial derivatives is divided by the diagonal element in the same row, the community matrix

(Levins 1968) results.

The computer program that implements this model assigns obligate generalists to resources with probabilities that equal resource frequencies. The frequency of resource j is its current abundance, which is its initial abundance or rate of supply less the present amount or rate of its use, divided by the sum of this quantity over all resources:

$$(A_j - \sum_{k=1}^{S} \alpha_{kj} n_{kj}) / \sum_{j=1}^{T} (A_j - \sum_{k=1}^{S} \alpha_{kj} n_{kj}).$$

Resources are ordered arbitrarily, their current abundance expressed as a cumulative frequency distribution, numbers in the unit interval are obtained from a pseudo-random number generator, and individuals are assigned to the resources to which these numbers correspond. The algorithm that assigns obligate specialists to resources chooses the one that they use most efficiently provided that it will sustain at least one individual. That is, the resource with the greatest associated e_{ij} is chosen provided that

$$(A_j - \sum_{k=1}^{S} \alpha_{kj} n_{kj}) \geq \alpha_{ij}.$$

Although the efficiencies with which species exploit alternative resources are not determined entirely by characteristics of the resources themselves, clearly they have an intrinsic component. Facultative specialists are assigned to the resource with the greatest unused capacity, assessed in individuals of that species,

$$(A_j - \sum_{k=1}^{S} \alpha_{kj} n_{kj}) / \alpha_{ij}.$$

3. EXPERIMENTAL DESIGN

In order to evaluate theoretically ideas about the evolution of efficiencies and strategies of resource use, one can simulate competition among species with different suites of these characteristics in different environments. If model species with appropriate characteristics fare better than others with alternative combinations, then the ideas have been corroborated; otherwise, they should be reconsidered. Model environments differing only in constancy were obtained for use in such studies by allowing the abundance of each resource to vary stochastically. Current resource abundances, the difference between initial abundances or rates of supply and present levels of use,

$$A_j - \sum_{k=1}^{S} \alpha_{kj} n_{kj},$$

were incremented by a quantity that transforms pairs of pseudo-random numbers into single random variables from a normal distribution with mean zero whose second moment can be specified. Depending on the magnitude of a third pseudo-random number, this quantity was either

$$\sqrt{(-2\ln X)} \ \sin(2\Pi Y)Z \quad \text{or} \quad \sqrt{(-2\ln X)} \ \cos(2\Pi Y)Z,$$

where X and Y are the pair of pseudo-random numbers and Z is the desired standard deviation (Chen 1971).

Three simulations were conducted in constant environments and, to maintain the errors within a few percentage points of their means, increasing numbers of simulations were conducted in more variable environments: four in environments in which the coefficients of variation of mean resource abundance were one percent, six in environments in which they were ten percent and fourteen in ones with coefficients of environmental variation of one hundred percent. There were five alternative resources, whose initial abundances or rates of supply were the same, and five competing species in each simulation. One of each of the obligate strategists, a generalist and specialist with even and uneven efficiencies of resource use, respectively, were represented, along with three facultative strategists, two with the same efficiencies of resource use as the obligate strategists and one with intermediate efficiencies. These model consumers were identical in all other characteristics (e.g., biotic potentials, initial abundances, carrying capacities). Parameter values are listed in the appendix.

4. RESULTS

Table 1 summarizes the results of simulated competition in environments differing in constancy. The columns refer to the five species: one is the generalist, five is the specialist and 2, 3 and 4 are the facultative strategists arrayed from most to least even efficiencies. The rows are treatments, the coefficients of variation of mean resource abundance ranging from 0 to 100 percent from top to bottom. The entries are mean frequencies and their standard errors, and the numbers of simulations on which each is based are shown in the last column. The most notable results are that generalists fare well relative to more selective consumers, predominating under constant conditions but being displaced by facultative specialists with even efficiencies as environments become increasingly variable.

5. DISCUSSION

The first result, that generalists fare well relative to more selective consumers irrespective of environmental variation, is an artifact of initially even resource-abundance distributions that were maintained more or less even by the facultative consumers, particularly species 2, but clearly the conventional wisdom

Table 1 Final Frequencies, $\overset{*}{N}_i / \sum_{k=1}^{S} \overset{*}{N}_k$, where $\overset{*}{N}_i = \sum_{j=1}^{T} [(A_j - \sum_{k \neq i}^{S} \alpha_{kj} n_{kj})/\alpha_{ij}]$, by

Species from Experiments in which the Coefficients of Environmental Variation Ranged from 0 to 100 Percent

Coefficient of Variation of Mean Resource Abundance	Species Final Frequencies					Sample Sizes
	1	2	3	4	5	
0	.27(0)	.22(.01)	.16(.00)	.17(0)	.17(0)	3
1	.26(.02)	.23(.01)	.17(.00)	.18(.00)	.18(.00)	4
10	.25(.02)	.25(.01)	.16(.00)	.18(.01)	.17(.00)	6
100	.23(.01)	.24(.01)	.17(.01)	.18(.01)	.18(.01)	14

Note: numbers in parentheses are standard errors; 0 indicates no error and .00 indicates an error smaller than .005, which would not round off to .01 .

that specialists outcompete generalists in heterogeneous environments in not universally true. The second result, that facultative consumers fare increasingly well relative to obligate ones in more variable environments, was one of the predictions made *a priori* (Glasser 1982). The only such prediction not corroborated by results of these simulations is that specialists should fare better than generalists in unsaturated environments: Their initial growth rates were comparable. However, this expectation is realized in simulations conducted in environments in which resource frequencies match the efficiencies with which specialists use them (Glasser and Price manuscript, experiment 6), and that is the only condition under which uneven efficiencies are likely to evolve. That is, species should become most efficient at using resources that they encounter most frequently, and the probabilities of encountering equally conspicuous resources are their frequencies.

Acknowledgments

The comments of participants at the International Conference on Population Biology were helpful in preparing the manuscript for publication, and Nancy Lyons provided a statistical reference that facilitated the research itself.

Appendix

Biotic potentials, $r_k = 0.01$ for all k
Number of resources, $T = 5$

Number of species, $S = 5$

Resources abundances or rates of supply, $A_j = 1000$ for all j

Individual rations, $R_k = 10$ for all k

Initial abundances, $N_k (t = 0) = 1$ for all k

Efficiencies of resource use, e_{kj} , for all k:

	Even	Intermediate	Uneven
e_{k1}	0.2	0.3	0.9
e_{k2}	0.2	0.25	0.04
e_{k3}	0.2	0.2	0.03
e_{k4}	0.2	0.15	0.02
e_{k5}	0.2	0.1	0.01

Thus, $\sum\limits_{j=1}^{T} e_{kj} = 1$ and $K_k = 100$ for all k

REFERENCES

Chen, E.H. (1971): A random normal number generator for 32-bit-word computers, *J. Amer. Stat. Assoc.* 66:400-403.

Glasser, J.W. (1982): A theory of trophic strategies: the evolution of facultative specialists, *Amer. Natur.* 119:250-262.

Glasser, J.W. and H.J. Price (manuscript): Competition for alternative resources.

Levins, R. (1968): *Evolution in Changing Environments: Some Theoretical Explorations*, Princeton Univ. Press, Princeton, N.J., 120 pp.

MacArthur, R.H. and R. Levins (1964): Competition, habitat selection and character displacement in a patchy environment, *Proc. Natl. Acad. Sci. U.S.* 51:1207-1210.

Vandermeer, J.H. (1972): Niche theory, *Ann. Rev. Ecol. Syst.* 3:107-132.

Verhulst, P.F. (1838): Notice sur la loi que la population suit dans son accroissement, *Correspondances Mathématiques et Physiques* 10:113-121.

COMPENSATORY STRATEGIES OF ENERGY
INVESTMENT IN UNCERTAIN ENVIRONMENTS

Jesús Alberto León

1. INTRODUCTION

What is the effect of randomly varying environments on the optimal partition of resources between effective reproduction and adult survival? Before attempting an answer, it is necessary to clarify the question, by stipulating (1) the life-history model envisaged (2) the pattern of uncertainty considered (3) the stage of the life history on which fluctuations are supposed to incide (4) the optimality criterion used.

(1) The basic life-history model used here is due to Charnov and Schaffer (1973). Let N(t) be the number of adults of a certain asexual genotype, just before reproduction occurs. Each individual produces thereafter B(t) offsprings. Each of these would reach maturity after a development period, with survival probability p(t). The adults survive without reproducing through an equivalent period with probability P(t). So the reproductive events are assumed to be synchronized, the inter-reproductive lapses having the same duration as the developmental periods. This duration is the time unit for the following difference equation, which governs the change in number of adults:

$$N(t+1) = N(t)\{B(t)p(t) + P(t)\} \tag{1}$$

The multiplicative growth rate λ of the genotype is therefore:

$$\lambda(t) = F(t) + P(t) \tag{2}$$

where the product B(t)p(t) has been called F(t), effective fecundity.

Since the initial question is about the effect of random change on the 'partition of resources', a variable portraying this partition must be introduced. Such a variable is the *reproductive effort* ε :   fraction of energy devoted by the adult to achieve effective fecundity F. F will be assumed to be an increasing concave function of ε, and $P(\varepsilon)$ decreasing concave. So the derivatives are:

$$\frac{\partial F}{\partial \varepsilon} > 0 \qquad \frac{\partial P}{\partial \varepsilon} < 0 \qquad \frac{\partial^2 F}{\partial \varepsilon^2} < 0 \qquad \frac{\partial^2 P}{\partial \varepsilon^2} < 0$$

(2) The pattern of uncertainty is as follows. The environment is a stochastic process independently and identically distributed over time units (specified above). That is, the process goes step wise; it remains constant during the time unit and can move to a new state (independently of where it was) when the change of

period occurs, i.e. just before the instantaneous reproductive episode. In one kind of models presented below, only two states can be adopted, each with probability 1/2. Also many-states models are considered.

(3) The stochastic process described before can affect either the juveniles or the adults. We call it X_j or X_A, according to the case. The respective survival probabilities become then functions of the stochastic process, $p(X_j)$ and $P(X_A)$.

(4) The optimality criterion preferred here is the gemometric mean of the λ's (see equation 2) corresponding to the different environmental states. This was used earlier by Cohen (1968) and Schaffer (1974) for simple life histories in random environments. Then Hasting and Caswell (1979) criticized its appropriateness. But I have challenged their criticisms elsewhere (León, manuscript), showing that the asexual genotype with maximal geometric mean of the yearly fitnesses λ (or its equivalent, the expectation of the logarithms of the λ's) comes to predominate in a population under a random environmental regime. Also Gillespie (1973) has established this in a genetical theory context.

2. JUMPING ENVIRONMENTS

Consider an environment whose quality jumps randomly between a good state $(\bar{X} + \Delta X)$ and a bad one $(\bar{X} - \Delta X)$, each state appearing with probability 1/2. Consider also an increasing continuous function $f(X)$ with value *one* at the mean $\bar{X}$, and values $f_G = f(\bar{X} + \Delta X)$, $f_B = f(\bar{X} - \Delta X)$. Define now $\delta_1 = 1 - f_B$ and $\delta_2 = f_G - 1$. Notice that if the function f is concave, $\delta_1 > \delta_2$. If it is linear, $\delta_1 = \delta_2$. If it is convex, $\delta_1 < \delta_2$.

(1) Suppose the uncertainty, is affecting only juvenile mortality. Then the fitnesses in good and bad years are:

$$\lambda_G = F_o \cdot f_F + P_o$$
$$\lambda_B = F_o \cdot f_B + P_o \tag{3}$$

where F_o and P_o are the effective fecundity and adult survivorship at the mean environment $\bar{X}$.

The geometric mean (squared) is:

$$\tilde{\lambda}^2 = F_o^2 \pi + F_o P_o \Sigma + P_o^2 \tag{4}$$

where $\pi = f_G f_B$, $\Sigma = f_G + f_B$.

If the environment was constant, with quality equal to $\bar{X}$, the multiplicative rate, i.e., the fitness of a genotype would be:

$$\lambda_o = F_o + P_o \tag{5}$$

Therefore, we can express the geometric mean as:

$$\tilde{\lambda}^2 = \lambda_o^2 + F_o^2(\pi-1) + F_o P_o (\zeta-2) \tag{6}$$

Notice also that

$$(\pi-1) = (\delta_2-\delta_1) - \delta_1\delta_2$$
$$(\zeta-2) = (\delta_2-\delta_1) \tag{7}$$

so that (6) could be rewritten in terms of the δ's.

We use the following method to assess the consequences of introducing random fluctuations. Were the environment constant at $\bar{X}$, the optimal reproductive effort would be a value $\hat{\epsilon}_o$ obtained by making $d\lambda_o/d\epsilon = 0$. That this is a maximum, and unique, is guaranteed by the concavity assumptions. We now calculate the derivative of the geometric mean $d\tilde{\lambda}/d\epsilon$ and evaluate it at ϵ_o. If this value is negative, $\tilde{\lambda}$ is already beyond its maximum, reached at $\hat{\epsilon}$, say, and so $\hat{\epsilon} < \hat{\epsilon}_o$. Conversely, a positive $d\tilde{\lambda}/d\epsilon$ at $\hat{\epsilon}_o$ implies $\tilde{\epsilon} > \hat{\epsilon}_o$.

The derivative of λ is:

$$(2\tilde{\lambda}) \frac{d\tilde{\lambda}}{d\epsilon} = (2\lambda_o) \frac{d\lambda_o}{d\epsilon} + 2F_o(\pi-1) \frac{dF_o}{d\epsilon} + (\zeta-2) \frac{d(F_o P_o)}{d\epsilon} \tag{8}$$

But since, at $\hat{\epsilon}_o$, $dF_o/d\epsilon = -dP_o/d\epsilon$ then

$$\frac{d(F_o P_o)}{d\epsilon} \bigg|_{\hat{\epsilon}_o} = (\hat{F}_o-\hat{P}_o) \frac{dP_o}{d\epsilon} \bigg|_{\hat{\epsilon}_o} \tag{9}$$

Therefore, the evaluation of (8) at $\hat{\epsilon}_o$, recalling (5), (7) and (9), leads to:

$$(2\tilde{\lambda}) \frac{d\tilde{\lambda}}{d\epsilon} \bigg|_{\hat{\epsilon}_o} = \{\hat{F}_o^2\delta_1\delta_2 - \hat{\lambda}_o(\delta_2-\delta_1)\} \frac{dP_o}{d\epsilon} \bigg|_{\hat{\epsilon}_o} \tag{10}$$

This equation deliniates the different kinds of effect that uncertainty in juvenile survivorship can have on optimal reproductive effort.

(i) 'Fluctuation factor' f concave ($\delta_2 < \delta_1$). The term between braces in the RHS of (10) is positive but $(dP_o/d\epsilon)$ is everywhere negative. Thus $d\tilde{\lambda}/d\epsilon < 0$ at $\hat{\epsilon}_o$. Therefore $\hat{\epsilon} < \hat{\epsilon}_o$. The same is of course true if f is linear. So these cases favor a *reduction* or reproductive effort as compensatory strategy when random variation impinges on effective reproduction.

(ii) 'Fluctuation factor' f slightly convex ($\delta_1\delta_2 > \delta_2-\delta_1$). Juvenile uncertainty will again cause a *reduction* optimal effort, $\hat{\epsilon} < \hat{\epsilon}_o$, except if $\hat{F}_o \ll \hat{P}_o$. This is so because the term between braces in (10) is positive insofar:

$$\frac{\hat{F}_o + \hat{P}_o}{2\hat{F}_o} < \frac{\delta_1 \delta_2}{\delta_2 - \delta_1} \tag{11}$$

(iii) 'Fluctuation factor' f strongly convex $(\delta_1 \delta_2 < \delta_2 - \delta_1)$. Now inequality (11) will *not* hold, except if $\hat{F}_o > \hat{P}_o$. Therefore, strongly convex uncertainty impinging on juveniles militates for an *increase* of reproductive effort, $\hat{\epsilon} > \hat{\epsilon}_o$, whenever the reference level $\hat{F}_o$ is not very high.

(2) If uncertainty affects only adult survival, the foregoing analysis can be easily paralleled. It is convenient in this case to use the survival effort S, fraction of energy devoted by the adult to its own survival during the time unit. Since we are not considering adult growth, $\epsilon + S = 1$. Again the investment S is *reduced* $(\hat{S} < \hat{S}_o)$ when the 'fluctuation factor' affecting P_o (g, say) is concave, linear or slightly convex (with $\hat{P}_o \ll \hat{F}_o$ bringing about possible exceptions to the very last case). And strong convexity in g elicits *increase* of survival effort $(\hat{S} > \hat{S}_o)$, whenever the reference level $\hat{P}_o$ is not higher than $\hat{F}_o$.

The models considered by Schaffer (1974) and Hastings and Caswell (1979) are particular instances of that presented here. Schaffer's model corresponds to the linear case, with $f_G = (1 + m)$, $f_B = (1 - m)$. Hastings and Caswell's model is in the borderline between slight and strong convexity, with $f_G = (1 + m)$, $f_B = \frac{1}{(1+m)}$

3. MANY-STATES ENVIRONMENTS

In order to extend our inquiry to environments adopting at random any of a variety of states, we approximate the expectations of the logarithm by the first two terms of a Taylor expansion.

$$E(\ln \lambda) \approx \ln E(\lambda) - \frac{V(\lambda)}{2E(\lambda)^2} \tag{12}$$

where E denotes expectation and V, variance. Analysis of the different cases gives results coinciding qualitatively with those obtained above. Only a sketch will be presented here.

For instance, if fluctuation affects only the juveniles:

$$E(\lambda) = F_o \cdot E(f) + P_o \qquad V(\lambda) = F_o^2 V(f)$$

Following the same procedure of calculating $dE(\ln \lambda)/d\epsilon$ and evaluating it at $\hat{\epsilon}_o$, gives:

$$\frac{d}{d\epsilon} E(\ell n\ \lambda)\Big|_{\hat{\epsilon}_o} \approx \frac{1}{\hat{E}(\lambda)} \left\{ -\left(1 + \frac{\hat{V}(\lambda)}{\hat{E}(\lambda)^2}\right) \Delta f + \frac{F_o}{\hat{E}(\lambda)}\ V(f) \right\} \frac{dP_o}{d\epsilon}\Big|_{\hat{\epsilon}_o} \tag{13}$$

where $\Delta f = E(f) - 1 = E(f) - f(\bar{X})$.

Again f concave ($\Delta f < 0$) or linear lead to a reduction of optimal effort ($\hat{\epsilon} < \hat{\epsilon}_o$). A similar result holds for f slightly convex, so far as the ratio ($\hat{P}_o/\hat{F}_o$) is not too high nor too low the variance $V(f)$. Strong convexity ($\Delta f > 0$ and large), on the other hand, bids for an increase in $\hat{\epsilon}$ ($\hat{\epsilon} > \hat{\epsilon}_o$). Exceptions now could be caused by a very low ratio ($\hat{P}_o/\hat{F}_o$) and a high $V(f)$.

Analogous results can be obtained for the survival effort when uncertainty affects adult mortality. A detailed examination of all these cases can be read in León (manuscript). Also situations in which both f and g fluctuate -- so that $V(t)$, $V(g)$ and $COV(fg)$ enter into the picture -- are considered there.

4. DISCUSSION

The models presented above contemplate only changes in intensity of unavoidable (or too costly to combat) mortality sources. That is, the *fraction* of effective fecundity (or of adult survival) gained when the environment ameliorates (δ_1)-- or lost when the latter worsens (δ_2) -- is independent of the energy fraction ϵ (or S) invested. Mortality factors of such a kind are easy to envisage: accidents and catastrophes, nor selective (or hard to avoid) predators, exogenously determined reductions in resource availability. They have in common its non-discriminatory nature: whatever the phenotype of the potential victim, the risk faced is the same. This of course depends not only on the nature of the environmental factor, but also on the peculiarities of the organism experiencing it.

Another way of saying all this is to note that the variance of the factor f (V_f) is independent of ϵ (Vg independent of S), depending only on the environmental variance σ^2. This is due to the unavoidable character of the mortality factors considered. Cases in which defense is possible and V_f or V_g are functions of ϵ, are of course conceivable. Models featuring this will be presented elsewhere (León, in preparation) and give results substantially different from those considered here.

We call *direct strategy* the change in optimal ϵ expected to supervene under natural selection as a response to uncertainty, when variance (V_f and/or V_g) -- being a function of ϵ -- is controllable by the organism. Conversely, we call *compensatory strategy* the change in $\hat{\epsilon}$ evoked by uncontrollable uncertainty, that is, fluctuation in unavoidable mortality.

Concavity or convexity (of f or g as functions of X) can be interpreted in several ways. A plausible suggestion is as follows. Concavity could be due to

a merely quantitative improvement in environment, which prompts diminishing returns
in fitness. Convexity, on the other hand, would be produced if the index X represents simultaneous improvement in quantity and quality, which beyond the mean environment brings in disproportionate advantages. Notice that concavity corresponds to
a decrement in the mean (of F or P) as compared with its value at the mean environment (F_0 or P_0). Convexity, on the contrary, brings as increment $E(F) - P_0 > 0$.
Only the linear case represents "pure uncertainty" in the sense of not bringing
about a change in the mean accompanying the variance. Therefore we call the three
cases "threatening"; "promising" and "neutral" uncertainty.

Two kinds of compensatory strategies are open as adaptive responses to uncontrollable uncertainty. One is *risk avoidance:* to reduce investment of resources
(ε or S) into that component of fitness (F or P) which is affected by random
fluctuations. The other is *risk incurrence*: to invest more there where uncertainty
impinges. Risk avoidance would occur when uncertainty is threatening, neutral or
only slightly promising. Risk incurrence would supervene whenever there is strongly
promising uncertainty, and the component of fitness affected by fluctuations is not
very high in the mean environment. Since neutral uncertainty evokes risk avoidance
just by introducing variance in the afflicted fitness component, the other strategies can be understood through the afore mentioned effects on the mean. Threatening uncertainty adds to variance a decrement of the mean. Slight promise ameliorates the mean, but not enough to overcome the variance effect. But the increase in
the mean brought about by strongly promising uncertainty is sufficient to merit risk
taking, unless the value around which fluctuations occur is high enough.

REFERENCES

Charnov, E.L., and W.M. Schaffer (1973): Life history consequences of natural
    selection, *Amer. Natur.* 107:791-793.

Cohen, D. (1968): A general model of optimal reproduction, *J. Ecol.* 56:219-228.

Gilespie, J.H. (1973): Natural selection with varying selection coefficients - a
    haploid model, *Genet. Res. Camb.* 21:115-120.

Hastings, A., and H. Caswell (1979): Role of environmental variability in the
    evolution of life history strategies, *Proc. Natl. Acad. Sci. USA.* 76:4700-
    4703.

León, J.A. (1982): Reproductive effort in random environments (I), (manuscript).

Schaffer, W.M. (1974): Optimal reproductive effort in fluctuating environments,
    Amer. Natur. 108:783-790.

A GEOMETRIC MODEL FOR OPTIMAL LIFE HISTORY

P.D. Taylor and G.C. Williams

ABSTRACT

Many authors (Williams, 1966; Gadgil and Bossert, 1970; Taylor et al., 1974; Schaffer, 1979) have considered models of optimal life history strategies. Most generally an organism at any particular age or size has a quantity of available resources which he can spend on maintenance, growth and/or reproduction, and his problem is to allocate these resources optimally. His objective is to maximize some measure of fitness or lifetime reproductive value. This optimization problem can be quite complex because decisions made at one stage may affect resources available at subsequent stages. Thus the mathematical analysis can be difficult and simple general patterns are hard to perceive. Our purpose is to consider a simple model which is quite general, and for which optimal allocation of resources can be determined by graphical analysis of suitable functions.

1. GENERAL ASSUMPTIONS

Assume that the resources available for growth and maintenance are a function solely of the size s of the organism, and that at any moment the organism of size s has two choices: grow or reproduce. If he decides to grow, his size increases according to the differential equation

$$\frac{ds}{dt} = sg(s) \tag{1}$$

where $g(s)$ is a fixed unit growth rate function. If he decides to reproduce he chooses an amount $s-s'$ to spend on reproduction, his size drops immediately to s' and he contributes an amount $k(s-s')$ to his offspring biomass, where k, his reproductive efficiency, is a fixed constant, $0 < k < 1$.

Assume that mortality is also a function $\mu(s)$ of size. We define the survivorship function $\ell(s',s'')$ to be the probability that an individual of size s' will survive to size s'' if he grows continuously till he reaches this size. Letting $s = s(t)$ be his size during this period of growth,

$$\ell(s,s'')\mu(s) = \frac{d}{dt}\,\ell(s,s'') = \frac{\partial}{\partial s}\,\ell(s,s'')\,\frac{ds}{dt} = \frac{\partial\ell}{\partial s}\,(s,s'')sg(s). \tag{2}$$

Hence

$$\frac{\partial}{\partial s}\,\ell(s,s'') = \ell(s,s'')\mu(s)/sg(s). \tag{3}$$

This equation will be useful later on.

We assume the functions $\mu(s)$ and $g(s)$ have graphs of the form depicted in Figure 1: μ and g are equal at points s_1 and s_2 , and $\mu < g$ at points between.

Our assumptions do not allow the organism to grow and reproduce simultaneously. But he can approach this behaviour by alternating growth and reproduction over a sequence of short time intervals, each time spending an infinitesimal amount on reproduction, and then growing up to or just past his previous size.

Our model is a continuous time model and takes no account of the effects of seasonality.

2. THE FORM OF AN OPTIMAL LIFE HISTORY STRATEGY

We now argue that the form of an optimal life history under our model is that all zygotes should be a fixed size x , organisms should grow continuously from size x to size y (maturity), reproduce and drop back to size z , grow back to y , reproduce as before, and continue this cycle until death. We allow z to be equal to y (in fact this will generally to be case) a situation which corresponds (by taking the limit as z approaches y) to spending resources on reproduction at rate $yg(y)$.

The argument is quite simple. Since growth and mortality depend on size only, any behaviour which is optimal for an individual at size s , will also be optimal when, if ever, he again attains this size. Thus if we simply choose y to be the first size after birth at which reproduction occurs under an optimal life history, and let z be the size immediately after reproduction, then the individual, at size z , will behave optimally by simply repeating his life history from his last occurrence of z . This creates a cycle.

3. THE OPTIMAL VALUES OF x , y , AND z

We assume a constant (over time) population size. In this case an optimal life history will simply maximize an individual's total number of expected offspring. Let x denote zygote size, y the size at reproduction, and z the size after reproduction. A zygote survives to his first reproduction with probability $\ell(x,y)$, and gets $k(y-z)/x$ offspring. He survives to get this many offspring a second time with probability $\ell(z,y)$. And so on. The total expected number of offspring of one zygote is

$$
\begin{aligned}
w(x,y,z) &= \frac{\ell(x,y)k(y-z)}{x} \left[1+\ell(z,y)+\ell(z,y)^2+\ldots\right] \\
&= \frac{k\ell(x,y)(y-z)}{x[1-\ell(z,y)]} .
\end{aligned}
\tag{4}
$$

To simplify this we introduce the function $L(s) = \ell(s,s_2)$ defined for $s \leq s_2$. Then for $s \leq y \leq s_2$,

$$L(s) = \ell(s,y)L(y) \tag{5}$$

so if we multiply top and bottom of (4) by L(y) we get

$$w(x,y,z) = \frac{kL(x)/x}{(L(y)-L(z))/(y-z)} \tag{6}$$

This expression has the form of the quotient of two slopes of secants to the graph of L(s), and suggests we maximize w with a geometric analysis of this graph. To find its concavity we differentiate to get

$$L''(s) = L(s)[R(s)^2+R'(s)], \tag{7}$$

where

$$R(x) = \mu(s)/sg(s) \tag{8}$$

A possible form of the graph of R is plotted in Figure 2 along with the graph of $1/s$. The concavity of the graph of L at s is found from the sign of $R(s)^2 + R'(s)$. To get ahold of this, observe that the family of solutions to the differential equation $f(s)^2 + f'(s) = 0$, is the function $f(s) = 1/s$ and all its horizontal translates $f(s) = 1/(s+c)$. Thus at each s we observe how R intersects the appropriate translate $f(s)$ of $1/s$. Thus at s_1, R cuts $1/s$ from below so $R' > f'$ and $L''(s_1) > 0$. Similarly $L''(s_2) < 0$. If we translate $1/s$ to the left in Figure 2 this intersection pattern persists until the curves separate with a final tangent at s_0. Thus, L(s) is concave down on $[s_1,s_0]$, concave up on $[s_0,s_2]$, and has a point of inflection at s_0. The graph of L is sketched in Figure 3.

Now let us maximize (6). The optimal zygote size x^* is that value which maximizes $L(x)/x$, the slope of the secant from the origin to the graph of L. This occurs where the secant is tangent to the graph and is depicted in Figure 3. Since x^* is a critical point of $L(s)/s$, and

$$\frac{d}{ds}\frac{L(s)}{s} = \frac{L(s)}{s}[R(s)-1/s] \tag{9}$$

we see that $R(x^*) = 1/x^*$, hence $x^* = s_1$.

Now look at the denominator of (6). For any fixed y, the best response z is that which minimizes the slope of the secant on $[z,y]$ to the graph of L. Such a (y,z) pair is depicted in Figure 3. The value y^* which minimizes all these minimum slopes is the inflection point s_0 marked in Figure 3. For this y^* we have $z^* = y^*$, the secant degenerated to a tangent, and the denominator of (6) becomes the derivative $L'(y^*)$.

We have shown that if the graph of R(s) has the form of Figure 2 then the optimum life history has $x^* = s_1$ and $y^* = z^* = s_0$. To get even a local optimum with $z^* \neq y^*$ requires a graph of R with the concavity structure of Figure 4.

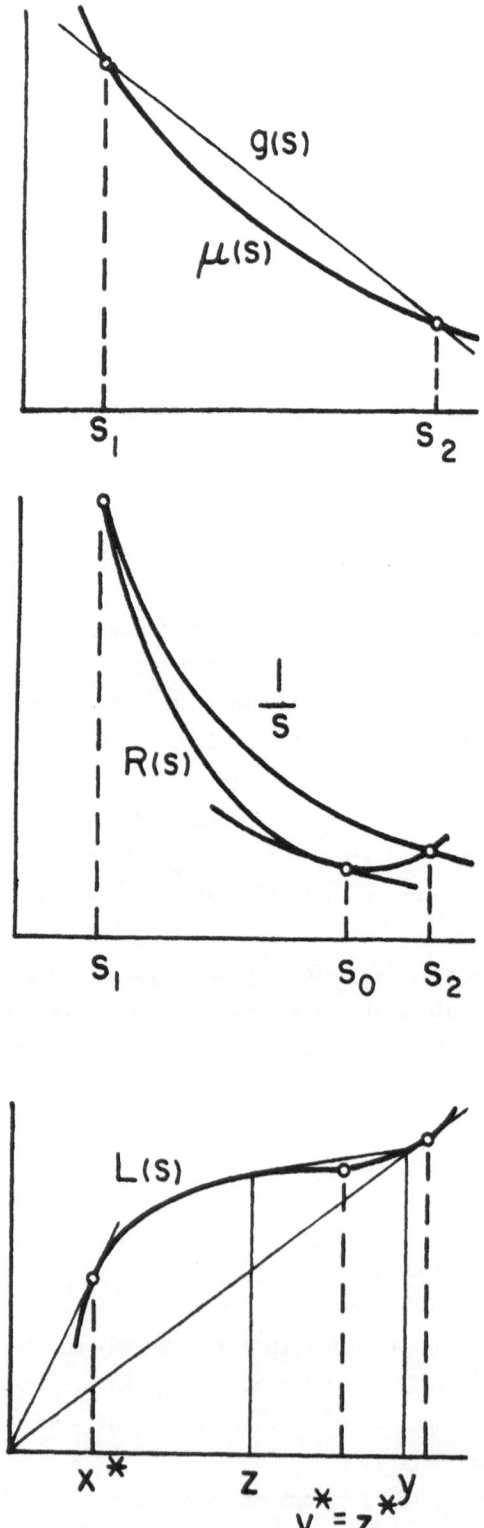

Figures 1, 2 and 3

For this graph the pair (y^*, z^*) is locally optimal in the sense that if (y,z) is a nearby (componentwise) pair then the secant on $[z,y]$ has a greater slope than the secant on $[z^*, y^*]$ and for any x, the strategy (x,y,z) is inferior to (x, y^*, z^*). However the pair (y^*, z^*) is not globally optimal, since the strategies at the two inflection points $z = y = s'$ and $z = y = s'''$ are both superior. This follows since the slopes of tangents at s' and s''' are less than that of the secant on $[z^*, y^*]$. The $R(s)$ graph corresponding to the $L(s)$ graph of Figure 4, will have the form given in Figure 5.

The above secant-tangent argument can be made quite general and leads to the result that no strategy with $z^* \neq y^*$ can be globally optimal. This follows from the geometric result that unless a graph is a straight line, there are always tangents with slope smaller than that of a given secant. In the framework of differential calculus, this is a consequence of the Theorem of the Mean.

4. DISCUSSION

We have made a number of major simplifying assumptions. First that when an organism reproduces he uses a block of resources which he has previously incorporated, thereby effectively reducing himself to a smaller size. By "effectively" we refer to the effects of mortality and growth (the ability both to find and incorporate new resources). For higher organisms it may usually be the case that the first pass through a size z (before reproduction) is quite different from the second pass. If mortality and growth interact differently on these two occasions, the qualitative conclusion of our model may be invalid, and we may find organisms which quite definitely alternate phases of growth and reproduction with substantial variation in "size" over one cycle.

Another major assumption is that environmental conditions are constant over time. Seasonality can be expected to impose a cyclical pattern on life history strategy and, as before, may override our conclusions. It may be that discrete time models provide the most appropriate way to account for these effects.

Thirdly, we have assumed no parental care. The effect of parental care is to change x from a discrete point to an interval of values, over which growth and mortality rates are altered by the intervention of the parent. The modelling of this important phenomenon requires new refinements of our model.

Fourthly, we have assumed the population is in a steady state, that is, population size is not changing. Populations which undergo periodic expansions and contractions will no doubt experience selective pressures for facultative modification of the values of x^* and y^* in response to these trends. The situation is complicated by the fact that those forces which drive the population cycles certainly influence mortality and probably even individual growth, so that our functions $\mu(s)$ and $g(s)$ are not time dependent.

With these assumptions, our conclusions are that an organism should grow to a

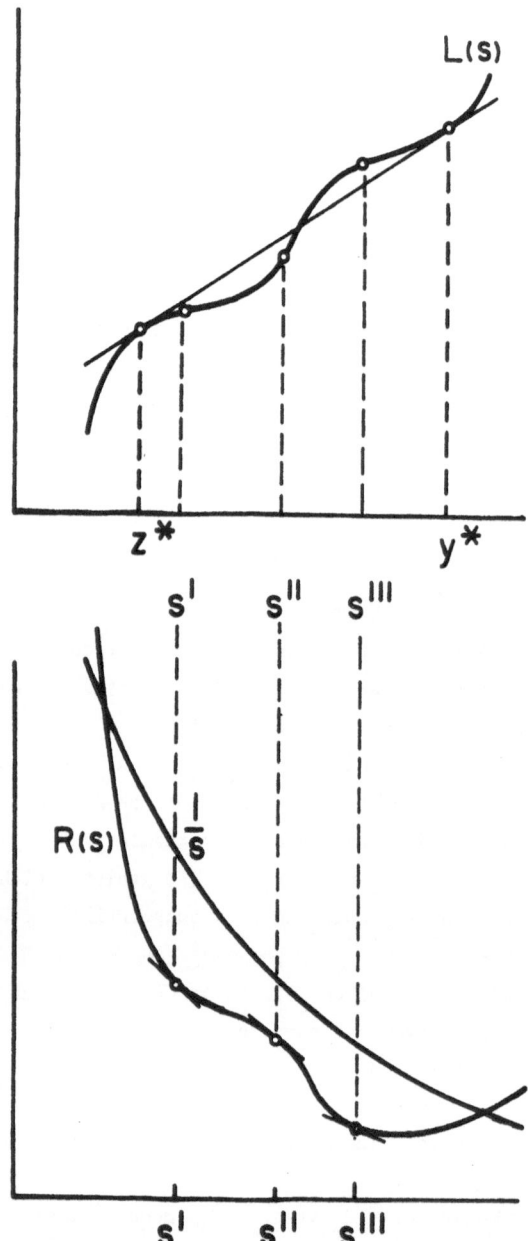

Figures 4 and 5

certain optimal size y (at which unit growth rate is somewhat greater than mortality) and then stay at that size for the rest of its life spending what extra
resources it can gather (over maintenance requirements) on reproduction. In higher
organisms this process of continuous reproduction might be best described as continuous parenting. The reproductive process may be complex and have several phases,
only one of which is the actual emergence of zygotes of size x. The organism can

be expected to time these phases so that expenditure levels are constant and its size and mortality do not vary noticeably over time.

The main advantage of our model is that it is simple enough to allow a geometric analysis. The advantage of such an analysis is that it allows both local and global optimization arguments to be easily made.

REFERENCES

Gadgil, M., and W. Bossert (1970): Life history consequences of natural selection, *Am. Nat.* 102:52-64.

Schaffer, W.M. (1979): Equivalence of maximizing reproductive value and fitness in the case of reproductive strategies, *Proc. Natl. Acad. Sci. USA* 76:3567-3569.

Taylor, H.M., R.S. Gourley, C.E. Lawrence, and R.S. Kaplan (1974): Natural selection of life history attributes: an analytical approach, *Theor. Popul. Biol.* 5:104-122.

Williams, G.C. (1966): Natural selection, the costs of reproduction, and a refinement of Lack's principle, *Am. Nat.* 100:687-690.

NECESSARY CONDITIONS FOR
AN INVASION PROOF STRATEGY

Thomas L. Vincent and Joel S. Brown

ABSTRACT

The evolutionarily stable strategy (ESS) as first formulated by Maynard Smith is a concept defined in terms of the pay-off functions of the "mutant" and one of the remaining "players". This paper demonstrates how this optimality concept may be extended to parametric games. Such games involve not only the payoff functions of every player but a model which puts constraints on the state of the system as well. The extended concept is then applied to a special class of "balanced" games. The balanced game not only greatly simplifies the necessary conditions for the extended ESS solution, but it is particularly applicable to ecological systems. Examples are given to illustrate the results.

1. INTRODUCTION

We turn our attention to static continuous parametric games and begin by defining the game. We designate control choices available to the players by u and the state which results from this control choice by x. For brevity in what follows we will not include inequality constraints on the state and/or control variables. A general analysis of continuous parametric games is given by Vincent and Grantham (1981).

DEFINITION 1.1 A vector payoff function $G(x,u) = [G_1,\ldots,G_r]^T$ whose state variables $x = [x_1,\ldots,x_n]^T$ and control variables $u = [u_1,\ldots,u_r]^T$ are related by equality constraints of the form $g(x,u) = [g_1,\ldots,g_r]^T = 0$ is said to be an r person *continuous static parametric game* if and only if for each $i \in (1,\ldots,r)$ there is associated a player who selects a control u_i seeking to maximize the ith component of the cost vector. Furthermore the equality constraints are required to satisfy $|\partial g(x,u)/\partial x| \neq 0$ for all u where x and u satisfy $g(x,u) = 0$.

Because of the requirements on the equality constraints, the implicit function theorem guarantees the existence of a function $\xi(\cdot)$ so that the solution to $g(x,u) = 0$ is given by $x = \xi(u)$. This allows us to define the payoff function as an explicit function of u.

$$\tilde{G}(u) \triangleq G(\xi(u),u) \tag{1.1}$$

The basic problem in game theory is to determine how the players should choose their controls. This is complicated by the fact that each player's payoff depends on the control choices made by every other player. There are many solution concepts associated with a game such as the Nash concept (Nash, 1951) based on the absence of coalitions.

DEFINITION 1.2 A point $\hat{u} \in R^r$ is a *Nash equilibrium solution (NES)* for a continuous static parametric game if and only if for each $i = 1,\ldots,r$

$$\tilde{G}_i(\hat{u}) \geq \tilde{G}_i(u_i,\hat{v}) \tag{1.2}$$

for all $u_i \in R$ where $u = [u_i,v]^T$ and R^r is the r dimensional set of real numbers.

Maynard Smith (1976) in examining solution concepts applicable in an evolutionary context defines an "evolutionary stable strategy" or ESS as follows.

DEFINITION 1.3 A strategy I is an *ESS* for all strategies J if either $E(I,I) > E(J,I)$ or $E(I,I) = E(J,I)$ and $E(I,J) > E(J,J)$ where $E(I,J)$ denotes the expected payoff to an individual playing I against an opponent using strategy J.

As Maynard Smith himself points out (Maynard Smith, 1982) "it [is] clear that there is some resemblance between an ESS and a Nash equilibrium". However this resemblance has not been made explicit and in the literature several authors have used the Nash equilibrium and the ESS more or less interchangeably (Auslander et. al., 1978; Mirmirani and Oster, 1978).

In order to formalize the notion of an ESS in continuous games we define an ESS like concept in such a way so that it is directly applicable to continuous static parametric games. We will call this new game solution an *invasion proof strategy* (IPS).

DEFINITION 1.4 A point $\bar{u} \in R^r$ is an *IPS* for a continuous static parametric game if and only if for each $i = 1,\ldots,r$ either

$$\tilde{G}_i(\bar{u}) > \tilde{G}_i(u_i,\bar{v}) \tag{1.3}$$

for all $u_i \in R$ with $u_i \neq \bar{u}_i$ or

$$\tilde{G}_i(\bar{u}) = \tilde{G}_i(u_i,\bar{v}) \tag{1.4}$$

for all $u_i \in R$ and for all $u = [\bar{u}_1+\delta, \bar{u}_2+\delta, \ldots, \bar{u}_r+\delta]^T \in R^r$

$$\tilde{G}_i(\bar{u}_i,v) > \tilde{G}_i(u) \tag{1.5}$$

where $u = [u_i, v]^T$ and δ is a nonzero scalar.

Neither the NES, definition 1.2 nor the IPS, definition 1.4 applied to a general continuous static parametric game would guarantee solutions for which all components of the u vector are equal or that the payoffs be equal. Some additional structure on the static parametric game is first required. We introduce the required additional structure in the next section via the concept of a balanced game.

2. THE BALANCED GAME

In order to provide a game setting more applicable to evolutionary biology, we now introduce some special structure to the continuous static parametric game. In particular, it is assumed that the environment presents the same conditions to each individual interacting with it.

DEFINITION 2.1 A point (x,u) is said to be *common* if and only if $x = \xi(u)$ is a solution to $g(x,u) = 0$ and u has all components equal, that is, $u = [s,s,\ldots,s]^T$ where s is a scalar.

DEFINITION 2.2 A continuous static parametric game is said to be *balanced* if and only if the following identities are satisfied

$$G_1(x,s) \equiv \cdots \equiv G_r(x,s) \tag{2.1}$$

$$\frac{\partial G_1(x,s)}{\partial x} \equiv \cdots \equiv \frac{\partial G_r(x,s)}{\partial x} \tag{2.2}$$

$$\frac{\partial G_1(x,s)}{\partial u_1} \equiv \cdots \equiv \frac{\partial G_r(x,s)}{\partial u_r} \tag{2.3}$$

$$\frac{\partial g(x,s)}{\partial u_1} \equiv \cdots \equiv \frac{\partial g(x,s)}{\partial u_r} \tag{2.4}$$

where use of s means that each component of the vector u is replaced by the scalar s.

The definition of a balanced game given here is similar to the symmetric game of Von Neumann and Morgenstern (1944) and Nash (1951). Since they are similar concepts, they place similar restrictions on the game. However, there are important differences which makes the balanced game more generally applicable to evolutionary problems than the symmetric game.

3. IPS SOLUTIONS

We begin by developing necessary conditions for a local IPS solution applicable to the game described by Definition 1.1.

THEOREM 3.1 If $\bar{u} \epsilon R^r$ is a local IPS for a continuous static parametric game and $\bar{x} = \xi(\bar{u})$ is a solution to $g(x,\bar{u}) = 0$ then for each $i = 1,\ldots,r$ there exists a vector $\lambda(i) \epsilon R^n$ such that

$$\frac{\partial L_i[\bar{x},\bar{u},\lambda(i)]}{\partial x} = 0 \tag{3.1}$$

$$\frac{\partial L_i[\bar{x},\bar{u},\lambda(i)]}{\partial u_i} = 0 \tag{3.2}$$

$$g(\bar{x},\bar{u}) = 0 \tag{3.3}$$

and either

$$\frac{\partial^2 \tilde{G}_i(\bar{u})}{\partial u_i^2} < 0 \tag{3.4}$$

or

$$\frac{\partial^2 \tilde{G}_i(\bar{u})}{\partial u_i^2} = 0 \tag{3.5}$$

and

$$\sum_{j=1}^{r} \frac{\partial^2 \tilde{G}_i(\bar{u})}{\partial u_i \partial u_j} \leq 0 \tag{3.6}$$

where $L_i = G_i - \lambda^T(i)g$ and $\tilde{G}$ is defined by (1.1).

Formal proof of this theorem will be given elsewhere. For now it is noted that from Definition 1.2 and conditions (1.4) and (1.5) of Definition 1.4 that an IPS point is also an NES point. Since u is an NES it follows from Definition 1.2 that $G_i(u)$ must take on a local minimum with respect to u_i. Thus (3.4) and (3.5) taken together are simply second order necessary conditions satisfying this requirement.

Note that the evaluation of $\partial^2 \tilde{G}_i/\partial u_i \partial u_j$ is not a trivial matter for the general case. An explicit expression for evaluating these derivatives in terms of L_i and g is given in Vincent and Grantham (1981).

Since conditions (3.4)-(3.6) involve second order partial derivatives, these conditions will not, in general, produce identities under the current definition of a balanced game. We therefore extend the concept of a balanced game to include

$$\frac{\partial^2 \tilde{G}_i(s)}{\partial u_i^2} \equiv \cdots \equiv \frac{\partial^2 \tilde{G}_r(s)}{\partial u_r^2} \tag{3.7}$$

$$\frac{\partial^2 \tilde{G}_i(s)}{\partial u_i \partial u_j} \equiv \cdots \equiv \frac{\partial^2 \tilde{G}_r(s)}{\partial u_r \partial u_j} \tag{3.8}$$

for all $j \in (1,\ldots,r)$.

4. IPS SOLUTIONS FOR A BALANCED GAME

We now apply the IPS conditions to an extended balanced game to obtain the following corollary.

COROLLARY 4.1 If $\bar{u} = [\bar{s},\bar{s},\ldots,\bar{s}]^T \in R^r$ is a common local IPS for an extended balanced continuous static parametric game and $\bar{x} = \xi(\bar{u})$ is a solution to $g(x,\bar{u}) = 0$, then for any $i \in (1,\ldots,r)$ and any $j \in (1,\ldots,r) \neq i$ there exists a vector $\lambda \in R^n$ such that

$$\frac{\partial L_i(\bar{x},\bar{s},\lambda)}{\partial x} = 0 \tag{4.1}$$

$$\frac{\partial L_i(\bar{x},\bar{s},\lambda)}{\partial u_i} = 0 \tag{4.2}$$

$$g(\bar{x},\bar{s}) = 0 \tag{4.3}$$

and either

$$\frac{\partial^2 \tilde{G}_i(\bar{s})}{\partial u_i^2} < 0 \tag{4.4}$$

or

$$\frac{\partial^2 \tilde{G}_i(\bar{s})}{\partial u_i^2} = 0 \tag{4.5}$$

and

$$\frac{\partial^2 \tilde{G}_i(\bar{s})}{\partial u_i \partial u_j} \leq 0 \tag{4.6}$$

where

$$L_i[x,u,\lambda(i)] \triangleq G_i(x,u) - \lambda^T(i)g(x,u) \tag{4.7}$$

and $\tilde{G}$ is defined by (1.1) and use of s means each component of the vector u is replaced by the scalar s.

The IPS necessary conditions applied to an extended balanced game are simpler than for the general game. In fact they are comparable in difficulty to optimizing a scalar function in a system with a single scalar control.

5. EXAMPLES

The following examples satisfy the requirements for an extended balanced game. In each case there are additional Nash solutions which lie on the boundary of the control set. The IPS solutions obtained are, of course, also Nash.

EXAMPLE 5.1 Let the payoff functions be given by

$$G_1 = 3u_1 - 5u_1u_2 - 2u_2 + 6 \tag{5.1}$$

$$G_2 = 3u_2 - 5u_1u_2 - 2u_1 + 6 \tag{5.2}$$

with constraints on the controls given by

$$0 \le u_i \le 1 \tag{5.3}$$

for i = 1,2. Since there are no equality constraints, the necessary conditions (4.1)-(4.3) simplify to $\partial G_i(s)/\partial u_i = 0$ for i = 1 or 2. This condition yields s = 3/5. Since both (4.5) and (4.6) are satisfied (Note in this case $\tilde{G} = G$) we conclude that s = 3/5 is a candidate for an IPS.

EXAMPLE 5.2 Let the payoff functions be given by

$$G_1 = 3u_1 - 5u_1u_2u_3 - 2(u_2+u_3) + 6 \tag{5.4}$$

$$G_2 = 3u_2 - 5u_1u_2u_3 - 2(u_1+u_3) + 6 \tag{5.5}$$

$$G_3 = 3u_3 - 5u_1u_2u_3 - 2(u_1+u_2) + 6 \tag{5.6}$$

with constraints on the controls given by

$$0 \le u_i \le 1 \tag{5.7}$$

for i = 1,2,3. Necessary conditions (4.1)-(4.3) are again simplified and yield s = $\sqrt{3/5}$. Again (4.5) and (4.6) are satisfied so that s = $\sqrt{3/5}$ provides a candidate for an IPS.

REFERENCES

Auslander, D.J., J.M. Guckenheimer, and G. Oster (1978): Random evolutionary stable strategies, *Theor. Pop. Biol.* 13:276-293.

Maynard Smith, J. (1976): Evolution and the theory of games, *Amer. Sci.* 64:41-45.

_____ (1982): *Evolution and the Theory of Games*, Cambridge Univ. Press, New York.

Mirimani, M., and G. Oster (1978): Competition, kin selection and evolutionary stable strategies, *Theor. Pop. Biol.* 13:304-339.

Nash, J.F. (1951): Non-cooperative games, *Ann. Math.*, 54, No. 2.

Vincent, T.L., and W.J. Grantham (1981): *Optimality in Parametric Systems*, John Wiley and Sons Inc., New York.

THE EVOLUTION OF STABLE STRATEGIES

James V. Whittaker

1. INTRODUCTION

The theory of games has been applied for nearly four decades to models of economic, social, and biological behaviour, largely through the notion of optimal strategy, but more recently through the notions of Nash equilibrium and evolution- arily stable strategy, the latter having been introduced in Maynard Smith (1974). The choice of strategy by the two agents in these models is governed by the assump- tion that the two agents are in competition for the advantages to be sought, and by the further assumption that their choices will be made with full knowledge and understanding of the mathematical analysis which lies behind the models. But the role of models where the strategies are arrived at through cooperation, instead of competition, between the two agents is greatly underestimated in some quarters, as the recent book by Boorman and Levitt (1980) will attest. Moreover, the assumption that the most desirable strategies will be chosen immediately by both agents, as soon as the mathematical analysis of the model is made manifest to them, sits rather uneasily alongside the sorts of assumptions found in other branches of modelling, where optimal solutions are reached gradually through a series of trials and errors. The original formulation of evolutionarily stable strategies tried to introduce a dynamical point of view into the static theory of games by explaining why we should expect deviations from stable strategies to be selected out of the population through evolution. However, a truly dynamic convergence to the desired strategies has only recently begun to emerge in the work of such authors as Hines (1980a,b).

In the present paper, we shall consider models of cooperative strategies and the dynamics of convergence to their stable limits. By cooperative strategies we shall mean that each agent is assumed to mould his latest strategy on the relative advantages conferred by the other agent's previous strategy, which is precisely Hines' (1980a,b) point of view. Of course, this sort of recursive choice of strat- egy implies a considerable amount of self-interest, for if we think of the agents as populations spanning many generations of individuals, each of whom exercises his own chosen strategy, then we can think of the collective strategy for any generation of the first agent as moulded on the relative advantages conferred by the other agent's strategy in the previous generation, and these relative advantages may be thought to be proportional to the number of offspring in the first agent's next generation who will practice the same strategy which begat them.

2. ASYMPTOTICALLY STABLE STRATEGIES

Suppose that we are given an m-by-m option matrix A whose entries are the advantages, measured in appropriate units, to the row agent after he and the column agent have chosen from among a fixed set of m options so as to form their strategies for survival, where the row agent belongs to some generation of a species competing for survival, and the column agent belongs to the next generation of that species. Suppose further that all agents are mortal, but that they beget offspring who may choose different strategies from those of their parents, depending upon how effective the parents' strategies were perceived to be. The strategies for successive generations of the species will be denoted by $P_0, P_1, \ldots, P_n, \ldots$, where $P_n = (p_1^{(n)} \ p_2^{(n)} \ \ldots \ p_m^{(n)})$ is the strategy of the $(n+1)$-th generation. If we assume that a spirit of cooperation prevails among successive generations, and that their common goal is to maximize the chance of survival for the entire population, then we might expect an arbitrary choice of strategy P_0 by the initial generation of the species to be followed by a strategy P_1 which was proportional to the relative advantages $P_0 A$ conferred by the various pure strategies available to the next generation of the species.

This analysis tacitly assumes that each generation knows precisely the strategy adopted by the previous generation and, indeed, that there was only one strategy followed by members of the previous generation. Let us consider the position of a member of the second generation more realistically. He is one of many in his generation, and his parent of the first generation is also one of many who has adopted an initial strategy P_0 which might, in fact, vary from one parent to another. What choice of P_1 ought a member of the second generation to make? He can not expect to encounter the same strategy P_0 from every member of the first generation who might come his way, so he is compelled to assume some sort of probability distribution for the various possible choices of P_0 which he might chance to meet. In the absence of any further information, the uniform distribution might seem to be the most natural one, but if such information were available, then some other distribution might be more appropriate. Thus, we should have to assume some probability density function $f_0(Q)$ to be defined at each point $Q = (q_1 \ q_2 \ \ldots \ q_m)$ in the region of euclidean m-space determined by $q_i \geq 0$ for each index i and $q_1 + q_2 + \ldots + q_m = 1$, a region which we immediately recognize as an $(m-1)$-dimensional simplex S. Now a member of the second generation might be assumed to adopt a strategy P_1 which was proportional to the relative expected advantages $E(QA)$ conferred by the various pure strategies available to the second generation with respect to the probability density function f_0. By this we mean that if the entries of A are denoted by a_{ij}, and a member of the second generation chooses the pure strategy which selects the j-th position, then his expected advantage can be expressed in terms of the j-th column A_j of A as

$$E(QA_j) = \int_S (a_{1j}q_1 + a_{2j}q_2 + \ldots + a_{mj}q_m) f_0(Q) dQ$$

$$= a_{1j} \int_S q_1 f_0(Q) dQ + a_{2j} \int_S q_2 f_0(Q) dQ + \ldots + a_{mj} \int_S q_m f_0(Q) dQ,$$

and if we let $E(Q) = \bar{Q}_0$ be the row matrix with entries $\int_S q_i f_0(Q)dQ$ for each index i, we then infer that $E(QA_j) = \bar{Q}_0 A_j$. If we arrange these entries in a row for each index j, we then conclude that the strategy P_1 ought to be proportional to $E(QA) = \bar{Q}_0 A$.

Now the uncertainties as to the precise strategy P_1 adopted by each member of the second generation will be found to be just as exigent for members of the third generation as they were to members of the second generation in regard to P_0, for the third generation can not be certain that all members of the second generation will adopt a strategy proportional to $\bar{Q}_0 A$, and it finds itself in the same position relative to its choice of P_2 as was just now described for the choice of P_1. That is to say, members of the third generation may be ignorant of the choice made by any given member of the second generation and are therefore compelled to assume some sort of probability distribution for the various possible choices of P_1 which they might happen to meet in the course of their lives. It is not at all clear that the same probability distribution should be assumed to govern the choice of P_1 as governed the choice of P_0, for the third generation has more information than was available to members of the second generation in forming their estimate of P_0, and this new information is simply the result of applying that same analysis used by the second generation so as now to form an estimate for P_1. Accordingly, we shall assume that the third generation adopts a probability density function $f_1(Q)$ governing the choice of P_1 which is proportional to the expected advantage conferred by the strategy Q upon members of the second generation against the expected strategy $\bar{Q}_0$ for the first generation, that is to say, $f_1(Q)$ should be proportional to $\bar{Q}_0 AQ'$, where Q' denotes the transpose of Q. Since the entries of $\bar{Q}_0 A$ are proportional to $P_1 = (p_1^{(1)} \; p_2^{(1)} \; \ldots \; p_m^{(1)})$, it follows that $f_1(Q)$ must be proportional to $p_1^{(1)} q_1 + p_2^{(1)} q_2 + \ldots + p_m^{(1)} q_m$. Now a member of the third generation might be assumed to adopt a strategy P_2 which was proportional to the relative expected advantages $E(QA)$ conferred by the various pure strategies available to the third generation with respect to the probability density function f_1. As in the case of the previous generation, we have $E(QA) = E(Q)A = \bar{Q}_1 A$, where the i-th entry of $\bar{Q}_1$ is $\int_S q_i f_1(Q)dQ$ for each index i, and P_2 will be proportional to $\bar{Q}_1 A$.

The conveyance of strategies from one generation to the next may be assumed to occur by means of some sort of haploid selection. In order to detect some pattern in this selection, we shall have to compare the strategy P_2 with its predecessor P_1 or, what amounts to the same thing, compare the expected strategy $\bar{Q}_1$ with its predecessor $\bar{Q}_0$. Now the i-th entry of $\bar{Q}_1$ is proportional to $\int_S q_i (p_1^{(1)} q_1 + p_2^{(1)} q_2 + \ldots + p_m^{(1)} q_m)dQ$, and from this form it is clear that we have to deal with essentially two types of terms in this integral, namely $\int_S q_i q_j dQ = u_m$ for all indices $i \neq j$, and $\int_S q_i^2 dQ = v_m$. In view of the symmetry of our region S of integration, the value of u_m is evidently independent of the choice of indices i,j, while v_m is independent of the choice of i. Our immediate aim will be to determine the relative sizes of u_m and v_m and to this end we observe that $\int_S dQ = w_m$ is the $(m-1)$-dimensional volume of S which could easily be computed from analytic geometry. However, this volume may also be computed as an iterated integral in which the final integration is with respect to q_1, and its integrand is proportional to the volume of an $(m-2)$-dimensional simplex with sides proportional to $1-q_1$, provided we assume that $m \geq 2$. Thus, we have

$$w_m = \int_S dQ = \int_0^1 c(1-q_1)^{m-2} dq_1 = c/(m-1),$$

so the constant of proportionality is $c = (m-1)w_m$, and our integrand can be thought of as the kernel $(m-1)w_m(1-q_1)^{m-2}$ for converting integrals in dQ to integrals in dq_1, so long as we are integrating functions of the single variable q_1. It is then an easy matter to evaluate $\int_S q_i dQ$, for this value must be independent of the index i, and from

$$\int_0^1 (1-q_1)(m-1)w_m(1-q_1)^{m-2} dq_1 = (m-1)w_m/m$$

we infer that $\int_S q_1 dQ = w_m - (m-1)w_m/m = w_m/m$, a result which we would have predicted from elementary geometric probability theory. Furthermore, from

$$\int_0^1 (1-q_1)^2 (m-1)w_m(1-q_1)^{m-2} dq_1 = (m-1)w_m/(m+1)$$

we infer that

$$\int_S q_1^2 dQ = (m-1)w_m/(m+1) - w_m + 2w_m/m = 2w_m/m(m+1),$$

and this is precisely the value of v_m.

In order to calculate u_m, we observe that

$$\begin{aligned}
w_m/m = \int_S q_1 dQ &= \int_S q_1(q_1 + q_2 + \ldots + q_m) dQ \\
&= \int_S q_1^2 dQ + \int_S q_1 q_2 dQ + \ldots + \int_S q_1 q_m dQ \\
&= v_m + (m-1)u_m,
\end{aligned}$$

and putting $v_m = 2w_m/m(m+1)$ into the relation $w_m/m = v_m + (m-1)u_m$ gives us the value $u_m = w_m/m(m+1)$. In particular, we have $v_m = 2u_m$, and this is all that we need in order to determine $\bar{Q}_1$, for we have already seen that the i-th entry of

$\bar{Q}_1$ is proportional to

$$u_m(p_1^{(1)} + \ldots + p_{i-1}^{(1)} + p_{i+1}^{(1)} + \ldots + p_m^{(1)}) + v_m p_i^{(1)} = u_m(1 + p_i^{(1)}).$$

If we arrange the entries of $\bar{Q}_1$ in a row for each index i, then we can see that $\bar{Q}_1$ is proportional to $E + P_1$, but since the row sum of $E + P_1$ is clearly $m + 1$, we can express this result in the form $\bar{Q}_1 = (E+P_1)/(m+1)$, where $E = (1\ 1\ \ldots\ 1)$.

3. HAPLOID SELECTION

The pattern for the conveyance of strategies through successive generations is by now fairly clear. Suppose that we have already determined a suitable probability density function f_n governing the choice of the strategy P_n. We must then compute $E(QA) = E(Q)A = \bar{Q}_n A$ with respect to f_n, and the next generation's strategy P_{n+1} ought to be proportional to $\bar{Q}_n A$. If we denote the row sum of $\bar{Q}_n A$ by s_n, then we can write $P_{n+1} = \bar{Q}_n A/s_n$, provided that $s_n \neq 0$. Moreover, the result of this computation is that $\bar{Q}_n = (E+P_n)/(m+1)$. Of course, members of the succeeding generation can not be sure that their predecessors will invariably follow the strategy P_{n+1}, so they will have to assume a probability density function $f_{n+1}(Q)$ governing the choice of P_{n+1} which is proportional to $\bar{Q}_n AQ'$ for each strategy Q. This completes our inductive definition of strategies and distributions for all indices n and all generations of our model species. Our concern in this analysis has gradually shifted away from the most likely strategy to be pursued by the n-th generation, and has begun to pay more attention to the form and nature of the probability distributions governing the choice of strategies by the n-th generation. Beyond that lies the question of how these strategies and distributions change with time as measured by the index n, and the ultimate form toward which they will tend through the passing of many generations, that is to say, as n tends to ∞.

In order to get some idea of the limiting forms of these strategies, we may simply combine the formulas $P_{n+1} = \bar{Q}_n A/s_n$ and $\bar{Q}_n = (E+P_n)/(m+1)$ for successive values of the index n. To begin with, we have

$$\bar{Q}_1 = E/(m+1) + P_1/(m+1) = E/(m+1) + \bar{Q}_0 A/s_0(m+1),$$

$$\bar{Q}_2 = E/(m+1) + P_2/(m+1) = E/(m+1) + \bar{Q}_1 A/s_1(m+1)$$

$$= E/(m+1) + EA/s_1(m+1)^2 + \bar{Q}_0 A^2/s_1 s_0(m+1)^2.$$

By now the pattern will be clear enough for us to write

$$\bar{Q}_{n+1} = E/(m+1) + EA/s_n(m+1)^2 + EA^2/s_n s_{n-1}(m+1)^3 + \ldots$$

$$+ EA^n/s_n s_{n-1} \cdots s_1(m+1)^{n+1} + \bar{Q}_0 A^{n+1}/s_n s_{n-1} \cdots s_1 s_0(m+1)^{n+1}$$

(1)

for each index n. As n tends to ∞, the expression for $\bar{Q}_n$ will tend to an infinite series whose convergence we shall now have to verify. In order to simplify
our work, we shall assume that the entries of A are non-negative and let r
denote the smallest row sum of A, while t denotes the largest row sum. Accordingly, we can write $rE' \leq AE' \leq tE'$, and repeated multiplication of this inequality by A leads to $r^k E' \leq A^k E' \leq t^k E'$ for every non-negative integer k.
In particular, we can multiply through on the left by E to obtain $mr^k \leq EA^k E' \leq$
mt^k, and for the same reason we infer that $r \leq \bar{Q}_k AE' = s_k \leq t$, as well as $r^k \leq$
$\bar{Q}_0 A^k E' \leq t^k$. Now every entry of every matrix appearing in (1) is non-negative, so
the infinite series which is the limit of $\bar{Q}_n$ as n tends to ∞ will converge in
case the series of entries from corresponding positions in all matrices of (1) is
dominated by a convergent series. Such a series of entries will clearly be dominated by the series of row sums which contain it, and we can get the series of row
sums simply by multiplying (1) through on the right by E'. If we apply the
inequalities already derived, we then find that

$$\bar{Q}_{n+1}E' \leq m/(m+1) + mt/s_n(m+1)^2 + mt^2/s_n s_{n-1}(m+1)^3 + \ldots$$

$$+ mt^n/s_n s_{n-1} \cdots s_1(m+1)^{n+1} + t^{n+1}/s_n s_{n-1} \cdots s_1 s_0(m+1)^{n+1}$$

$$\leq m/(m+1) + m(t/r)/(m+1)^2 + m(t/r)^2/(m+1)^3 + \ldots$$

$$+ m(t/r)^n/(m+1)^{n+1} + (t/r)^{n+1}/(m+1)^{n+1}.$$

Evidently the latter series will converge as n tends to ∞, provided we assume
that $t/r < m + 1$, and this assumption implies, in particular, that $r \neq 0$. We
shall denote the limit of $\bar{Q}_n$ as n tends to ∞ by $\bar{Q}$ and define the limit P
of P_n by means of the equation $P = \bar{Q}A/s$, where $s = \bar{Q}AE'$. Finally, the limiting
probability density function will be denoted by f(Q) and is defined to be proportional to $\bar{Q}AQ'$.

If our assumption $t/r < m + 1$ is satisfied, then the continuity of the
relations $\bar{Q}_n = (E+P_n)/(m+1)$ and $P_{n+1} = \bar{Q}_n A/s_n$ implies not only that $P = \bar{Q}A/s$
but also that $\bar{Q} = (E+P)/(m+1)$. When these two equations are combined, we then find
that

$$\bar{Q} = E/(m+1) + P/(m+1) = E/(m+1) + \bar{Q}A/s(m+1),$$

and so $\bar{Q}((m+1)I-A/s) = E$ implies that $\bar{Q} = E((m+1)I-A/s)^{-1}$. We shall inscribe
this result in the following theorem.

THEOREM 1 *Suppose that* A *is an* m × m *option matrix with non-negative entries,
and* r *is its smallest row sum, while* t *is its largest row sum. Let* A *govern
the evolutionary process whose* n-th *generation has the expected strategy* $\bar{Q}_n$

defined by the probability density function f_n *in such a way that* $f_{n+1}(Q)$ *is proportional to* $\bar{Q}_n AQ'$ *for every strategy* Q *and arbitrary initial density function* f_0. *If* $t/r < m + 1$, *then this process will converge as* n *tends to* ∞ *in the sense that* $\bar{Q}_n$ *tends to the limit* $\bar{Q} = E((m+1)I-A/s)^{-1}$, *where* $s = \bar{Q}AE'$, *and* f_n *tends to the density function* $f(Q)$ *which is proportional to* $\bar{Q}AQ'$ *for each strategy* Q, *regardless of the initial density function* f_0. *The expected value of* Q *with respect to* $f(Q)$ *is* $\bar{Q}A/s$.

We can compute error bounds for the difference $\bar{Q} - \bar{Q}_{n+1}$ most conveniently in case all the row sums of A are equal, that is to say, $r = s_k = s = t$, by observing that if u is the largest column sum of A, then we can write $EA \leq uE$ and, after further iteration, $EA^k \leq u^k E$ for every positive integer k. Now we know from the derivation of Theorem 1 that

$$\bar{Q} = E/(m+1) + EA/s(m+1)^2 + \ldots + EA^n/s^n(m+1)^{n+1} + \ldots . \tag{2}$$

If we subtract equation (1) from (2) and apply our hypothesis $s_k = s$, then

$$|\bar{Q}-\bar{Q}_{n+1}| = |EA^{n+1}/s^{n+1}(m+1)^{n+2} + EA^{n+2}/s^{n+2}(m+1)^{n+3} + \ldots - Q_0 A^{n+1}/s^{n+1}(m+1)^{n+2}|$$

$$\leq E((u/s)^{n+1}/(m+1)^{n+2} + (u/s)^{n+2}/(m+1)^{n+3} + \ldots)$$

$$= E(u/s(m+1))^{n+1}/(m+1-u/s),$$

provided $u/s < m + 1$, but this follows from the fact that each entry of A is $\leq s$, so $u \leq ms$. We shall record this result in the following corollary.

COROLLARY 1 *With the same hypotheses as in Theorem 1, if all row sums of* A *are equal to* s, *and* u *is the largest column sum of* A, *then* $|\bar{Q}-\bar{Q}_{n+1}| \leq$ $E(u/s(m+1))^{n+1}/(m+1-u/s)$.

COROLLARY 2 *With the same hypotheses as in Theorem 1, if all column sums of* A *are equal to* u, *then* $\bar{Q} = E/m$ *and* $s = u$.

PROOF. Our hypothesis says that $EA = uE$, so we have $E((m+1)I-A/s) = E(m+1-u/s)$. If we multiply both sides of this equation by $((m+1)I-A/s)^{-1}$, the result is then

$$E = E((m+1)I-A/s)^{-1}(m+1-u/s) = \bar{Q}(m+1-u/s),$$

so $\bar{Q} = E/(m+1-u/s)$. Thus, $\bar{Q}$ is a scalar multiple of E with row sum 1, so $\bar{Q} = E/m$. Comparing $1/m$ with $1/(m+1-u/s)$, we infer that $s = u$.

The latter is a useful formula for s when the column sums of A are all equal, but the row sums need not be so, for generally s is not easy to find unless we know that $r = t$. Furthermore, $\bar{Q}$ is not generally an evolutionarily stable strategy, for we may choose to take the option matrix

$$\begin{pmatrix} 0 & 1 & -1 \\ -1 & 0 & 1 \\ 1 & -1 & 0 \end{pmatrix}$$

for the game of rock-scissors-paper which is well known to have no evolutionarily stable strategy, provided we add 1 to all entries so as to make it non-negative, for this operation does not alter the evolutionarily stable strategies. Evidently $A = \begin{pmatrix} 1 & 2 & 0 \\ 0 & 1 & 2 \\ 2 & 0 & 1 \end{pmatrix}$ satisfies the hypotheses of Theorem 1, and Corollary 2 says that $\bar{Q} = (1/3 \ \ 1/3 \ \ 1/3)$ is the limiting strategy.

A somewhat different approach to evolutionary processes will be found in the work of Hines (1980b). Instead of our relation $\bar{Q}_{n+1} = E/(m+1) + \bar{Q}_n A/s_n(m+1)$, he writes (using the notation of the present paper) $\bar{Q}_{n+1} = \bar{Q}_n + \bar{Q}_n A C_n$, where C_n is a covariance matrix, and proceeds to investigate various properties of the limit $\bar{Q}$ on the assumption that C_n is nearly constant for large values of n.

REFERENCES

Boorman, S.A., and P.R. Levitt (1980): *The Genetics of Altruism*, Academic Press, New York.

Hines, W.G.S. (1980a): Three characterizations of population strategy stability, *J. Appl. Prob.* 17:333-340.

_____(1980b): Strategy stability in complex populations, *J. Appl. Prob.* 17:600-610.

Maynard Smith, J. (1974): The theory of games and the resolution of animal conflicts, *J. Theoret. Biol.* 47:209-221.

_____(1976): Evolution and the theory of games, *Amer. Scientist* 64:41-45.

PART III:

POPULATION GROWTH

FORECASTING THE HUMAN POPULATION

Nathan Keyfitz

1. IMPOSSIBILITY OF FORECASTING

Population numbers change in accord with other changes in the social and natural world. A true forecast of population, say for the year 2000, would require
that we know what will happen between now and then in the way of epidemics, bio-
medical discoveries, war and peace, economic development in poor countries, prosperity and depression in rich countries. The vast cultural change of the last two
decades summarily described as women's liberation has had a major unanticipated
effect on the birth rate; nothing precludes further such changes in the decades
ahead.

If population is tied to other social and economic variables, then it cannot
be forecast without taking account of them. Yet merely to cite the unexpected
changes of the last decades is to show how difficult forecasting is. Who could now
claim to have foreseen the rising divorce rate of the past twenty years by which
divorces in the United States increased from under 400,000 per year in 1960 to 1.2
million in 1981? Who, observing that men's incomes were rising, and hence that they
were more able than before to support their wives and children, would have expected
the movement of women, including a large fraction of married women with young children, into the labor force?

Similar unpredicted changes have occurred in mortality. In North America
life expectancy for men of 65 remained at just about 13 years during the 1950s
and 1960s, so this came to be seen as an upper limit; during the 1970s it resumed
the increase of earlier times and is now close to 14.5 years. We think of improved
health as associated with economic growth, but the difficult 1970s saw 50 percent
more improvement for U.S. males of all ages than the prosperous 1950s and 1960s
taken together. China's income has risen slowly, but its expectation of life at
birth has increased spectacularly, form 46 years in 1950 to 67 in 1975, according to figures published by the United Nations (1981, p. 90). In the Soviet Union,
on the other hand, there have been recent increases in mortality for many segments
of the population. The United Nations (1981, p. 100) shows a drop of 0.8 years
from 1970-75 to 1975-80, and 1975-80 is actually below all previous figures as far
back as 1960-65.

While men's expectation of life everywhere rises slowly, that of women ac-
celerates, so that the expectation of life for a girl just born is 8 years more,
at current death rates, than for a boy, where in 1920 the difference was only 1
year (*Stat. Abstr.*, 1981, p. 69). None of these changes is understood even after
the fact, and even less could it have been forecast before it occurred.

We cannot even explain the past, a simpler task than predicting the future. To ask, "Can Prediction Become a Science?" (Lilley, 1946) is asking whether there is genuine novelty in human affairs. The study of past attempts at prediction gives the answer: There ia genuine novelty, and anticipating it can never be a science. If birth, death, and migration changes are tied into a complex set of equations including independent social, economic, and biological variables, we are at least two removes from effective forecasting. In the first place we do not know what will happen to the many independent variables -- whether there will be peace or war, prosperity or depression. In the second place, even if we could foretell these perfectly, we would not be able to translate their changes into the dependent variable of population. The linkages between the variables are either not known or are known too loosely to be applied.

2. ESSENTIALITY OF FORECASTING

Despite all this one cannot walk away from the problem of prediction. Any present policy or action depends on whatever knowledge of the future we can secure. No one really cares what the situation was in the past; it is the reigning circumstances of the future that will govern the success or failure of what we do now. The point of gathering statistics on the past is that they will enable us to be more knowledgeable regarding the future. Whereas most economic forecasts are for months ahead, and weather forecasts for hours or days, population forecasts are needed for decades ahead. Action implies assumptions about what will happen in the future, and it is better to make those assumptions explicitly.

We have to live with the fact that forecasting is both impossible and unavoidable. Population projection or forecasting has adapted to these circumstances by distancing itself from other changes in society. Beyond a formal caution that the forecast assumes no major upheavals like wars, most forecasts work from little beyond previous population figures. Economists would like to endogenize population, but in practice economic models suppose a future population trajectory that is fed into the model from outside. The population forecast is best described as an extrapolation, though we must understand that it is often a very elaborate and sophisticated extrapolation.

3. EXTRAPOLATING TOTAL POPULATION

For a long time the extrapolation was of the total population. A curve was put through the population curve of the last decades, and its extension into the future gave the forecast. At times an exponential has been used; at other times the more appropriate logistic. Yet it is now some 40 years since practitioners became disillusioned with such curves. One reason is that past data seem to resist selection among possible curves. A simple example of this will suffice.

If we fit a logistic,

$$P_t = \frac{a}{1+be^{-rt}} \, ,$$

where a,b, and r are constants whose evaluation constitutes the fitting, and P_t is the population at time t, to the succession of counts of the United States population from 1800 to 1960, we find that it projects towards a horizontal asymptote of 256 million; after some time in the 21st century the population comes close to this value and never exceeds it. It we project with a hyperbola,

$$P_t = \frac{8500}{2009 - t}$$

we reach an asymptote also, but this time vertical, at the year 2009, when the population goes to infinity. The standard error of fit for the 200-year series is about half as great again with the hyperbola as with the logistic, but with short series the hyperbola fits as well as the logistic, and for most countries we have much less than 200 years of population count. The point is not that one would consider a hyperbola, but data unable to distinguish between a reasonable curve and an absurd one are even less likely to distinguish between two reasonable curves. Thus a cumulative normal curve and an inverse tangent fit about as well as a logistic.

If too much is changing below the surface for the effective selection of a method of extrapolation for the whole population, and still bearing in mind that we have to do something, what about extrapolating the components, birth, death, and migration? That is indeed what demographers now do. Especially effective methods have been developed for mortality, and we start with it.

4. MORTALITY: MINIMUM PARAMETER REPRESENTATION

The extrapolation of mortality rates by age demands a compact representation of the age schedule. Think of deaths for each five-year age group for 1975 and 1980, say, and suppose extrapolation of each age (by straight lines or by ratios) to 1985, 1990, etc. Within as little as five or ten years the resultant curve would have lost the characteristic bathtub shape; some rates might even be negative -- though this latter could be prevented by projecting logarithms. Recognizing separate five-year age intervals amounts to using 18 parameters. In the corresponding parameter space all possible life tables would be contained in a very small island. Any age-by-age extrapolation, say from 1975-80 to some future year, would quickly run outside that island. We need to get down to a smaller parameter space.

An extreme way of doing this is to index life tables on their expectation of life at age zero, $\mathring{e}_0$, and for each $\mathring{e}_0$ use an average of all known tables. The United Nations followed this method for the first set of what are called model tables (U.N. 1956). The approach escapes the difficulty that extrapolating would take one out of the island of possible life tables; between the expectations of 20 and 80 any extrapolation simply works through successive tables of the set. The trouble is the disregarded variation among mortality schedules with a given expectation at age zero.

A two-dimensional set, perhaps indexed on the ratio of adult to child mortality, as well as on $\overset{\circ}{e}_0$, seems better. John Pollard (1973, 1979, and elsewhere) has proposed a number of analytical formulas, in a tradition that goes back to Gompertz (1825), and finds 6 or more parameters essential to adequate description.

A particularly promising approach is the relational method due to William Brass (1971). Brass showed that if the probability of surviving from age zero to age x, $\ell(x)$, is transformed by the logit function:

$$Y(x) = \frac{1}{2} \log_e \left(\frac{1-\ell(x)}{\ell(x)} \right)$$

then practically any life table can be represented as a function of a standard life table. If $Y_s(x)$ is the logit of the survivorship of the standard then only two constants are needed to portray any given life table:

$$Y(x) = \alpha + \beta Y_s(x).$$

In application, suppose that we have constants α_{1950}, β_{1950} for the 1950 life table of a given country (either in terms of a universal standard, or of its own 1940 life table), and similarly for later years, then we need only project the two series

$$0, \ \alpha_{1950}, \ \alpha_{1960}, \ \alpha_{1970}, \ \ldots,$$

$$1, \ \beta_{1950}, \ \beta_{1960}, \ \beta_{1970}, \ \ldots,$$

and we have our answer. (For the standard table $\alpha = 0$ and $\beta = 1$.) The two conditions that make the process acceptable are

1) The simple curve remains within the island of possible life tables.
2) The succession of values of each of the two series is a reasonably simple curve, for instance a straight line.

On this process the projection cannot easily escape from the island of possible life tables. Experimentation due to Brass (1974) and others suggests that the two conditions. In some circumstances further constants are needed to deal with twists of the life table at the youngest and oldest ages, and a generalizing of the logistic to a transformation involving as many as four constants has been proposed by Stoto and de Leon (1981).

5. THE FERTILITY COMPONENT

A similar approach applied to fertility (Brass, 1974), and the curve of child-bearing by age, transformed by a cumulative Gompertz or other curve, can be projected by the same relational method. Unfortunately the eccentric movements of the fertility tranjectory remain to plague the user of this as of any other method that in its nature cannot recognize turning points before they occur.

Fertility is on the one hand more difficult to forecast than is mortality and on the other makes a larger difference to the future population. An error of 1 birth has no greater immediate effect than an error of 1 death. But the birth forecasts are subject to much more error.

The standard error of estimate of the rate of increase of the population is $\sqrt{\sigma_b^2 + \sigma_d^2}$, where σ_b^2 is the birth-rate error variance and σ_d^2 the death-rate error variance. We will see that σ_d^2 is about 10 times as great as σ_d^2.

Table 1 supposes that we are projecting the birth rates and the death rates by four simple methods, starting with calling the future year the same as the last past year, $Y_t = Y_{t-5}$, then going on to suppose arithmetic increase, geometric increase, and a parabola. The results differ relatively little among methods. We note that

1) Geometric progression (c) is not conspicuously superior to arithmetic (b).
2) Even putting a zero degree curve through, i.e., supposing this moment the same as five years back, as in (a), is not much worse than higher degree curves.
3) The standard error for births is at least three times as great as that for deaths.
4) On the whole the standard error of the rate of increase is in the neighborhood of 0.0025; we can bet two to one odds that the forecast rate of increase will come within 0.25 percentage points of the realized rate.

If one adds an allowance for migration to the 0.0025 of the bottom of Table 1, the result is not very different from the 0.3 or 0.4 percentage points that represent the errors of actual forecasts made in the past. We proceed to study these empirical errors in estimates of total populations.

6. EMPIRICAL MEASURE OF PRECISION IN FORECASTS BY COUNTRY

Rather than pursue further the error of naive forecasts (e.g. by adding in the effect of migration), we proceed directly to comparing actual forecasts made in the past with the subsequently realized population, with some 90 countries for which past forecasts are to be found.

The first problem is to find a metric for the departure of the forecast from the realization that will be comparable between countries of different sizes and for different projection spans. An error of 1 million for the United States 10 years

Table 1 Standard Error of Births and Deaths, and of Birth, Death and Natural
Increase Rates, from Data for the United States, 1800-1960, with 4 Simple
Estimates

Absolute Births and Deaths (Thousands)

	σ_B	σ_D	σ_B/σ_D
a) Y_t as Y_{t-5}	372.6	116.8	3.19
b) Y_t as $2Y_{t-5} - Y_{t-10}$	461.0	77.9	5.92
c) Y_t as Y_{t-5}^2/Y_{t-10}	455.7	90.1	5.06

Crude Rates of Birth and Death

	σ_b	σ_d	σ_b/σ_d
a) Y_t as Y_{t-5}	0.002412	0.000654	3.69
b) Y_t as $2Y_{t-5} - Y_{t-10}$	0.002508	0.000599	4.19
c) Y_t as Y_{t-5}^2/Y_{t-10}	0.002370	0.000653	3.63
d) Y_t as $3Y_{t-1} - 3Y_{t-2} + Y_{t-3}$	0.002124	0.000324	6.56

Effect of Crude Birth and Death Rates Combined

	$\sqrt{\sigma_b^2 + \sigma_d^2}$
a) Y_t as Y_{t-5}	0.002499
b) Y_t as $2Y_{t-5} - Y_{t-10}$	0.002579
c) Y_t as Y_{t-5}^2/Y_{t-10}	0.002458
d) Y_t as $3Y_{t-1} - 3Y_{t-2} + Y_{t-3}$	0.002149

ahead would be remarkably good; an error of 1 million for Alberta 5 years ahead
would be poor. A metric for error that is at least roughly comparable across countries and projection spans is the difference between the implied annual rate of
increase of the forecast and that of the realization (McNees, 1981; Keyfitz, 1981;
Stoto, 1981). If the forecast series is F_t and the realized series is P_t, then
the metric is

$$e = \left(\frac{F_t}{F_0}\right)^{1/t} - \left(\frac{P_t}{P_0}\right)^{1/t}.$$

7. PRINCIPAL CONCLUSIONS OF THE EMPIRICAL STUDY

1) A round figure that summarizes the error of past projections is that they have a root-mean-square difference from the subsequent performance of 0.4 percentage points of increase per year.

2) The error is about 0.6 percentage points for the fastest growing populations against 0.3 or less for the slowest.

3) The skill and care with which the calculation was done seems less important than the time when the forecast was made. Forecasts tended to be low in the 1940s for developed countries because the baby boom was not foreseen, and for the less developed countries because of the subsequent fall in mortality.

4) Whether the country is rich or poor, as measured by national income per capita, has almost no relation to accuracy for countries of given rate of increase.

5) A most important constancy is with the projection span: on the measure used (forecast mean rate of increase less rate in subsequent performance) the forecast five years ahead has the same range of error as the forecast 20 years ahead. Extrapolation of this constancy permits statements about the error of forecasts 30 and more years ahead, on which we have no direct experience.

6) More recent forecasts have been closer to realization than those of the 1950s, and those of the 1950s closer than the 1930s and 1940s. Methods have somewhat improved, but the degree to which this affects the matter, and the degree to which the discontinuities of the 1970s have been less that those earlier, is hard to say. It is probably over-cautious to suppose that the error of the present forecasts is equal to the average error of all past ones; one may prefer to take it that present forecasts will attain the precision of the more recent past forecasts.

7) Forecasts for areas within countries have been subject to more than twice as much error as forecasts for entire countries, to judge from the states of the U.S. in the 1970s, or from forecasts of urban population in 30 countries. Internal migration adds a large amount of unpredicted change.

8) Not enough is known (perhaps not much is knowable) about the extent to which errors of national estimates balance out in larger aggregates. One can say little on the error of present estimates of the world population in the year 2000 and after, except that the agreement of a number of writers is not conclusive evidence; there have been past cases of agreement among seemingly independent calculations for a particular country, all of which later proved much too low or too high.

8. ASSESSMENT OF A PREDICTION AFTER THE EVENT

An incidental use of the methods here developed is to assess the merits of individual predictions. Abraham Lincoln, in his annual message to Congress of

December 1982, estimated the U.S. population of 1900 at 103,208,000. Was that a good or a bad forecast, given that the 1900 census counted only 75,995,000? It shows the delicate character of such work that even after we know the outcome we cannot immediately say whether the forecaster should have done better. But with the above results as background, a judgment can be made on the quality of Lincoln's demographic analysis. The implied annual rate of increase of Lincoln's forecast to 1900 was 3.0 percent from the 1860 census figure of 31,444,000, the latest data available to Lincoln. In actuality the 1900 population of 75,995,000 showed a rate of increase from 1860 of only 2.2 percent. The departure of about 0.8 percentage points is only slightly greater than the 0.6 found for rapidly increasing populations. Thus Lincoln's estimate was decidely high, but his error was not far out of line with errors made since World War II using much more sophisticated methods.

We could make the error seem even less important by reminding ourselves that Lincoln could not have anticipated the fall of fertility that started in the 1870s throughout the English-speaking world, but once we start looking for reasons why forecasts went wrong the field for speculation is much too wide. (I am indebted to William Kruskal for the excerpt from Lincoln's message.)

9. THE MEANING OF THE VARIANTS

It has been the custom to present not one projection variant but three or more. In recent years less stress has been laid on the high and low variants, partly because they are so hard to interpret. No one has ever thought they are absolute upper and lower bounds on the possible future populations.

One way of interpreting them is to see in what portion of cases they straddled the realization. For instance, out of 5 estimates of the 1965 population of the United States made subsequent to 1947, 2 straddled; out of 6 estimates for 1970, 4 straddled; out of 9 estimates of 1975, 2 straddled; out of 6 estimates of 1980, none straddled. In all the record shows 26 estimates from 1947 to 1972, mostly with four variants each, and in 8 cases the highest and the lowest straddled the population that eventuated. The Population Investigation Committee of London made 16 estimates of the population of Great Britain in 1977, and 15 of these turned out to be low; the highest of the 16 was about right.

It is not easy to translate the range of assumptions into probabilities. The present official estimates for the United States suppose lifetime births per woman of 1.7 for the low and 2.7 for the high variant. The 1980 births are somewhat below the lower figure, and much of the time from 1945 to 1965 the realized births were above the higher figure. Thus although the range of 1.7 to 2.7 might appear to be wide enough to be safe, yet in fact it is somewhat daring. It sets a window for the 1980s and 1990s that fertility might easily exceed or fall below.

Better than such speculation is to note that the offical high variant for the

United States in the year 2000 is 283 million, the low is 246 million. For the 1980 figure of 226 million the implied annual rate of increase is 1.13 percent down to 0.42 percent, a half-range of 0.35 percent which seems close to the root-mean-square that we found. One can conclude that the published range is approximately a two-thirds confidence interval.

LONG-TERM FORECASTS ALL BUT USELESS

The most important single discovery is that the error remains constant with variation in the time span or horizon of projection. Thus we find for the United Nations forecasts made with data up to about 1955 (U.N., 1958) that the error in percentage points on our metric was

0.589 for 1960,

0.553 for 1965,

0.541 for 1970,

0.536 for 1975.

Supposing that this constancy holds for longer periods permits a statement of error for various time spans.

Table 2 Upper and Lower Bounds Calculated From Root-Mean-Square Error

	Lower Bound	Forecast	Upper Bound
1975	22.7	22.7	22.7
2000	31.3	33.7	36.2
2025	28.8	33.4	38.7
2050	28.0	35.0	43.8
2075	26.0	35.0	47.2
2100	24.1	35.0	50.8

Suppose that for Canada the root-mean-square error is 0.3 percentage points, and suppose that someone makes the forecasts shown in the middle column of Table 2. (Those for 2000 and 2025 are from the United Nations (1981, p. 35), while those for later years are hypothetical.) Then the two thirds confidence intervals are as given; for instance the low variant for 2100 is

$$(22.7)\left(\left(\frac{35.0}{22.7}\right)^{1/125} - 0.003\right)^{125} = 24.1.$$

Note that up to the year 2000 the range is narrow; beyond that it is too wide for the figures to be useful.

In application the metric is very different for fast- and slow-growing countries. Dividing the 90 countries of over 1,000,000 population in the U.N. 1958 forecasts gives in percentage points:

> 0.288 for the 30 slowest growing,
> 0.478 for the middle 30,
> 0.604 for the 30 fastest growing.

This is a real defect in the measure of error here used that one hopes is rectifiable. A much more sophisticated approach is due to Silvert (1981) that no one has yet applied in demography.

11. POPULATION VS. ECONOMIC FORECASTS

How does the accuracy of population forecasts compare with that of other attempts to foresee the future? The most readily available material is for Gross National Product, percentage of the labor force unemployed, the Consumer Price Index, and other economic facts. These are intrinsically more volatile and harder to predict than population, so it is not surprising that the error comes out to be larger. Economists are wise enough to stay with short-term estimates, which are more feasible and more useful than long-term.

Employing the same measure that we have used, the departure of the forecast annual rate from the realized rate, Stephen McNees (1981) has studied the record of 13 economic forecasters. McNee's tables end at 8 quarters, or two years, but being on an annual basis they are comparable with our estimates. One result is 1.5 percentage points for the root-mean-square error of GNP in money terms and about 0.8 percentage points for the deflated GNP. These are harder to forecast than the increase in real goods and services. Fully 7 percentage points is the best anyone has done for investment in residential building. Nonresidential fixed investment did much better, at about 3 percentage points.

Compared with these numbers the 0.4 percentage points average for population forecasts is not bad. But mostly what this represents is the fact that the population of each year overlaps more than 98 percent with the population of the previous year. There is no such help in forecasting the economy.

REFERENCES

Brass, William (1971): On the scale of mortality, in W.Brass (ed.), *Biological Aspects of Demography*, pp. 69-110.

_____(1974): Persepctives in population prediction, illustrated by the statistics of England and Wales, *Journal of the Royal Statistical Society*, Series A, 137:532-583.

Gompertz, Benjamin (1825): On the nature of the function expressive of the law of human mortality, in D. Smith and N. Keyfitz (eds.) (1977) *Mathematical Demography: Selected Papers*, pp. 279-288. Berlin: Springer-Verlag.

Jöckel, Karl-Heinz, and Peter Pflaumer (1981): Some remarks on errors in population forecasting (Manuscript).

Keyfitz, Nathan (1981): The limits of population forecasting, *Population and Development Review* 7, 4:579-593.

_____(1982): Choice of function for mortality analysis, *Theoretical Population Biology* (in press).

Lilley, S. (1946): Can prediction become a science? *Discovery*, November: 336-340. Reprinted in B. Barber and W. Hirsch (eds.), *The Sociology of Science*. New York: Free Press of Glencoe.

McNees, Stephen K. (1981): The recent record of thirteen forecasters, *New England Economic Review*, Sept./Oct.: 5-21.

Pollard, John H. (1973): *Mathematical Models for the Growth of Human Populations*, London: Cambridge University Press.

_____(1979): Factors affecting mortality and the length of life, *Population in the Service of Mankind*: 53-79. Liege: International Union for the Scientifc Study of Population.

Silvert, William (1981): The formulation and evaluation of predictions, *International Journal of General Systems* 7:189-205.

Stoto, Michael A. (1979): The accuracy of population projections. Working Paper WP-79-75. Laxenburg, Austria: International Institute for Applied Systems Analysis.

United Nations (1956): *Manual III: Methods for Population Projections by Sex and Age*. Sales No. 1956.XIII.3. New York: United Nations.

_____(1981): *World Population Prospects as Assessed in 1980*. ST/ESA/SER. A/78 New York: United Nations

U.S. Bureau of the Census (1981): *Statistical Abstract of the United States: 1981* (102d edition). Washington, D.C.: U.S. Government Printing Office.

PERSISTENCE IN UNCERTAIN ENVIRONMENTS

Zvia Agur

ABSTRACT

The colonization-extinction process in uncertain environments is studied. A model is described for the dynamics of populations with a complex life-cycle, in which the non-reproducing life-stage is resistant to environmental disturbances inflicted on the adult population. The basic equation is of a deterministic logistic type, with a time-delay in the growth term. Envrionmental uncertainties, in the form of a coloured noise ("telegraphic noise"), are superimposed on the deterministic equation. It is suggested that persistence in uncertain environments is mostly influenced by the duration of the resistant stage, and that a universal colonizer of such environments should originate from relatively stable environments.

1. INTRODUCTION

Persistence, which is at the roots of all population dynamics studies, is a static rather than a dynamic entity. Being delimited by colonization and extinction, persistence can be conceived through these two processes.

The fundamental attempt to specify the general prerequisites for colonization success was forwarded by MacArthur and Wilson (1967). Their conjectures were based on a model in which the growth of a finite population was taken as the difference between the processes of stochastic birth and death, while the environment was assumed to be constant. It seems though that in many colonization events the number of founding propagules is big enough for such demographic stochasticity to be of a marginal importance only. In the marine intertidal, for example, once the right conditions prevail in a spot, a new colony may be established there by a large number of larvae drifted with an occasional current. Moreover, in some systems, environmental stochasticity, not considered in MacArthur and Wilson's model, may have a decisive role in determining persistence chances of newly formed colonies. Whether there exists a global pattern to the relation between environmental stochasticity and colonization success is not clear yet. May (1975) suggests that higher stability of the environment is linked with a weaker ability of recolonization. Loya (1976) shows contradictory results, namely that colonization rate of hermatypic corals is much lower in a relatively unpredictable reef flat than in a neighboring, relatively stable, reef flat.

Environmental disturbances have sometimes a harsh prolonged effect such that susceptible life-stages become totally extinct, and the recovery of a local population depends on the ability of a resistant life-stage to survive throughout the unfavourable period. It is not evident that the estimation of the intrinsic rate of increase, r, suggested by Safriel and Ritte (1980) as a universal correlate of colonizing ability, may be sufficient for systems of this kind.

This paper studies the colonization-extinction process in uncertain environments. It describes a model for the dynamics of populations with a complex life-cycle, in which the non-reproducing life-stage is resistant to environmental disturbances affecting the adult population. Demographic stochasticity is ignored in the model, upon assuming a large founding population and a harsh environmental stochasticity. Optimal strategies of colonization are discussed, when the optimal strategy is defined here as the duration of the resistant-stage which minimizes the chances of extinction of its bearer when competing with other types in a given environment.

2. THE COLONIZATION - EXTINCTION MODEL

The model is intended to describe population dynamics in the marine intertidal, some insect populations, and other populations having the following properties. The life-cycle consists of two principal stages: 1) birth of juveniles, 2) recruitment of juveniles to the adult population. The juvenile stage is immune to environmental uncertainties inflicted on adults. This phase is also the only mobile form in the life history, and a time delay in reproductive activity. As such its role is complex and in some environments its optimal duration may be determined by contradicting factors.

In order to derive the optimal duration of the juvenile stage, it is assumed that the founding population is made of M competing types, differing in their juvenile longevity, τ. If we assume that there exists a single optimal strategy, then one must find $\tau = \tau^*$, such that all $\tau + \epsilon$ are at a disadvantage for any small ϵ. That is, τ will be the optimal strategy, τ^*, is its frequency of extinction, F_τ, is lower than the frequency of extinction of all $\tau + \epsilon$, $F_{\tau+\epsilon}$. Subsequently, the advantage of a given strategy, τ, with regard to $\tau + \epsilon$, will be measured by the net frequency of extinction, so that:

$$\tau = \tau^* \quad \text{if} \quad F_{\tau+\epsilon} - F_\tau > 0 \ \forall \epsilon. \tag{1}$$

Local population dynamics will be described in the following way. If $J(t-\tau)$ denotes a cohort of juveniles, born at time $t - \tau$, during one time interval, then $J(t-\tau) = \lambda X(t-\tau)$, where λ is the birth rate and $X(t-\tau)$ is the number of adults at time $t - \tau$. The dynamics of this cohort is given by:

$$dJ/dt = -\mu'J$$

(μ' is the mortality rate of juveniles) and

$$J(t) = \lambda X(t-\tau)e^{-\mu'\tau}. \tag{2}$$

If τ is the duration of the juvenile stage, then (2) is the number of juveniles, born at time $t - \tau$ that are ready to be recruited at time t. The model assumes that recruitment is the only density dependent process in the life history: the number of recruited individuals at time t will be then

$$X(t-\tau)e^{-\mu'\tau}(1-X(t)/K)$$

where K is the environmental limiting factor. Adding now the deterministic term of adult mortality, μ, the dynamics of the adult population will be:

$$dX/dt = \lambda X(t-\tau)e^{-\mu'\tau}(1-X(t)/K) - \mu X(t). \qquad (3)$$

It was chosen here to employ the simplistic logistic-type form of population dynamics. A situation of density dependent growth and density independent death term was previously considered by Horn (1968). Linear analysis of this equation (Agur and Deneubourg, manuscript), shows that the nontrivial steady state, $X = K(1-\mu e^{\mu'\tau}/\lambda)$, is a stable point.

In order to describe as simply as possible a habitat like the patchy intertidal zone, the environment is assumed to be made of a row of N patches, located on a boundless line, with a distance, S, between neighbouring patches. The mechanism responsible for the arrival of the founding propagule to one of the patches of the target (i.e., the mechanism of its long range dispersal) is not discussed here. As demographic stochasticity is ignored in the present model the initial population size should play no role in determining the chances of persistence. For this reason this size was assumed to be a constant. Juveniles born in a patch move randomly in their linear medium according to Fick's law, and eventually settle in an encountered patch within the habitat. Their movement is described by the function $g(\tau,S)$:

$$g(\tau,S) = 1/\sqrt{4\pi d\tau}\ e^{-S^2/2d\tau}$$

where d is the diffusion coefficient. For the justification and limitations of the use of such a function see Pielou (1977) and Okubo (1980).

The complete dynamics of the population in a constant environment will be:

$$dX/dt = \sum_{j=1}^{N} \lambda X_{j,k}(t-\tau_k)e^{-\mu'\tau_k}g(S_{ji},\tau_k)(1 - \sum_{k=1}^{M} \frac{1}{K} X_i,\tau_k) - \mu X_{i,k} \qquad (4)$$

where $X_{i,k}$ is the population size of type k in the island i.

Harsh environmental disturbances were incorporated in the model in the following form. The times between the end of one disturbance and the beginning of the next one are independent, exponentially-distributed random variables with a mean $1/\phi$. During these intervals the local population grows according to Equation

(4). When a disturbance occurs the adult population is wiped out and the subsequent recruitment of juveniles is prevented during a constant period of δ time units. Such a form of noise is conventionally called "telegraphic noise". Disturbances strike each patch independently. The renewal of a local population depends on the availability of juveniles in the spot when it becomes habitable again.

Analytical solution of this population dynamics is impossible. Thus a computer simulation was performed instead, using the following procedure. Equation (4) was integrated by the Euler method. At the same time a stochastic generator was activated in order to determine the times of the initiation of the disturbance. The time unit of integration was 1/50 - 1/100 of τ. Simulation experiments were carried out over 5000 time units (30-80 generations). Population was defined as extinct when its numbers remained zero for a period equal to τ.

3. RESULTS AND DISCUSSION

Simulation experiments of Equation (4) were carried out in the manner described above for the competition between couples of invading types in the range: τ = 60 - 180. This range is assumed to represent the physiological capacities of the population. Results of the competition, following definition (1), are brought in Figure (1) and Figure (2).

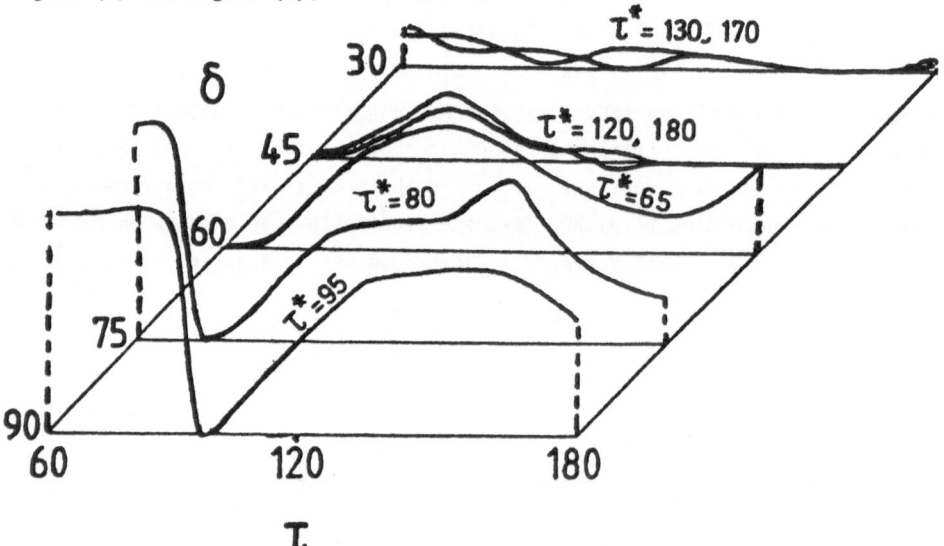

Figure 1   The optimal strategy of colonization as a function of the duration of the disturbance, δ. Curves show the net frequency of extinction $F_\tau - F_{\tau^*}$, calculated from simulation of Equation (4), for competition between couples of types in the region: τ = 60 to 180. λ = .1, μ' = 0.0, μ =.01, N = 4, S = 11.25, δ = 30 to 90. ϕ = 1/30, d = .1, K = 100, initial population size of each type is 40. Note that τ values were taken in intervals of 15 units.

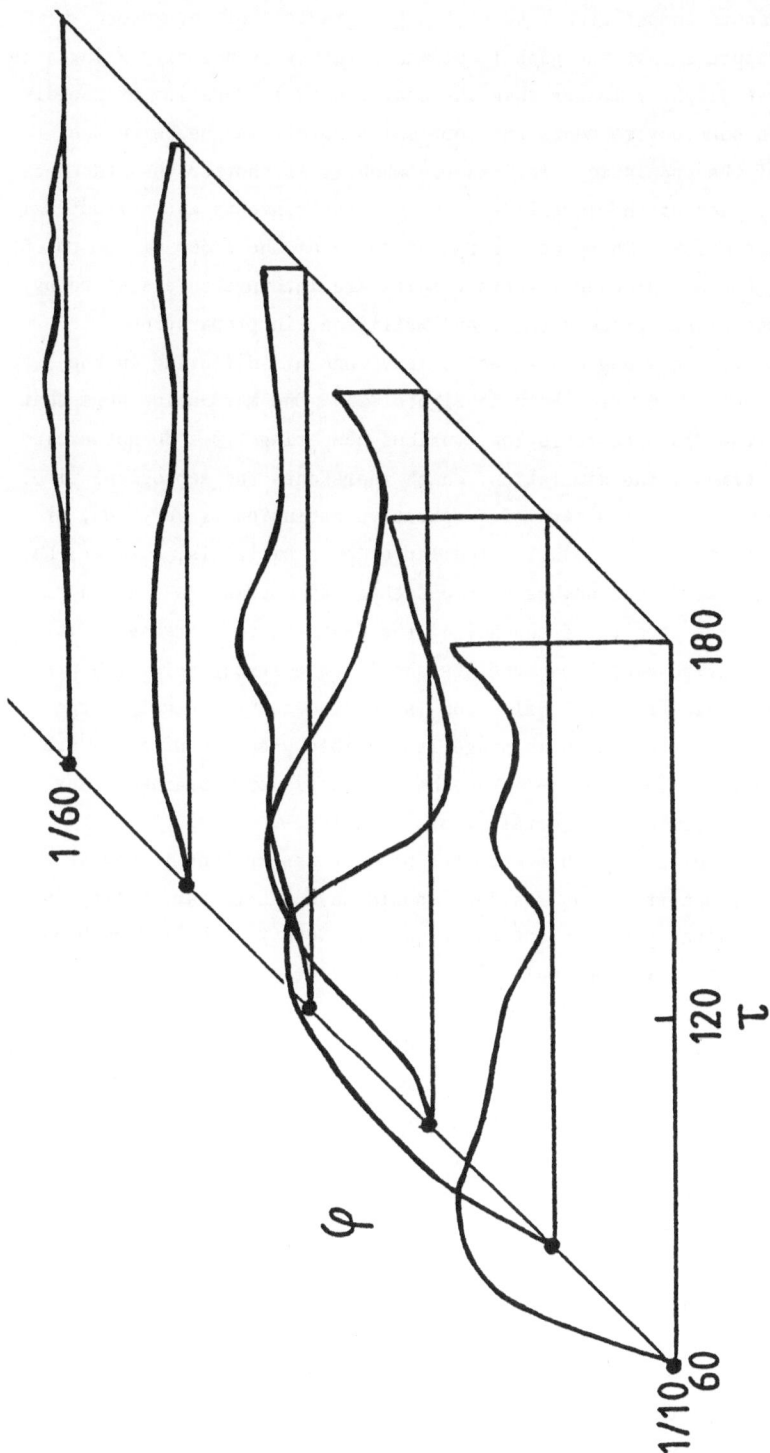

Figure 2 The net frequency of extinction as a function of $1/\phi$. Simulation
results of Equation (4) for competition between $\tau = 63$ and all other
types in the region $\tau = 60 - 180$. $\delta = 60$, other parameters like in Figure 1.

Figure (1) presents competition results along a gradient of environments differing in δ. It appears that the global optimal strategy is normally a juvenile longevity which is just slightly longer than the characteristic duration of the disturbance. However, in some environments this optimal strategy may be below the physiological range of the population, for example when it is shorter than the necessary developmental period of the juveniles. In such environments another optimal strategy will be selected for. This local strategy being of the order of 3 to 5 times the duration of the disturbance. These results are interpreted elsewhere by the features of a pure Poisson process (Agur and Meilijson, in preparation).

Results of competition along a gradient of environments differing in the frequency of the disturbance, φ, are shown in Figure (2). One may notice here that when the frequency of the disturbance is low, various competing types do not exclude each other during the time of the simulation, which represents the ecological time. This result implies either that the rate of competitive exclusion is very low, or that a mixed optimal strategy may prevail. The latter possibility is, however, beyond the scope of this paper. The number of types that can coexist in a given environment becomes smaller when the frequency of the disturbance increases. It seems then that a high environmental uncertainty should pose strong selection pressures for a precise optimal strategy. Selection is relaxed and relatively large variation in the duration of the juvenile stage is possible when disturbances are rare. It should be kept in mind, though, that these results were obtained under the conditions of very low deterministic mortality of juveniles.

Following these results and subject to the present assumptions I suggest that a universal colonizer of uncertain environments should have a high variability in the duration of the resistant stage in the source area. Such variability enables selection in the target areas for optimal τ's which are determined by the parameters of the local stochastic regimes. Relaxation of the selection for a specific optimal τ, in environments with low frequency of disturbance (and low mortality of juveniles) will lead to high variability of this trait. It is conjectured, therefore, that the universal colonizer in uncertain environments will originate from relatively stable environments, where stability is measured by the frequency of the disturbance. Support for this hypothesis may be found in Martins and Jain (1979) and Mayr (1965).

The influence of spatial heterogeneity of the habitat on the optimal strategy was also studied. The results (not shown here for reasons of brevity of the text) indicate the following conclusions:

1. τ, just slightly larger than the duration of the disturbance is the global optimal strategy of an individual patch.
2. Persistence of an archipelago is determined by the difference between the rate of patch extinction and the rate of recolonization.
3. The rate of recolonization depends on the probability of crossing from an

inhabited to an empty patch, and also on the number of dispersers. The latter factor is a function of the environmental limiting factor of an individual patch, K_i, in systems in which a steady state is reached early in the favourable period.

4. For a given archipelago there is a threshold, K_{ic}, specific to the population, above which there will be at least one propagule crossing from an inhabited patch to an empty patch during one favourable period.

5. Once the optimal strategy is reached, the stability of the system is increased when patches are further divided. This will be so as long as $K_i \geq K_{ic}$.

It can be concluded that in regimes of harsh long term disturbances τ is the biological parameter which mostly influences persistence. Optimal τ's will be smaller in dense archipelagos with shorter disturbances and longer in dispersed archipelagos with larger disturbances. Hence, unlike MacArthur and Wilson (1967), and Safriel and Ritte (1983) the present model does not suggest a systematic relation between the persistence of colonizing populations and their intrinsic rate of increase.

The simulation results brough forward above, are sensitive to three of the basic assumptions of the model: juveniles being immune to environmental uncertainties imposed on adults; Poisson distribution of intervals between disturbances; low mortality of adults in the intervals between disturbances. A constant juvenile stage and a constant duration of the disturbance were assumed here in order to keep the model as simple as possible. Further results imply that these assumptions are not crucial and that the same properties are manifested under the conditions of a small variance in these parameters.

This paper suggests to treat colonization systems within the framework of the stochastic regimes in the different geographical areas. A more complete understanding of local persistence chances requires the estimation of the parameters of the environmental disturbances, as well as those of the flow between the patches within each habitat.

Acknowledgements

Discussions with D. Cohen and J.L. Deneubourg were very rewarding.

REFERENCES

Agur, Z., and J.L. Deneubourg (manuscript): The role of non-reproducing resistant life-stages in uncertain environments.

Horn, H.S. (1968): Regulation of animal numbers: A model counter-example, *Ecology* 49(4):776-778.

Loya, Y. (1976): Recolonization of Red Sea corals affected by natural catastrophes and man-made perturbations, *Ecology* 57:278-289.

MacArthur, R.H., and E.O. Wilson (1967): *The Theory of Island Biogeography*, Princeton, N.J.

Martins, P.S., and S.K. Jain (1979): Role of genetic variation in the colonizing ability of rose clover (Trifolium hirtum All.), *Am. Nat.* 114:591-595.

May, R.M. (1975): The tropical rainforest, *Nature* 257:737-738.

Mayr, E. (1965): Summary. In: *The Genetics of Colonizing Species* (H.G. Baker and G.L. Stebbins, eds.), pp. 553-562. Academic Press, New York.

Okubo, A. (1980): Diffusion and ecological problems: mathematical models, *Biomathematics*, 10, Springer-Verlag.

Pielou, E.C. (1977): *Mathematical Ecology*, Wiley and Sons, New York.

Safriel, U.N., and U. Ritte (1980): Criteria for the identification of potential colonizers, *Biol. J. Linn. Soc. Lond.*, 13:287-297.

_____(in press): Universal correlates of colonizing ability. In: *The Ecology of Animal Movement* (I. Swingland and P. Greenwood, eds.), Oxford Univ. Press.

FORMULATING POPULATION MODELS WITH
DIFFERENTIAL AGING

S.P. Blythe, R.M. Nisbet and W.S.C. Gurney

1. INTRODUCTION

Throughout the course of this conference the importance of age-structure effects upon population dynamics has been emphasised frequently, and some solutions to the difficulties of formulating realistic and tractable age-structure models proposed. In particular, Gurney and Nisbet (these Proceedings) showed how it is possible to formulate analytically and computationally tractable models incorporating age-structure by dealing with an arbitrary number of biologically identifiable developmental stages, within each of which all individuals are functionally identical, so that a transition between stages occurs when some critical parameter value (such as "time-in-stage", or mass) is reached, and where this value is the same for all individuals. In this paper we address the problem of model formulation when the assumption of functional identity of individuals within each stage is relaxed, for the particular case where the natural variation among individuals in aging rate is allowed for, and where the between-stage transitions are age-dependent. We consider only the case where there are two well-defined developmental stages - immatures and adults - so that differential aging is expressed as the variation in the total transit-time through the immature stage, i.e. the "maturation period".

Conventionally, equations used to model adult population dynamics under differential aging assume some distribution of maturation periods, and are of the form

$$\dot{N}_A(t) = F(N_A(t) , \int_0^\infty u(a)N_A(t-a)da) \qquad (1)$$

where $F(\cdot,\cdot)$ is a usually non-linear function of the contemporary adult population size $N_A(t)$ and of a weighted average ($u(a)$ a weighting function) of past population sizes (e.g. May 1974, Oster 1976, or MacDonald 1978). Such equations were originally formulated to describe predator-prey or herbivore-vegetation dynamics (e.g. Volterra 1927), and whatever their merits in those fields, equations of the form Equation (1) are as completely inapplicable in age-structure modelling as is their near relative, the time-delayed logistic, as we shall demonstrate in section 3.

We shall first illustrate what we believe to be the correct general formalism for a two developmental-stage population model with differential aging, and then extract an analytically and computationally tractable equation describing adult population dynamics under conditions of adult-only competition, through the judicious choice of a certain generalised weighting function of realistic form.

2. GENERAL FORMULATION

We first note that because individuals age at different rates, it is possible for two individuals born at time t to be in different developmental stages at some time $t + x$ in the future. Hence the "age" of an individual is not necessarily a good indication of its functional "place" in the population, and instead we shall use the "time-to-date" spent in each developmental stage, r and s for the immature and adult stages respectively. Thus we denote by $f_I(r,t)dr$ and $f_A(s,t)ds$ the numbers of immatures and adults in the small duration-intervals $(r,r+dr)$ and $(s,s+ds)$, respectively, and by $\delta_I(r,t)$ and $\delta_A(s,t)$ the per capita immature and adult death rates respectively, all at time t. We also define a per capita maturation rate $\phi(r)$, such that $\phi(r)dr$ is the average fraction of the immatures in the duration-interval $(r,r+dr)$ who mature into adults during that interval, and a per capita fecundity, $\beta(s,t)$. The entire population dynamics can now be written as a pair of partial differential equations (each akin to the MacKendrick (1926) equation), and their associated renewal conditions, as follows,

$$\frac{\partial f_I(r,t)}{\partial t} + \frac{\partial f_I(r,t)}{\partial r} = -\phi(r)f_I(r,t) - \delta_I(r,t)f_I(r,t) \tag{2}$$

$$f_I(0,t) = \int_0^\infty \beta(s,t)f_A(s,t)ds \tag{3}$$

$$\frac{\partial f_A(s,t)}{\partial t} + \frac{\partial f_A(s,t)}{\partial s} = -\delta_A(s,t)f_A(s,t) \tag{4}$$

$$f_A(0,t) = \int_0^\infty \phi(r)f_I(r,t)dr. \tag{5}$$

By integrating equations (2) and (4) over all r and s respectively, we arrive at a pair of balance equations for the total adult population $(N_A(t))$ and the total immature $(N_I(t))$ sub-population,

$$\left. \begin{array}{l} \dot{N}_I(t) = B(t) - M(t) - D_I(t) \\ \dot{N}_A(t) = R(t) - D_A(t), \end{array} \right\} \tag{6a, 6b}$$

where

$$N_I(t) \equiv \int_0^\infty f_I(r,t)dr, \tag{7}$$

$$N_A(t) \equiv \int_0^\infty f_A(s,t)ds, \tag{8}$$

$$B(t) \equiv f_I(0,t), \tag{9}$$

$$R(t) = M(t) \equiv f_A(0,t), \tag{10}$$

$$D_I(t) = \int_0^\infty \delta_I(r,t)f_I(r,t)dr \tag{11}$$

and

$$D_A(t) = \int_0^\infty \delta_A(s,t)f_A(s,t)ds. \tag{12}$$

Little progress can be made in analysing equations (6) in this general form, so instead we consider a particular laboratory competition regime, that of "adult-only" competition, and try to extract analytically and computationally tractable equations.

3. ADULT COMPETITION

We assume that the adults compete for food to produce eggs, and that adult density effects are much more important than aging, so that the per capita fecundity and adult death rates are functions only of $N_A(t)$, and hence

$$B(t) = B(N_A(t)) \tag{13}$$

and

$$D_A(t) = D_A(N_A(t)). \tag{14}$$

The immatures do not compete, so that density effects are negligible, and thus the per capita immature death rate is a function only of r,

$$D_I(t) = \int_0^\infty \delta_I(r)f_I(r,t)dr. \tag{15}$$

Given these assumptions, we can write down the solution to equation (2) for $t > r$ (and if the experimental initial conditions are as stated by Gurney and Nisbet (these Proceedings), i.e. an empty system into which new adults, in this case, are introduced, then equation (16) is the only solution)

$$f_I(r,t) = B(N_A(t-r))\exp\left\{-\int_0^r [\phi(x)+\delta_I(x)]dx\right\}. \tag{16}$$

The adult population balance equation (6b) decouples from that for the immatures (6a), and can be written as

$$\dot{N}_A(t) = \int_0^\infty w(r)B(N_A(t-r))dr - D_A(N_A(t)), \tag{17}$$

where

$$w(r) = \phi(r)\exp\left\{-\int_0^r [\phi(x)+\delta_I(x)]dx\right\}. \tag{18}$$

It is clear that equation (17) is not a special case of the conventional "distributed-delay" model, equation (1). In fact, the latter can only be of the correct form when the per capita birth rate is constant, and where there is no cross-multiplication of "delayed" and "undelayed" terms in $F(\cdot,\cdot)$. Unfortunately, most theoretical attention seems to have been paid to models of the generalised "time-delayed logistic" form (e.g. Barclay and van den Driessche 1975, Kazarinoff, Wan and van den Driessche 1978, MacDonald 1978), which has precisely this character of cross-multiplication, and so is of no help in analysing equation (17).

There is little difficulty in dealing with the density-dependent $B(\cdot)$ and $D_A(\cdot)$ terms in equation (17), as the biological constraints upon the functional forms of these functions, and their effects upon population dynamics, have been extensively studied for the uniform aging case ($w(r)$ a delta-function at some positive value of r)-Blythe, Nisbet and Gurney (1982). It is thus with $w(r)$ itself that we must be concerned if equation (17) is to be made fully tractable.

4. THE WEIGHTING FUNCTION

At first sight, equation (18) is not encouraging, as even with known functions for $\phi(r)$ and $\delta_I(\dot{r})$, the analytic form of $w(r)$ will almost certainly be spectacularly intractable. However, inspection of equations (17) and (18) makes it clear that the biological meaning of the weighting function can be expressed as follows for any synchronously laid group (cohort) of eggs

$$w(r)dr = \frac{\text{number of individuals maturing in interval } (r,r+dr)}{\text{initial cohort size}} . \qquad (19)$$

Integrating over all r gives

$$S \equiv \int_0^\infty w(r)dr = \frac{\text{total number ever maturing}}{\text{initial cohort size}} \leq 1 \qquad (20)$$

so that S is the experimentally easily measureable quantity, the "egg-to-adult survival". A comparison of equations (19) and (20) suggests that a third quantity, $\psi(r)$, can be defined, such that

$$\psi(r)dr = \frac{1}{S} w(r)dr$$

$$= \frac{\text{number of individuals maturing during interval } (r,r+dr)}{\text{total number ever maturing}} , \qquad (21)$$

that is the frequency distribution of maturation periods.

We have thus replaced the intractable equation (18) by the much simpler form,

$$w(r) = S\psi(r), \qquad (22)$$

where S is a constant, and $\psi(r)$ a well-behaved function. It only remains then
to *choose* some functional form for $\psi(r)$ that will characterise experimentally
observed distributions, and which also ensures the analytic and computational
tractability of equation (17).

In Figure 1 (a,b) we have plotted two examples of maturation period distributions: (a) for the blowfly *Phaenicia sericata* (Ash and Greenberg 1975), where
maturations occur according to an extremely asymmetric distribution, starting about
10 days after the eggs were laid, and (b) for the damselfly *Pyrrhosoma nymphula*
(Lawton 1970, and 1982 pers. comm.), where the distribution is very much more
symmetrical, and occurs very much longer (about 2 years!) after egg laying. The
pattern illustrated by these two examples, namely of a minimum maturation period
(τ_1) followed by a reasonably well-defined distribution (summarised in Figure 1(c))
appears to be common, although of course far from universal. We thus choose a
general function for $\psi(r)$ which is capable of reproducing a great variety of
variations on the theme of the $w(r)$ illustrated in Figure 1, such that

$$w(r) = Sg(r-\tau_1;p)H(r-\tau_1),\tag{23}$$

where $H(\cdot)$ is the Heaviside step function (zero for $r < \tau_1$, and unity for
$r \geq \tau_1$), and where

$$g(x;p) = \frac{c^{p+1}}{p!}\, x^p e^{-cx} \quad c > 0, \quad p = 1,2,\ldots\tag{24}$$

is the gamma distribution of integer order. Equation (24) is often used where a
general "humped" function with a peak of adjustable width is required (e.g. Lewis
1977, MacDonald 1978, Cushing 1980): the addition of the Heaviside function permits
very narrow distributions to be specified without using large values of p, which
not only permits distribution "width" and symmetry to be controlled using different
parameters, but also greatly facilitates numerical analysis, as we indicate in
section 5. Curves of $w(r)$ as given by equation (23), fitted using a simple method
of moments with p taken to the nearest integer, are superimposed on the observed
distributions of Figure 1 (a,b). Clearly while the statistical fit (especially for
the *P. sericata* data) is not overly impressive, the essential characteristics of the
two distributions are well represented.

5. ANALYSIS

The choice of equation (23) as the weighting function in equation (18) also
fulfils the requirements for analytic and computational tractability of the integro-
differential equation. A full account will be published elsewhere (Blythe, Nisbet
and Gurney, in preparation), and we will merely mention some important results here.

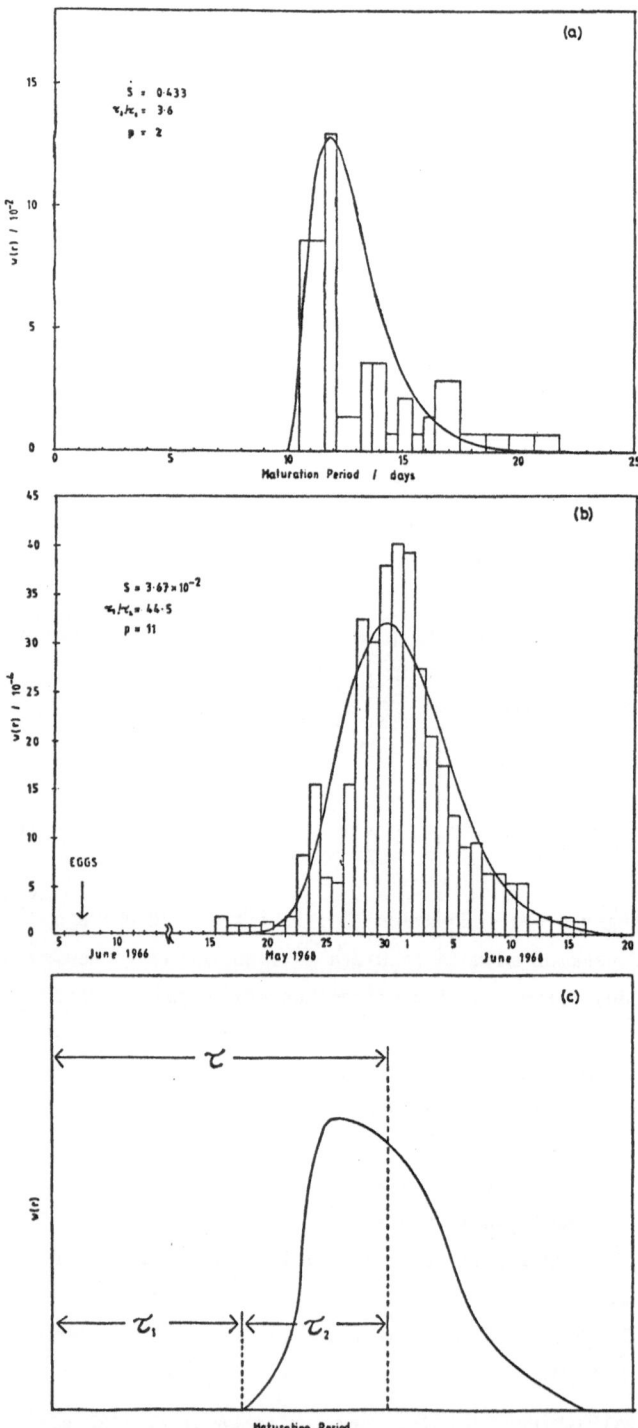

Figure 1 Observed and fitted w(r) for (a) the blowfly *Phaenicia sericata* (Ash and Greenberg 1975), and (b) the damselfly *Pyrrhosoma nymphula* (Lawton 1970, 1982 pers. comm.). (c) The important features of w(r): τ_1 the minimum maturation period; τ the average maturation period; $\tau_2 = \tau - \tau_1$.

(i) The linear equation (describing the fate of small perturbations from the steady state) associated with equation (17) has a characteristic equation that, with a little cunning, readily yields up parametric equations for the local stability boundary, for general τ_1/τ_2 and p (parameters controlling the shape of w(r)). With increasing τ_1/τ_2 and/or p, the local stability boundary tends to the limiting (uniform aging) case.

(ii) The numerical analysis of equation (17), with w(r) given by equation (23), is also straightforward, and is achieved using an extension of the "linear chain trick" of MacDonald (1978). By defining a set of p + 1 auxiliary variables,

$$V_j(t) = \int_{\tau_1}^{\infty} g(r-\tau_1;p+1-j)B(N_A(t-r))dr, \quad j = 1,2 \ldots p+1 \tag{25}$$

and differentiating each with respect to time, the integro-differential equation (17) can be replaced by the set of p + 2 equations

$$\dot{N}_A(t) = SV_1(t) - D_A(N_A(t))$$
$$\dot{V}_j(t) = c(V_{j+1}(t)-V_j(t)), \quad j = 1,2 \ldots p \tag{26}$$
$$\dot{V}_{j+1}(t) = c(B(N_A(t-\tau_1))-V_{p+1}(t))$$

(where c is the constant from equation (24)), with the associated set of initial values, and initial history,

$$N_A(t) = 0 \quad -\tau_1 \leq t \leq 0$$
$$V_j(0) = 0, \quad j = 1,2 \ldots p+1 \tag{27}$$
$$R(t) = SV_1(t) + I(t)$$

where I(t) is zero for all t except a small interval just after t = 0, and represents the initial "innoculation" of the population by newly emerged adults (see Gurney and Nisbet, these Proceedings). The system of equations (26) is straighforward to solve numerically (we have used the algorithim of Maas, Nisbet and Gurney (1982)), and because p need not be large, requires relatively little computing time.

6. CONCLUSIONS

In this paper we have derived a single, tractable, integro-differential equation describing adult population dynamics in a population where only adults compete, and where individuals require different lengths of time to pass through the immature developmental stage. We emphasize that this equation is fundamentally different from the conventionally applied "distributed-delay" models. The choice of a gamma distribution with a shift in origin, a realistic form for the distribution of maturation periods, permits a full local stability analysis to be performed for the general equation, and greatly facilitates numerical analysis.

REFERENCES

Ash, N. and B. Greenberg (1975): Developmental temperature responses of the sibling species. *Phaenicia sericata* and *Phaenicia pallescens*. *Annals. Entom. Soc. America* 68: 197-200.

Barclay, H.B. and P. van den Driessche (1975): Time lags in ecological systems. *J. Theoretical Biology* 51: 347-356.

Blythe, S.P., R.M. Nisbet and W.S.C. Gurney (1982): Instability and complex dynamic behaviour in population models with long time delays. *Theor. Pop. Biology* 22: 147-176.

Cushing, J.M. (1980): Model stability and instability in age-structured populations. *J. Theoretical Biology* 86: 709-730.

Gurney, W.S.C. and R.M. Nisbet (this Proceedings): The systematic formulation of delay-differential models of age or size structured populations.

Kazarinoff, N.D., Y.-H. Wan and P. van den Driessche (1978): Hopf bifuscations and stability of periodic solutions of differential-difference and integro-differential equations. *J. Inst. Math. and Its Applications* 21: 461-477.

Lawton, J.H. (1971): A population study on larvae of the Damselfly *Pyrrhosoma nymphula* (Sulzer) (Odonata: Zygoptera) *Hydrobiologia* 36: 33-52.

Lewis, E.R. (1977): Linear population models with stochastic time delays. *Ecology* 58: 738-749.

Maas, P., R.M. Nisbet and W.S.C. Gurney (1982): SOLVER, an adaptable program template for initial value problem solving. *Applied Physics Industrial Consultants*, University of Strathclyde, Glasgow.

MacDonald, N. (1978): *Time Lags in Biological Models*. (Lecture Notes in Biomathematics 27) Springer-Verlag, Berlin/Heidelberg/New York.

MacKendrick, A.G. (1926): Applications of mathematics to medical problems. *Proc. Edinburgh Math. Soc.* 44: 98-130.

May, R.M. (1974): *Stability and Complexity in Model Ecosystems*. Princeton U.P., Princeton.

Oster, G.F. (1976): Lectures in Population Dynamics pp 149-190 *in* R.C. DiPrima (Ed.) *Modern Modelling of Continum Phenomena*. American Mathematical Soc. Providence (Rhode Island).

Volterra, V. (1927): Teoria de les functionels y de las ecuacions integrales e integro-differentiales. Conferencias explicades en la Facultad de la Ciencias, de la Universitad, 1925, redactndas por L. Fanteppie, Madrid (in Spanish). Reprinted (1959) *Theory of Functionals and of Integral and Integro-Differential Equations*. Dover.

POPULATION MODELLING BY HIGHER ORDER DIFFERENTIAL EQUATIONS

George Bojadziev

ABSTRACT

Modelling in ecology by the means of higher order differential equations is discussed and compared to the established practice of using first-order equations. The study is based on the premise that a process of population growth responds not only to its present level, but also to the rate of change of that level.

1. INTRODUCTION

It is well known that under some simplifying assumptions the growth of a single population is modelled by a first order ordinary differential equation of the type $\dot{x} = G(x)$, $\dot{x} = dx/dt$, called the first order model. Here x represents the population density (number of individuals per unit area) or biomas and $G(x)$ is a known function of x. Dealing with this equation is equivalent to assuming that the behaviour of a species can be adequately represented by a single variable. It specifies the behaviour of the rate of growth $\dot{x}$ as a function of the population density x. One of the most popular models is the logistic growth equation $\dot{x} = rx(1-x/K)$. When x is small the growth is almost exponential and as t increases, x approaches a steady value K without oscillations.

Only on rare occasions has the second order derivative been used in modelling the dynamics of growth of a single population. Clark (1971) was the first to discuss the significance of population modelling by the means of second order differential equations. Innis (1972) has criticized some controvertial points in Clark's paper. However, he has failed to appreciate its contribution. Freedman (1980) briefly considers Clark's work and discusses a rather general model described by a second order differential equation.

The aim of this paper is to discuss the philosophical aspects of using higher order differential equations in population dynamics (higher order models), their meaning and significance. To facilitate this task the author reviews Clark's and Freedman's second order models, introduces more general models, and as illustration considers some particular cases.

2. SECOND ORDER EQUATIONS

History of science provides numerous examples of ideas developed to investigate physical and mechanical problems which later have been used in more complicated sciences. For instance that is the case with Clark's paper. Its starting point is Newton's second law of motion $F = m\ddot{x}$ describing how a particle with position x

and mass m reacts to a force F. A knowledge of x alone in the sense of how x
changes is not sufficient to determine the future position of x. In addition, we
also need to know how the velocity $v = \dot{x}$ changes. Thus x and v together with
their initial conditions $x(t_0) = x_0$ and $v(t_0) = v_0$ are sufficient to determine
the particle's motion. That makes clear why Clark has used a second order differential equation $\ddot{x} = G(x,\dot{x})$ in order to model second order processes that responds
not only to its present level x, but also to the rate of change $\dot{x}$ of that level.
In other words this equation specifies the functional behaviour of $\ddot{x}$ (the rate of
growth of $\dot{x}$) in terms of x and $\dot{x}$. As Freedman notes, the second derivative $\ddot{x}$,
acceleration in physics, may be thought of as representing the "life-force" of the
population. Perhaps the Newton's equation of a moving body $m\ddot{x} = -S(x) - T(x,\dot{x})$,
where S(x) is a restoring force, and $T(x,\dot{x})$ is damping force motivates Clark's
assumption that the life force is a sum of two forces, and then his modelling
equation becomes

$$\ddot{x} = G_E(x) + G_H(x,\dot{x}). \tag{1}$$

The first force G_E if a force of restoration to an equilibrium position K which
is related to the existence of a carrying capacity of the environment. According to
Clark the second force G_H is connected to the recent previous history of the
population, which is influenced by the rate of growth $\dot{x}$ of the population.

To summarize, we note that a first order model involves only one variable
x (the density of one species) and is described by a first order differential
equation, while a second order model involves two variables x and $\dot{x} = v$ and
leads to a second order differential equation or to a system of two first order
equations $\dot{x} = v$, $\dot{v} = G(x,v)$. Both, the first and second order models predict the
population growth of a single species.

3. SOME SECOND ORDER MODELS

To illustrate his idea, Clark (1971) has considered a simple second order
model for which $G_E = a(K-x)$, $G_H = -b\dot{x}$, where a and b are positive constants.
Then equation (1) takes the form

$$\ddot{x} = a(K-x) - b\dot{x}. \tag{2}$$

The term G_E shows that a population is to be affected more strongly the farther it
is away from its equilibrium position K. The force G_H reflects the recent
history of the population which is assumed to depend on the rate of growth $\dot{x}$ of
the population. Actually G_H is a damping force reflecting the internal resistance within the population to fast changes in terms of population size. Equation (2)
expresses the influence of the forces G_E and G_H on the life-force $\ddot{x}$, that is
the faster a population increases, the less able it is to increase any faster.

Since a and b are positive, the equilibrium position (K,0) in the phase plane $(x, \dot{x} = v)$ is a stable node if $b^2 - 4a \geq 0$, and a stable focus if $b^2 - 4a < 0$.

Freedman (1980) has discussed a more general model

$$\ddot{x} + f(x)\dot{x} + g(x) = 0 \qquad (3)$$

which is of the type (1) with $G_E = -g(x)$ and $G_H = -f(x)\dot{x}$. Here $g(x)$ has the following properties:

(i)   There exists an equilibrium $K > 0$ such that $g(K) = 0$;
(ii) $(x-K)g(x) > 0$ for $x \neq K$.

It is pointed out that (3) is a form of the Lienard equation which has been extensively studied in the literature (say Cesari 1971).

Now we consider a specific second order model with $G_E = -c(x-K)$ and $G_H = -r(2x/K-1)\dot{x}$, where c, r, and K are positive constants. Then equation (1) takes the form

$$\ddot{x} + r\left(\frac{2}{K} x - 1\right)\dot{x} + c(x - K) = 0. \qquad (4)$$

Setting c = 0 in (4) and integrating gives the logitic equation plus a constant of integration which can be made zero if the initial conditions of (4) are selected appropriately. In that sense (4) is a generalized logistic equation.

The equilibrium position (K,0) of (4) in the phase plane $(x,\dot{x})$ is a stable node if $r^2 \geq 4c$ and a stable focus if $r^2 < 4c$, hence it attracts the trajectories of equation (4). Depending on the values of r and c, and the initial conditions several types of behaviour of x close to the equilibrium K are possible. It is worth noticing the case of damped oscillations ($r^2 < 4c$) which are not exhibited by the logistic equation.

4. THIRD AND HIGHER ORDER MODELS

Generalizing Clark's approach we use a third order differential equation $\dddot{x} = G(x,\dot{x},\ddot{x})$ to model "third order" processes that respond not only to their present level x and its rate of change $\dot{x} = v$, but also to the rate of change of the life-force $\ddot{x}$. This equation specifies the functional behaviour of the rate of growth $\dddot{x}$ of the life-force. The function G can be expressed as a sum of three forces $G = G_E + G_H + G_L$, where $G_E = -g(x)$ and $G_H = -f(x,\dot{x})$ have the same meaning as in the second order model, i.e. G_E is a force of restoration and G_H is a damping force. The third force G_L involves $\ddot{x}$ and hence it might be called "life-force". Then the third order model reads

$$\dddot{x} + \phi(x,\dot{x},x) + f(x,x) + g(x) = 0. \qquad (5)$$

As an illustration we consider a linear model

$$\dot{x} + c\dddot{x} + b\ddot{x} + a(x-K) = 0 \tag{6}$$

with forces $G_E = a(K-x)$, $G_H = -b\dot{x}$, $G_L = -cx$, where a, b, and c are positive constants. While G_E and G_H have the same meaning as in Clark's model (2), G_E is to account for the life-force of the population. The characteristic equation of (6) has at least one real negative root since a, b, and c are positive. If in addition bc > a, then the remaining roots are negative or have a real negative part. Hence the equilibrium (K,0,0) of (6) in the space $(x,\dot{x},\ddot{x})$ is strictly stable.

Besides by direct generalization as above we can derive third order equations of the type (5) modelling certain elastic systems with internal friction and relaxation. The motion of such a system can be described by the following equations $m\ddot{x} + \sigma = 0$, $\sigma + R(\dot{\sigma}) = S(X) + T(x,\dot{x})$, where x is the deformation, m is mass, σ is stress, $S(x)$ is the elastic resistance of material, $T(x,\dot{x})$ is internal friction, and $R(\dot{\sigma})$ is resistance of the relaxation. The ecological interpretation of the quantities in these equations may be thought of as follows: x is the population density, σ is stress due to environmental and intrapopulation forces, $S(x)$ is a force of restoration to an equilibrium position, and $R(\dot{\sigma})$ is the relaxation function which dampers the stress.

In general, the study of third order equations of the type (5) (see for example Reissig et al. 1974), except in particular cases like (6), is a difficult task.

Analogically to the second and third order model we introduce an "n-th order" model $x^{(n)} = G(x,\dot{x},\ldots,x^{(n-1)})$ where G is a sum of n-forces.

5. DISCUSSION

A first order model is based on the assumption that the behaviour of a single population x can be adequately represented by a single variable. The first order modelling equation specifies the rate of change $\dot{x}$ of the population density x. It requires one initial condition $x(t_0) = x_0$.

A second order model is based on the more elaborated assumption that the behaviour of a single population x can be adequately represented by two variables, x and $v = \dot{x}$. The modelling equation (1) specifies the life-force, the rate of change $\ddot{x}$ of v. In addition two initial conditions $x(t_0) = x_0$ and $v(t_0) = \dot{x}_0$ are required. Of course one may use a system of two first order differential equations instead of (1).

In comparison to a first order model, which is often a crude one, a second order model involves more information about the population and is more elaborated. Taking into account more patterns of behaviour gives the possibility for describing more precisely the growth of the population. The results obtained from a second order model are richer and more detailed, both from qualitative and quantitative point of view in comparison to those obtained by a first order model. We cannot expect simple models to accurately predict population growth.

However, the fact that more information is required to construct a second order model than a first order model has its own shortcoming. Since observations of population changes are limited, sometimes it might be difficult to acquire the necessary information in order to design a meaningful model. In addition one takes into account that a second order model (meaning a second order differential equation) is more difficult to be investigated or solved than a first order differential equation.

Let us note that instead of $\dot{x}$ as a second variable one can select a more suitable variable for a specific problem at hand. Maynard Smith (1974) has indicated that the density of a species is usually not sufficient to determine its behaviour and considered as an additional variable a density with delayed argument. Another example is the model of the population genetics of cystic fibrosis in the United States presented by Beck (1982), $\dot{x} = \beta x(1-x/K)$, $\dot{K} = \gamma K(1-K/L)$, where the second variable is the carrying capacity K.

Note also the well known connection between second (and higher) order linear differential equations and integral equations of Volterra's type.

REFERENCES

Beck, K. (1982): A model of the population genetics of cystic fibrosis in the United States. *Math. Biosciences* 58: 243-257.

Cesari, L. (1971): *Asymptotic Behaviour and Stability Problems in Ordinary Differential Equations*. Springer Verlag, New York.

Clark, J.P. (1971): The second derivative and population modelling. *Ecology* 52: 606-613.

Freedman, H.I. (1980): *Deterministic Mathematical Models in Population Ecology*. Marcel Dekker, New York, 252 p.

Innis, G. (1972): The second derivative and population modelling: another view, *Ecology* 53: 720-723.

Maynard Smith, J. (1974): *Models in Ecology*. Cambridge University Press, 146 p.

Reissig, R., G. Sansone and R. Conti (1974): *Non-linear Differential Equations of Higher Order*. Noordhoff International Publishing, Leyden, The Netherlands, 669 p.

CONCERNING THE TIME-DEPENDENCE OF POPULATION VITAL RATES

C. Scott Findlay, Robert F. Rockwell and Fred Cooke

Deterministic models are useful for gaining insight into the mechanisms of population dynamics. However, they are not generally useful for the practical problem of estimating real populations.

- G. Oster (1981)

A solution to the population process ... can usually be obtained only when the age-specific survival probabilities are constant over time and independent of population size ... Its importance does not lie in the fact that natural populations frequently satisfy these conditions: clearly, they do not in general. Rather, many of the more basic properties of more complex types of age-structured populations can be understood by reference to this case.

- B. Charlesworth (1980)

All is flux, nothing is stationary.

- Heracleitus (502 B.C.)

1. Population models come in two flavours: deterministic and stochastic. While deterministic models of population growth have been bandied back and forth for some time, the interest in stochastic models is comparatively recent (e.g. Bartlett 1960, Ludwig 1974, May 1974). The reasons for the historical apathy towards stochastic models are unclear. Of some importance, however, is the fact that the field has traditionally been the domain of the applied mathematician. To these researchers, it is often not only mathematical tractability that counts, but also mathematical elegance. And in this regard, stochastic models are usually far outdistanced by their deterministic counterparts.

Mathematical elegance is, however, a rather flimsy yardstick with which to measure a model's value. Oster (1981) suggested that models could be classified as either general ("models which address general phenomena, but are not designed to explain any particular set of data") or special ("models ... designed to answer a particular question - or organize the data - for a specific experiment or data set"). As such, the value of a special model is primarily predictive. The value of a general model is, on the other hand, largely heuristic.

Clearly then, we should assess models not according to the complexities of their mathematical viscera, but rather on the basis of their ecological relevance. Models are, after all, abstractions based (albeit sometimes rather loosely) on empirical observation. This is true of both the general and special cases, though at qualitatively different levels. Yet formulating the appropriate model from a set of observations is not an easy task. Even for the most rudimentary ecological systems, it may not be immediately obvious whether a stochastic or deterministic representation is more appropriate. May (1976) and May and Oster (1976) have shown that simple deterministic models may exhibit extremely complicated behaviour. Indeed, under several biologically reasonable conditions, such systems can describe dynamic trajectories that are virtually indistinguishable from the chaos generated by truly stochastic processes (May 1974, Mackey and Glass 1977, Sparrow 1980). We are thus faced with the inescapable conclusion that what is observed may belie the nature of the underlying process(es): pattern need not imply determinism, nor chaos stochasticity.

The stochastic-deterministic debate is vaguely reminiscent of the chance-necessity controversy. It may, on occasion, provide lively entertainment at somnolent academic gatherings. But little is gained by engaging in *a priori* speculation as to which class of models provides the better representation of reality. Lest we forget, fitting a model requires at least a modicum of data. Hence, only through empirical observation can we hope to formulate appropriate models of ecological systems.

Leslie (1945) first introduced the basic deterministic model for an age-structured population in discrete time. It consisted of an age census at time t, $Y(t)$, and a (usually) linear operator $x(t)$ which mapped $Y(t) \rightarrow Y(t+1)$. The operator is a $k \times k$ projection matrix having as its first row the age-specific fecundities at time t, $m_k(t)$, $k = 1,\ldots,K$, and the age-specific survivorships $0 \leq s_k \leq 1$, $k = 1,\ldots,K-1$ as the sub-diagonal. Cohen (1979a) summarized three classes of theorems pertaining to the long run behaviour of demographic models. The strong ergodic theorem holds that $x(t)$ is invariant over time. The weak ergodic theorem assumes that $\{x(t)\}$ is a determinate sequence. Finally, stochastic ergodic theorems assume that $\{x(t,\omega)\}$ is a sample path of a (stationary or non-stationary) stochastic process generating $x(t)$ from a set of potential operators. Thus, we see that the formal distinctions among the various classes of demographic models are based on the presence or absence (as well as the nature) of time-dependent vital rates.

There are, however, few natural populations for which vital rates are truly time-independent. Most populations exploit environments which are, at least to some extent, temporally heterogeneous. Insofar as vital rates are correlated with the state of the environment, we expect to find some level of time-dependence. The question is thus not *whether* vital rates are time-dependent, but rather, *to what extent* are they time-dependent?

There are some populations whose vital rates show relatively little temporal variability. In other cases, however, the amount of temporal variation appears to be greater than can be safely accommodated with the framework of the strong ergodic theorem. For example, Figure 1 shows annual variation in the mean clutch size of Great *(Parus major)* and Blue *(P. caerulescens)* Tits at Marley Wood, U.K. Figure 2

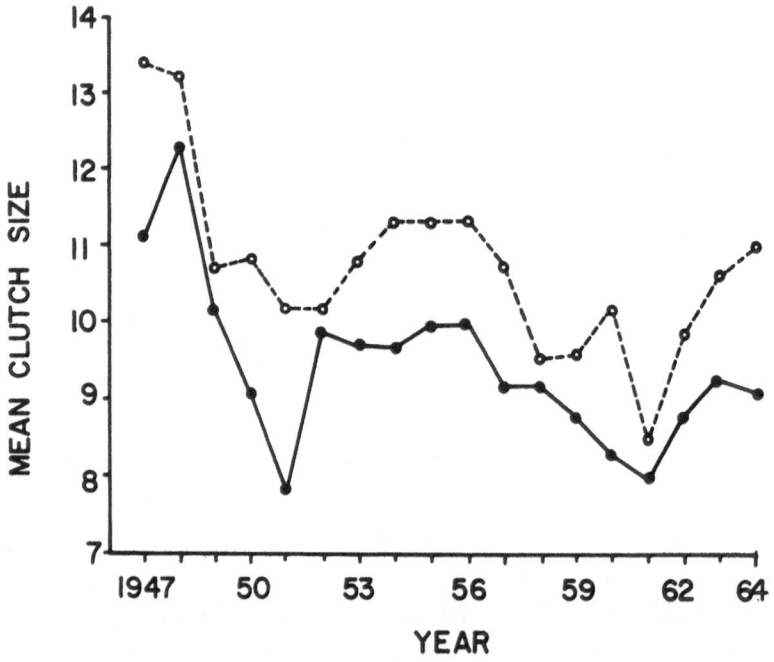

Figure 1 Annual variation in mean clutch size of Great Tits (●) and Blue Tits (o)
at Marley Wood, U.K. After Lack (1966).

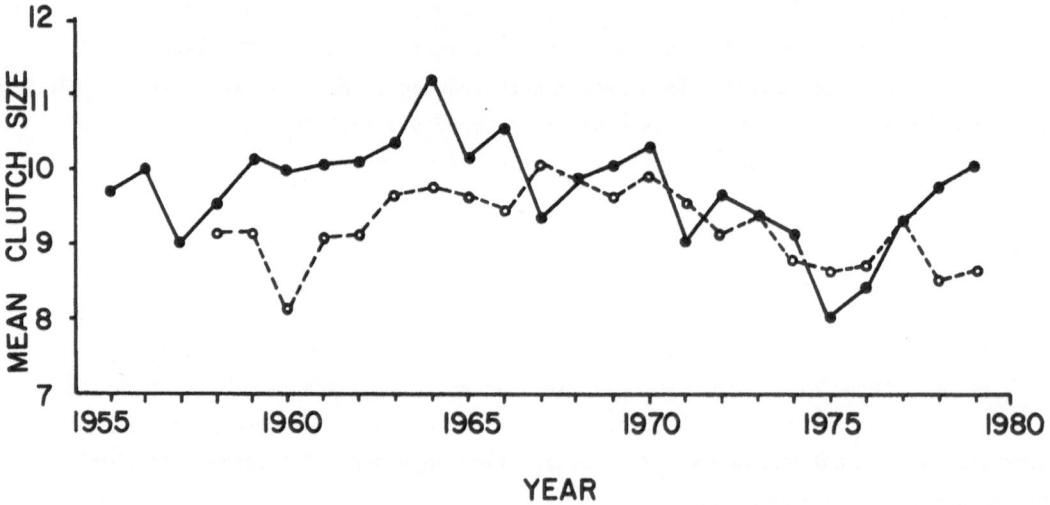

Figure 2 Annual variation in mean clutch size of Great Tits in Hoge Veluwe (●) and
Vlieland (o), Netherlands. After van Noordwijk (1981).

provides similar information from two populations of *P. major* studied by Kluijver, van Balen and co-workers in the Netherlands. All four populations show considerable temporal variation in mean fecundity (clutch size) over the study period (CV × 100: 1(●) = 11.32, 1(o) = 25.90, 2(●) = 5.52, 2(o) = 6.91). Figure 3 documents a similar phenomenon in a population of Lesser Snow geese *(Anser caerulescens caerulescens)* breeding at La Perouse Bay, Manitoba.

Temporal variation in vital rates need not be confined to the first row of the projection matrix. Age-specific survivorships may also vary over time. A good example is the study by North and Morgan (1979) of survivorship among Grey Herons *(Ardea cinerea)* in the U.K. (Figure 4). First-year survivorship showed considerable variation between seasons, ranging from near zero in 1962 to approximately 0.75 in 1960 (CV × 100 over all years of the study = 38.3). This phenomenon of time-specific avian survival probabilities has been the subject of both general (e.g. Seber 1970, Brownie and Robson 1976, Brownie et. al. 1978) and special (e.g. North and Morgan 1979) models.

There are several conditions under which there exists at least the potential for significant time-dependence of population vital rates. These include: (1) populations for which age-specific fecundities and/or survivorships are closely linked to extrinsic conditions (e.g. weather conditions, season phenology, etc.) in a temporally heterogeneous environment; (2) populations whose vital rates are density-dependent; (3) populations whose vital rates are cohort-specific; or (4) populations for which the proportion of surviving adults who breed varies from season to season. To illustrate the latter case, we refer to the basic matrix model for population growth in discrete time:

$$Y(t+1) = x(t+1)Y(t), \qquad (1)$$

where the total number of individuals in age class 1 at time t+1 is given by

$$Y_1(t+1) = \sum_{k=1}^{K} m_k(t+1)Y_k(t), \qquad (2)$$

and

$$Y_k(t) = s_{k-1}(t)Y_{k-1}(t-1), \qquad (3)$$

(modified from Cohen 1979) where $s_{k-1}(t)$ is "the survival proportion per unit time" (Cohen 1979), i.e. the probability that an individual age $k-1$ at time $t-1$ will survive to age k at time t. Substituting equation (3) into (2), we have

$$Y_1(t+1) = \sum_{k=1}^{K} m_k(t+1)s_{k-1}(t)Y_{k-1}(t-1). \qquad (4)$$

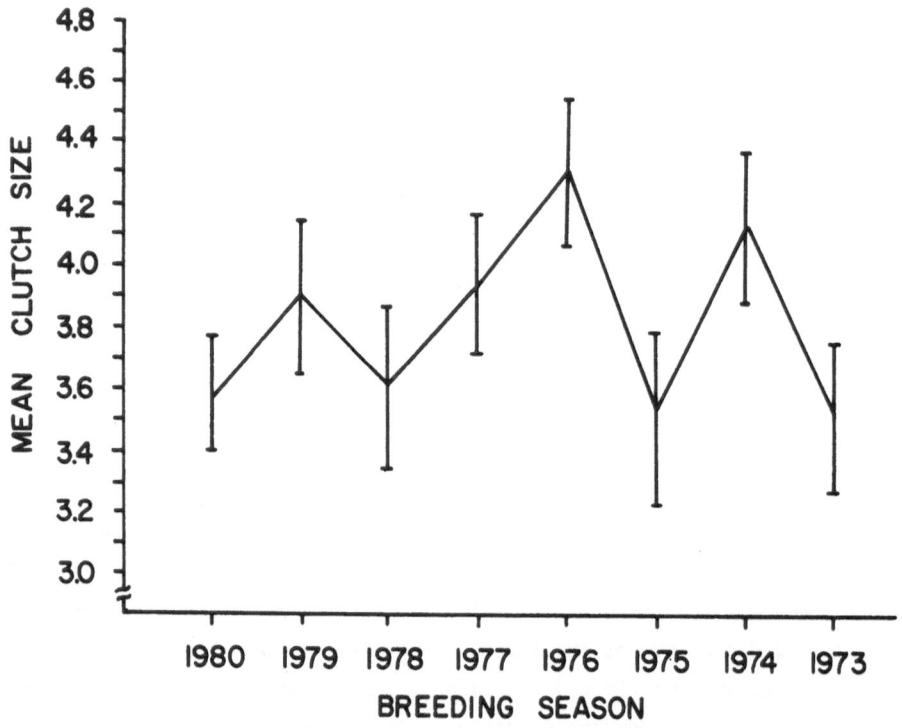

Figure 3 Annual variation in mean clutch size of female Lesser Snow Geese at La Perouse Bay, Manitoba. Vertical bars represent 2 SE. After Rockwell et al. (1983).

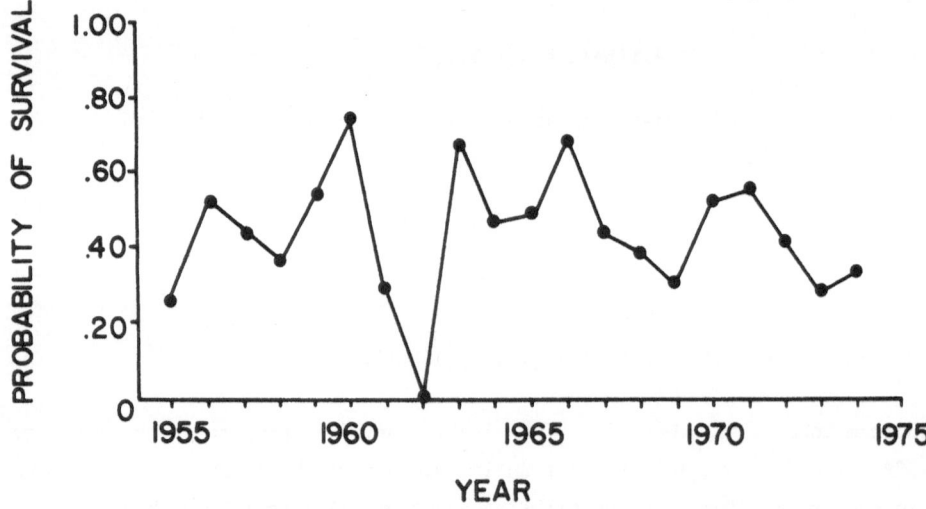

Figure 4 First-year survivorship of Grey Herons in Britain, 1955-1974. After North and Morgan (1979).

From (4), we see that the model implicitly assumes that *all* individuals who survive the interval $t-1$ to t (or age $k-1$ to k) subsequently *breed at time* t. If this is not the case, then $Y_k(t)$ represents the age-structure of the total *surviving* population, but *not* the age-structure of the *breeding* population. Strictly speaking then, the subdiagonal elements of $x(t)$ should represent not only the probability that an individual age $k-1$ at time $t-1$ survives to age k at time t, but also that having survived, it then breeds ($\gamma_k(t)$). If $\gamma_k(t) \neq \gamma_k(t+1) \neq$ ..., $k = 1, \ldots, K$, then once again population trajectories may not conform to demographic predictions based on the strong ergodic theorem. As an example, we refer again to the breeding biology of Snow Geese. Using data collected at the La Perouse Bay colony, Sulzbach (1975) estimated the proportion of surviving females breeding at age k for several different census years (Table 1). Here, we see that, in general, $\gamma_2(t) < \gamma_3(t) < \gamma_4(t)$. More importantly, there is considerable temporal variation in γ for some age classes, e.g. $\gamma_2(1971) = 0.127$, whereas $\gamma_2(1974) = 0.520$.

Table 1 Estimates of $\gamma_k(t)$ for Female Lesser Snow Geese at La Perouse Bay, Manitoba. After Sulzbach (1975).

	t			
k	1971	1972	1973	1974
2	0.127	0.120	0.470	0.520
3		0.480	0.470	0.520
4			0.744	0.708
5				1.000

Temporal variation need not, however, be confined solely to individual γ_k's. Of greater importance (from a demographic perspective) is the possibility that the entire vector $[\gamma]$ may vary over time. This phenomenon of aperiodic non-breeding has been investigated in several arctic-breeding birds (e.g. Bertram et al. 1934), but has received comparatively little attention elsewhere. Goose biologists are, however, well aware of the "boom-bust" breeding syndrome characteristic of many high arctic colonies. In years of particularly harsh conditions, less than 30% of the surviving adult population may breed. Under more benign conditions, this figure may jump to 80-90%.

2. CONCLUSIONS

We have cited several examples of natural populations whose vital rates tend to exhibit relatively large fluctuations over time. Hence, demographic models which assume no time-dependence may be inappropriate. Indeed, - depending on the magnitude of temporal variation, - such models may not generate even qualitatively accurate predictions of population growth. In such instances, stochastic models incorporating time-dependent vital rates may well provide the only useful method for deriving empirically relevant population projections.

Acknowledgements

The authors thank Dolf Harmsen and Joan Geramita for their critical reading of earlier drafts of this paper. Continuing research into the population biology of the Lesser Snow Goose is funded, in part, by the Canadian Wildlife Service, the Natural Sciences and Engineering Research Council of Canada, and the Central and Mississippi Flyway Councils.

REFERENCES

Bartlett, M.S. (1960): *Stochastic Population Models in Ecology and Epidemiology*, Methuen, London.

Bertram, G.C.L., D. Lack, and B.B. Roberts (1934): Notes on East Greenland birds with a discussion of the periodic non-breeding among arctic birds, *Ibis* 76: 816-831.

Brownie, C., D.R. Anderson, K.P. Burnham and D.S. Robson (1978): *Statistical Inference from Band Recovery Data - a Handbook*, U.S. Depart. of the Interior Fish and Wildlife Service, Resource Publ. 131.

Brownie, C., and D.S. Robson (1976): Models allowing for age-dependent survival rates for band return data, *Biometrics* 32:303-323.

Charlesworth, B. (1980): *Evolution in Age-Structured Populations*, Camb. Studies in Mathematical Biology 1, Cambridge Univ. Press, Cambridge.

Cohen, J. (1979): Ergodic theorems in demography, *Bull. Amer. Math. Soc.* 1(2): 275-295.

Lack, D. (1966): *Population Studies of Birds*, Clarendon, Oxford.

Leslie, P.H. (1945): On the uses of matrices in certain population mathematics, *Biometrika* 35:213-245.

Ludwig, D. (1974): *Stochastic Population Theories*, Lecture Notes in Biomathematics, Vol. 3, Springer-Verlag, New York.

May, R.M. (1974): Biological populations with non-overlapping generations: stable points, stable cycles, and chaos, *Science* 186:645.

_____ (1976): Simple mathematical models with very complicated dynamics, *Nature* 261:459-467.

May, R.M., and G. Oster (1976): Bifurcations and dynamic complexity in simple ecological models, *Am. Nat.* 110:573-599.

Mackey, M.C., and L. Glass (1977): Oscillation and chaos in physiological control systems, *Science* 197:287-289.

North, P.M., and B.J.T. Morgan (1979): Modelling heron survival using weather data, *Biometrics* 35:667-681.

Oster, G. (1981): Predicting populations, *Amer. Zool.* 21:831-844.

Rockwell, R.F., C.S. Findlay and F. Cooke (1983): Life history studies of the Lesser Snow Goose (*Anser caerulescens caerulescens*). I. The influence of age and time on fecundity, *Oecologia* (in press).

Seber, G.A.F. (1970): Estimating time-specific survival and reporting rates for adult birds from band returns, *Biometrika* 57:313-318.

Sparrow, S. (1980): Bifurcations and chaotic behavior in simple feedback systems, *J. Theor. Biol.* 83:93-105.

Sulzbach, D.S. (1975): A study of the population dynamics of a nesting colony of the Lesser Snow Goose (*Anser caerulescens caerulescens*). Unpubl. M.Sc. Thesis, Queen's Univ., Kingston, Ontario.

van Noordwijk, A.J. (1981): Genetic and environmental variation in clutch size of the Great Tit, *Neth. J. Zool.* 31:342-372.

EXTINCTION PROBABILITIES IN STOCHASTIC
AGE-STRUCTURED MODELS OF POPULATION GROWTH

Lev R. Ginzburg and Andrea Pugliese

Stochastic age-structured models of population growth have received significant attention over the past years, both from a purely theoretical point of view,
and with an interest towards applications. This interest sprang from the consideration that many species are iteroparous and that vital parameters vary unpredictably
from year to year.

In a variety of applications, from environmental impact assessment and
fisheries management problems to problems of life history evolution, two distinct
problems receive the greatest attention. One is the evaluation of the rate of
growth, which is the trend in population dynamics and analogue of 'fitness' in
evolutionary contexts. The other is the evaluation of extinction probabilities.

Though both problems are relevant to many ecological and evolutionary questions, this paper will be devoted mainly to the latter one. We start with a short
review of the mathematical background on the stochastic age-structured model and on
the extinction problem for other models. Then, the two topics will be put together,
showing the difficulties of computing extinction probabilities for the age-structured
model, together with an approximate method for resolving them. We will end with an
overview of possible applications.

A distinction has to be made between density-dependent and density-independent
models. In general, researchers have avoided the study of stochastic density-
dependent models with an explicit age-structure. Reed (1982) recently considered
such a model to study the dependence of the variance of fluctuations on the intensity
of harvesting. The method there is to linearize equations around the deterministic
equilibrium, assuming that fluctuations are small.

Here we will restrict to linear (density-independent) models. When the model
is linear, $N_{k+1} = A_k N_k$, where N_k is the vector representing age-structured population at time k, and A_k are Leslie matrices, some results about the asymptotic
structure of the population have been proved (Cohen 1976, 1977; Lange and Holmes
1981; Tuljapurkar and Orzack 1980), owing especially to a paper by Furstenberg and
Kesten (1960).

The more revelant ones are the following:
- the probability distribution of the normalized vector representing the
  age structure converges to a stationary one.
- the logarithmic growth rate converges, with probability one, to its
  average according to that stationary distribution.
- the logarithm of any linear combination of the age classes converges in
  distribution, when opportunely scaled, to a normal variable.

Unfortunately, a direct computation of these quantities require extensive numerical computations, that probably become unfeasible when the dimension of the problem is high (the organism being modelled is extendedly iteroparous). A method for obtaining approximations, asymptotically true as the fluctuations go to zero, has been introduced by Tuljapurkar (1982a, 1982b). This method has been used by Ginzburg et al. (1982b) to study the probabilities of 'quasi-extinction'. By 'quasi-extinction' we mean the trajectory crossing some (low) threshold. The study of the probabilities of such events (equivalent to a consideration of the distribution of first passage times) has been introduced in Capocelli and Ricciardi (1974) as a way of studying extinctions in a model in which the population size will be positive at any finite time, with probability one. Discussion on how to model extinction and alternative approaches can be found in Keiding (1975), Ludwig (1976), Turelli (1978) and Sawyer and Slatkin (1981).

In an age-structured model, it is also necessary to choose a scalar quantity that is considered to be the most relevant one for this problem; this might be total population size, adult population size, total biomass, etc. Whichever it is, in what follows we consider it fixed (let it be (b, N_t) where b is a positive vector) and call it 'size'.

Explicit formulae for the first passage times probabilities exist only, to our knowledge, for diffusion processes with constant coefficients (in which case the density function at time t of first passages through the level N_c, if the position at time 0) was N_o, the infinitesimal drift and variance are a and σ^2, is given by

$$\ln(N_o/N_c)\exp\{-(\ln N_o/N_c + at)^2/(2\sigma^2 t)\}/(t\sqrt{2\pi\sigma^2 t}) \qquad (1)$$

and few other Markov processes (like simple random walks). Since the probability distribution of population size asymptotically becomes undistinguishable from the one resulting from a diffusion process with constant coefficients, the crossing level probabilities for the latter (1) were used for the former.

However, age-structure induces a correlation between subsequent increments of population size (whether or not the underlying noise in the parameters is auto-correlated): it is clear that stochastic processes with the same asymptotic variance but different autocorrelation structure will have noticeably different probabilities of crossing any threshold.

One way of approaching the problem is to use the idea that population size can be approximately described as a one-dimensional process with correlated increments (Ginzburg et al., 1982b). Other ways of reducing a multidimensional process to a one-dimensional one are used by Chesson and Werner (1981), Deriso (1980) and Horwood and Shepherd (1981). In this context, it is interesting to note that the autocorrelation depends strongly on the 'size' measure chosen, through the vector b, as shown in Figure 1 and Figure 2.

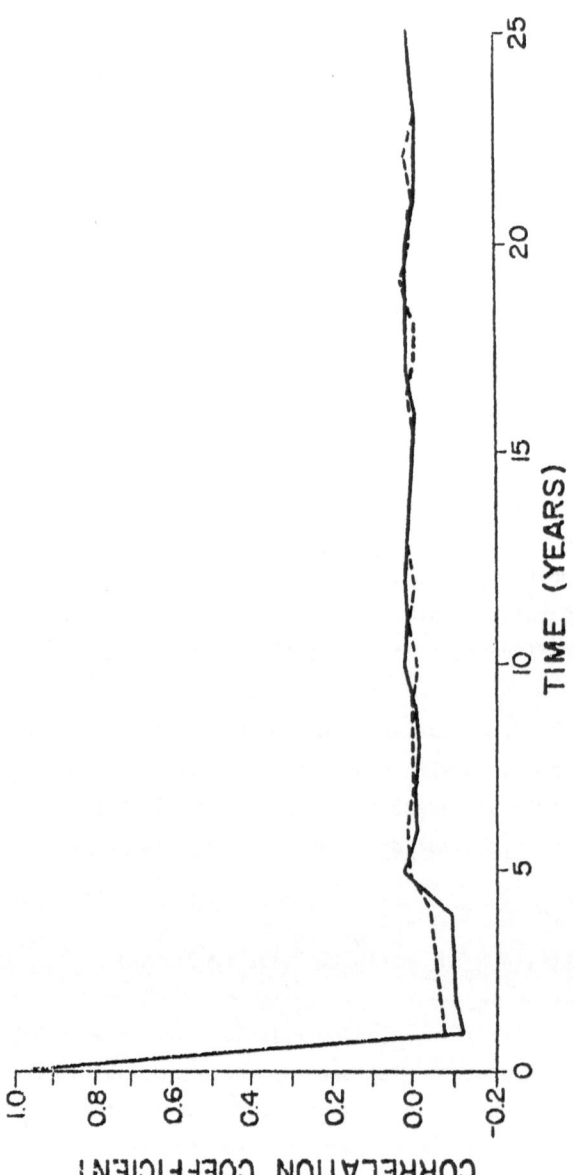

Figure 1

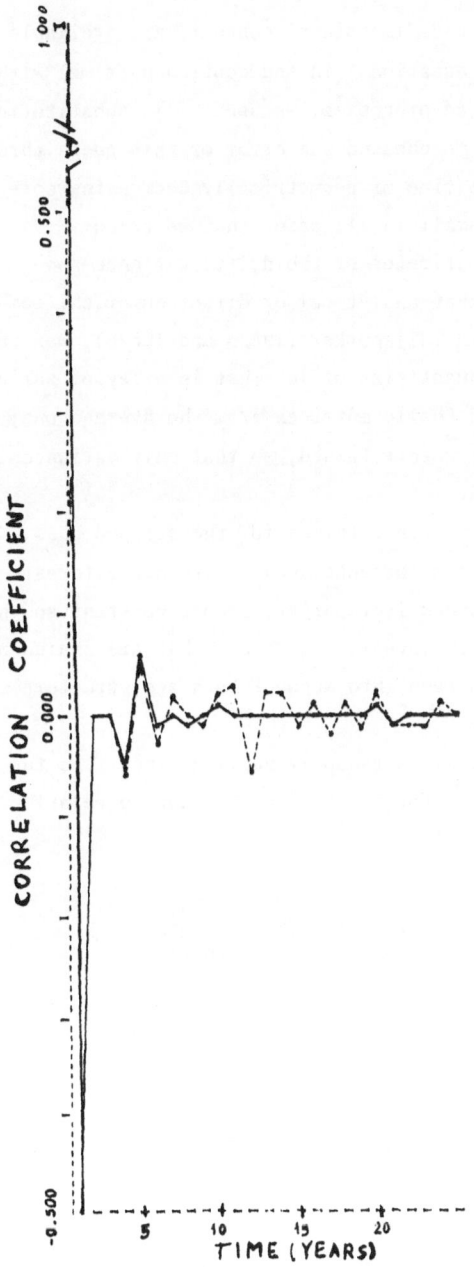

Figure 1 and 2   The autocorrelation for the variables   log(b,N[t+n+1])
                 - log(b,N[t+n])   considered as function of   n.   In Figure 1, b   is
                 such as to count the total number of individuals 5 to 20 years old.
                 In Figure 2   b'   is   (0,1,0 ...).   In both figures the solid line is
                 computed from formulae (3) and (4), the broken line is an average
                 over the simulations and over   t   between   450   and   500   years.

No analytical results, even in the one-dimensional context, are available for random walks (or solution of differential equations, in the continuous time) with memory. As a first guess for autocorrelated processes, we used (1) substituting $V(t)$, its variance at time t, for $\sigma^2 t$. We checked the error of this guess through computer simulations of processes with one-time or geometrically decreasing correlations and it proved to be relatively small in all cases that we tried.

The step that was left was to get estimates of the drift, variance and autocorrelation. We already pointed out that analytical or direct numerical computations are unpractical. The idea, as in Tuljapurkar (1982a and 1982b), was to formally develop the expressions for the quantities of interest in a Taylor series in terms of the deviations of the realized Leslie matrices from the average one; then only the terms up to the second order were retained, so that only variances and covariances of vital rates were needed.

We used the method outlined above on data obtained for the striped bass population of the Hudson River. We chose that because, beyond its own interest, it exemplifies many life history traits (extended iteroparity, almost constant and high adult survival, very low and highly variably juvenile survival) that are shared by many organisms and compel investigators to take into account both age structure and stochasticity.

Assuming that only first year survival is randomly varying simplifies the algebra somewhat. In this case the formulae for a, σ^2, and the autocorrelation $\phi(\cdot)$ become

$$a \approx \log \lambda - \frac{(CV)^2}{2(U,V)^2} \qquad \sigma^2 = \phi(0) + 2 \sum_{k=1}^{\infty} \phi(k) \approx \frac{(CV)^2}{(U,V)^2} \qquad (2)$$

$$\phi(0) \approx \frac{(CV)^2}{(U,V)^2} + \frac{(CV)^2}{(b,V)^2} \sum_{\beta \neq 1} \sum_{\alpha} \frac{(b,V_\alpha)(b,V_\beta)\lambda_\alpha \lambda_\beta}{(U_\alpha,V_\alpha)(U_\beta,V_\beta)(\lambda^2 - \lambda_\alpha \lambda_\beta)} (1 - \lambda_\beta/\lambda) \qquad (3)$$

$$\phi(k) \approx - \frac{(CV)^2}{(b,V)^2} \sum_{\beta \neq 1} \sum_{\alpha} \frac{(b,V_\alpha)(b,V_\beta)\lambda_\alpha \lambda_\beta}{(U_\alpha,V_\alpha)(U_\beta,V_\beta)(\lambda^2 - \lambda_2 \lambda_\beta)} (1 - \lambda_\beta/\lambda)^2 (\lambda_\beta/\lambda)^{k-1} \qquad (4)$$

for $k \geq 1$.

where CV is coefficient of variation of first year survival, $\lambda = \lambda_1, \ldots, \lambda_s$ are the eigenvalues of the average Leslie matrix, ordered by the absolute value, U_α and V_α are the corresponding left and right eigenvectors.

To check that these approximations were reasonable at the level of variation we considered, we extended the expansion of a and σ^2 to the fourth order (without appreciable difference, as shown in Ginzburg et al., (1982b)) and checked the computed autocorrelation against the one estimated from simulations, for various age groups. Some results are illustrated in Figures 1 and 2.

Finally, in Figure 3, the extinction probabilities, as resulting from Monte Carlo simulations, are shown together with the expectations predicted from (1), with and without the autocorrelation induced by the age structure, using the procedure outlined above. It is clear that consideration of autocorrelation improves the predictions

The use of the crossing level probabilities in finite times as the most reasonable way of studying extinction in wildlife management, has been suggested in Ginzburg et al. (1982a). Of course, the problem is the sensitivity of the result to inaccurate estimation of parameters. This sensitivity, unfortunately, seems to be quite high, since the expected probabilities depend on all the eigenvalues of the average Leslie matrix.

Having explicit formulae for extinction probabilities could prove to be useful also for answering evolutionary questions. Local extinction and recolonizations have been thought to be important in the genetic cohesiveness of a species (Grant 1980), in the coexistence of predator-prey systems (Huffaker 1958) and are fundamental to the process of group selection (see for example Maynard Smith 1976). Estimates for the probability of extinction are relevant to the debate about the importance of these factors.

Another way of assessing the occurrence of group selection is through a study of life history parameters. If local extinctions were a usual event, it is likely that the life history that is observed now minimizes extinction risks, while an individual selection argument would lead to the prediction that average growth rate (in logarithmic scale) has been maximized. These predictions might be so different that an empirical study will be able to distinguish between them.

As a one-dimensional example, we will consider a model introduced by Cohen (1966) of an annual plant growing in a random environment; the seeds are capable of surviving dormant for many years. Cohen computed the percentage of the seed bank that should germinate at the beginning of a year, in order to maximize the growth rate. Mountford (1971), using a branching process model, noted that the percentage would be different if the aim were to minimize extinction risks. In Figure 4, the logarithmic growth rate and the probabilities of 'quasi-extinction' at various times, for Cohen's model with arbitrary values of the parameters, are compared. It is clear that the shorter the average time between extinctions, the lower will be the germination percentage selected and, in any case, it is lower than the one optimal for growth rate.

Acknowledgements

This work was supported by Environmental Protection Agency grant number R807885010 to Lev R. Ginzburg and is contribution 453 from the Department of Ecology and Evolution of the State University of New York and Stony Brook. The conclusions represent the view of the authors and do not necessarily reflect opinions, policies or recommendations of the Environmental Protection Agency.

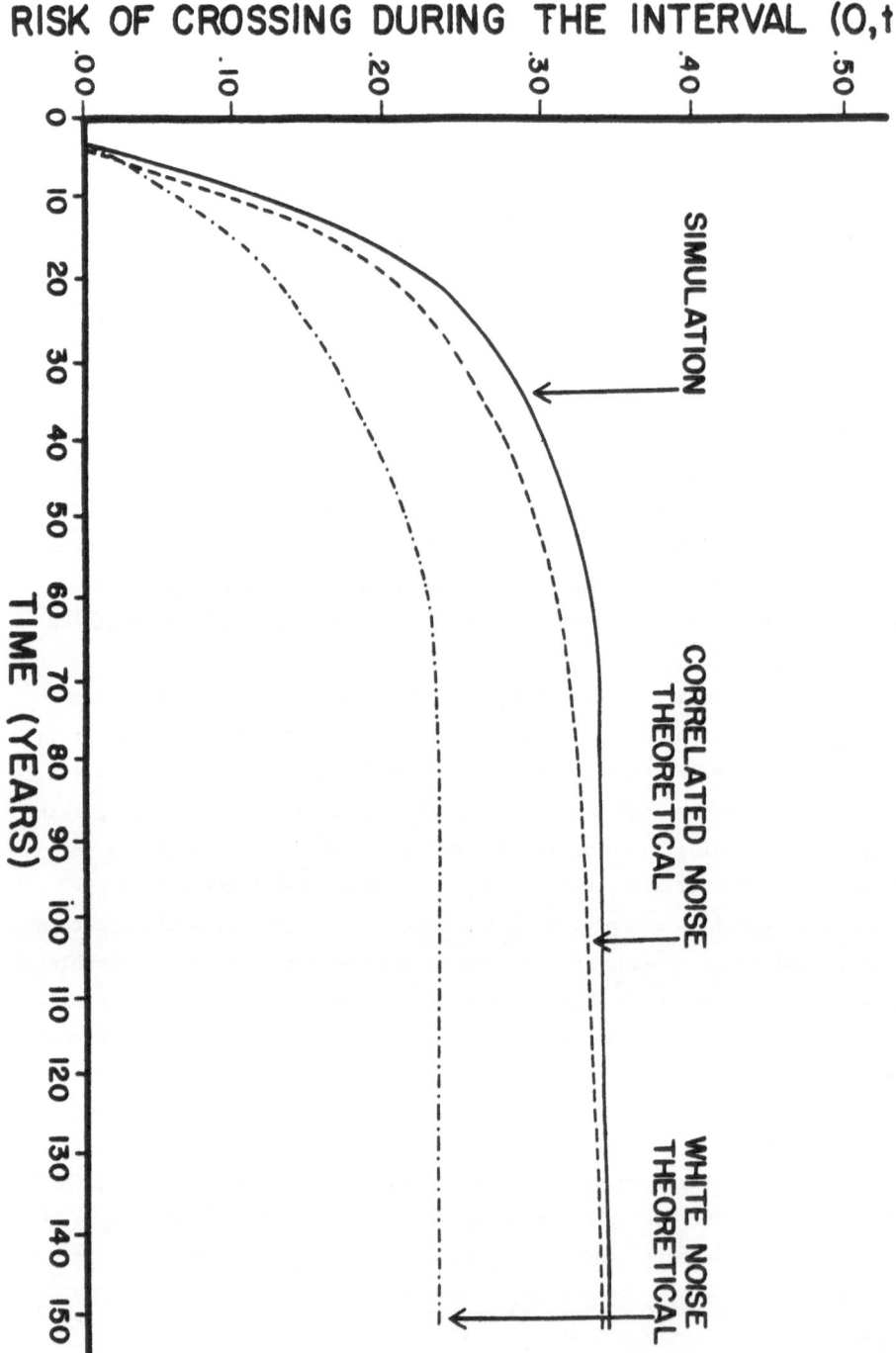

Figure 3 The risk of crossing the level $N_c = .8N_o$ at least once in the interval $(0,t)$ as functions of t. The white noise curve is computed through (1) and (2). The correlated noise using (1), (2), (3) and (4) as outlined in the text.

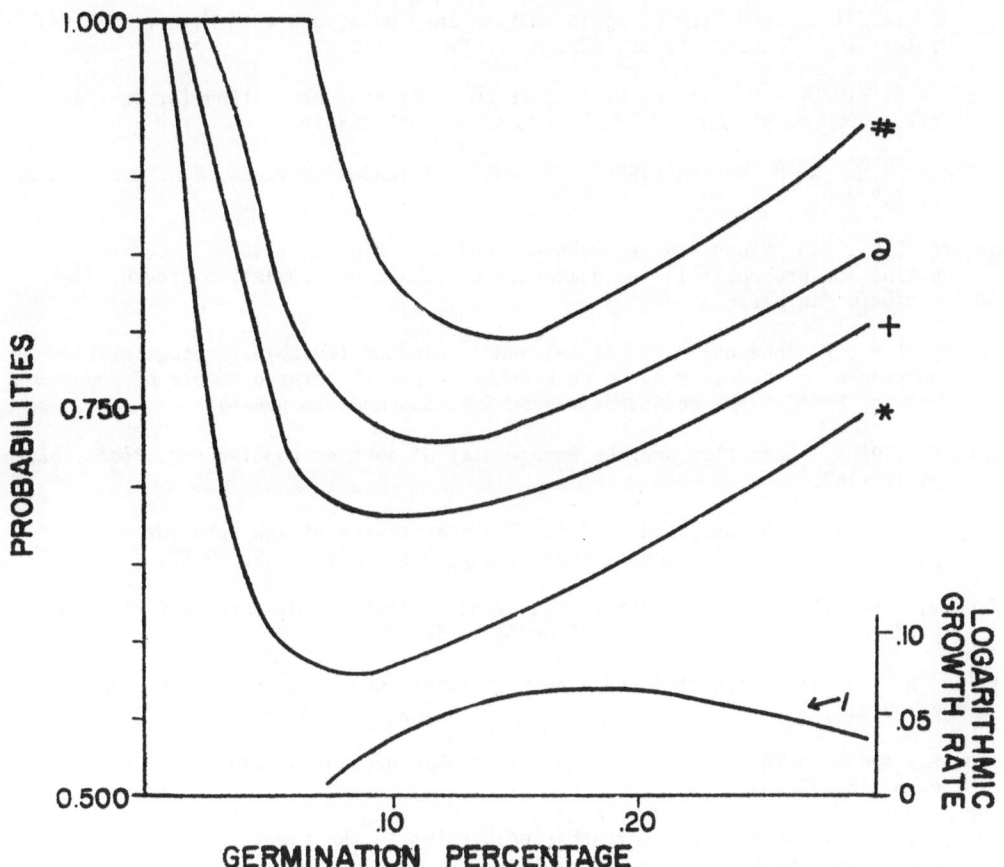

Figure 4   The logarithmic growth rate (1) and the probabilities of quasi-extinction
within 25 (*), 50 (+), 100 (∂) years and ever (#) versus germination
percentage for Cohen's model with Y = 50, p = .2, V = .75. The critical
level was chosen as .135 of the initial population size. The scale for
probabilities is on the left side of the figure, the growth rate is on
the right side.

REFERENCES

Capocelli, R.M., and L.M. Ricciardi (1974): A diffusion model for population growth
in a random environment, *Theor. Pop. Biol.* 5:28-41.

Chesson, P.L., and R.R. Warner (1981): Environmental variability promotes coexistence in lottery competitive systems, *Am. Nat.* 117:923-943.

Cohen, D. (1966): Optimizing reproduction in a randomly varying environment, *J.
Theor. Biol.* 12:119-129.

Cohen, J.E. (1976): Ergodicity of age structure in populations with Markovian vital
rates, I. Countable states, *J. Amer. Stat. Ass.* 71:335-339.

_____(1977): Ergodicity of age structure in populations with Markovian vital rates, II. General states, *Adv. Appl. Prob.* 9:18-37.

Deriso, R.B. (1980): Harvesting strategies and parameter estimation for an age-structured model, *Can. J. Fish. Aquat. Sci.* 37:268-282.

Furstenberg, H. and H. Kesten (1960): Products of random matrices, *Ann. Math. Stat.* 31:457-469.

Ginzburg, L.R., L.B. Slobodkin, K. Johnson, and A.G. Bindman (1982a): Quasi-extinction probabilities as a measure of impact on population growth, *Risk Analysis* (in press).

Ginzburg, L.R., K. Johnson, A. Pugliese, and J. Gladden (1982b): Ecological risk assessment methodology based on stochastic age-structured models of population growth, *Proc. Symp. on Statistics in Env. Science* (in press).

Grant, V. (1980): Gene flow and the homogeneity of species populations, *Biol. Zbl.* 99:157-169.

Horwood, J.W., and J.A. Shepherd (1981): The sensitivity of age-structured populations to environmental variability, *Math. Biosci.* 57:59-82.

Huffaker, C.B. (1958): Experimental studies on predation: dispersion factors and predator-prey oscillations, *Hilgardia* 27:343-383.

Keiding, N. (1975): Extinction and growth in random environments, *Theor. Pop. Biol.* 6:199-216.

Lange, K., and W. Holmes (1981): Stochastic stable population growth, *J. Appl. Prob.* 18:325-334.

Ludwig, E. (1976): A singular perturbation problem in the theory of population extinction, *SIAM-AMS Proc.* 10:87-104.

Maynard Smith, J. (1976): Group selection, *Qua. Rev. Biol.* 51:277-283.

Mountford, M.D. (1971): Population survival in a variable environment, *J. Theor. Bio.* 32:75-79.

Reed, W.J. (1982): Recruitment variability and age structure in harvested animal population (manuscript).

Sawyer, S., and M. Slatkin (1981): Density-independent fluctuations of population size, *Theor. Pop. Biol.* 19:37-51.

Tuljapurkar, S.D. (1982a): Population dynamics in variable environments, II. Correlated environments, sensitivity analysis and dynamics, *Theor. Pop. Biol.* 21:114-140.

_____(1982b): Population dynamics in variable environments, III. Evolutionary dynamics of r-selection, *Theor. Pop. Biol.* 21:141-165.

Tuljapurkar, S.D., and S.H. Orzack (1980): Population dynamics in variable environments, I. Long-rum growth rate and extinction, *Theor. Pop. Biol.* 18:314-342.

Turelli, M. (1978): A reexamination of stability in random varying versus deterministic environments with comments on the stochastic theory of limiting similarity, *Theor. Pop. Biol.* 13:244-267.

THE SYSTEMATIC FORMULATION OF DELAY-DIFFERENTIAL
MODELS OF AGE OR SIZE STRUCTURED POPULATIONS

W.S.C. Gurney and R.M. Nisbet

1. INTRODUCTION

The vital rates of individual organisms usually depend strongly on their
chronological or physiological age. Thus the average vital rates which characterize
the dynamics of a whole population of such organisms will remain constant over time
only so long as the population retains a constant age-profile. Despite almost universal agreement on the evident corollary of this; namely that proper modelling of
population fluctuations must require due attention to the dynamics of changes in the
population age-distribution, the great majority of theoretical studies take little
or no account of such changes.

It is our belief that the primary reason for this omission lies in the failure of any currently popular age-structure description to provide a modelling tool
combining a useful degree of realism with an acceptable level of mathematical difficulty. The partial differential equation formalism originally due to McKendrick
(1926) and repopularised by von Foerster (1959) is the most rigorous and elegant
available, but poses such horrendous technical difficulties that few analysts
succeed in wringing from it any significant biological insight. The more popular
matrix formalism (Leslie 1945) poses fewer technical problems but is seriously
flawed as a modelling tool because its necessarily rigid assumption of equal length
age classes precludes the realistic identification of such age classes with groups
of functionally similar individuals (e.g. insect instars). The manifest failure of
these two mathematically rigorous descriptions to come up with the biological goods
has led in turn to the use of a variety of *ad hoc* models based on sets of heuristi-
cally formulated delay-differential equations (see for example MacDonald 1978).
Such models, although appealingly tractable, all carry the inherent flaw that their
lack of any mathematically rigorous foundation makes it hard to distinguish real
(i.e. biologically real) dynamic subtlty from irrelevant artifacts of erroneous
formulation.

In this paper we wish to argue that it is possible to set out a prescription
which, under well defined simplifying assumptions, allows one to systematically
formulate models which combine the rigour of the von Foerster approach with the
mathematical docility of heuristic delay-differential models. The key element in
this prescription is the realization that within most age-structured populations
there are readily identifiable sub-populations whose members can, without serious
error, be regarded as functionally identical (and thus, in particular, as having
identical vital rates). If we write the number of individuals in the ith such

functional class at time t as $N_i(t)$ and think of a closed population (so as to exclude immigration and emigration) then it is self evident that the dynamics of the total population must be described by a set of differential equations of the general form

$$\dot{N}_i(t) = \text{recruitment} - \text{maturation} - \text{deaths}. \qquad (1)$$

The process of model formulation now focusses upon calculating the three terms on the right-hand side of this equation. The two difficult terms are the recruitment and maturation rates which describe moults, emergencies, pupation and so on, but it turns out that if any inter-class transition is triggered by a critical value of a single parameter (such as age, physiological age, size etc.) then the equivalent recruitment and maturation rates can be obtained rigorously from the von Foerster equation or one of its more complex cousins. In the remainder of this paper we shall concentrate on two particular cases: firstly inter-class transitions which take place at a critical age, and secondly transitions governed by body size (which is thought to be the critical factor in triggering many insect moults). However we would like to draw your attention also to some work which Stephen Blythe will be describing in these Proceedings, in which he examines the possibility of similarly rigorous formulation of "distributed delay" models in which individual members of sub-populations differ in their aging rate.

2. AGE-DEPENDENT TRANSITIONS

We consider first the class of models in which transitions out of class $i-1$ into class i take place at age a_i, and transitions out of class i into class $i+1$ occur at age a_{i+1}, so that the duration of age class i is

$$\tau_i = a_{i+1} - a_i. \qquad (2)$$

We seek to calculate the magnitudes of the terms in the population balance equation (1) which we re-write more formally as

$$\dot{N}_i(t) = R_i(t) - M_i(t) - D_i(t). \qquad (3)$$

The death term is easy; we are assuming that all members of age class i have the same instantaneous per capita death rate $\delta_i(t)$ so

$$D_i(t) = \delta_i(t)N_i(t). \qquad (4)$$

The formal derivation of the recruitment and maturation terms (see Gurney, Nisbet and Lawton 1982) is considerably more complex, but it is rather easy to get a feel for the results with the aid of George Oster's analogy between the process of aging

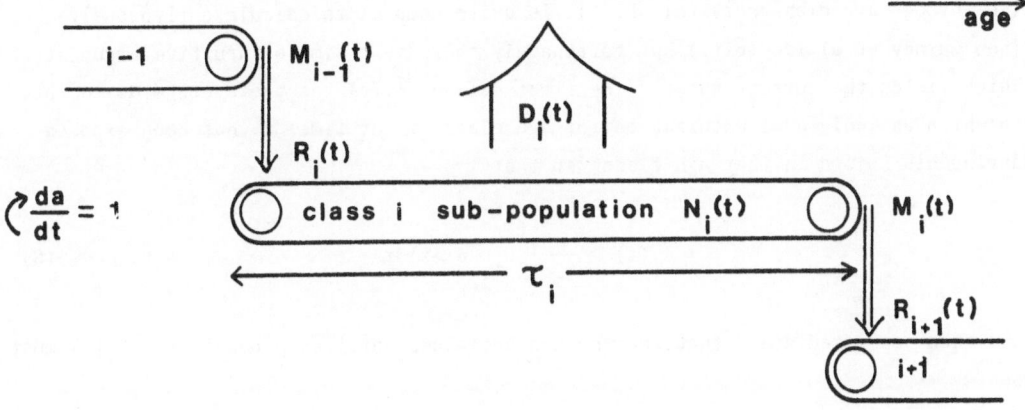

Figure 1   Age dependent transitions

and the motion of a conveyor belt. By picturing age class i as a belt of length
τ_i turning at constant unit speed (see Figure 1) it is elementary to see that

1. The recruitment rate to the first age class (i = 1 say) is just the total popu-
    lation reproduction rate, and thus, since all individuals in age class i at
    time t are deemed to have the same instantaneous per capita fecundity $\beta_i(t)$,
    it must be given by

    $$R_1(t) = \sum_i \beta_i(t)N_i(t). \tag{5}$$

2. For age classes other than the first, new recruits get onto the start of the
    belt by falling off the end of the previous one, so the rate of recruitment into
    class i must be exactly equal to the rate of maturation out of class i - 1

    $$R_i(t) = M_{i-1}(t) \quad i > 1. \tag{6}$$

3. Once on the belt individuals get off it only by dying (evaporating?) or falling
    off the end into class i + 1. Thus if $P_i(t)$ is the probability that an
    individual falling onto the start of the belt at time $t - \tau_i$ is still alive to
    fall off the end at time t, then the rate of maturation out of age class i at
    time t is

    $$M_i(t) = R_i(t-\tau_i)P_i(t) \tag{7}$$

The through age class survival $P_i(t)$ is quite complex to calculate rigorously (see Gurney et al loc. cit.) but fortunately there is a simple intuitive argument which yields the correct answer. We first define $\bar{\delta}_i(t)$ as the average death rate to which an individual maturing out of age class i at time t has been exposed during his sojourn in that age class, so that

$$\bar{\delta}_i(t) = \frac{1}{\tau_i} \int_{t-\tau_i}^{t} \delta_i(x)dx \tag{8}$$

and then see immediately that the through age-class survival probability $P_i(t)$ must be

$$P_i(t) = \exp\{-\tau_i\bar{\delta}_i(t)\} = \exp\left\{-\int_{t-\tau_i}^{t} \delta_i(x)dx\right\}. \tag{9}$$

3. PRACTICAL LUMPED AGE STRUCTURE MODELLING

In common with other delay-differential systems, the dynamic behaviour of lumped age-structure models is most readily investigated by judiciously selected numerical integrations. We have found that the evaluation of such integrals is noticeably facilitated by re-stating the integral equation (9) as a delay-differential equation plus an initial condition, and hence obtaining a model description consisting of pairs of delay-differential equations of the form

$$\dot{N}_i(t) = R_i(t) - R_i(t-\tau_i)P_i(t) - \delta_i(t)N_i(t) \tag{10a}$$

$$\dot{P}_i(t) = P_i(t)[\delta_i(t-\tau_i)-\delta_i(t)] \tag{10b}$$

where

$$R_i(t) = R_{i-1}(t-\tau_{i-1})P_{i-1}(t) \quad i > 1 \tag{11a}$$

$$R_1(t) = \sum_i \beta_i(t)N_i(t) \tag{11b}$$

and

$$P_i(0) = \int_{-\tau_i}^{0} \delta_i(x)dx \tag{12}$$

Once a suitable initial history has been constructed the system of equations can be integrated rapidly and effectively by a simple predictor-corrector algorithm such as that described by Maas, Nisbet and Gurney (1982). However, at this point we encounter an apparent impass, because the validity of equation (11a) hinges upon the entire historical record being itself a valid solution of equations (10) and (11). Thus we appear not be be able to solve the equations until we've constructed a prior history - by solving the equations!! The remedy for this difficulty lies in the fact that a totally empty system (all $N_i = 0$) is always a stationary solution of (10) and (11), so if we set

$$N_i(t) = 0 \quad \text{for all} \quad i \quad \text{and} \quad t \geq 0 \tag{13}$$

then we have constructed a valid, if apparently uninteresting, initial history. Moreover, such an initial history is much less boring than it seems, because it corresponds to the actual prior state of a laboratory experiment which is initiated by the innoculation (at $t = 0$) of a previously empty culture vessel or enclosure. It thus only remains to model the innoculation process. If we insist that the innoculated population shall have an arbitrary age distribution this is a very complex undertaking, but by restricting ourselves to "newley qualified" immigrants (e.g. newly laid eggs, newly emerged adults etc. etc.) we can reduce the problem to one which can be solved simply by modifying equation (11) to take account of an immigration rate $I_i(t)$ thus

$$R_i(t) = R_{i-1}(t-\tau_{i-1})P_{i-1}(t) + I_i(t) \tag{14a}$$

$$R_1(t) = \sum_i \beta_i(t)N_i(t) + I_1(t) \tag{14b}$$

and then setting all the I_i's to zero except during some short "innoculation period" just after $t = 0$.

4. GROWTH-DEPENDENT TRANSITIONS

In many species transitions between clearly defined functional classes (e.g. insect instars) occur not at a critical age but at a critical value of body size, body weight, or some other physiological parameter (Beddington et al. 1976). Provided that only a single critical factor is involved we can rather easily extend the formalism of section 2 to cover this case. As before the population balance equation for each functional class must of necessity take the general form

$$\dot{N}_i(t) = R_i(t) - M_i(t) - \delta_i(t)N_i(t) \tag{15}$$

and our task is to discover the relationship between recruitment, maturation and reproduction. The formal mathematical derivation of the results we require (see Nisbet and Gurney 1983) from the general formalism due to Sinko and Streiffer (1967) is even more intimidating than before, but again Oster's conveyor belts come to our rescue. We now visualize distance along the belt as representing body size, so that (Figure 2) we regard size class i as a single belt spanning a size increment Δm_i and turning at a rate, $\dot{m}_i = g_i(t)$, determined by the instantaneous growth rate of the current population of the size class. By simply examining the picture we deduce at once that provided all newborns have the same weight (say m = 0):

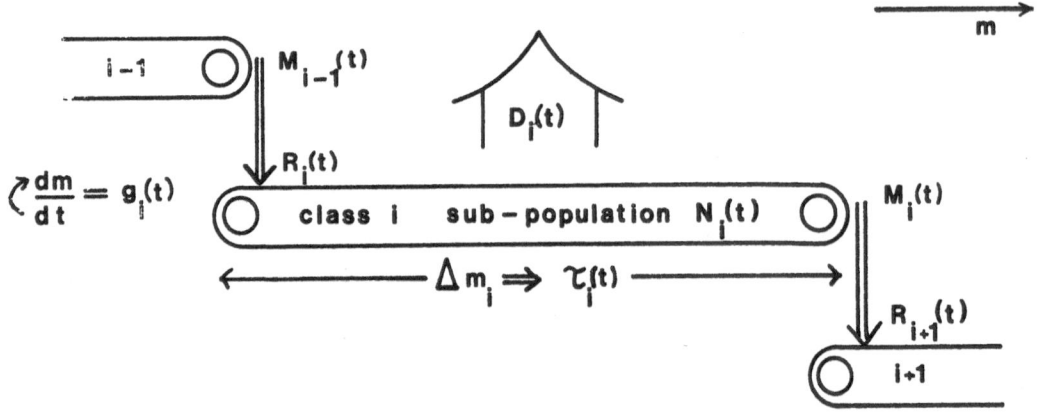

Figure 2   Size dependent transitions

1.   Recruitment to the first size class is just the total population reproduction
     rate

$$R_1(t) = \sum_i \beta_i(t)N_i(t) \tag{16a}$$

     and

2.   Recruitment to subsequent classes occurs only by maturation out of the previous
     class

$$R_i(t) = M_{i-1}(t) \quad i > 1. \tag{16b}$$

Connecting the maturation rate out of class i to the recruitment rate into it is
rather more complex in this case because the belt doesn't turn at a constant speed.
To assist us in thinking about the problem we first define $\rho_B(t)$ and $\rho_E(t)$ to be
the densities of individuals to be found at time t on the beginning and the end of
the belt respectively.   Now, since at time t, the belt is turning at speed $g_i(t)$
we know at one that

$$M_i(t) = g_i(t)\rho_E(t) \tag{17}$$

and

$$\rho_B(t) = R_i(t)/g_i(t) \tag{18}$$

Clearly if individuals maturing one of class i at time t entered it at a time
which for reasons that will become clear in a moment we write as $t - \tau_i(t)$, then
ρ_B and ρ_E must obey the relation

$$\rho_E(t) = \rho_B(t-\tau_i(t))P_i(t) \tag{19}$$

where the through age-class survival $P_i(t)$ is now defined as

$$P_i(t) \equiv \exp\left\{-\int_{t-\tau_i(t)}^{t} \delta_i(x)dx\right\}. \tag{20}$$

This (equation (19)) in turn enables us to combine equations (17) and (18) to yield a relationship between M_i and R_i

$$M_i(t) = g_i(t) \frac{R_i(t-\tau_i(t))}{g_i(t-\tau_i(t))} P_i(t). \tag{21}$$

which becomes meaningful as soon as we can calculate the delay $\tau_i(t)$. But $\tau_i(t)$ simply represents the time taken by an individual maturing out of class i at time t to traverse that class (i.e. to achieve a size gain Δm_i). Thus, although it has now itself become a dynamically varying quantity, it is always defined by the requirement that

$$\Delta m_i = \int_{t-\tau_i(t)}^{t} g_i(x)dx \tag{22}$$

As we found in section 3, we can facilitate numerical analysis of this system by restating the two integral equations (20) and (22) as differential equations plus initial conditions. This yields a model description composed of sets of three delay-differential equations of the form

$$\dot{N}_i(t) = R_i(t) - \frac{g_i(t)}{g_i(t-\tau_i(t))} R_i(t-\tau_i(t))P_i(t) - \delta_i(t)N_i(t) \tag{23a}$$

$$\dot{P}_i(t) = P_i(t)\left[\frac{g_i(t)\delta_i(t-\tau_i(t))}{g_i(t-\tau_i(t))} - \delta_i(t)\right] \tag{23b}$$

$$\dot{\tau}_i(t) = 1 - \frac{g_i(t)}{g_i(t-\tau_i(t))} \tag{23c}$$

where, again as before, we modify the inter-class links to allow for "newly qualified" immigrants thus

$$R_1(t) = \sum_i \beta_i(t)N_i(t) + I_1(t) \tag{24a}$$

$$R_i(t) = R_{i-1}(t-\tau_{i-1}(t)) \frac{g_{i-1}(t)}{g_{i-1}(t-\tau_{i-1}(t))} P_{i-1}(t) + I_i(t) \tag{24b}$$

and where the initial states of the auxiliary variables P_i and τ_i are defined by

$$P_i(0) = \exp\left\{-\int_{-\tau_i(0)}^{0} \delta_i(x)dx\right\},$$ (25a)

$$\int_{-\tau_i(0)}^{0} g_i(x)dx = \Delta m_i.$$ (25b)

As before we normally adopt the empty system $(N_i(t) = 0$ for all i and $t \leq 0)$ as our assumed prior history and start off the culture with a burst of immigration just after $t = 0$.

5. THE POPULATION DYNAMICS OF DAMSELFLY THEORETICA

To demonstrate the use of the formalism we have just developed, and also to illustrate the potential of "mixed mode" models in which some stages are size-controlled while others are age-controlled, we now formulate a strategic model of a generalized damselfly (D. theoretica). This model was suggested to us by some very elegant experiments of Lawton et al. (1980) which showed that the damselfly *Ishnura elegans* can exhibit hugh variations in instar duration in response to changes in food supply. The model, illustrated in Figure 3, is intended to explore the possibility of dynamic control mediated only by changes in instar duration, and postulates an organism with two life-history stages, larvae and adults, whose populations we shall write as $L(t)$ and $A(t)$ respectively. The population dynamics of the organism are as follows:

i. *Food supply*. larval food is supplied at a constant rate ϕ into a "pool" of size $F(t)$, and is eaten by the larvae at a *per capita* rate E which depends on the availability of food, i.e. $F(t)$. Thus we write

$$F(t) = \phi - L(t)E(F(t)).$$ (26)

ii. *Larvae*.

(a) We assume that larvae grow at a rate directly proportional to their food intake rate, and mature into adults when they reach a critical size. Thus, from (23c) the dynamics of the development delay are

$$\dot{\tau}(t) = 1 - E(F(t))/E(F(t-\tau(t))).$$ (27)

(b) We assume that the larval death rate is a density independent constant (per unit time), so that the through age-class survival $P(t)$ is

$$P(t) = \exp\{-\tau(t)\Delta\}$$ (28)

(c) From equation (23a) we can see that if the adults produce eggs (i.e. newborn larvae!) at a rate $R(t)$ then the larval population dynamic must be

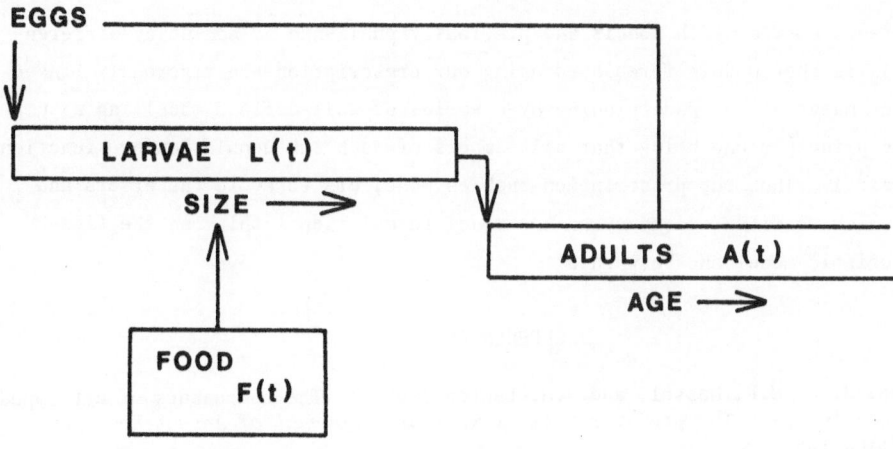

Figure 3   Life cycle of Damselfly Theoretica

$$\dot{L}(t) = R(t) - \frac{E(t)}{E(t-\tau(t))} \cdot R(t-\tau(t)) \cdot P(t) - \Delta \cdot L(t) \tag{29}$$

iii.   *Adults*. We assume that adults have unlimited food supplies (or do not feed) so that they have a constant *per capita* fecundity and death rate β   and δ respectively. Thus

$$R(t) = \beta A(t) \tag{30}$$

and

$$\dot{A}(t) = \frac{E(t)}{E(t-\tau(t))} \cdot R(t-\tau(t))P(t) - \delta A(t) \tag{31}$$

The dynamics of this model are fully discussed in Nisbet and Gurney (1983). Here it is sufficient to say that notwithstanding its postulation of entirely density *independent* birth and death *rates* the model (when equipped with a sensible functional form for E(F)) exhibits stable stationary states over large regions of its parameter space. The reason for this is very easy to understand: the density dependent length of the larval stage, combined with the loss of a constant fraction of the larval population per unit time, implies strong density dependence of egg to adult survival, and this "implicit density dependence" is just as capable of population stabilization as straightforwardly density dependent vital rates.

5.   CONCLUSIONS

We have constructed a framework within which age and/or size structured populations can be modelled with a high degree of mathematical rigour using simple and comprehensible delay-differential equations. It cannot be over-emphasized that the difference between such models and previously published *ad hoc* delay-differential models is that models formulated using our prescription are rigorously linked to an exact mathematical underpinning by a series of well-defined modelling assumptions, the principle one being that all members of each sub-population are functionally identical. Thus our prescription enables modellers to avoid the errors and pitfalls which so often accompany *ad hoc* model formulation - think on the time-delayed logistic model and be warned!

REFERENCES

Beddington, J.R., M.P. Hassel, and J.H. Lawton (1976): The components of arthropod predation II. The predator rate of increase, *Journal of Animal Ecology* 45:165-185.

von Foerster, H. (1959): Some remarks on changing populations, in *The Kinetics of Cellular Proliferation* (Ed. F. Stohlman, Jr.) Frame and Stratton, New York, pp. 382-407.

Gurney, W.S.C., R.M. Nisbet, and J.H. Lawton (1983): The systematic formulation of tractable single species population models incorporating age structure, *Journal of Animal Ecology*, in press.

Lawton, J.H., B.A. Thompson, and D.J. Thompson (1980): The effects of prey density on survival and growth of damselfly larvae, *Ecolog. Ent.* 5:39-51.

Leslie, P.H. (1978): On the use of matrices in certain population mathematics, *Biometrika* 33:183-212.

Maas, P., R.M. Nisbet, and W.S.C. Gurney (1982): Solver-an adaptable program template for initial value problem solving, *Applied Physics Industrial Consultants*, University of Strathclyde, Glasgow.

McDonald, N. (1978): *Time Lags in Biological Models*, Springer-Verlag. Lect. notes in biomath., Vol. 27, New York.

McKendrick, A.G. (1926): Applications of mathematics to medical problems, *Proc. Edin. Math. Soc.* 44:98-130.

Nisbet, R.M., and W.S.C. Gurney (1983): The systematic formulation of population models for insects with dynamically varying instar duration, *Theor. Pop. Biol.*, in press.

Sinko, J.W., and W. Streiffer (1967): A new model for the age-size structure of a population, *Ecology* 48:910-918.

LOTKA DISTRIBUTION FOR A FINITE MIXTURE
OF HUMAN POPULATIONS I

P. Krishnan and N.M. Lalu

ABSTRACT

It is shown that the dominant root and the associated stable vector of a
convex mixture of Leslie matrices is approximately equal to the convex mixture of
the dominant roots and their stable vectors of the component matrices.

1. INTRODUCTION

The Lotka distribution has been studied extensively by demographers, mathematicians, and others. In the last several years, various generalizations have been
suggested by incorporating migration (Lalu and Krishnan, 1979; Lalu, 1980;
MacFarland, 1969), non-linear forms of vital rates (Goel, et al., 1971; Gopalswamy,
this Proceedings) etc. In demographic literature, the quasi-stable form when
mortality is allowed to decline over time and the multi-regional analogue have been
the most popular forms of generalization (Coale, 1963, 1972; Rogers, 1975). The
properties of the Lotka distribution are put to use to estimate the vital rates of
developing countries, where the data collection system is defective and incomplete.

An issue of a different nature arises in the developed countries. The population of a developed country (this is true of any country, as a matter of fact) is
not homogeneous by any criterion. People have different religious affiliations,
belong to different ethnic groups, speak different languages etc. It is known that
health, mortality, and fertility of a population are affected by ethno-cultural and
social factors as well. While the vital rates are known for the country/province as
a whole and for certain major population groups, such data are hard to come by for
smaller groups of population. Since the population of a country/province is conceivable as a mixture of population groups (based on a criterion under consideration), a study of the properties of Lotka distribution of a finite mixture of populations may help find some means to estimate the vital rates of some of the smaller
groups.

2. METHODOLOGY

Since this is an unexplored area in mathematical demography, we have used
computer simulation to derive the patterns of results. No formal proofs will be
attempted here.

The following conditions are imposed.

a) We consider only a mixture of two populations.

b) We use the discrete time approach suggested by the Leslie scheme.

c) The female population only is considered.

d) No interaction between the population groups, through inter-marriage, is assumed.

The Leslie matrix can be written as:

$$L = \left[\begin{array}{c|c} M & 0 \\ \hline A & B \end{array} \right].$$

where M is the so-called projection matrix (see, Keyfitz, 1968, p. 37). Then we known that M is primitive. Let λ_1 be the dominant root of M and Z_1 the associated stable vector (or the stable age distribution).

DEFINITION 2.1 The population of a country/province is called the *main* population.

DEFINITION 2.2 The populations of the groups comprising the main population, according to a given criterion, are called the *components*.

2.3 *Projection matrix of the mixture.* Let M^1 and M^{11} be the M-matrices associated with the two components. If α is the proportion of the first component, then $M = \alpha M^1 + (1-\alpha)M^{11}$ ($1 \geq \alpha \geq 0$), a convex mixture of M^1 and M^{11}.

2.4 *Dominant component.* If α is large (i.e. close to unity), we say M^1 (or the first component) *dominates* the other.

The questions to be answered are:

i) how is λ_1 related to λ_1^1 and λ_1^{11}, the dominant roots of M^1 and M^{11} respectively?

ii) how is Z_1 related to Z_1^1 and Z_1^{11}, the stable vectors of M^1 and M^{11}?

3. RESULTS FROM COMPUTER SIMULATION

Since there are no theorems, as far as we know, in Matrix Algebra that present the relationship between the characteristic roots and vectors of a convex mixture and its components, we resorted to simulation to explore the patterns that may emerge. Real data have been employed in the simulation experiments.

3.1 *Data used for simulation.* The date (Statistics Canada, 1978) employed for the simulation exercises are given in Table 1. The 1926 and 1976 rates of fertility are indicative of high and low conditions respectively in both Alberta and Quebec, Canada. Only the mortality schedules of 1976 were input into the simulation. For Bangladesh, the world fertility survey results of 1976 (see Lightbourne et al.,

Table 1 Fertility and Mortality, Alberta, Bangladesh, and Quebec

Age Group	1926	1976	Bangladesh 1973-75	Quebec 1926	1976
15 - 19	37.1	44.5	253	24.2	20.4
20 - 24	175.7	130.7	302	152.8	100.1
25 - 29	188.6	140.8	250	· 216.2	137.1
30 - 34	146.3	66.5	198	199.5	70.0
35 - 39	99.9	21.0	120	179.7	22.4
40 - 44	49.6	4.2	53	79.6	4.4
45 - 49	9.0	0.4	1	9.5	0.3
Total Fertility	3526	2040	5885	4307	1774
Life Expectancy at Birth	--	77.7	45.0*	--	76.5

Source: Vital Statistics of Canada, 1976 and World Fertility Survey

*West Model Level 11

Table 5, 1982) were used. Life expectancy at birth for females in Bangladesh, according to the United Nations Demographic Year Book for 1979, is 46.6. Hence West Model Level 11 (Coale and Demeny, 1966), yielding a life expectancy of 45 years, was taken as a proxy for the female mortality conditions in that country.

Various combinations of fertility and mortality used in the simulation are as follows:

Case I   1976 levels of fertility and mortality for both Quebec (M^1) and Alberta (M^{11})

Case II   1926 fertility and 1976 mortality levels for Quebec (M^1) and 1976 fertility and mortality for Quebec (M^{11})

Case III   1976 fertility and West Model Level 11 for Bangladesh (M^1) and 1976 fertility and mortality levels for Quebec (M^{11})

The dominant roots for these cases are given below:

	M^1	M^{11}
Case I	0.97391	0.99920
Case II	1.12900	0.97390
Case III	1.14690	0.97390

3.2 *Simulation results.* The M^1 and M^{11} matrices were combined linearly with the mixing parameter α ranging from 0.5 to 0.95. The dominant roots and the stable vectors were computed by raising the convex mixture $\alpha M^1 + (1-\alpha)M^{11}$ matrices to high powers till stability is attained. In Table 2 are shown the different values of α, the corresponding dominant roots of the linear mixture, and the convex combinations of the roots of M^1 and M^{11}. From this table, it is clear that the dominant root of the mixture is very closely approximated by the convex combination of the dominant roots of the components. It is interesting to note that this approximation is on the conservative side. Also as α approaches unit the dominant root of M is very close to that of M^1.

Table 2 Dominant Roots of the Mixture and the Mixture of the Dominant Roots

| | CASE I | | CASE II | | CASE III | |
| | Root (s) | | Root (s) | | Root (s) | |
α	Mixture	Weighted Average	Mixture	Weighted Average	Mixture	Weighted Average
.5	.98665	.98656	1.0683	1.0515	1.0790	1.0604
.6	.98412	.98403	1.0822	1.0670	1.0946	1.0777
.7	.98158	.98150	1.0950	1.0825	1.1091	1.0950
.8	.97903	.98897	1.1070	1.0980	1.1226	1.1123
.9	.97647	.97644	1.1183	1.1135	1.1351	1.1296
.95	.97519	.97517	1.1237	1.1212	1.1411	1.1383

In Tables 3A and 3B, the female stable age distributions truncated at age 50 for the three cases (for $\alpha = .5$ and $.9$) are shown. The indices of dissimilarity reveal that the distributions, one derived for the convex mixture and the other convex sum of the distributions associated with M^1 and M^{11}, are indeed close. Again as $\alpha \to 1$, Z_1 is very close to Z_1^1.

On the bases of the simulation results, we can state the following theorems.

THEOREM 3.1 *If* $(M_i, \lambda_1^{(i)})$ *is a projection (primitive matrix and the associated dominant root of the* i*th population group and* $M = \sum_{i=1}^{k} \alpha_i M_i$ $(1 \geq \alpha_i \geq 0,$ $\Sigma \alpha_i = 1)$, *then the dominant root* λ_1 *of the convex mixture* M *is approximately equal to* $\sum_{i=1}^{k} \alpha_1 \lambda_1^{(i)}$.

THEOREM 3.2 *If* $Z_{1_k}^{(i)}$ *is the stable vector associated with the dominant root* $\lambda_1^{(i)}$ *of* M_i *and* $M = \sum_{i=1}^{k} \alpha_i M_i$ $(0 \leq \alpha_1 \leq 1, \Sigma \alpha_i = 1)$, *then the stable vector* Z_1, *associated with the dominant root* λ_1 *of* M *is approximately given by* $Z_1 = \sum_{i=1}^{k} \alpha_i Z_1^{(i)}$.

THEOREM 3.3 *If* $\alpha_i \to 1$, M_i *dominates* M.

Table 3A Stable Age Distribution, Females (For α = 0.5)

| | CASE I | | | CASE II | | |
Age Group	Mixture	Convex Sum	$\|\Delta\|$	Mixture	Convex Sum	$\|\Delta\|$
0 - 4	9.54	9.54	0.01	13.37	12.70	0.67
5 - 9	9.64	9.65	0.01	12.48	11.85	0.63
10 - 14	9.76	9.76	----	11.67	11.13	0.54
15 - 19	9.87	9.87	----	10.90	10.50	0.40
20 - 24	9.98	9.97	0.01	10.17	9.95	0.22
25 - 29	10.08	10.08	----	9.50	9.49	0.01
30 - 34	10.18	10.18	----	8.86	9.10	0.24
35 - 39	10.27	10.28	0.01	8.26	8.75	0.49
40 - 44	10.33	10.34	0.01	7.68	8.44	0.76
45 - 49	10.34	10.36	0.02	7.10	8.13	1.03
Index of Dissimilarity	-----	-----	0.035	-----	-----	1.995

| | CASE III | | |
Age Group	Mixture	Convex Sum	$\|\Delta\|$
0 - 4	14.99	14.38	0.61
5 - 9	13.38	12.61	0.71
10 - 14	12.24	11.53	0.71
15 - 19	11.19	10.62	0.47
20 - 24	10.19	9.84	0.35
25 - 29	9.25	9.17	0.08
30 - 34	8.38	8.61	0.23
35 - 39	7.55	8.14	0.59
40 - 44	6.79	7.74	0.95
45 - 49	6.06	7.40	1.34
Index of Dissimilarity	-----	-----	3.02

4. REMARKS

The simulation results encompass a wide range of fertility and mortality conditions and hence the statements made here are fairly general. The results developed here have two aspects:

a) For primitive matrices, some interesting results regarding the dominant root and stable vector of a convex mixture have emerged. Mathematicians are invited to prove these results formally.

Table 3B Stable Age Distribution, Females (For $\alpha = 0.9$)

Age Group	CASE I Mixture	CASE I Convex Sum	$|\Delta|$	CASE II Mixture	CASE II Convex Sum	$|\Delta|$
0 - 4	9.09	9.09	----	15.87	15.67	0.20
5 - 9	9.29	9.29	----	14.16	13.97	0.19
10 - 14	9.50	9.50	----	12.64	12.48	0.16
15 - 19	9.70	9.70	----	11.27	11.16	0.11
20 - 24	9.91	9.91	----	10.05	10.00	0.05
25 - 29	10.12	10.12	----	8.97	8.99	0.02
30 - 34	10.33	10.33	----	8.00	8.08	0.08
35 - 39	10.54	10.54	----	7.13	7.27	0.14
40 - 44	10.71	10.71	----	6.33	6.54	0.21
45 - 49	10.82	10.83	0.01	5.58	5.87	0.29
Index of Dissimilarity	-----	-----	.005	-----	-----	0.725

Age Group	CASE III Mixture	CASE III Convex Sum	$|\Delta|$
0 - 4	18.89	18.69	0.20
5 - 9	15.57	15.33	0.24
10 - 14	13.41	13.20	0.21
15 - 19	11.55	11.39	0.16
20 - 24	9.88	9.79	0.09
25 - 29	8.40	8.39	0.01
30 - 34	7.11	7.19	0.08
35 - 39	5.99	6.17	0.18
40 - 44	5.02	5.29	0.27
45 - 49	4.55	4.55	0.37
Index of Dissimilarity	-----	-----	0.905

b) The results can be used to generate meaningful estimates of the birth/death parameters for certain population segments in a country (eg. Catholics in Canada), if estimates for the main population and all but the component under consideration are known. Several applications are under study.

REFERENCES

Coale, A.J. (1963): Estimates of various demographic measures through the quasi-stable age distribution, in *Emerging Techniques in Population Research*, New York: Milbank Memorial Fund, pp. 175-195.

_____(1972): *The Growth and Structure of Human Populations,* Princeton University Press, Princeton.

Coale, A.J., and P. Demeny (1966): *Regional Model Life Tables and Stable Populations,* Princeton University Press, Princeton.

Gopalaswamy, K (this Proceedings): Nonlinear age dependent population system with periodic vital rates.

Goel, N.S., S.C. Maitra, and E.W. Montroll (1971): *On the Volterra and Other Nonlinear Models of Interacting Populations,* Academic Press, New York.

Keyfitz, N. (1968): *Introduction to the Mathematics of Population,* Addison-Wesley, Reading (Mass.).

Lalu, N.M. (1980): The impact of migration on the stability of a population: a simulation study, *Modeling and Simulation* 11:1429-1433.

Lalu, N.M., and P. Krishnan (1979): Lotka distribution open to migration. Paper presented at the Conference of the Federation of Canadian Demographers, Montreal.

Lightbourne, R., S. Singh, and C. Green (1982): The world fertility survey: charting global childbearing, *Population Bulletin* 37:1-54.

McFarland, D.D. (1969): On the theory of stable population: a new and elementary proof of the theorems under weaker assumptions, *Demography* 6:301-322.

Rogers, A. (1975): *Introduction to Multiregional Mathematical Demography,* John Wiley and Sons, New York.

Statistics Canada (1978): *Vital Statistics of Canada 1976,* Statistics Canada, Ottawa.

A PROBLEM IN NONLINEAR

AGE DEPENDENT POPULATION DIFFUSION

Michel R. Langlais

1. INTRODUCTION

We are interested in a mathematical model of an age-dependent population
moving in a limited environment. We focus our attention on properties coming from
the nonlinearity in the diffusion mechanism designed to provide an anti-crowding
model; the birth-death process is chosen to be linear.

Let $u(t,a,x)$ denote the age-space structure of a single species population
moving in a bounded domain Ω in $\mathbf{R}^N$, $N = 1,2$ or 3; here $t > 0$ is time, a is
age: $0 < a < A$ where A is the maximum life expectancy of the species, and $x \in \Omega$
is the position. Integrating over all ages yields the space structure or total
population

$$P(t,x) = \int_0^A u(t,a,x)da \qquad (1)$$

We first assume that the flux of individuals of age a by spatial diffusion
is given by $-P \cdot \nabla u$: it lies in the direction of decreasing density as in the linear
theory but it is weighted by the total population (∇ stands for the gradient in
$\mathbf{R}^N$). Next we assume that the supply to Ω of individuals of age a is primarily
due to death and is given by $-\mu \cdot u$ where $\mu(t,a,x) \geq 0$ is the death-modulus.

Studying under those assumptions the rate of change of the class of age a
with the method described in Gurtin (1973) leads to the law of population balance of
individuals of age a;

$$u_t + u_a - \mathrm{div}(P \cdot \nabla u) + \mu u = 0 \qquad (2)$$

We furthermore assume that no individual leaves Ω through its boundary $\partial \Omega$

$$P \cdot \frac{\partial u}{\partial \eta} = 0 \qquad (\frac{\partial}{\partial \eta} = \text{normal derivative}) \qquad (3)$$

and that the birth-law takes the form

$$u(t,0,x) = \int_0^A \beta(t,a,x)u(t,a,x)da \qquad (4)$$

where β is the maternity function. Starting with an initial distribution

$$u(0,a,x) = u_0(a,x) \geq 0 \qquad (5)$$

our problem is to determine the evolution of u as a solution of (1)-(5).

To get a better understanding we perform a preliminary computation: assuming either A = +∞ or A finite and u(t,A,x) = 0 let us integrate (2)-(5) with respect to a from 0 to A. We obtain

$$P_t - div(P \cdot \nabla P) = \int_0^A (\beta-\mu)uda \qquad (6)$$

$$P(0,x) = P_0(x) = \int_0^A u_0(a,x)da \geq 0 \qquad (7)$$

$$P \frac{\partial P}{\partial \eta} = 0 \qquad (8)$$

Hence P is solution of a nonlinear and degenerate parabolic equation for which an extensive literature is available (see Alikakos and Rostamian, 1981; Aronson, 1970; Diaz-Diaz, 1979; Di Benedetto, 1981, and their references). We can expect P to be continuous but if P_0 has a compact support in Ω then P(t,) has a compact support in Ω for small t and P has discontinuous derivatives. This lack of smoothness in the diffusion coefficient of (2) shows that we must look for functions satisfying (1)-(5) in a weak sense.

2.   In the next sections u_0 satisfies at least the following assumptions:

$$u_0(a,x) \geq 0, \quad \int_{(0,A)\times\Omega} u_0^2(a,x)dadx < +\infty;$$

$$P_0(x) = \int_0^A u_0(a,x)da \quad \text{is continuous on} \quad \bar{\Omega};$$

$$\int_\Omega |\nabla P_0^2|^2 dx < +\infty.$$

3.   DEFINITION

Let T be a finite real number; a weak solution of (1)-(5) defined in (0,T) × (0,A) × Ω is any nonnegative and integrable u such that P in (1) is continuous on [0,T] × $\bar{\Omega}$,

$$\int_{(0,T)\times(0,A)\times\Omega} P|\nabla u|^2 dtdadx + \int_{(0,T)\times(0,A)\times\Omega} u^2 dtdadx < +\infty,$$

and satisfying for any smooth φ vanishing at t = T and about a = A the integral identity

$$\int_{(0,T)\times(0,A)\times\Omega} [-(\frac{\partial\phi}{\partial t} + \frac{\partial\phi}{\partial a})u + \mu u\phi + P\nabla u \cdot \nabla\phi]dtdadx =$$

$$= \int_{(0,A)\times\Omega} u_0(a,x) \cdot \phi(0,a,x)dadx + \int_{(0,T)\times\Omega} \int_0^A \beta uda \cdot \phi(t,0,x)dtdx.$$

This is derived from (1)-(5) upon integrating by parts (2)-(5). The integrals appearing in this relation make sense when μ and β satisfy the properties listed in Theorems 1 and 2 below.

4. RESULTS

To illustrate some features of our model we present two typical examples in which a solution exists and offers various qualitative properties depending on whether A is finite or not, on u_0 and on μ and β. More details are found in Langlais (manuscript).

4.1. We first assume $A = +\infty$ and we take μ and β to be independent of the variable a, nonnegative and smooth with respect to the variables t and x. We can rewrite equations (6)-(8) as:

$$P_t - \text{div}(P \cdot \nabla P) + (\mu - \beta)P = 0 \qquad (9)$$

$$P(0,x) = p_0(x), \qquad P \frac{\partial P}{\partial \eta} = 0 \qquad (10)$$

and we note that u does not explicitly appear in this set of equations. When age-dependence is neglected similar equations for P are derived in Gurney and Nisbet (1975) and in Gurtin and MacCamy (1977).

Therefore (9)-(10) can be solved separately and provides a solution labelled $\bar{P}$; substituting $P = \bar{P}$ in (2)-(5) (note that (4) becomes $u(t,0,x) = \beta\bar{P}$) and solving the resulting linear equations gives a solution $\bar{u}$. To conclude we must show that (1) holds, namely the solution $\bar{P}$ to (9)-(10) and the integral with respect to the variable a of $\bar{u}$ are identical; this last step is obtained upon checking that both of them are solutions of the same linear and possibly degenerate parabolic equations namely:

$$R_t - \text{div}(\bar{P} \cdot \nabla R) + \mu R = \beta\bar{P}, \qquad R(0,x) = p_0(x), \qquad \bar{P} \frac{\partial R}{\partial \eta} = 0.$$

This can be mathematically established and we have:

THEOREM 1 *Assume* $A = +\infty$, μ *and* β *are nonnegative, independent of the variable* a *and smooth with respect to* t *and* x. *Then there exists a unique solution of* (1)-(5) *and* P *satisfies* (9)-(10) *in a suitably weak sense.*

Now we take advantage of the equations satisfied by P to derive qualitative properties of u and P. First using comparison techniques and results of Diaz-Diaz (1979) we obtain

COROLLARY 1

    i) *If* $P_0(x)$ *has a compact support in* Ω *then for small* t, $P(t,\cdot)$ *has a*
    compact support in Ω.

    ii) *If* $0 < m_0 \leq P_0(x)$ *then* $0 < m_1 \leq P(t,x)$ *in* $[0,T] \times \bar{\Omega}$.

    The case i) means that a population which is initially confined in a compact subregion of Ω spreads out at a finite speed (note that if for some t_0 $P(t_0,x) = 0$ then $u(t_0,a,x) = 0$ for any $a > 0$). In the case ii) the regularity of P follows closely that of P_0.

    The large time behaviour can be investigated by using comparison techniques and results of Alikakos and Rostamian (1981).

COROLLARY 2

    i) *If* $\mu(t,x) - \beta(t,x) \geq c_0 > 0$ *then* $\lim\limits_{t \to +\infty} P(t,x) = 0$.

    ii) *If* $\mu(t,x) = \beta(t,x)$ *then* $\lim\limits_{t \to +\infty} P(t,x) = \dfrac{1}{\text{mes } \Omega} \int_\Omega P_0(x)dx$.

    iii) *If* $\mu(t,x) - \beta(t,x) \leq c_0 < 0$ *then* $\lim\limits_{t \to \infty} P(t,x) = +\infty$.

    In ii) and iii) the whole domain Ω is ultimately populated while in i) the population dies out and it may happen for suitable u_0 that only a closed subdomain of Ω is ultimately populated: see Gurtin and MacCamy (1977).

4.2. In this second example we take A finite and we assume that μ and β depend only on the variable a. A finite means that $\lim\limits_{a \to A} u(t,a,x) = 0$ and this is obtained by requiring $\int_0^A \mu(a)da = +\infty$, namely μ is rapidly increasing at a = A.

    Clearly P (resp. u) explicitly appears in the equation for u (resp. P) and our problem is much more complicated than in the previous section. To handle it we treat together the system of nonlinear equations (2)-(5) and (6)-(8), by a method involving a delay, and obtain a solution $(\bar{u},\bar{P})$. As above to show that (1) holds we check that $\bar{P}$ and $\int_0^A \bar{u}(t,a,x)da$ satisfy the same equation. Under suitable assumptions we are able to prove the existence of at least a solution.

THEOREM 2 *Assume* A *finite and that* β *is nonnegative bounded and depends only on the variable* a. *Let* μ *be nonnegative, depend only on* a, *satisfy* $\int_0^A \mu(a)da = +\infty$ *but* $\mu \in C^1([0,A))$. *If for some* $p > \dfrac{N}{2} + 1$ *we have* $\mu^p \mu_0 \in L^2((0,A) \times \Omega)$ *and*

$$\mu^2 - p \frac{du}{da} \geq \lambda_1 \mu + \lambda_2, \quad \lambda_1, \lambda_2 \quad \textit{real constants} \tag{11}$$

then there exists at least a solution to (1)-(5)

The main difficulty here lies in the unboundedness of μ; but under condition (11) it is possible to show that the right hand side of (6) belongs to L^p and the result proved in Di Benedetto (1981) ensures that P is continuous. Condition (11) is fulfilled by various functions satisfying the divergence condition $\int_0^A \mu(a)da = +\infty$ and has already been used in Langlais (to appear).

In this second example the qualitative properties of the solution may differ from those obtained in the first example. Let $[a_1,a_2]$ be the support of β and assume that the support of u_0 lies in $[a_2,A] \times \bar{\Omega}$; then one can check that $t = A$ is an extinction time: $u(t,a,x) = 0$ for any $t > A$. A typical example is $\beta = 0$ for which $t = A$ is an extinction time whatever u_0 can be. Therefore the analogue of ii) in Corollary 1 is no more true. To avoid an extinction time some relations must exist between u_0, μ and β.

COROLLARY 3 *Assume that there exist* λ *a real number and* α *in* $C^1([0,A])$ *such that* $0 \leq \alpha(a) \leq 1$, $\alpha(0) = 1$, $\alpha(A) = 0$ *and*

$$0 \leq \beta(a) - [\mu(a)+\lambda]\alpha(a) + \frac{d\alpha}{da} \tag{12}$$

$$\int_0^A \alpha(a)u_0(a,x)da \geq m_0 > 0 \tag{13}$$

Then the weak solution obtained in Theorem 2 satisfies $P(t,x) \geq m_1 > 0$.

We may notice that (13) implies $p_0(x) \geq m_1 > 0$ and that (12) is fulfilled if β is not identically 0. Finally (12) and (13) are simultaneously realized where the initial population is young enough. u_0 and β must be large in the same interval of ages: the initial population must be able to produce enough offsprings.

On the other hand the anti-crowding effect still holds: namely the initial population still spreads out at a finite speed (Corollary 1, i)).

Due to the boundedness of β and the unboundedness of μ the assumptions ii) and iii) in Corollary 2 are irrelevant.

REFERENCES

Alikakos, N.D., and R. Rostamian (1981): Large time behaviour of Neumann boundary value problem, *Indiana Univ. Math. J.* 30:378-412.

Aronson, D.G. (1970): Regularity properties of flow through porous media: a counterexample, *SIAM J. Appl. Math.* 19:299-307.

Diaz-Diaz, I. (1979): Solutions with compact support for some degenerate parabolic problem, *J. Nonlinear Analysis* 3:831-847.

Di Benedetto, E. (1981): Continuity of weak solutions to a general porous media equations, *M.R.C. Tech. Rep.* #2182, Univ. of Madison, Wisconsin.

Gurney, W.S.C., and R.M. Nisbet (1975): The regulation of inhomogeneous population, *J. Theor. Biol.* 52:441-452.

Gurtin, M.E. (1973): A system of equations for age dependent population diffusion, *J. Theor. Biol.* 40:389-392.

Gurtin, M.E., and R.C. MacCamy (1977): On the diffusion of biological population, *Math. Biosc.* 38:35-49.

Langlais, M. (to appear): On some linear age dependent population model, *Quart. Appl. Math.*

_____(manuscript): A problem for age dependent population with nonlinear diffusion.

PART IV:

COMPETITIVE SYSTEMS

COEXISTENCE OF COMPETITORS IN A STOCHASTIC
ENVIRONMENT: THE STORAGE EFFECT

Peter L. Chesson

The pre-reproductive, or juvenile stages of an organism, are often the most precarious. These are the stages in the life cycle that are most subject to the vagaries of the environment, while potentially reproductive adults tend to have both higher average survivorship and more predictable survivorship. It follows that recruitment to the adult population is often much more variable than adult survivorship. This difference in variability is increased in some species by a tendency of adults to reduce reproductive effort when conditions are poor thus increasing their own survivorship or enhancing later reproductive success (Murdoch 1966, Goodman 1974, Nichols et al. 1976, Tyler and Dunn 1976).

High variability in recruitment, together with both less variable and high adult survivorship is a characteristic feature of many populations of commercially exploited fishes (Gulland 1982). The low adult death rates mean that the adult population can be maintained for a long time by a single large recruitment event. Essentially, the adult population stores up recruitment events subject to a yearly discount equal to the adult death rate. As a consequence, the adult stocks are not nearly so variable as the sporadic recruitment might suggest. The interaction between variable recruitment and high, less variable, adult survivorship that leads to these population characteristics will be called the *storage effect*. The storage effect is present whenever recruitment is variable and generations overlap, but the effect is strongest for long-lived species.

Variable recruitment and long-lived adults are also features of perennial plant populations (Harper 1977, Grubb 1977, Hubbell 1980) but the consequences of the storage effect for these populations is not nearly so well documented.

For a guild of competing species, the storage effect can have quite surprising consequences: it can be the mechanism of coexistence. Chesson and Warner (1981) investigated the storage effect in the lottery model of competition for space among reef fishes. Only in the presence of both variable recruitment and overlapping generations was coexistence possible; that is, the storage effect was found to be essential for coexistence. Is this result merely a special feature of the lottery model or can it apply more generally? To answer this question I shall discuss a general mathematical model incorporating the storage effect. I shall then go on to consider a variety of examples that may be useful in applications, and which illustrate the generality of situations where the storage effect may promote coexistence of competing species. A companion paper (Warner and Chesson, manuscript) discusses the biological implications of the storage effect, and suggests procedures for testing the hypothesis that storage is indeed the mechanism of coexistence in field situations.

1. THE GENERAL MODEL

To model the storage effect, let $X_i(t)$ be the population density of adults of species i at time t; δ_i is the adult death rate and $R_i(t)$ is the per capita recruitment rate during $(t,t+1)$. It follows that

$$X_i(t+1) = (1-\delta_i)X_i(t) + R_i(t)X_i(t). \tag{1}$$

In this equation $R_i(t)$ will generally depend on $X_i(t)$, the other species present in the system, and the state of the environment, $E(t)$, during $(t,t+1)$. Thus we have

$$R_i(t) = f_i(E(t),X_1(t),\ldots,X_k(t)), \tag{2}$$

where f_i is some function. For simplicity the death rate δ_i is assumed to be constant, but, as can be seen from Chesson and Warner (1981) and Chesson (1982), δ_i may vary with the environment without substantially altering the conclusions below. Moreover the death rates can be made to depend on population densities to a limited degree. However, in general we shall think of the adult death rates as small and relatively constant while the recruitment rates are highly variable.

To analyze the model we define

$$\rho_i(t) = R_i(t)/\delta_i. \tag{3}$$

Species i increases when $\rho_i(t) > 1$ and decreases when $\rho_i(t) < 1$. Persistence of species i depends on its mean instantaneous growth rate at low density, Δ_i, which is given by the formulae

$$\Delta_i = E \log X_i(t+1)/X_i(t)$$
$$= E \log \{1 + \delta_i[\rho_i(t)-1]\}, \tag{4}$$

evaluated for $X_i(t) = 0$. If $\Delta_i > 0$ then species i persists in the sense of invasibility (Turelli 1978), i.e., species i tends to increase from low values. Invasibility often implies persistence in the more satisfactory sense of stochastic boundedness or "s.b. persistence" (Chesson 1978, 1982). S.b. persistence means that $X_i(t)$ is stochastically larger than some positive random variable U, uniformly in t, in symbols

$$P(X_i(t) > x) \geq P(U > x).$$

To determine invasibility by calculating Δ_i it is usually assumed that $(E(t), X_1(t), \ldots, X_k(t))$ approaches some stationary process when $X_i(t) = 0$; and the expected value is taken for the stationary distribution, or more commonly, for some approximating distribution. In this study the stationary distribution is not always available but we do find stationary lower bounds to $\rho_i(t)$ which allows us to deduce conservative conditions for invasibility. If these conservative conditions are satisfied, and the lower bound to $\rho_i(t)$ is an independent and identically distributed (i.i.d.) function of $E(t)$, then Theorem 5.1 of Chesson (1982) can be modified to prove s.b. persistence for the regular examples given below. However, in general, we shall be content to show that a species satisfies the invasibility criterion.

To determine invasibility, we note from Chesson (1982) that

$$\Delta_i \geq \delta_i E[\log \rho_i | \rho_i > 1] P(\rho_i > 1) + \log(1 - \delta_i) P(\rho_i < 1). \qquad (5)$$

It follows that Δ_i will be positive whenever

$$E[\log \rho_i | \rho_i > 1] \geq c(\delta_i) P(\rho_i < 1)/P(\rho_i > 1) \qquad (6)$$

where

$$c(\delta_i) = -\log(1 - \delta_i)/\delta_i \qquad (7)$$

and $c(\delta_i)$ is close to 1 for small δ_i.

The expression (6) provides a highly conservative yet very useful criterion for persistence in the sense of invasibility. Note that the criterion does not depend on the magnitude of ρ_i, the ratio of the recruitment rate and death rate, during periods of population decrease. Referring to periods of population increase and decrease respectively as favorable and unfavorable, we see that the criterion (6) depends only on the mean magnitude of $\log \rho_i$ during favorable periods and on the frequency of these periods relative to unfavorable periods. Thus a species can persist provided only that some of its favorable periods are sufficiently favorable (as measured by $\log \rho_i$) relative to their frequency of occurrence. This is the storage effect in operation: large recruitment events are stored in the adult population reducing the significance of the actual magnitude of $\log \rho_i$ during unfavorable periods. A general tendency for population increase can be maintained by favorable periods alone.

The magnitude of $\log \rho_i$ during unfavorable periods becomes important when criterion (6) is not satisfied. If the unfavorable periods are not greatly unfavorable then persistence can occur with substantially poorer favorable periods than suggested by (6). Precise relationships between mean favorability, variance in favorability and adult death rate, can be found from an obvious reinterpretation of the figures in Chesson and Warner (1981).

The fact that a species may persist by having only occasional good recruitment periods has important consequences for a set of interacting species. If the interactions are negative, so that not all species can have favorable periods simultaneously, coexistence may nevertheless occur. Each species simply needs to experience favorable periods that are sufficiently favorable relative to their frequency of occurrence. Such periods may be brought on by fluctuations in the environment, or by fluctuations, which may or may not be environmentally induced, in other species.

We now illustrate these ideas with a variety of specific examples. Additional examples can be found in Ellner (1983) and Shmida and Ellner (1983).

(i) The multispecies lottery model.

This first example shows that the storage effect may promote coexistence not just in a two-species system, as shown by Chesson and Warner (1981), but in systems with many species.

In the lottery model of Chesson and Warner each adult individual holds a unit of space, called a "home" which may be a defended territory (e.g., some reef fishes) or simply room to grow (e.g., trees or bushes). The total number of homesites available is assumed fixed and sites become available to new recruits only by death of adults. Thus the amount of space becoming available during $(t,t+1)$ is $\Sigma \delta_i X_i(t)$. The $\beta_i(t)X_i(t)$ juveniles of species i compete for this space with juveniles of all species. The outcome is assumed to be a random or biased random division of space among the individuals of the k species. In the slightly simpler completely random case we have

$$R_i(t)X_i(t) = \sum_{j=1}^{k} \delta_j X_j(t) \cdot \frac{\beta_i(t)X_i(t)}{\sum_{j=1}^{k} \beta_j(t)X_j(t)} \tag{8}$$

The parameters $\beta_i(t)$ representing birth and survivorship to the age of recruitment, are assumed functions of the environment $E(t)$. The process $E(t)$ is assumed to be stationary in all examples, and for proofs of stochastic boundedness we make the stronger assumption that it is an i.i.d. process. Formula (8) involves the implicit assumption that the total number of juveniles exceeds the available space, and hence that all space is filled at each recruitment period. Measuring density as the proportion of space that a species occupies, it follows that $\Sigma_{i=1}^{k} X_i(t) = 1$.

For species i at zero density, $\rho_i(t)$ $(= R_i(t)/\delta_i)$ is given by

$$\rho_i(t) = \frac{\sum_{j \neq i} \delta_j X_j(t)}{\delta_i} \cdot \frac{\beta_i(t)}{\sum_{j \neq i} \beta_j(t) X_j(t)} \tag{9}$$

$$\geq \frac{\min_{j \neq i} \delta_j}{\delta_i} \cdot \frac{\beta_i(t)}{\max_{j \neq i} \beta_j(t)} \tag{10}$$

The general result (6) implies that species i will persist if $\rho_i(t)$ takes on values that are sufficiently large relative to their frequency of occurrence. The inequality (10) thus shows that the species will coexist if each species has periods when its birth rate, $\beta_i(t)$, is sufficiently superior to the birth rates of the other species. Thus coexistence results from sufficient variability in the birth rates of each species relative to the others. Moreover since (10) is independent of the population densities it is an i.i.d. process when $(\beta_1(t), \ldots, \beta_k(t))$, $t = 0, 1, \ldots$ is an i.i.d. sequence of random vectors. It follows that each species will be s.b. persistent when there is sufficient variability in birth rates.

Although inequality (10) helps show that coexistence will occur with sufficient variability it does not give a very useful indication of the amount of variation necessary. It suggests that the necessary variation increases quite sharply as the number of species increases, however the equality (9), which gives a much better indication, suggests a substantially milder increase in the necessary variability. For $\rho_i(t)$ to be large in (9), $\beta_i(t)$ has to be large relative to, not the maximum birth rate for the other species, but to a weighted average of their birth rates, the weights being the species' densities.

Coexistence results from a sufficient storage effect in the multispecies lottery model; the storage effect is also necessary for coexistence. When generations are nonoverlapping, equation (6) of Chesson and Warner (1981) shows that the species with the largest value of $E \log \beta_i(t)$ eventually dominates the system. In a constant environment, the species with the largest value of β_i/δ_i, is the only one that persists.

(ii) The lottery model with vacant space.

The application of the previous model may be restricted by the assumption that the total number of juveniles surviving to the recruitment stage always exceeds the available space. To remove this assumption for the two-species case we let $Y(t) = 1 - X_1(t) - X_2(t)$, the amount of unoccupied space. The recruitment term, $R_i(t)X_i(t)$, of the lottery model is modified to

$$\min \left\{ \beta_i(t) X_i(t), \; [Y(t) + \delta_1 X_1(t) + \delta_2 X_2(t)] \frac{\beta_i(t) X_i(t)}{\beta_1(t) X_1(t) + \beta_2(t) X_2(t)} \right\}. \tag{11}$$

In this new model, if there are more juveniles than the space can hold, allocation is at random, otherwise the density of new recruits of species i is simply the density of juveniles.

For $X_i(t) = 0$ we have

$$\rho_i(t) = \min\left\{1, \frac{Y(t) + \delta_j X_j(t)}{\beta_j(t) X_j(t)}\right\} \frac{\beta_i(t)}{\delta_i} \tag{12}$$

$$\geq \min\left\{1, \frac{\delta_j}{\beta_j(t)}\right\} \frac{\beta_i(t)}{\delta_i} . \tag{13}$$

From (13) we see that coexistence will occur if each species has periods when the ratio of its birth rate to its death rate is large both absolutely and relative to the ratio for the other species. This is not much different from the lottery model because the lottery model really only makes sense when $\beta_i(t) \geq \delta_i$. Consequently, when $\beta_i(t)/\delta_i$ is large relative to $\beta_j(t)/\delta_j$, it must also be large absolutely. However the above criterion for coexistence is conservative when there is empty space. Indeed (12) suggests that coexistence will occur under more general conditions, especially if the amount of empty space can be large.

Although we make no attempt to explore more general conditions for coexistence, one thing is clear: the storage effect is essential. To see this consider first the case of a constant environment. For species i to persist $\beta_i/\delta_i > 1$. Thus if both species persist, all space is filled in finite time, and as in the lottery model, the species with the smaller value of β_i/δ_i becomes extinct. When generations are not overlapping

$$X_1(t+1)/X_2(t+1) = (\beta_1(t)/\beta_2(t))X_1(t)/X_2(t). \tag{14}$$

This equation also holds for the lottery model and it follows that the species with the smaller value of $E \log \beta_i(t)$ becomes extinct. Thus the interaction between variable recruitment and overlapping generations (the storage effect) is essential to coexistence.

(iii) The lottery model with other kinds of competition.

Competition need not be restricted to a single time in the life of an organism. For example in addition to lottery competition for space at settlement, competition among adults may affect reproductive success. Provided this competition among adults does not affect adult death rates, it still fits easily into the general model above. Another interesting possibility involves competition among juveniles before they reach the settlement stage. We model just this latter possibility because the former situation, involving adult competition, can be modeled as a minor modification with very similar conclusions.

To add presettlement competition we do not need to change equation (8) of the lottery model; we just need to make the $\beta_i(t)$ density dependent as follows, for the two-species case.

$$\beta_i(t) = B_i(t)f_i(B_i(t)X_i(t),B_j(t)X_j(t)). \tag{15}$$

Here $B_i(t)$ is the birth rate and so $B_i(t)X_i(t)$ is the number of juveniles before presettlement competition, while the function f_i represents the fractional reduction in $B_i(t)$ due to this competition. The result, $\beta_i(t)$, is the potential recruitment rate because it is the number that would recruit, per adult, if the space were available.

The interpretation of the model is aided by the following definition of generalized competition coefficients (c.f., Abrams 1975).

$$\alpha_{ij}(\ell_i,\ell_j) = -\frac{\partial}{\partial \ell_j} \log f_i(\ell_i,\ell_j), \quad i \neq j, \tag{16}$$

and α_{ii} is defined analogously. These competition coefficients represent the relative depression of species i juveniles by each additional juvenile of species j and species i respectively. Making the assumptions $f_i(0,0) = 1$, $X_i(t) = 0$, we now obtain

$$\rho_i(t) = \frac{\delta_j B_i(t)}{\delta_i B_j(t)} e^{\int_0^{B_j(t)} \alpha_{jj}(\ell,0)-\alpha_{ij}(0,\ell)d\ell}. \tag{17}$$

Inspection of (17) shows that coexistence must always occur if birth rates vary sufficiently in the following sense. Let p and L be fixed positive constants and

$$P(B_i(t) > M, B_j(t) < L) \geq p, \tag{18}$$

$i \neq j$; $i, j = 1,2$. By increasing M, variability is increased in the sense of an increase in the difference between the small and large values that the birth rates take on. Expression (18) also insists that some periods of low birth rate for each species coincide with some periods of high birth rate for the other species. With sufficiently high variability (high M) coexistence occurs in this model of general presettlement competition with lottery competition at settlement.

Although we have used a very general definition of high variability it does not cover all cases in which variability may lead to coexistence, especially for particular forms of the model where quite different definitions of high variability can be adequate. The present definition has the property that $EB_i(t)$ will be large when variability is large, however $E \log B_i(t)$ need not be large. Chesson (1982) discusses the relative merits of different ways of defining high variability.

To see what happens in a constant environment we use the general definition of coexistence in terms of an attractor block given by Armstrong and McGehee (1980). In a constant environment we say that the species coexist if there is a positive number ε such that both species densities rise above ε and remain there, given any positive initial densities. With this definition, a necessary and sufficient condition for coexistence is $\rho_i > 1$, $i = 1,2$. Inspection of (17) is sufficient to give the general features of coexistence in a constant environment. We consider four cases.

(A) $\alpha_{ij} < \alpha_{jj}$, $i \neq j$, $i, j = 1,2$ (intraspecific competition exceeds interspecific competition). Coexistence occurs if $\delta_2 B_1 / B_2 \delta_1 = 1$, but also in many other situations. In general, coexistence is favored by any of the following: $\delta_2 B_1 / B_2 \delta_1$ is near 1; the B_i are large in absolute value; the differences $\alpha_{jj} - \alpha_{ij}$ are large.

(B) $\alpha_{ij} > \alpha_{jj}$, $i \neq j$, $i, j = 12$, (interspecific competition exceeds intraspecific competition). Coexistence never occurs for this case.

(C) $\alpha_{ij} < \alpha_{jj}$, $\alpha_{ji} > \alpha_{ii}$ (species i is a superior competitor). Coexistence can occur if the presettlement competitive advantage of species i is balanced by a large value of $\delta_i B_j / \delta_j B_i$ giving species j an advantage with respect to birth and adult survivorship rates. However it is possible for this ratio to be too large so that species i does not persist.

(D) $\alpha_{ij} \equiv \alpha_{jj}$, $i \neq j$, $i, j = 1,2$ (interpsecific competition equals intraspecific competition). The case is equivalent to the lottery model and so coexistence cannot occur in a constant environment.

Coexistence in a constant environment occurs for subcases of only two of the above cases, however coexistence will always occur in a sufficiently variable environment, as discussed above. Thus a stochastic environment broadens the range of situations in which coexistence can occur.

Although not obvious from our analysis so far, it is not just the stochastic environment that is important but its combination with overlapping generations, i.e., the storage effect. Without overlapping generations coexistence remains impossible in case (D), because this case is equivalent to the lottery model, while in case (B) $\Delta_1 + \Delta_2 = E \log \rho_1 + E \log \rho_2 < 0$, so that it is impossible for both species to satisfy the invasibility criterion. In addition, although it is possible that a stochastic environment could lead to coexistence without overlapping generations for some instances of (A) and (C), these instances necessarily belong to a subset of cases where the storage effect yields coexistence. This follows from the general result for the lottery model (Chesson and Warner 1981), which applies here also, that smaller death rates favor coexistence.

(iv) The storage effect in a Lotka-Volterra model.

In addition to the storage effect, the above models have all involved lottery competition. To see that lottery competition is not essential to coexistence by the storage effect we now consider a model in which there is no lottery competition. Instead, competition of the Lotka-Volterra kind occurs among juveniles and among adults.

Let $L_i(t)$ be the number of juveniles born to the $X_i(t)$ adults of species i in $(t,t+1)$, then the recruitment term takes the form

$$R_i(t)X_i(t) = \frac{\Theta_i(t)L_i(t)}{1+\alpha_{ii}(t)L_i(t)+\alpha_{ij}(t)L_j(t)} \cdot \tag{19}$$

This equation is the hyperbolic form of Lotka-Volterra competition in discrete time (Leslie 1958). It represents the reduction in juvenile survivorship due to competition among juveniles. The parameter $\Theta_i(t)$ is per capita juvenile survivorship in the absence of competition; $\alpha_{ii}(t)$ and $\alpha_{ij}(t)$ are the competition coefficients. Although these parameters are possibly time dependent they are assumed to be stationary stochastic processes not involving adult or juvenile densities.

Birth rates of juveniles are affected by competition among adults and the model is completed by the equation

$$L_i(t) = \frac{B_i(t)X_i(t)}{1+\gamma_{ii}(t)X_i(t)+\gamma_{ij}(t)X_j(t)} \tag{20}$$

with the same assumptions as applied to (19) on the time varying parameters.

For species i at 0 density we obtain

$$\rho_i(t) = \frac{\Theta_i(t)B_i(t)/\delta_i}{[1+\gamma_{ij}(t)X_j(t)]\left[1+\dfrac{\alpha_{ij}(t)B_j(t)X_j(t)}{1+\gamma_{jj}(t)X_j(t)}\right]} \cdot \tag{21}$$

The behavior of $\rho_i(t)$ depends on the behavior of $X_j(t)$ when $X_i(t) = 0$. Assuming $X_i(t) = 0$, (19) and (20) imply

$$X_j(t+1) \leq (1-\delta_j)X_j(t) + \Theta_j(t)/\alpha_{jj}(t). \tag{22}$$

It follows that $X_j(t)$ is bounded by the random variable

$$\xi_j(t) = (1-\delta_j)^t X_j(0) + \sum_{s=0}^{t-1} (1-\delta_j)^s \Theta_j(t-s)/\alpha_{jj}(t-s). \tag{23}$$

The significant feature of (23) is that the birth rates are not functionally involved although this does not deny statistical dependence. Substituting in (21) we conclude

$$\rho_i(t) \geq \frac{\Theta_i(t)B_i(t)/\delta_i}{[1+\gamma_{ij}(t)\xi_j(t)]\left[1+\dfrac{\alpha_{ij}(t)B_j(t)\xi_j(t)}{1+\gamma_{jj}(t)\xi_j(t)}\right]} . \tag{24}$$

It follows that if $B_1(t)$ and $B_2(t)$ are large at different times, both $\rho_1(t)$ and $\rho_2(t)$ can take large values. Thus coexistence can occur by the storage effect.

In a constant environment with species j alone, $X_j(t)$ converges to the equilibrium value

$$(\Theta_j B_j/\delta_j - 1)/(\gamma_{jj}+\alpha_{jj}B_j) . \tag{25}$$

To see when species i can invade, this equilibrium value is substituted in expression (21) for ρ_i (values of $\rho_i > 1$ indicate invasibility). Inspection of the resulting formula reveals that coexistence can occur if interspecific competition coefficients α_{ij} and γ_{ij} are small while the B_i are large. On the other hand, if the α_{ij} are large, the species cannot coexist for any values of the B_i. While no attempt has been made to delineate precise regions of coexistence, it is clear that the storage effect does indeed lead to a broader range of situations of coexistence: with the storage effect coexistence can occur for any values of the interspecific competition coefficients.

2. SUMMARY

It is not uncommon for the rate of recruitment to an adult population to be much more variable than adult survivorship. If adult survivorship is high, while recruitment is quite variable, the average growth rate of the population will be dependent mostly on the strength of good recruitments and little dependent on the strength of poor recruitments. It is possible that the population is maintained by infrequent strong recruitment events. The interaction between variable recruitment and low adult death rates, that leads to these population properties, is called the storage effect. Because the storage effect can permit a species to have a positive average growth rate if its good recruitments are sufficiently beneficial, independently of the paucity of recruitment at other times, it can promote the coexistence of species that compete quite strongly. For coexistence, each species simply needs periods when its recruitment rate is sufficiently high relative to the frequency of these periods.

Several different competition models are given which illustrate the variety of situations in which the storage effect promotes coexistence.

Acknowledgements

The development of the ideas in this paper has benefited from discussions with a great many people, but especially I wish to thank Robert Warner.

REFERENCES

Abrams, P. (1975): Limiting similarity and the form of the competition coefficient, *Theoret. Pop. Biol.* 8:356-375.

Armstrong, R.A. and R. McGehee (1980): Competitive exclusion, *Am. Nat.* 115:151-170.

Chesson, P.L. (1978): Predator-prey theory and variability, *Annu. Rev. Ecol. Syst.* 9:323-347.

_____ (1982): The stabilizing effect of a random environment, *J. Math. Biol.* 15: 1-36.

Chesson, P.L. and R.R. Warner (1981): Environmental variability promotes coexistence in lottery competitive systems, *Am. Nat.* 117:923-943.

Ellner, S.P. (1983); Stationary distributions for some stochastic difference equation models, *J. Math. Biol.* To appear.

Goodman, D. (1974): Natural selection and a cost ceiling on reproductive effort, *Am. Nat.* 108:247-268.

Grubb, P.J. (1977): The maintenance of species richness in plant communities: the importance of the regeneration niche, *Biol. Rev.* 52:107-145.

Gulland, J.A. (1982): Why do fish numbers vary? *J. Theor. Biol.* 97:69-75.

Harper, J.L. (1977): *Population Biology of Plants*, Academic Press, London, 892 pp.

Hubbell, S.P. (1980): Seed predation and the coexistence of tree species in tropical forests, *Oikos* 35:214-229.

Leslie, P.H. (1958): A stochastic model for studying the properties of certain biological systems by numerical methods, *Biometrika* 45:16-31.

Murdoch, W.W. (1966): Population stability and life history phenomena, *Am. Nat.* 100:5-11.

Nichols, J.D., W. Conley, B. Batt, and A.R. Tipton (1976): Temporally dynamic reproductive strategies and the concept of r- and K- selection, *Am. Nat.* 110:995-1005.

Schmida, A. and S. Ellner (1983): Coexistence of trophically equivalent plant species, *Vegetatio*, in press.

Turelli, M. (1978): Does environmental variability limit niche overlap? *Proc. Natl. Acad. Sci. USA* 75:5085-5089.

Tyler, A.V. and R.S. Dunn (1976): Ration, growth, and measures of organ condition in relation to meal frequency in winter flounder, *Pseudo-pleuronectes americanus*, with hypotheses regarding population homeostasis, *J. Fish. Res. Board Canada* 33:63-75.

MODELS OF COMPETITION FOR A SINGLE RESOURCE

Paul Waltman*

1. INTRODUCTION

This lecture presents a survey of recent work of the author and several colleagues on competition models, centering around the chemostat. The chemostat is a device for culturing microorganisms, and since the common mathematical assumptions made in ecological modeling are best met by simple organisms, one anticipates good agreement between theory and experiment here. In its simplest form the chemostat is a piece of laboratory apparatus -- a constant concentration of nutrient liquid (with all necessary nutrients except one in abundance) is pumped into a well mixed culture vessel at a fixed rate, the volume of the culture vessel being kept constant by allowing the contents of overflow into a collection vessel. The overflow is a mixture of nutrient and the microorganisms growing in the vessel. This apparatus produces a continuous supply of microorganisms and hence the term continuous culture is sometimes used to contrast the process with the more familiar batch culture technique. The importance of the chemostat in ecology is well documented in several survey articles, Frederickson and Stepanopoulos (1981), Jannash and Mateles (1974), Veldcamp (1977), Waltman et al. (1980), to which the reader is referred for background.

2. THE SIMPLE CHEMOSTAT

The equations governing the population interactions in the chemostat are fairly well accepted and go back as far as Monod (1942). For two competitors, they take the form

$$S' = (S^{(0)}-S)D - \frac{m_1}{Y_1}\frac{xS}{a_1+S} - \frac{m_2}{Y_2}\frac{yS}{a_2+S}$$

$$x' = x\left(\frac{m_1 S}{a_1+S} - D\right)$$

$$y' = y\left(\frac{m_2 S}{a_2+S} - D\right)$$

$$S(0) = S_0 \geq 0, \quad x(0) = x_0 > 0, \quad y(0) = y_0 > 0.$$

* Research supported by NSF Grant MCS-8120380.

The quantity $S^{(0)}$ is the input concentration of the nutrient (S is used since the equations are those for enzyme kinetics and the nutrient is thought of as a substrate) and D is the dilution rate -- both of these quantitites are under the control of the experimenter. γ_i is a yield constant, m_i is the maximal growth rate, and a_i is called the Michaelis-Menten (or half saturation) constant. All of these parameters can be viewed as properties of the organism and are readily measured in the laboratory; in fact, this is an important feature in joining the mathematical model and the biology -- all the parameters in the model can be determined by growing each organism alone on the nutrient.

Mathematically it is best to work with nondimensional variables. The equations can be rescaled to the form

$$S' = 1 - S - \frac{m_1 xS}{a_1+S} - \frac{m_2 yS}{a_2+S}$$

$$x' = x\left(\frac{m_1 S}{a_1+S} - 1\right) \qquad (2.1)$$

$$y' = y\left(\frac{m_2 S}{a_2+S} - 1\right)$$

where, of course, m_i and a_i have changed their meaning. The controls, S^0 and D, and the yield constants, γ_1 and γ_2, have been "scaled out." If one observes further that the quantity $\Sigma = S + x + y$ satisfies the differential equation

$$\Sigma' = 1 - \Sigma$$

then it follows that $\lim_{t\to\infty} \Sigma(t) = 1$. In the language of dynamical systems, the omega limit set of any trajectory of (2.1) lies in the plane $S + x + y = 1$. This has the effect of reducing the problem to a consideration of the planar system

$$x' = x\left(\frac{m_1(1-x-y)}{1+a_1-x-y} - 1\right)$$

$$y' = y\left(\frac{m_2(1-x-y)}{1+a_2-x-y} - 1\right). \qquad (2.2)$$

THEOREM 2.1 *Let* $\lambda_i = a_i/m_i - 1$. *If* $m_1 \leq 1$ *or* $m_1 > 1$ *and* $\lambda_1 \geq 1$, *then* $\lim_{t\to\infty} x(t) = 0$. *If* $m_2 \leq 1$ *or if* $m_2 > 1$ *and* $\lambda_2 \geq 1$, $\lim_{t\to\infty} y(t) = 0$.

This theorem may be proved by simple differential inequalities. The result is competition independent and may be interpreted as a case of "inadequate competitors" -- the competitors cannot survive individually in this chemostat.

THEOREM 2.2 *Suppose* $m_i > 1$, $i = 1,2$ *and* $0 < \lambda_1 < \lambda_2 < 1$. *Then*

$$\lim_{t \to \infty} x(t) = 1 - \lambda_1, \quad \lim_{t \to \infty} y(t) = 0.$$

This theorem shows that the competitive exclusion principle holds -- at most one competitor survives the competition. Proofs have been given in Hsu (1978), Hsu et al. (1977), Powell (1958), Stewart and Levin (1973), but a rigorous, simple geometric argument can be made by establishing that the phase plane diagram shown in Figure 1 holds, and using the fact that if the omega limit set contains a (locally) asymptotically stable critical point it must be that point. It is also clear from the figure that the argument can be made to apply to more general differential equations -- the isoclines need not be straight lines. It is necessary only that they intersect the axes, not intersect each other, and that the two critical points on the boundary have the appropriate stability properties.

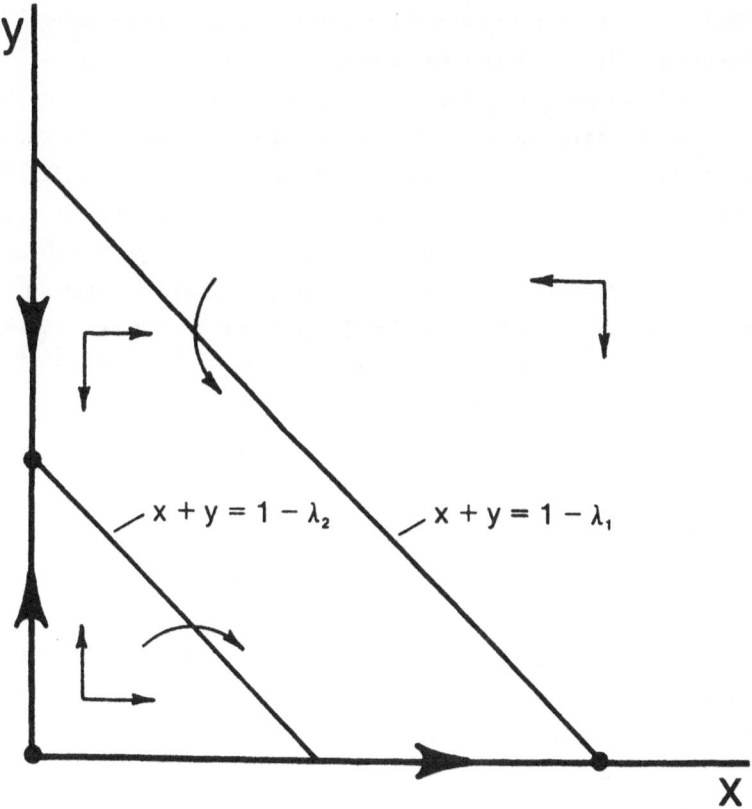

Figure 1   Phase plane for (2.2).

A similar theorem -- competitive exclusion -- holds with n competitors, Hsu (1978), Hsu et al. (1978a), provided

$$0 < \lambda_1 < \lambda_2 \leq \lambda_3 \leq \quad \leq \lambda_n \qquad (H)$$

while if the λ's are equal the competitors coexist.

The immediate question is how well do the theorems fit the actual chemostat? Experiments were described in Hansen and Hubbell (1980) which tested the validity of the above "λ-criterion." If competitive exclusion does occur alternative possibili-ties are that the organism with the largest m or with the smallest a should win the competition. Three experiments were performed as shown in Figure 2 reproduced from Hanson and Hubbell (1980). In the first experiment the m's are approximately equal while $a_1 < a_2$, and in the second the a's are equal and $m_1 > m_2$. The run parameters are shown in the table while the predicted time course is shown in dashed lines and the fit to data is shown in solid lines in Figure 2. First of all the predicted competitor -- the one with the lowest value of λ -- won the competition. Secondly, for biological data, the fit is good. Two discrepancies are immediate, however. Where the theory predicted a monotone approach to steady state the ex-periment showed damped oscillations. Moreover, the losing competitor was eliminated faster than predicted. The organisms used in the second experiment had growth rates which were affected by a chemical (nalidixic acid). A graph of this sensitivity is shown in the Figure 2(c); with an increase in concentration, m is relatively even for one and decreases rapidly for the other. (The ordinate of the graph is $r = m - D$; since D is the same for both organisms this is effectively a change in m.) By a proper choice of concentration of nalidixic acid it was possible to achieve $\lambda_1 = \lambda_2$ where coexistence is predicted. This is shown in the final graph. Overall, this is excellent agreement between the mathematics and the biology, and confirms that it is the relative sizes of λ_1 and λ_2 which determines the outcome. (Note that $J = \lambda$, $\mu = m$, and $K = a$ in the table in Figure 2.)

3. COMPETITION FOR A RENEWABLE RESOURCE

The interesting question in competition theory is "who survives?" The previous section can be summarized by saying that the principle of competitive exclusion -- one survivor on one resource -- holds in the chemostat. In Sections 3 and 4 results are presented which show circumstances where the principle is violated. A mathematical example of the failure of competitive exclusion was given in McGehee and Armstrong (1977) and some numerical examples were provided in Koch (1974). Coexistence -- the opposite of competitive exclusion -- was shown to occur in quite general circumstances in Butler (1980). Sections 3 and 4 present models which yield coexistence in the form of a limit cycle.

We leave the chemostat momentarily and consider competition for a renewable resource instead of a constant-input resource. We take the equation to be of the

Experiment No.	Bacterial strain	Auxotrophic for tryptophan					Other run parameters		
		Yield (cell/g)	K_s (g/liter)	μ (per hour)	r (per hour)	J (g/liter)	S_0 (g/liter)	D (per hour)	Volume (ml)
1	C-8*	2.5×10^{10}	3.0×10^{-6}	0.81	0.75	2.40×10^{-7}	1×10^{-4}	6.0×10^{-2}	200
	PAO283†	3.8×10^{10}	3.1×10^{-4}	0.91	0.85	2.19×10^{-5}			
2	C-8 nal^{s}spec^s	6.3×10^{10}	1.6×10^{-6}	0.68	0.61	1.98×10^{-7}	5×10^{-6}	7.5×10^{-2}	200
	C-8 nal^{s}spec^r	6.2×10^{10}	1.6×10^{-6}	0.96	0.89	1.35×10^{-7}			
3‡	C-8 nal^{s}spec^s	6.3×10^{10}	1.6×10^{-6}	0.68	0.61	1.98×10^{-7}	5×10^{-6}	7.5×10^{-2}	200
	C-8 nal^{s}spec^r	6.2×10^{10}	0.9×10^{-6}	0.41	0.34	1.99×10^{-7}			

*Escherichia coli.   †Pseudomonas aeruginosa.   ‡Nalidixic acid added (0.5 μg/ml).

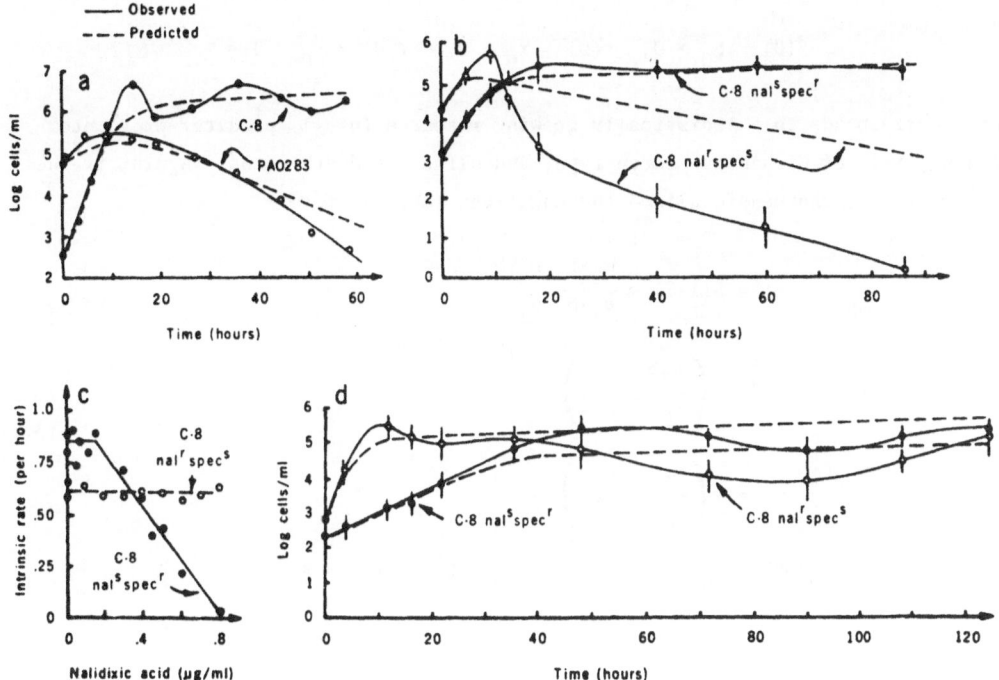

Figure 2   (a) Experiment 1: Strains differ principally in their half-saturation constants for tryptophan, and PAO283 loses to C-8 as predicted. (b) Experiment 2: Strains differ in their intrinsic rates of increase, but not in their half-saturation constants, and C-8 nal^{r}spec^s loses to C-8 nal^{s}spec^r as predicted.   (c) Effect of nalidixic acid on intrinsic rate of increase of strains C-8 nal^{s}spec^r and C-8 nal^{r}spec^s.   (d) Experiment 3: Strains differ in the half-saturation constants and in their intrinsic rates of increase, but nevertheless have identical J parameters, and the strains coexisted for the duration of the experiment, as predicted. In each experiment, the predicted curves were obtained by numerical integration of Equation 1.   Bars around points in experiments 2 and 3 (b and c) indicate ranges of three replicate values.

form

$$S' = \gamma S(1 - \frac{S}{K}) - \frac{m_1 xS}{a_1 + S} - \frac{m_2 yS}{a_2 + S}$$

$$x' = x\left(\frac{m_1 S}{a_1 + S} - D_1\right)$$

$$y' = y\left(\frac{m_2 S}{a_2 + S} - D_2\right)$$

$$S(0) = S_0 > 0, \quad x(0) = x_0 > 0, \quad y(0) = y_0 > 0,$$

(3.1)

which corresponds to a logistically growing resource (prey), predator-prey interactions still of Michaelis-Menten type, and differing death rates. Again, if the
variables are nondimensionalized the equations take the form

$$S' = S(1-S) - \frac{m_1 xS}{a_1 + S} \frac{m_2 yS}{a_2 + S}$$

$$x' = x\left(\frac{m_1 S}{a_1 + S} - D_1\right)$$

$$y' = y\left(\frac{m_2 S}{a_2 + S} - D_2\right)$$

$$S(0) = S_0 > 0, \quad x(0) = x_0 > 0, \quad y(0) = y_0 > 0.$$

(3.2)

We note that solutions of (3.2) are bounded and that the positive cone is invariant. If one defines $b_i = m_i/D_i$ and $\lambda_i = a_i D_i/m_i - D_i = a_i/b_i - 1$ then the case of
"inadequate competitors" takes the form that if $b \leq 1$ or if $b > 1$ and $\lambda_1 \geq 1$,
then $\lim_{t\to\infty} x(t) = 0$. Similarly, for $y(t)$ if $b_2 \leq 1$ or $b_2 > 1$ and $\lambda_2 \geq 1$.
Thus we assume:

$$b_i > 1, \quad i = 1,2$$

$$0 < \lambda_1 < \lambda_2 < 1.$$

(H)

This corresponds to (H) of Section 2.

Circumstances for competitive exclusion are contained in the following result.

THEOREM 3.1 (Hsu, Hubbell and Waltman, 1978a, 1978b). *If* (H) *holds and if*
$b_1 \geq b_2$ *then* $\lim_{t\to\infty} y(t) = 0.$

The impact of this theorem is that the omega limit set of trajectories of

(3.2) is two-dimensional. Specifically it consists of trajectories of

$$S' = S(1-S) - \frac{m_1 xS}{a_1 + S}$$

$$x' = x\left(\frac{m_1 S}{a_1 + S} - D_1\right).$$

(3.3)

The relevant features of solutions of (3.3) are as follows (assuming $\lambda_1 < 1$), Cheng (1981), Hsu et al. (1978a,b).

(i) If $a + 2\lambda_1 > 1$, (3.3) has a globally asymptotically critical point

$$\left(\lambda_1, \frac{\lambda_1(1-\lambda_1)}{D_1}\right).$$

(ii) If $a_1 + 2\lambda_1 < 1$, (3.3) has a unique limit cycle which is a global attractor of noncritical orbits in the positive quadrant.

Under the hypotheses of Theorem 3.1, all solutions of (3.2) then tend to a critical point or to a periodic orbit. Thus $b_1 < b_2$ is a necessary condition for coexistence of the predators. ($b_1 < b_2$ and $\lambda_1 < \lambda_2$ forces $a_1 < a_2$.)

THEOREM 3.2 (Butler and Waltman, 1981). *Let* m_1, a_1, D *be fixed so that* $m_1 > D_1$, $a_1 + 2\lambda_1 < 1$. *Fix* m_2 *and* D_2 *such that* $b_2 > b_1$. *There exists a number* $a_2^* > a_1$ *such that if* $a_2 < a_2^*$, $a_2^* - a_2$ *sufficiently small, then* $\lambda_1 < \lambda_2$ *and* (3.2) *has a periodic solution arbitrarily near the* S-x *plane.*

Note that if $(S(t), x(t))$ is the unique limit cycle of the system (3.3), then $(S(t), x(t), 0)$ is a periodic solution of (3.2). Theorem 3.2 describes the bifurcation of that limit cycle into two limit cycles, one which is planar (remains in the S-x plane) and one which is three dimensional. The biological interpretation of this limit cycle in S-x-y space is coexistence of the two predators feeding on a single prey. Note that this is possible only in an oscillatory manner.

Using asymptotic techniques, the stability of the bifurcation was determined in Smith (1982) when $\lambda_2 - \lambda_1$ is small. The solution which bifurcates out of the S-x plane can be shown numerically to continue and to collapse into the S-y plane, reversing the competitive outcome. For $\lambda_2 - \lambda_1$ small this continuation was proved, using asymptotic methods, in Keener (to appear).

4. COMPETING PREDATORS IN THE CHEMOSTAT

Suppose now that in a chemostat, as described in the introduction, there is only one organism growing on the constantly input nutrient but two predators feeding exclusively on this organism are added to the culture vessel. Such a system might consist of a sugar, a type of bacteria, and two different types of ciliates. Using the corresponding nondimensional variables as in Section 2, the equations take the form

$$S' = 1 - S - \frac{m_1 xS}{a_1 + S}$$

$$x' = x \left[\frac{m_1 S}{a_1 + S} - 1 - \frac{m_2 y}{a_2 + x} - \frac{m_3 z}{a_3 + x} \right]$$

$$y' = y \left[\frac{m_2 x}{a_2 + x} - 1 \right] \tag{4.1}$$

$$z' = z \left[\frac{m_3 x}{a_3 x} - 1 \right]$$

$$S(0) = S_0 \geq 0, \quad x(0) = x_0 > 0, \quad y(0) = y_0 > 0, \quad z(0) = z_0 > 0.$$

This system has been investigated in detail in Butler et al. (to appear). This is a three level system, S, a nutrient at the lowest level, x, a prey population growing on the nutrient at the intermediate level, and y and z, competing predator populations, feeding on the prey, occupy the top level. As before, m_i and $\lambda_i = a_i/m_{i-1}$, if $m_i > 1$, are the key parameters. If $m_1 \leq 1$ or if $m_1 > 1$ and $\lambda_1 \geq 1$, then $\lim_{t \to \infty} x(t) = 0$ (and, of course, $\lim_{t \to \infty} y(t) = 0$ and $\lim_{t \to \infty} z(t) = 0$). There are corresponding results for $y(t)$ and $z(t)$ using the pairs m_2 and λ_2 and m_3 and λ_3, respectively. The assumption corresponding to (H) of Section 2 is

$$m_i > 1, \quad \lambda_i < 1, \quad i = 1,2,3$$

$$\lambda_2 < \lambda_3. \tag{H-1}$$

The last inequality is merely a choice of labeling for the predators.

THEOREM 4.1 (Butler, Hsu, and Waltman, (to appear)). *Let* (H-1) *hold. If* $m_2 \geq m_3$ *then* $\lim_{t \to \infty} z(t) = 0$.

This shows that the principle of competitive exclusion holds if $m_2 \geq m_3$. In this case the omega limit set consists of trajectories of the system

$$S' = 1 - S - \frac{m_1 xS}{a_1 + S}$$

$$x' = x \left[\frac{m_1 S}{a_1 + S} - 1 - \frac{m_2 y}{a_1 + x} \right] \tag{4.2}$$

$$y' = y \left[\frac{m_2 x}{a_2 + x} - 1 \right]$$

$$S(0) = S_0 > 0, \quad x(0) = x_0 > 0, \quad y(0) = y_0 > 0.$$

(4.2)

This system represents a food chain, a model of which is of interest in its own right. It has been studied in Canale (1969, 1970), Sell (1977), and related experiments may be found in Jost et al. (1976) and Tsuchiya et al. (1972).

If $\Sigma = S + x + y$, then $\Sigma' = 1 - \Sigma$ or the omega limit set of any trajectory of (4.2) lies in the plane $S + x + y = 1$. Thus one variable may be eliminated and one need study only trajectories of

$$x' = x \left(\frac{m_1(1-x-y)}{1+a_1-x-y} - 1 - \frac{m_2 y}{a_2 + x} \right)$$

$$y' = y \left(\frac{m_2 x}{a_2 + x} - 1 \right).$$

(4.3)

The essential mathematical results for (4.3) may be summarized as follows (Butler, et al. (to appear), Sell 1977).

(i) If $m_1 \leq 1$ or $m_1 > 1$ and $\lambda_1 \geq 1$, $\lim_{t \to \infty} x(t) = 0$. If $m_2 \leq 1$ or $m_2 > 1$ and $\lambda_2 \geq 1$, $\lim_{t \to \infty} y(t) = 0$. If $\lambda_1 + \lambda_2 \geq 1$, $\lim_{t \to \infty} y(t) = 0$.

(ii) If $\lambda_1 + \lambda_2 < 0$, there exists a critical point (x_c, y_c), $x_c > 0$, $y_c > 0$. This critical point is globally asymptotically stable if

$$\frac{y_c}{m_2 \lambda_2^2} < \frac{m_1 a_1}{(1+a_1 - \lambda_2 - y_c)^2}$$

(4.4)

(iii) If the inequality in (4.4) is reversed, the critical point is unstable and there exists a least one periodic solution of (4.3).

If hypothesis (H-1) holds then the omega limit set of any trajectory is the critical point (x_c, y_c) or contains a periodic orbit. Coexistence of the top level predators then is possible only if $m_2 < m_3$ (which forces $a_2 < a_3$ since $\lambda_2 < \lambda_3$).

In case (iii) above, it is not known whether the limit cycle is unique (numerical evidence supports this conclusion) but certainly if more than one exists (there is at most a finite number) the inner one is stable from the inside and the outer one is stable from the outside. Thus there must be at least one stable limit cycle -- unfortunately we need slightly more than that and assume

If (4.4) is reversed, there is a limit cycle of (4.3)
and the linearization about it has one Floquet multiplier   (H-2)
(strictly) inside the unit circle.

The remaining Floquet multiplier is equal to one since the system has a periodic solution. (H-2) eliminates the possibility that the remaining multiplier is equal to one in the parameter range of interest. (It is necessarily less than or equal to one since it may be taken to be asymptotically stable.)

The coexistence result may now be stated.

THEOREM 4.2. *Let* a_i, m_i $i = 1, 2$ *be fixed so that* $m_i > 1$, $\lambda_1 + \lambda_2 < 1$, (H-2) *holds. Fix* $m_3 > m_2$. *Then there exists a number* $a_3^* > a_2$ *such that for* $a_3 < a_3^*$, $a_3^* - a_3$ *sufficiently small,* $\lambda_2 < \lambda_3$ *and* (4.1) *has a periodic orbit in the positive cone which is arbitrarily close to (but not in) the plane* $S + x + y = 1$, $z = 0$.

With additional restrictions, the stability and continuability of this orbit can be established (Keener, private communication).

Theorem 4.2 predicts a range of parameters where an experiment would show coexisting predators feeding exclusively on a single prey. The entire system, of course, would be oscillatory.

REFERENCES

Butler, G.J. (1980): Coexistence in predator prey systems, in: *Modeling and Differential Equations in Biology*, T. Burton, Ed., Marcel Dekker, New York.

Butler, G.J., S.B. Hsu, and P. Waltman (to appear): Coexistence of competing predators in a chemostat, *J. Math. Biology*.

Butler, G.J., and P. Waltman (1981): Bifurcation from a limit cycle in a two prey one predator ecosystem modeled on a chemostat, *J. Math. Biology* 12:295-310.

Canale, R.P. (1969): Prey relationships in a model for activated process, *Biotechnology and Bioengineering* 11:887-907.

_____ (1970): An analysis of models describing predator prey interaction, *Biotechnology and Bioengineering* 12:353-378.

Cheng, K.S. (1981): Uniqueness of limit cycle for a predator-prey system, *SIAM J. Math. Anal.* 12:541-548.

Drake, J.F., and H.M. Tsuchiya (1976): Predation of Escherichia Coli by Colpoda Stenii, *Appl. and Envr. Microbiology* 31: 870-874.

Fredrickson, A.G., and G. Stephanopoulos (1981): Microbial competition, *Science* 213:972-979.

Hansen, S.R., and S.P. Hubbell (1980): Single-nutrient microbial competition: Agreement between experimental and theoretical forecast outcomes, *Science* 207:1491-1493.

Hsu, S.B. (1978): Limiting behavior for competing species, *SIAM J. Appl. Math.* 34:760-763.

Hsu, S.B., S. Hubbell, and P. Waltman (1977): A mathematical theory for single-nutrient competition in continuous cultures of microorganisms, *SIAM J. Appl. Math.* 32:366-383.

_____(1978a): Competing predators, *SIAM J. Appl. Math.* 4:617-625.

_____(1978b): A contribution to the theory of competing predators, *Ecol. Mongr.* 48:337-349.

Jannash, H.W., and R.T. Mateles (1974): Experimental bacterial ecology studied in continuous culture, *Advances in Microbial Physiology* 11:165-212.

Jost, J.L., S.F. Drake, A.G. Fredrickson, and M. Tsuchiya (1976): Interaction of tetrahymena pyriformis, escherichia coli, azotobacter vinelandii and glucose in a minimal medium, *J. Bacteriology* 113:834-840.

Keener, J.P. (to appear). Oscillatory coexistence in the chemostat: A codimension two unfolding, *SIAM J. Appl. Math.*

_____: Private communication.

Koch, A.L. (1974): Competitive coexistence of two predators utilizing the same prey under constant environmental conditions, *J. Theor. Biol.* 44:378-386.

McGehee, R., and R.A. Armstrong (1977): Some mathematical problems concerning the ecological principle of competitive exclusion, *J. Diff. Eq.* 23:30-52.

Monod, J. (1942): *Reserches sur la Croissance des Cultures Bacteriennes,* Herman Paris.

Powell, E.O. (1958): Criteria for the growth of contaminants and mutants in continuous culture, *J. Gen Microbiology*, 18:259-268.

Sell, G. (1977): What is a dynamical system?, in: *Studies in Ordinary Differential Equations,* J. Hale, Ed., MAA Studies in Mathematics No. 14.

Smith, H.L. (1982): The interaction of steady state and Hopf bifurcation in a two-predator-one-prey competition model, *SIAM J. Appl. Math.* 42:27-43.

Stewart, F.M., and B.R. Levin (1973): Partitioning of resources and the outcome of interspecific competition; a model and some general considerations, *Amer. Nat.* 107:171-198.

Tsuchiya, H.M., S.F. Drake, J.L. Jost, and A.G. Fredrickson (1972): Predator-prey interactions of dictoyostelium discordeum and escherichia coli in continuous culture, *J. Bacteriology* 110:1147-1153.

Veldcamp, H.(1977): Ecological studies with the chemostat, *Advances in Microbial Ecology* 1:59-95.

Waltman, P., S.P. Hubbell, and S.B. Hsu (1980): Theoretical and experimental investigations of microbial competition in continuous culture, in: *Modeling and Differential Equations in Biology,* T. Burton, Ed., Marcel Dekker, New York.

COMPETITIVE PREDATOR-PREY SYSTEMS AND COEXISTENCE

G.J. Butler

1. INTRODUCTION

When two predator populations compete exploitatively for the same prey population, one possible outcome is the extinction of both predator populations owing to
their inability to predate successfully. Another possible outcome is the extinction
of one of the predator populations due to its being out-competed by the other which
remains to survive in some stable configuration with the prey population, either in
equilibrium or in some oscillating node. This is an example of the "competitive
exclusion principle" (Levin, 1970; May, 1973). Although such competitive exclusion
is a common occurrence in nature, there are studies (Koch, 1974; MacArthur, 1958;
Stewart and Levin, 1973) indicating that under suitable circumstances, both predator
populations can coexist. One mechanism suggested for this is that one population is
more suited to growing at low resource (prey) levels, while the other population is
better equipped for growing at high resource levels (MacArthur, 1958, 1960;
MacArthur and Wilson, 1967).

The earliest mathematical models of two predator-one prey systems were of
Lotka-Volterra type and seemed to support the "competive exclusion principle". In
important theoretical papers in 1976 and 1977, Armstrong and McGehee put this into
clearer perspective by demonstrating that if the specific predator growth rates are
affine functions of prey density (as is the case in the Lotka-Volterra model), then
coexistence is impossible. They were, however, able to construct a model with nonlinear specific growth rates in which coexistence could occur (Armstrong and McGehee,
1976; McGehee and Armstrong, 1977). Since then, other examples have been given
(Grasman, 1980; Zicarelli, 1975). Indeed there are examples of coexistence amongst
n   predator populations competing for a single prey species (Zicarelli, 1975).

Such examples, created with the intention of illustrating coexistence, may be
somewhat contrived. There are, however, models that arise quite naturally, that
appear to have the potential for exhibiting coexistence. One such model is that in
which the prey population has a logistic growth rate and a Holling-type functional
response is assumed for the predator populations. Koch (1974) carried out numerical
studies with this model, which suggested that for certain ranges of parameter values
appearing in the model, coexistence might be occurring. Hsu, Hubbell and Waltman
(1978) made a qualitative study of the model, obtaining a complete global analysis
for a large region of the parameter space. While leaving the question of coexistence unresolved, they also conjectured that it would occur for suitable parameter
values. This was verified by the author (Butler, 1980) and subsequently it was
shown that such coexistence would be exhibited as limit-cycle behaviour (Butler and
Waltman, 1981; Keener, to appear; Smith, 1982; Wilker, to appear).

The object of this paper is to present coexistence phenomena as a rather common feature of predator-prey models.

2. THE MODEL

Let $S(t)$ denote the prey density and $x_1(t)$, $x_2(t)$ denote the predator densities, at time t. The model is described by the system of ordinary differential equations

$$S'(t) = \gamma S(t)g(S(t),K) - c_1 x_1(t)p(S(t),a_1) - c_2 x_2(t)p(S(t),a_2)$$

$$x_1'(t) = x_1(t)(-D_1+p(S(t),a_1)) \tag{2.1}$$

$$x_2'(t) = x_2(t)(-D_2+p(S(t),a_2))$$

We shall also have occasion to consider the reduced system that occurs if the predator x_{3-i} is absent:

$$S'(t) = \gamma S(t)g(S(t),K) - c_i x_i(t)p(S(t),a_i)$$

$$x_i'(t) = x_i(t)(-D_i+p(S(t),a_i)) \tag{2.2}$$

γg and p are the per capita prey growth rate and predation response functions, respectively, and we shall make the following assumptions:

$$g(S,K) \text{ and } p(S,a) \text{ are smooth functions of } (S,K) \text{ and } (S,a)$$
$$\text{respectively,} \tag{2.3}$$

$$g(0,K) = 1, \quad \partial g/\partial S < 0 < \partial^2 g/\partial S\partial K \text{ for all } S,K \geq 0, \text{ and}$$
$$\lim_{K\to\infty} \frac{\partial g}{\partial S} = 0 \tag{2.4}$$

$$(K-S)g(S,K) > 0 \text{ for all } S \geq 0, \quad S \neq K \tag{2.5}$$

$$p(0,a) = 0, \quad \frac{\partial p}{\partial S} > 0 \text{ and}$$
$$\frac{\partial p}{\partial S} < \frac{p(S,a)}{S} \text{ for all } S,a > 0. \tag{2.6}$$

$$\frac{\partial p}{\partial a} < 0 \text{ for all } S,a \geq 0. \tag{2.7}$$

Apart from the smoother assumptions, (2.3)-(2.7) are quite natural within the context of predator-prey interactions. γ is the maximum growth rate of the prey and K plays the role of a carrying capacity.

One other assumption that we shall use later is that except for a closed, nowhere dense set of parameter 5-tuples (γ,K,c_i,a_i,D_i), all critical points and periodic orbits of $(2.2)_i$ are hyperbolic, i.e. the characteristic roots or multipliers associated with such solutions do not have zero real part. This is a generic assumption on the parametrized vector fields defined by $(2.2)_i$, hence we shall refer to this assumption by saying that $(2.2)_i$ is *generic*.

3. PRELIMINARY ANALYSIS OF THE MODEL AND EXTINCTION RESULTS

The equilibria for (2.1) are given by $x_i = 0$ or $p(S, a_i) = D_i$. Since $\partial p / \partial S > 0$, $p(S, a_i) = D_i$ has at most one solution, $S = \lambda_i$, say. If $x_1 = x_2 = 0$, we must have $Sg(S, K) = 0$ which implies that $S = 0$ or $S = K$. Thus we have the two equilibria $E_0(0,0,0)$ and $E_3(K,0,0)$ in the non-negative (S, x_1, x_2) octant, with additional equilibria $E_1(\lambda_1, \hat{x}_1, 0)$, $E_2(\lambda_2, 0, \hat{x}_2)$, where $\hat{x}_i = \gamma \lambda_i g(\lambda_i, K)/c_i D_i$, provided that D_i is in the range of $p(S, a_i)$ (so that λ_i is defined) and that $\lambda_i \leq K$ (so that $\hat{x}_i \geq 0$). Interior equilibria in the positive octant can only occur in the nongeneric case that $\lambda_1 = \lambda_2$, in which case there is a line of critical points. Henceforth we shall exclude this case from consideration. (For a discussion of this case for the Hsu-Hubbell-Waltman model, see Wilker, to appear.

Let $B_i = \lim\limits_{S \to \infty} p(S, a_i)$ $(0 < B_i \leq \infty)$ and let $b_i = B_i/D_i$.

As a preliminary result, we need to know that (2.1) is a well-posed system:

LEMMA 3.1 *Given initial conditions in the non-negative octant, solutions of are unqiuely defined for all time, vary continuously with initial data, and are bounded and remain in the non-negative octant for* $t \geq 0$.

PROOF. This is straightforward and we omit it.

In order that E_i be defined and lie in the non-negative octant with $\hat{x}_i > 0$, we must have $b_i > 1$ and $\lambda_i < K$. Without either one of these conditions, x_i cannot survive in the absence of x_{3-i} (in system $(2.2)_i$). Not surprisingly, this is also the case for the full system (2.1), and this is expressed in the lemma below. We omit the proof, which is similar to that of the corresponding result given in Hsu et. al. (1978).

LEMMA 3.2 *A necessary condition for* x_i *to survive in the system* (2.1) *is that* $b_i > 1$ *and* $0 < \lambda_i < K$.

From Lemma 3.2, we have the following result for the extinction of both predators:

THEOREM 3.3 *If* $b_1 \leq 1$ *or* $\lambda_1 \geq K$ *and* $b_2 \leq 1$ *or* $\lambda_2 \geq K$, *then for all solutions of* (2.1) *with initial conditions in the positive octant, we have* $\lim\limits_{t \to \infty} S(t) = K$, $\lim\limits_{t \to \infty} x_i(t) = 0$, $i = 1, 2$.

From now on we shall assume that $b_1 < 1$ and $\lambda_1 < K$, and without loss of generality $\lambda_1 < \lambda_2$ if λ_2 is defined.

Stability of the equilibria. First consider the equilibria as restricted to the system $(2.2)_i$. The stability of E_i is determined by the sign of the partial derivative with respect to S of $Sg(S,K)/p(S,a_i)$, evaluated at $S = \lambda_i$ (Cheng et al., 1981).

Taking the full 3-dimensional system (2.1) into account, it is easily verified that E_1 has a 1-dimensional stable manifold transverse to the (S,x_1)-plane and E_2 has an unstable 1-dimensional manifold transverse to the (S,x_2)-plane. E_0 and E_3 are unstable. Summarizing these remarks, we have

LEMMA 3.4 *Assume that* $b_1 < 1$ *and* $0 < \lambda_1 < K$, *and* $\lambda_1 < \lambda_2$ *if* λ_2 *is defined. Then* E_0, E_2 *(if it is defined) and* E_3 *are unstable.* E_1 *is asymptotically stable or unstable according as* $Sg(S,K)/p(S,a_1)$ *has a negative or positive derivative with respect to* S *at* $S = \lambda_1$.

LEMMA 3.5 E_1 *is unstable for* K *sufficiently large.*

PROOF. (2.4) implies that $(\partial g/\partial S)(\lambda_1,K) \to 0$ as $K \to \infty$, and a simple computation using (2.6) shows that $\dfrac{\partial}{\partial S} \left(\dfrac{Sg(S,K)}{p(S,a_1)} \right)_{S=\lambda_1} > 0$ for K large. The result now follows from Lemma 3.4.

REMARK. As K increases, there will be a critical value K_1 at which a Hopf bifurcation occurs giving rise to a periodic solution in the (S,x_1)-plane. (The requisite transversality condition is assured by (2.4)). A similar analysis with respect to the (S,x_2)-plane applies to E_2.

The following result determines some conditions under which x_1 will survive and x_2 will become extinct. The proof is along the same lines as that given in Hsu et. al. (1978).

THEOREM 3.6 *Let* $0 < \lambda_1 < K$. *If any of the following holds, we shall have*
$$\lim_{t \to \infty} S(t) > 0, \quad \lim_{t \to \infty} x_1(t) > 0, \quad \lim_{t \to \infty} x_2(t) = 0:$$

(i) $b_2 \leq 1$; (ii) $\lambda_2 \geq K$; (iii) $0 < \lambda_1 < \lambda_2 < K$ *and there exists* $\xi > 0$ *such that*

$$\xi\left(1 - \frac{p(S,a_2)}{D_2}\right) - \left(1 - \frac{p(S,a_1)}{D_1}\right) < 0$$

for all S *with* $0 < S \le K$.

4. COEXISTENCE

We have seen that a_1, D_1 can be chosen so that as K increases, there is a value K_1 of K at which E_1 undergoes a Hopf bifurcation resulting in periodic solutions in the (S,x_1)-plane.

Under suitable conditions, it may be shown that these bifurcating periodic solutions are exponentially asymptotically stable for the system $(2.2)_1$.

THEOREM 4.1 *Let* $f(S,K) = Sg(S,K)$. *The bifurcating periodic solutions will be exponentially asymptotically stable and exist for* $K > K_1$ *if*

$$\frac{\partial f}{\partial S}(\lambda_1,K_1)\left[\frac{\partial^3 f}{\partial S^3}(\lambda_1,K_1) - f(\lambda_1,K_1)\frac{\partial^3 p}{\partial S^3}(\lambda_1,a_1)\right] < 0. \tag{4.1}$$

In particular, if E_1 *is globally asymptotically stable for* $K < K_1$, *the bifurcating periodic solutions will be asymptotically stable provided the generic condition*

$$\frac{\partial f}{\partial S}(\lambda_1,K_1)\left[\frac{\partial^3 f}{\partial S^3}(\lambda_1,K_1) - f(\lambda_1,K_1)\frac{\partial^3 p}{\partial S^3}(\lambda_1,a_1)\right] \ne 0 \tag{4.2}$$

holds.

PROOF. A somewhat tedious computation shows that (4.1) is equivalent to the "vague attractor" condition that guarantees asymptotic stability of the bifurcating periodic solutions.

REMARK. Cheng et. al. (1981) have given general conditions for E_1 to be globally stable.

The periodic solutions in the (S,x_1)-plane may now be destabilized by adjusting the death-rate D_2 of predator x_2. In this way, we may produce a periodic solution of (2.1) lying in the positive (S,x_1,x_2)-octant. This may be achieved for a non-empty open subset of parameter space.

THEOREM 4.2 *Let* (2.3)-(2.7) *hold and assume that* $(2.2)_i$ *is generic* (i=1,2). *Then there is a non-empty open subset of the full parameter space for which* (2.1) *possesses a periodic solution lying in the positive* (S,x_1,x_2)-octant.

PROOF. Let γ,c_1,a_1,D_1 be given such that $b_1 > 1$. By Lemma 3.5, if K is sufficiently large, E_1 will be defined and will be unstable, a Hopf bifurcation at $K = K_1 = K_1 (\gamma, c_1, a_1, D_1)$ resulting in a periodic solution in the (S, x_1)-plane. The instability of E_1 for $K > K_1$ implies via the Poincare-Bendixsson theory that $(2.2)_1$ will have at least one non-trivial stable periodic solution. The assumption that $(2.2)_1$ is generic implies that this solution is in fact exponentially stable. Standard theory shows that this periodic solution will vary continuously over a sufficiently small open region of $(K, \gamma, c_1, a_1, D_1)$-space where it will remain exponentially asymptotically stable.

Let $(S(t), x_1(t))$ be this periodic solution for fixed values of K, γ, c_1, a_1, D_1, and let its period be T. Let c_2, a_2 be given.

Let $D^* = \frac{1}{T} \int_0^T p(\bar{S}(t), a_2) dt$.

The Floquet exponents for the linearization of (2.1) about the periodic solution $(\bar{S}(t), \bar{x}_1(t), 0)$ are $0, -\mu, \nu$, where $-\mu < 0$ and $\nu = D^* - D_2$ (see Lemma 4.1 and 4.2 of Butler and Waltman, 1981 for the case of Holling kinetics). Now $\frac{\partial \nu}{\partial D_2} = -1$, so that as D_2 decreases through D^*, the Poincaré map associated with $(\bar{S}(t), x_1(t), 0)$ has an eigenvalue that crosses the unit circle transversally. As in Butler and Waltman (1981), we find that a one-parameter family of positive periodic solutions of (2.1) bifurcate from $(\bar{S}(t), \bar{x}_1(t), 0)$.

5. EXAMPLES

Predation functions $p(S, a)$ that are found in the literature include the following:

$$\frac{S}{S+a} \qquad \text{(Holling-type)} \qquad (5.1)$$

$$1 - e^{-as} \qquad \text{(Ivlev, 1961)} \qquad (5.2)$$

$$S^q \qquad (0 < q < 1) \quad \text{(Rosenzweig, 1971)} \qquad (5.3)$$

$$\frac{S^2}{S^2 + a} \qquad \text{(May, 1973).} \qquad (5.4)$$

If we take $g(S, K)$ to correspond to logistic growth, $g(S, K) = 1 - S/K$, or the growth rate suggested by Goel, Maitra and Montroll (1971), $g(S, K) = 1 - (S/K)^\alpha$ $(0 < \alpha < 1)$, it is easily verified that with any one of the predation functions (5.1)-(5.4), the conditions (2.3)-(2.7) and (4.1) hold. Thus we have coexistence possible for any model based on these functions.

REFERENCES

Armstrong, R.A., and R. McGehee (1976): Coexistence of two competitors on one resource, *J. Theor. Biol.* 56:499-502.

Butler, G.J. (1980): Coexistence in predator-prey systems. In: *Modeling in Differential Equations and Biology*, T.A. Burton (Editor), Marcel Dekker, New York, pp. 199-207.

Butler, G.J., and P. Waltman (1981): Bifurcation from a limit cycle in a two predator-one prey ecosystem modeled on a chemostat, *J. Math. Biol.* 12:295-310.

Cheng, K.-S., S.-B. Hsu, and S.-S. Lin (1981): Some results on global stability of a predator-prey system, *J. Math. Biol.* 12:115-126.

Goel, N.S., S.C. Maitra, and E.W. Montroll (1971): On the Volterra and other non-linear models of interacting populations, *Rev. Mod. Phys.* 43:231-276.

Grasman, W. (1980): The existence of a periodic solution to a model of two predators and one prey. In: *Differential Equations and Biology*, T.A. Burton (Editor), Marcel Dekker, New York.

Hsu, S.-B., S.P. Hubbell, and P. Waltman (1978): Competing predators, *SIAM J. Appl. Math.* 35:617-625.

_____ (1978): A contribution to the theory of competing predators, *Ecol. Monogr.* 48:337-349.

Ivlev, V.S. (1961): *Experimental Ecology of the Feeding of Fishes*, Yale University Press, New Haven, Connecticut.

Keener, J.P. (to appear): Oscillatory coexistence in the chemostat: a codimension two unfolding, *SIAM J. Appl. Math.*

Koch, A.L. (1974): Competitive coexistence of two predators utilizing the same prey under constant environmental conditions, *J. Theor. Biol.* 44:378-386.

Levin, S. (1970): Community equilibria and stability, and an extension of the competitive exclusion principle, *Amer. Nat.* 104:413-423.

MacArthur, R.M. (1958): Population ecology of some warblers of northeastern coniferous forests, *Ecology* 39:599-619.

_____ (1960): On the relative abundance of species, *Amer. Nat.* 94:25-36.

MacArthur, R.M., and E.O. Wilson (1967): *The Theory of Island Biogeography*, Princeton University Press, Princeton, New Jersey.

May, R.M. (1973): *Stability and Complexity in Model Ecosystems*, Princeton University Press, Princeton, New Jersey.

McGehee, R., and R.A. Armstrong (1977): Some mathematical problems concerning the ecological principles of competitive exclusion, *J. Differential Equations*, 23:30-52.

Rosenzweig, M.L. (1971): Paradox of enrichment: destabilization of exploitation ecosystems in ecological time, *Science* 171:385-387.

Smith, H.L. (1982): The interaction of steady state and Hopf bifurcation in a two-predator-one-prey competition model, *SIAM J. Appl Math.* 42:27-43.

Stewart, F.M., and B.R. Levin (1973): Partitioning of resources and the outcome of interspecific competition: a model and some general considerations, *Amer. Nat.* 107:171-198.

Wilken, D.R. (1982): Some remarks on a competing predator problem, *SIAM J. Appl. Math.* 42:895-902.

Zicarelli, J. (1975): Mathematical analysis of a population model with several predators on a single prey, *Ph.D. Thesis,* University of Minnesota, Minneapolis, Minnesota.

COEXISTENCE OF MANY COMPETING SPECIES

Jean Coste

1. INTRODUCTION

We consider here dynamical systems of the form:

$$\dot{x}_i = x_i G_i (\{x_j\}) \quad i,j \in [1,\ldots N] \tag{1}$$

This type of vector field is of interest in ecological matters as well as in various other fields. It is characterized by the existence of N invariant planes $x_i = 0$. Therefore some variables may go to zero in the course of evolution, and the dimensionality of the attractors may be decreased by the vanishing of those variables. We want to study this phenomenon in the limit of large N.

Several papers have been devoted to this question in the last ten years, essentially of two kinds. i) Those based upon the study of linear stability of the field's fixed points, which gave only partial and contradictory results (Gilpin 1975, May 1972, Roberts 1974). ii) Those which try to give some general persistence arguments. In addition to our own work we must mention the results of McGehee and Armstrong (1977). They considered systems of the type:

$$\dot{x}_i = x_i U_i (z_1 \ldots z_k) \quad i \in [1,\ldots N]$$

$$\dot{z}_j = r_j (x_1 \ldots x_N) \quad j \in [1,\ldots k]$$

when N species feed on k resources which, in turn, depend on the N population levels. McGehee and Armstrong were able to show that no persistent solution is available if $k < N$ (and if the U_i are linear functions of the z_j).

We consider here the more general case $k = N$ where nothing general has been proved, and where many examples of persistent solutions are known. We propose to show that, in the case of Lotka-Volterra (or L.V.) fields (and for an open class of fields in the neighborhood of L.V.) persistent solutions are highly improbable in the limit of large N. The argument if probabilistic.

The basic matter of the present work was first briefly outlined by Coste et al (1978).

2. ABOUT THE CONCEPT OF PERSISTENCE

We remember the usual definition of a persistent vector field of type (1): there exists an attractor in $\text{int}(\mathbb{R}_N)$ (cf. McGehee and Armstrong 1977). Stated equivalently: there exist orbits such that $x_i > \varepsilon$ (ε fixed arbitrarily small) $\forall i$, $\forall t$.

Let us now consider the vector field of May and Leonard (1975), which is of the L.V. type, and whose attractor C has the following phase portrait:

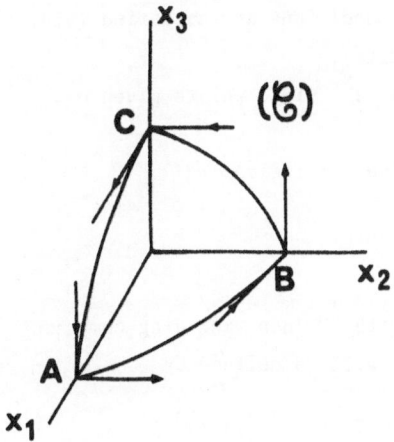

Figure 1

C is the union of three 2-dimensional arcs lying in the coordinate planes. Can such a system be thought of as 2-species persistent? As $t \to \infty$ an orbit point solution, converging toward C, approaches closer and closer to the unstable fixed points A,B,C, staying longer and longer in their neighborhood. Therefore 2 variables among the 3 are almost always extincted, and the system, for large t, is almost always found with only one non vanishing species. Then the system could hardly be called 2-species persistent, in spite of the geometry of its attractor.

At this point, it is worth remarking that, in most biological or physical systems, there exists a threshold value for any x_i below which the continuous model becomes meaningless and x_i usually vanishes (for instance under the fluctuations of environment). Thus the evolution of a May-Leonard system actually ends up at some finite time on one of A,B or C fixed points.

The above remarks lead us to look for the number of variables "almost always extincted" along an orbit, and to introduce the following definition.

DEFINITION: A system will be said to be "not persistent of order (N-p)" if, along any orbit and in the limit $t \to \infty$, one can almost always find a set of p variables $\{x_i\}$ such that $x_i < \varepsilon$, ε given arbitrarily small. We shall say that, in this case p variables are almost always extincted, or "A.A. extincted".

3. THE LOTKA-VOLTERRA CASE

Let

$$G_i \equiv b_i - \sum_j a_{ij}x_j, \qquad i,j \in [1,\dots N]$$

We assume the boundedness of solutions, a condition to be satisfied by any accept-able model. It is verified, in particular, if $0 < b_i < M$ and $a_{ij} > K > 0$, $\forall i,j$. We suppose here by simplicity that these particular conditions are satisfied (although they are not essential for obtaining our results).

Let us first remark that the vector field has 2^N fixed points given by:

$$
\begin{cases}
\xi^k = 0 & k \in \{p\}, \text{ any set of indices} \\
\sum_j a_{ij}\xi^j = b_i & i,j \in [1,\ldots,N] - \{p\}
\end{cases}
$$

and we shall denote by ξ_N the unique fixed point with N non vanishing components. $\xi_{N-1}(k)$, one of the $(N-1)$ components-fixed points, will be defined by

$$\sum_{i,j \neq k} a_{ij}\xi^j_{N-1}(k) = b_i; \quad \xi_{N-2}(k,\ell) \quad \text{by} \quad \sum_{i,j \neq k,\ell} a_{i,j}\xi^j_{N-2}(k,\ell) = b_i, \text{ etc}\ldots$$

Time averaging the equations of motion written in the form $\dfrac{\dot{x}_i}{x_i} = b_i - \sum_j a_{ij}x_i$, we obtain:

$$\frac{1}{T} \text{Log} \frac{x_i}{x_i^0} = b_i - \sum_{j=1}^{N} a_{ij}\langle x_j \rangle_T, \tag{2}$$

where $\langle\,\rangle_T = \dfrac{1}{T}\displaystyle\int_0^T dt'$. (2) can be rewritten as:

$$\langle x^i \rangle_T = \xi^i_N - \frac{1}{T}\sum_j \lambda_{ij} \text{Log}\left(\frac{x_j}{x_j^0}\right) \tag{3}$$

where $\underset{\approx}{\lambda} = \underset{\approx}{a}^{-1}$ ($\underset{\approx}{a}$ assumed invertible).

Now the persistence of the N variables $(x_i > \varepsilon, \forall_i, \forall t)$ implies, by Equation (2), that, in the limit $T \to \infty$, $\langle x \rangle = \xi_N$, and, therefore, that ξ_N is positive (ξ^i_N positive $\forall_i$). If ξ_N is not positive, at least one of its components, say ξ^ℓ_N, is negative and Equation (3) written for $i = \ell$ shows that, as $T \to \infty$, at least one component of x, say x_k must go to zero. We want now to show that if $\xi_{N-1}(k)$ is not positive $\forall k$, 2 populations at least are "A.A. extincted".

Indeed, if x_k is the only vanishing variable at time T, $\dfrac{x_j}{x_j^0} > \varepsilon$ $\forall j \neq k$ and, if $|\lambda_{ij}| < \dfrac{\lambda}{N}$

$$\rho_\ell = \xi^\ell_N - \frac{1}{T}\sum_{j \neq k} \lambda_{\ell j} \text{Log}\left(\frac{x_j}{x_j^0}\right) < \xi^\ell_N + \frac{\lambda}{T} \text{Log}\left(\frac{1}{\varepsilon}\right)$$

therefore, there exists $\rho > 0$ and $T_0 = \dfrac{Log(1/\varepsilon)}{\rho + \xi_N^\ell}$ such that if $T > T_0$ $\rho_\ell < -\rho < 0$,

and the positivity of $\langle x \rangle_\ell$ implies that $x_k < x_k^0 e^{-\frac{\rho}{\lambda_{\ell k}} T}$ (variable x_k being such that $\lambda_{\ell k} > 0$).

Let us now take T as time origin $(\langle \ \rangle_t = \int_T^{T+t} dt')$. Since $\dfrac{\dot{x}_k}{x_k} < b_k$, we have

$$\left.\begin{array}{c} x_k(t) \\ \langle x_k \rangle_t \end{array}\right\} < x_k^0 e^{-\frac{\rho}{\lambda_{\ell k}} T} e^{b_k t} \quad \text{and} \quad \left.\begin{array}{c} x_k(t) \\ \langle x_k \rangle_t \end{array}\right\} < \varepsilon \quad \text{for} \quad 0 < t < \tau, \text{ where } b_k \tau = \dfrac{\rho}{\lambda_{\ell k}} T -$$

$Log\left(\dfrac{x_k^0}{\varepsilon}\right)$.

Let $\xi_{N-1}^i(k)$ be one of the negative components of $\xi_{N-1}(k)$. Equation (3) can be rewritten as:

$$\langle x_i \rangle_t = \left\{ \xi_{N-1}^i(k) - \sum_{j \neq k} \lambda_{ij} a_{jk} \langle x_k \rangle_t \right\} - \frac{1}{t} \sum_{j \neq k} \lambda_{ij} Log\left(\dfrac{x_j}{x_j^T}\right) \tag{4}$$

If x_k is the only "extincted" species for $t \in (0,\tau)$, $\dfrac{x_j(t)}{x_j^T} > \varepsilon$ in this time

interval. If $\left| \sum_{j \neq k} \lambda_{ij} a_{jk} \right| < \alpha_i$ we have:

$$\langle x_i \rangle_j < \xi_{N-1}^i(k) + \alpha_i \varepsilon + \frac{\lambda}{t} Log\left(\frac{1}{\varepsilon}\right)$$

If ε is chosen small enough in order that $\xi_{N-1}^i + \alpha_i \varepsilon < 0$, T can always be chosen large enough in order that $\langle x_i \rangle_t$ must become negative in $(0,\tau)$, which is impossible. Therefore, one more species at least must become small in $(0,\tau)$. Let us assume that there is only one, say x_p, and let us put:

$$\nu_i = \xi_{N-1}^i(k) - \sum_{j \neq k} \lambda_{ij} a_{jk} \langle x_k \rangle_t - \frac{1}{t} \sum_{j \neq k,p} \lambda_{ij} Log\left(\dfrac{x_j}{x_j^T}\right)$$

then $\nu_i < \xi_{N-1}^i(k) + \alpha_i \varepsilon + \frac{\lambda}{t} Log(1/\varepsilon)$ and there exists $\eta > 0$ and

$t_0 = -\dfrac{\lambda Log(1/\varepsilon)}{\xi_{N-1}^i(k) + \alpha_i \varepsilon + \eta}$ such that for $t > t_0$, $\nu_i < -\eta$ $(\eta > 0)$. Finally, $\langle x_i \rangle_t > 0$

implies that $x_p < x_p^T e^{-\frac{\eta}{\lambda_{ip}} t}$ $(\lambda_{ip} > 0)$. We see that $x_p < \varepsilon$ for $t > t_1 =$

$\dfrac{\lambda_{ip}}{\eta} Log\left(\dfrac{x_p^T}{\varepsilon}\right)$. Let $t' = \sup(t_0, t_1)$. We see that x_p and x are both smaller than ε over the "extinction interval" $t_c = \tau - t'$ and the relative time of simultaneous extinction is $\frac{\tau-t'}{\tau}$, which obviously goes to unity when $T \to \infty$. We conclude that at least 2 species are "A.A. extincted" on any orbit as $T \to \infty$. According to our definition the system is not persistent of order (N-1). Obviously this argument can be carried out one step further in the same way and would lead to conclude that the non-positivity of $\xi_{N-2}(k,p)V_{k,p}$ implies non persistence of order (N-2), etc. In general: *the non-positivity of all* ξ_{N-r} *for* $r = 0,1 \ldots m$, *implies non-persistence of order* (N-m).

According to our necessary conditions of persistence, we must investigate the possibility of finding a positive solution of a set of n linear equations. We consider this problem from a probabilistic point of view; the field parameters a_{ij}, b_i being chosen at random according to some given probability law, what is the probability $P(n)$ of finding a positive solution (in the limit of large n)? Two preliminary remarks are the following: i) $P(n) \approx 1$ if matrix $\underset{\approx}{a}$ is "nearly diagonal" ($a_{ii} \gg a_{ij\ i,j}$). Indeed, the species are then nearly independent, a situation of obvious coexistence. Therefore, dynamical regression concerns only the variables which are effectively interacting. An interesting example of multi-connected variables is provided by a set of similar species (indeed we have in this case $a_{ij} \approx a_{ii}\forall i,j$, since mutual and self interactions are nearly the same). ii) Since the L.V. field exhibits 2^N fixed points, a necessary condition of strong dynamical regression (most of the species get A.A. extincted) is that $P(m) \underset{\sim}{<} 2^{-m}$.

Now the general problem of probability we have to solve is by no means an easy one, and only partial results (unpublished) are available at present time. One of them concerns a system of similar species for which

$$\begin{cases} a_{ij} = a(1 + \varepsilon_{ij}) \\ \\ b_i = b(1 + \varepsilon_i) \end{cases} \quad \begin{cases} \varepsilon_{ij} \\ \varepsilon_i \end{cases} < \varepsilon \ll 1 \tag{5}$$

It can be shown in this case that $P(n) < (n-1)^{-(n-1)/2}$. We have also performed numerical calculations, solving sets of linear equations whose coefficients were pseudo random numbers (taken in $(0,1)$ or $(-1,1)$), which seem to confirm that $P(n)$ decreases faster than 2^{-n}.

4. PERTURBATION OF THE LOTKA-VOLTERRA FIELD

Let $G_i = b_i - \sum_j a_{ij}x_j + \phi_i(\{x_k\})$ where the ϕ_i are arbitrary non singular, non linear functions. If the solutions are still bounded, it may be possible to state that $|\phi_i(\{x_k\})| < \frac{\varepsilon}{N} \forall i, \forall t$. Time averaging the equations of motion gives:

$$\frac{1}{T} \text{Log} \frac{x_i}{x_i^0} = b_i - \sum_{j=1}^{N} a_{ij} \langle \alpha^j \rangle + \langle \phi_i \rangle$$

or

$$\langle x^i \rangle_T = \xi_N^i - \sum_j \lambda_{ij} \left[\frac{1}{T} \text{Log} \left(\frac{x_j}{x_j^0} \right) - \langle \phi_j \rangle \right]$$

where ξ_N and λ_{ij} refer to unperturbed L.V. fields. N-persistence and the positivity of the $\langle x^i \rangle$'s implies that: $\xi_N^i + \sum_j \lambda_{ij} \langle \phi_j \rangle_T > 0$. We therefore obtain the necessary conditions of N-persistence: $\xi_N^i > -\lambda\varepsilon, \forall i$. Now the probability $P(N,C)$ of finding a solution ξ_N of the linearized equilibrium equations such that $\xi > C$ is obviously a continuous function of C around $C = 0$. We conclude that there exists an open set of vector fields around L.V. for which previous qualitative conclusions still hold.

5. REMARKS ON THE DYNAMICS OF THE REGRESSION. EFFECT OF FLUCTUATIONS

How much time does it take for most of the species getting extinct (in a case of strong regression)? If the field's coefficients are of the same order of magnitude, "extinction time" t_c is expected to be of the order of the average inverse growth rate, which is confirmed by numerical integrations of the equations of many species motion. Things are different in the interesting case of a system made of similar species (when inequalities (5) are obeyed). Then it may be conjectured that t_c is a decreasing function of ε. Indeed, for $\varepsilon = 0$, any solution $\sum x_j = b/a$ is an equilibrium solution. Then, for ε small but finite, this solution becomes quasi-stationary and its life time goes to infinity when $\varepsilon \to 0$. Let us also remark that, in the case of small ε, the selection process is determined by small differences between the field parameters. Therefore, the fluctuations of environment could likely make these parameters to fluctuate. What would be the effect on dynamical regression? We have performed a numerical experiment where we choose the a_{ij}'s and b_i's randomly in $[0,1]$ at random time intervals t_i. A non trivial situation occurs when the average time interval $\langle t_{i+1} - t_i \rangle$ is much smaller than the average selection time in the non fluctuating field. The evolution is sketched on Figure 2a (without fluctuations) and 2b (with fluctuations), for a system of 20 species, showing a strong selection in both cases. This result suggests that dynamical regression is a robust phenomenon with respect to fluctuations.

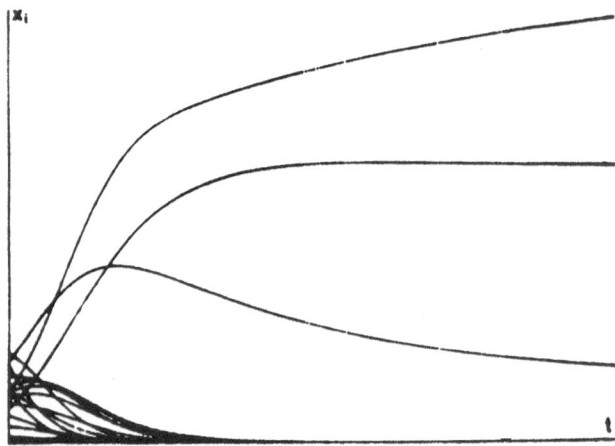

Figure 2a

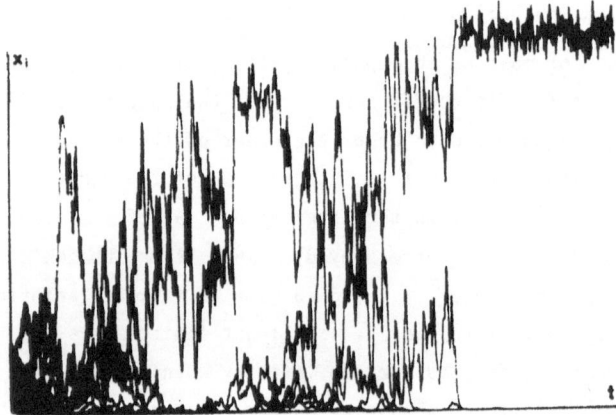

Figure 2b

REFERENCES

Coste, J., J. Peyraud, and P. Coullet (1978): Does complexity favor the existence of persistent ecosystems?, *J. Theor. Biol.* 73:359-362.

Gilpin, M. (1975): Stability of a feasible predator-prey system, *Nature* 254:137-139.

May, R. (1972): Will a large complex system be stable?, *Nature* 238:413-414.

May, R., and W. Leonard (1975): Non linear aspects of competition between three species, *SIAM J. Appl. Math.* 29:243-253.

McGehee, R., and R.A. Armstrong (1977): Some mathematical problems concerning the ecological principle of competitive exclusion, *J. Diff. Equations* 23:30-52.

Roberts, A. (1974): The stability of feasible random ecosystems, *Nature*, 251:607-608.

COMPETITION AND AGE-STRUCTURE

J.M. Cushing and M. Saleem*

1. INTRODUCTION

The theory of two species competing for the same resources plays an extremely important role in present day ecology, both applied and theoretical. The role played by age structure in competitive interactions has not, however, been investigated to the extent that it has been for predator-prey interactions. The purpose of this work is to provide some insight into the dynamics of competitive interactions with age structure. To make the model simple and tractable we consider a two-species competition model where only one species has age-structure. This is not purely a mathematical simplification but has some biological justification as well. For example, for some populations of birds and lizards competing for insects only the lizards have a significant age-structure (we thank R.M. May who suggested this biological application in a personal conversation). The existence and stability of the equilibria of the general model equation (2.3) is studied as it depends on the inherent net reproductive rate n of the age-structured population.

2. MODEL EQUATIONS

As in Cushing and Saleem (1982), the model equations are derived from McKendrick and Pai (1910) equations. The equation for total population size $P(t)$ of age-structured species is (see Cushing and Saleem, 1982)

$$\frac{dP}{dt} = -\mu P + \int_0^\infty f_a(a,t)P(t-a)e^{-\mu a}da,$$

where μ is the (constant) per unit death rate and f is the (per unit density) fecundity rate of P. f is assumed to be of the form

$$f = b\beta(a)h(P,Q).$$

The function $\beta(a)$ is assumed to satisfy for $a \geq 0$

$$\beta(0) = 0, \quad \beta(a) \geq 0, \quad \int_0^\infty e^{-\mu a}\beta(a)ada < +\infty \tag{2.1}$$

while the function h is assumed to be defined and twice continuously differentiable on $\mathbf{R}^2$ and to satisfy

*Research supported by National Scholarship for Study Abroad (1980) from the Government of India.

$$h(0,0) = 1, \quad h(P,Q) \geq 0, \quad h_p < 0, \quad h_Q < 0 \quad \text{for} \quad P, \quad Q \geq 0. \qquad (2.2)$$

The dynamics of the non-age structured species is assumed to be governed by the equation

$$\frac{dQ}{dt} = rQg(P,Q).$$

Thus the model equations for this paper are

$$\frac{dP}{dt} = -\mu P + bh(P,Q) \int_0^\infty \beta'(a)P(t-a)e^{-\mu a}da$$

$$\frac{dQ}{dt} = rQg(P,Q), \qquad (2.3)$$

where $r > 0$ is a constant and g is assumed to be defined and twice continuously differentiable on $\mathbf{R}^2$ and to satisfy

$$g(0,0) = 1, \quad g_p < 0, \quad g_Q < 0 \quad \text{for} \quad P, \quad Q \geq 0$$

$$g(P,Q) < 0 \quad \text{for all large} \quad P^2 + Q^2, \quad p \geq 0, \quad Q \geq 0. \qquad (2.4)$$

We define $n = b\beta^*(\mu)$, where $\beta^*(\mu) = \int_0^\infty e^{-\mu a}\beta(a)da$, to be the inherent net reproductive rate of P for our future reference.

3. AN EXAMPLE

A simple example (but yet one which motivates the general results for the model equations) is presented in this section. To describe it, we consider

$$\beta(a) = \frac{1}{m^2} ae^{-a/m}; \quad h(P,Q) = [1 - c_{11}P - c_{12}Q]_+, \quad g(P,Q) = 1 - c_{21}P - c_{22}Q, \quad (3.1)$$

where c_{ij}, $m > 0$ (constants). $[\cdot]_+$ means $[x]_+ = x$ if $x \geq 0$ and $-x$ if $x < 0$. For (3.1), the model equations (2.3) have the following equilibria

$$E_0 = [0,0], \quad E_1 = \left[\frac{1 - \frac{1}{n}}{c_{11}}, 0 \right], \quad E_2 = \left[0, \frac{1}{c_{22}} \right], \quad \text{and}$$

$$E_3 = \left[\frac{c_{22}(1 - \frac{1}{n}) - c_{12}}{\Delta}, \frac{c_{11} - c_{21}(1 - \frac{1}{n})}{\Delta} \right]$$

where $\Delta = c_{11}c_{22} - c_{12}c_{21}$ and $n = b/(\mu m+1)^2$. The existence and stability or instability of the above non-negative equilibria E_0, E_1, E_2 and positive equilibrium E_3 are shown in the following pictures given in Figures 1(a,b). In these graphs, the magnitude $|E-E_0|$ is plotted against the inherent net reproductive rate n. A "u" represents instability while an "s" stability. $n_{cr} = c_{22}/(c_{22}-c_{12})$, $\overline{n_{cr}} = c_{21}/(c_{21}-c_{11})$.

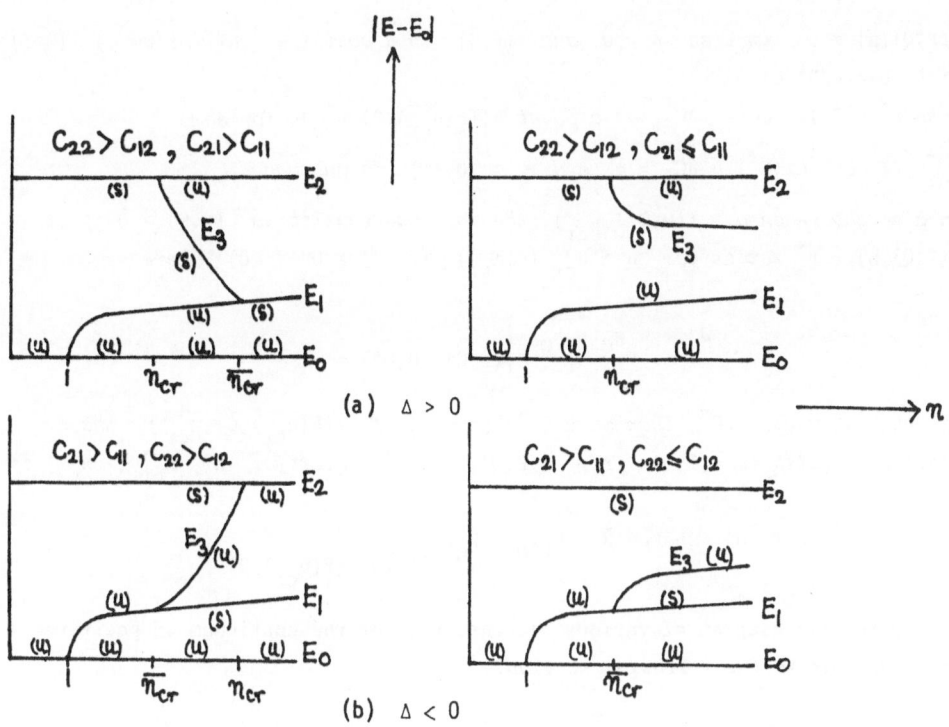

Figure 1

4. EXISTENCE OF EQUILIBRIA

Non-negative equilibria of (2.3) are solutions of

$$P[-1+h(P,Q)n] = 0, \qquad Qg(P,Q) = 0. \tag{4.1}$$

These equations have at least two non-negative equilibria $E_0 = (0,0)$ and $E_1 = (P(n),0)$ if $1 < n < n^* \leq +\infty$. We assume that there also exists an equilibrium $E_2 = (0,Q^*)$. By monotonicity conditions (2.2) and (2.4) these are the only equilibria on the axis. To show the existence of positive equilibria we assume the following.

Let $\mathbf{R}$ and $\mathbf{R}^+$ denote the set of reals and non-negative reals respectively. Let $\mathbf{C}^+ = \mathbf{R}^+ \times \mathbf{R}^+$ denote the positive quadrant. A continuum is a closed connected set. If $E^+ \subset \mathbf{C}^+ \times \mathbf{R}$ is a set and $c\ell(E^+)$ is a continuum, then the *spectrum* of E^+ is defined to be the interval $\sigma = \{n: ((P,Q),n) \in E^+$ for some $(P,Q) \in \mathbf{C}^+\}$. The following theorem can be proved using the global bifurcation results of Rabinowitz (1971).

THEOREM 1 *Assume* $h(0,Q^*) > 0$. *There exists a set* $E^+ \subset C^+ \times \mathbf{R}$ *whose closure* $c\ell(E^+)$ *is a continuum, with the following properties:*

(a) $((0,Q^*),n_{cr}) \in c\ell(E^+)$, $n_{cr} = 1/h(0,Q^*)$.

(b) $((P,Q),n) \in E^+$ *implies* $n > 0$ *and* (P,Q) *is a positive equilibrium of* (2.3) *with* $b = n/\beta^*(u)$.

(c) *Either* $c\ell(E^+)$ *meets* ∂C^+ *at a point* $((P(\overline{n_{cr}}),0),\overline{n_{cr}})$, *for some* $\overline{n_{cr}} > 1$ *or* $c\ell(E^+) \cap \partial C^+ = \phi$ *and* $\sigma \subset (0,+\infty)$ *is unbounded.*

(d) *In a neighborhood of* $((0,Q^*),n_{cr})$, *the positive equilibria* $(P,Q) > 0$ *from* $((P,Q),n) \in E^+$ *exist for* $n > n_{cr}$ $(< n_{cr})$ *if* $J^* > 0$ (< 0) *where*

$$J^* = h_P g_Q - h_Q g_P \big|_{(P,Q)=(0,Q^*)} \quad .$$

(e) *If* $c\ell(E^+)$ *meets* ∂C^+, *then in a neighborhood of* $((P(\overline{n_{cr}}),0),\overline{n_{cr}})$, *the positive equilibria* (P,Q) *from* $((P,Q),n) \in E^+$ *exist for* $n < \overline{n_{cr}}$ $(> \overline{n_{cr}})$ *if*

$$\bar{J} > 0 \; (< 0) \quad where \quad \bar{J} = h_P g_Q - h_Q g_P \big|_{(P,Q)=(P(\overline{n_{cr}}),0)} \quad .$$

A systematic diagram of various alternatives for the continuum of positive equilibria appears in the following section.

5. STABILITY

In this section , the results about the stability or instability of non-negative equilibria of (2.3) will be given. We found that $E_0 = (0,0)$ is always unstable. $E_2 = (0,Q^*)$ is (locally) asymptotically stable (unstable) if $n > n_{cr}$ $(< n_{cr})$. Equilibrium $E_3 = (P(n),0)$ is (locally) asymptotically stable (unstable) for $n > \overline{n_{cr}}$ $(< \overline{n_{cr}})$. For the stability of positive equilibria $(P,Q) > 0$ from the branch E^+ (Theorem 1) near the critical values n_{cr} and $\overline{n_{cr}}$ we give the following theorem which can be proved by a linearized stability analysis.

THEOREM 2 *Assume* (2.2) *and* (2.4) *and* $h(0,Q^*) > 0$.

(a) *If* $J^* \neq 0$ *then the positive equilibria from* E^+ *are (locally) asymptotically stable (unstable) if* $n \approx n_{cr}$ *and* $n > n_{cr}$ $(< n_{cr})$.

Assume in addition alternative (c) of Theorem 1.

(b) *If* $\bar{J} \neq 0$ *then the positive equilibria form* E^+ *are (locally) asymptotically stable (unstable) if* $n < \overline{n_{cr}}$ $(> \overline{n_{cr}})$.

Note that by Theorem 1, the stability of the positive equilibrium at least near the critical values n_{cr} and $\overline{n_{cr}}$ is determined by the sign of the Jacobians

J* and J̄.

Pictorially, the results of this section are summarized in Figure 2 below.

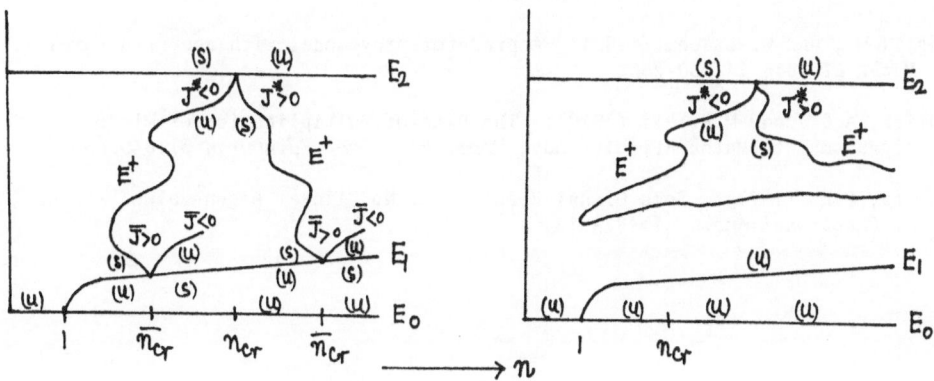

Figure 2

These graphs are just a systematic representation to show how many equilibria are
there for a certain value of n. Figure 2(a) represents the first alternative in
(c) of Theorem 1, while Figure 2(b) the second alternative in (c). In this figure
only two of many of orientations for E⁺ are shown in each of the alternatives of
Theorem 1.

6. CONCLUSIONS

     We point out the following interesting features from Figure 1 for the special
case (3.1).

1. The theory of two species competition where one species has age structure mimics,
    at least with regard to existence and stability of equilibria, that of the
    classical Lotka-Volterra theory in the sense that positive equilibrium exists
    only for a restricted set of parameter values and its stability depends on the
    sign of Δ.
2. In order to survive the competition with a non-age-structured species, the age-
    structured species must have an inherent net reproductive rate n which lies
    in a specified critical interval $(n_{cr}, \overline{n_{cr}})$.

     From Figure 2 it is clear that more or less the same features as exist in the
special case (3.1) are present for the general system (2.3). Namely: for
(2.3) there exist two values n_{cr} and $\overline{n_{cr}}$ of n which play a critical role in
the sense that stability is determined by whether n is greater or smaller than
these values. *Furthermore* instead of the sign Δ, it is now the sign of the

Jacobian J at the equilibrium that determines the stability or instability of equilibria.

REFERENCES

Cushing, J.M., and M. Saleem (1982): A predator-prey model with age-structure, *J. Math. Biology* 14:230-250.

McKendrick, A.G., and M.R. Pai (1910): The Rate of Multiplication of Micro-organisms: A Mathematical Study, *Proc. Roy. Soc. Edinburgh* 31:649-655.

Rabinowitz, P.H. (1971): Some Global Results for Non-linear Eigen-value Problems, *J. Functional Analy.* 7:487-513.

SELECTION OF MOLECULAR SPECIES
BY COMPETITION FOR A COMMON SOURCE

M.I. Granero-Porati and A. Porati

1. INTRODUCTION

One of the main problems modern biology is faced with is the one of morpho-
genesis. It is well known that cells of different tissues of the same organism have
the same "potentiality of information", and it is also known that the expression of
this common potentiality differs from cell to cell.

The attempts of a rational explanation of this fundamental phenomenon must
necessarily account for two different facts. The first one is the existence of the
so-called "positional information" (Wolpert, 1969) (a cell "knows" its position).
The second is the existence of a molecular selection, a necessary prerequisite for
differentiation.

The problem of positional information has been analyzed, from the theoretical
point of view, by several authors. We recall, in particular, a well-known model
proposed by Gierer (Gierer and Meinhardt, 1972) and based on lateral inhibition. It
has been shown, in the case of this model, the possibility of a spontaneous estab-
lishment of a gradient of concentration of chemical substances (morphogens), starting
from a nearly equal distribution (Granero et al. 1976; Haken and Olbrich, 1978).

For the second point, as far as we know, very few theoretical attempts have
been made.

Nevertheless, an increasing number of experimental observations (Puglisi and
Algeri, 1971; Pollak and Sutton, 1980; Ferrero et al. 1981) indicate that in several
eucaryotic organisms the differential expression of the nuclear genetic complement
is controlled by products (in particular, enzymes) of mitochondrial macromolecular
synthesis.

The experimental observation that mitochondrial protein synthesis plays a
precise role in the differentiation process, lead to express the hypothesis that the
modulation of the activity of the nuclear gene complement relies on macromolecular
signals, also of mitochondrial origin, whose concentration is a function of energy
availability (i.e. ATP concentration).

One can in fact suppose that the ATP variations give both a qualitative
and quantitative selection of the synthesized enzymes. The ATP level would modu-
late the amount and the type of polypeptides synthesized in function of their parameters (number of residues, turn-over, functioning threshold and so on).

We present here a simple mathematical model on the role that ATP level may
play on the modulation of protein synthesis.

2. THE MODEL

We start with the simplest mathematical model suggested by the previous biological evidence. We construct a model with the aim of investigating the possibility that, owing to competition among molecular species due to the consumption of a
common energetic material, we can have different spectra for the concentrations,
when the quantity of common material has different values.

The variables of the model are:

x_i = concentration of macromolecules of the i-th species (i = 1,2,...,k)

A = concentration of molecules of ATP

Φ = input flux of A

τ_i = mean life of the i-th macromolecule

h_i = efficiency parameter (roughly proportional to $1/n_i$, where n_i is the number
of sub-units of the i-th molecule).

The differential equations proposed are:

$$\dot{x}_i = -x_i/\tau_i + h_i A(t) \quad (i = 1,2,...,k)$$

$$\dot{A} = \Phi(t) - \sum_{j=1}^{k} h_j A(t)$$

(1)

The first k equations describe the temporal evolution of the k macromolecular
concentrations as a function of their decay rates and their formation by means of
ATP. The last equation describes the temporal evoluation of A(t) and means a
simple balance between input and output.

It must be noted that a system of this kind has been recently proposed
(Ebeling and Schmelzer, 1980; Schmelzer and Ebeling, 1980) in the context of competing predators. The main difference of the present model is the absence of
autocatalytic terms: this choice is due to the fact that proteins are not self-
replicating units.

Let us now consider some particular case.

2.1 *The Case of Constant Input Flux*

In the first case we consider the input flux, $\Phi(t)$, as a constant Φ_0.
This hypothesis is near to the experimental conditions: in fact (see for example
Kaniuga et al. 1969) it is possible, with some specific inhibitor of ATP produc-
tion, to keep the input flux constant, but at different levels.

System (1) is now a linear system of k + 1 ordinary differential equations
of the first order. It is easily recognised that the steady state (s.s) is
asymptotically stable for every value of the parameters. In fact, after translation into the origin of the equilibrium point, system (1) is of the form:

$$\dot{\underline{\xi}} = \Gamma \underline{\xi} \tag{2}$$

and the characteristic equation is:

$$(\sum_{j=1}^{k} h_j + \lambda) \prod_{j=1}^{k} (\frac{1}{\tau_j} + \lambda) = 0 \tag{3}$$

The eigenvalues (all negative) are:

$$\lambda_i = \frac{-1}{\tau_i} \quad (i = 1,2,\ldots,k)$$

$$\lambda_{k+1} = -\sum h_j \tag{4}$$

The general solution of (1), is, in this case:

$$A(t) = \frac{\phi_0}{\Sigma h_j} + \left[A(o) - \frac{\phi_0}{\Sigma h_j} \right] e^{-\Sigma h_j t}$$

$$x_i(t) = \left[x_i(0) - \frac{h_i \tau_i \phi_0}{\Sigma h_j} - \frac{h_i(A(o) - \phi_0/\Sigma h_j)}{1/\tau_i - \Sigma h_j} \right] e^{-t/\tau_i} \tag{5}$$

$$+ \frac{h_i[A(o) - \phi_0/\Sigma h_j]}{1/\tau_i - \Sigma h_j} + \frac{h_i \tau_i \phi_0}{\Sigma h_j}$$

In Figure 1 the temporal behaviour of the concentrations of two molecular species is illustrated for the particular case $A(o) = 1$, $\phi_0 = 1$. From this simple example, one can see that, if some threshold for the functionality of proteins exists, (and this is realistic from biological evidence), during the time one can have a "selection" of molecular species, and therefore, a differentiation.

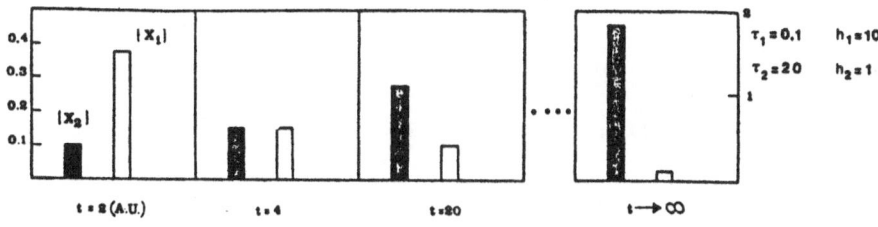

We have also considered a more simple case, that is the one in which the ATP level is constant for $0 \le t < t^*$ and suddenly varies at $t = t^*$ assuming a new constant level for $t > t^*$. The result is summarized in Figure 2: the ratio between the concentrations of two molecular species is plotted as a function of time, with A_o (level of ATP) as a parameter. It must be noted that, for high level of ATP the ratio has a slow variation in time. When the ATP level is lowered, the ratio exhibits a peak (and the sharpness of the peak is enhanced for lower values of ATP level). This behaviour shows that, under competition of molecular species for a common energetic source, the spectra of relative concentrations are different for different values of ATP level.

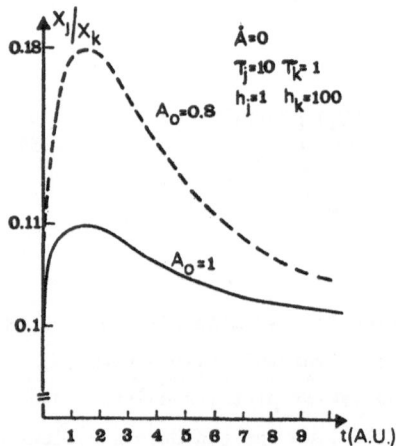

Figure 2

2.2 *The Case of Oscillating Flux*

In this case we consider the input flux of ATP as a periodic function of time, assuming a termporal behaviour of this kind:

$$\Phi(t) = k \cos \omega t + \Phi_0$$

This hypothesis is more similar to the "in vivo" conditions: in fact in the cell, the ATP level is under the control of internal mechanisms that can originate temporal oscillations (as, for example, glycolytic oscillations (Boiteux and Hess, 1974)).

In this case system (1) is a linear non homogeneous system whose general solution is:

$$A(t) = \tilde{A}e^{-\Sigma h_j t} + \alpha \cos \omega t + \beta \sin \omega t + \phi_0/\Sigma h_j$$

$$x_i(t) = \tilde{x}_i e^{-t/\tau_i} + c_i e^{-\Sigma h_j t} + a_i \cos \omega t + b_i \sin \omega t + \frac{h_i \tau_i \phi_0}{\Sigma h_j}$$

(6)

where:

$$\tilde{A} = \left[A(o) - \frac{k\Sigma h_j}{\omega^2 + (\Sigma h_j)^2} - \frac{\phi_0}{\Sigma h_j} \right]; \quad \alpha = \frac{k\Sigma h_j}{\omega^2 + (\Sigma h_j)^2}; \quad \beta = \frac{k\omega}{\omega^2 + (\Sigma h_j)^2}$$

$$\tilde{x}_i = x_i(0) - \frac{h_i}{1/\tau_i + \Sigma h_j} [A(o) - \alpha - \phi_0/\Sigma h_j] - \frac{h_i}{1/\tau_i^2 + \omega^2} (\alpha/\tau_i - \beta\omega) - \frac{h_i \tau_i \phi_0}{\Sigma h_j}$$

$$c_i = \frac{h_i}{1/\tau_i + \Sigma h_j} [A(o) - \alpha - \phi_0/\Sigma h_j]; \quad a_i = \frac{h_i k(\Sigma h_j/\tau_i - \omega^2)}{(1/\tau_i^2 + \omega^2)[\omega^2 + (\Sigma h_j)^2]};$$

$$b_i = \frac{h_i k(\omega \Sigma h_j + \omega/\tau_i)}{(1/\tau_i^2 + \omega^2)}$$

One can see that, for $t > t_0$, the behaviour of $x_i(t)$ is of the kind:

$$x_i(t) \cong a_i \cos \omega t + b_i \sin \omega t + \text{const.}$$

(the exponential terms can be disregarded after a sufficiently long time). The relative influence of the terms "sinus" or "cosinus" is determined by the coefficients a_i and b_i, as illustrated in Figure 3.

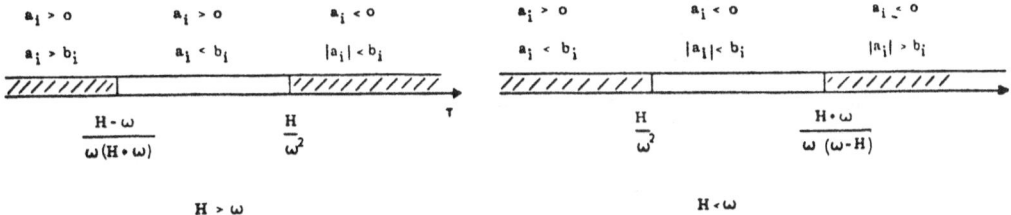

Figure 3

For example, if one has two molecular species with different mean lives (and we know that, for enzymes, the mean life ranges from minutes to days), it can happen that the behaviour of the first is cosinus-like, and that of the second is sinus-like. This means that the oscillations of the two concentrations are out of phase: when the concentration of a molecular species is increasing, that of the other is decreasing. A particular case of this kind is summarized in Figure 4.

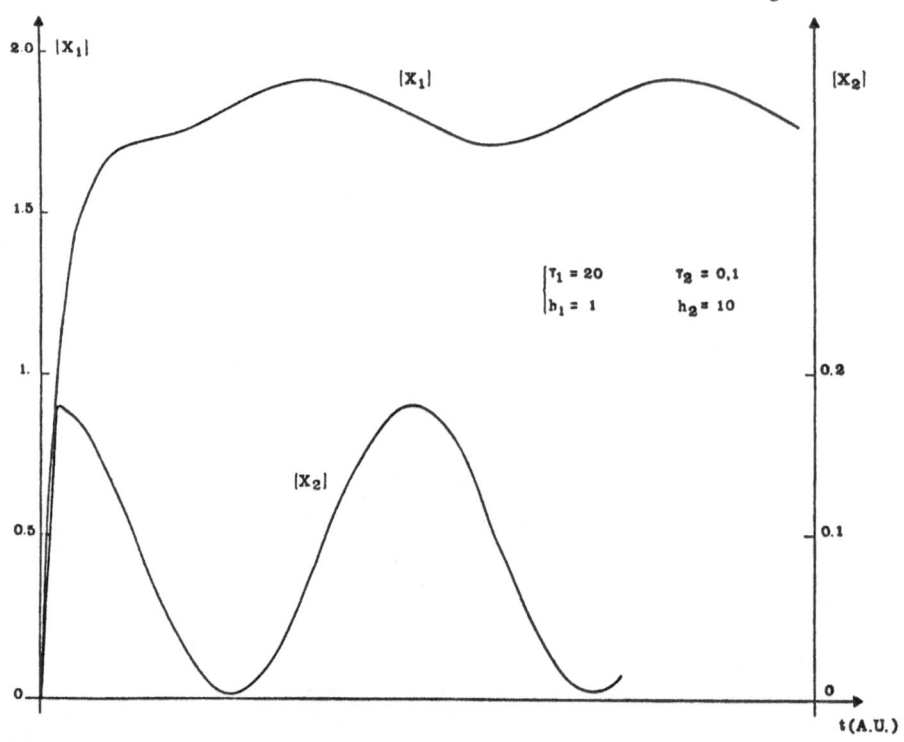

Figure 4

REFERENCES

Boiteux, A., and B. Hess (1974): *Faraday Symposia of the Chemical Society* 9:202-214.

Ebeling, W., and J. Schmelzer (1980): *Zeitschrift für Physikalische Chemie* 261:677-696.

Ferrero, I., C. Rossi, N. Marmiroli, C. Donnini, and P.P. Puglisi (1981): *Antonie van Leeuwenhoek* 47:311-323.

Gierer, A., and H. Meinhardt (1972): *Kybernetik* 12:30-59.

Granero, M.I., A. Porati, and D. Zanacca (1977): *Journal of Mathematical Biology* 4:21-27.

Haken, H., and H. Olbrich (1978): *Journal of Mathematical Biology* 6:317-331.

Kaniuga, Z., J. Bryla, and E.C. Slater (1969): In *Inhibitors - Tools in Cell Research,* Springer-Berlin.

Pollak, J.K., and R. Sutton (1980): *Trends in Biochemical Science,* 23-27.

Puglisi, P.P., and A.A. Algeri (1971): *Molecular and General Genetics* 110:110-117.

Schmelzer, J., and W. Ebeling (1980): *Studia Biophysica* 80:12-31.

Wolpert, L. (1969): *Journal of Theoretical Biology* 25:1-47.

AN ANALYSIS OF COMPETITION BETWEEN THE LARVAE OF A DEPOSIT-FEEDING
AND A FILTER-FEEDING SPECIES OF CHIRONOMID (DIPTERA), USING A MECHANISTIC MODEL

Joseph B. Rasmussen

Density manipulations performed in field enclosures on larvae of *Chironomus riparius* Meigen and *Glyptotendipes paripes* (Edwards) (Figure 1) have shown that increased larval density decreases growth rate in a concave-upward hyperbolic manner. No density effects could be demonstrated at densities below 1 larva/cm^2. The density manipulations were carried out with 1:9 mixtures of the species, and with 9:1 mixtures, and in both cases, the minority species was less affected by competition. This indicates that the larval competition will likely contribute to stable coexistence rather than extinction of either species. My analyses of gut contents, behavioral observations, and food addition experiments, have demonstrated that *G. paripes* larvae filter-feed and *C. riparius* larvae deposit-feed on fine material on the mud surface. Both species live in mud-mucous tubes within the top 2-3 cm. of mud, with no tendency toward microhabitat segregation, and the life-cycles are nearly identical with regard to timing of emergence, oviposition, and larval growth.

Papers giving the details of the experiments and analyses described above are in preparation. In this paper I will discuss the possible mechanisms involved in the density effects observed, and develop some competition models based on the proposed mechanisms. I will then outline some criteria for coexistence under various food regimes.

Tubiculous benthic chironomid larvae are known to utilize only a small area around the tube mouth (even if more is available) and to space themselves regularly at high densities and so minimize or remove overlap between the territories utilized (Edgar and Meadows 1969; McLachlan 1977; Wiley 1980). This would account for the absence of density effects at low densities and their abrupt onset. The reciprocal of the threshold density would serve as an estimate of the area utilized by a larva. Once the territories begin to abut, a sharp increase in competition occurs. Behavioral observations indicate that this results mainly from territory overlap rather than from reduction in territory size. If a is the territory area, and F is the rate at which assimilatable material becomes available within it, then assuming perfect spacing of the larvae, and that the turnover time is long enough to allow complete food depletion by one larva, then

$$A = Fa/N \quad N/a \geq 1 \quad (A = Fa \text{ if } N/a < 1) \tag{1}$$

where A = the amount of food assimilated by each larva per unit time, and N/a = the average number of larvae sharing the territory. Wiley (1980) has shown that when individuals of *Cricotopus bicinctus* directly contact their neighbors, as they do infrequently, the resulting avoidance reaction will detract from feeding time. This will however not likely prove to be a factor significant enough to prevent A from being limited by food availability per individual. Equation 1 produces a strongly curved hyperbola, in fact, much more strongly curved than that observed in the density manipulation experiments (Figure 1). It is likely that turnover time of the food will be short enough so that food depletion will be greater at high N/a, with a diminishing returns mechanism being present. If Ψ is the encounter rate between an individual larva and food particles when N/a = 1, then $e^{-\Psi(N/a)}$ will represent the proportion of food particles not found at a given larval density (P_0 term of a Poisson distribution), and $1 - e^{-\Psi(N/a)}$ will represent the proportion of food consumed. Thus equation 1 can be modified to

$$A = \frac{Fa(1 - e^{-\Psi(N/a)})}{N} \qquad N/a \geq 1 \qquad (2)$$

A more detailed derivation of this diminishing returns mechanism has been given by Hassell (1978) and Schoener (1973, 1978). Equation 2 can be given more realism in the manner outlined by Schoener (1973, 1974) who formulated mechanistic models for population growth of the form

$$(1/N) \ dN/dt = R[A(N)-M(N)] \qquad (3)$$

R being the proportionality constant between growth of individuals (G = A - M) and per capita rate of increase in numbers, and M = costs of metabolism in biomass units. We can write

$$G = \frac{Fa(1 - e^{-\Psi(N/a)})}{N} - C \qquad N/a \geq 1 \qquad (4)$$

for a purely food-limited growth vs density model, where C = the metabolic rate of the larvae in biomass units, or

$$G = \frac{Fa(1 - e^{-\Psi(N/a)})}{N} - C - \gamma N/a \qquad N/a \geq 1 \qquad (5)$$

to incorporate increased metabolic costs due to interactions. γ = the increase in metabolic costs per individual. Here it is assumed that larvae only influence each other's metabolic rate when territories overlap. The γ-effects will be termed interference effects after Schoener (1973). The importance of interference was established for larvae of *G. paripes* in a laboratory experiment where dissolved O_2 uptake per individual was found to increase sharply with density, when density exceeded 1 larva/cm^2 (Figure 2). Equations 4 and 5 are similar in theme to III and III' of Schoener (1973). Another model outlined by Schoener is obtained by

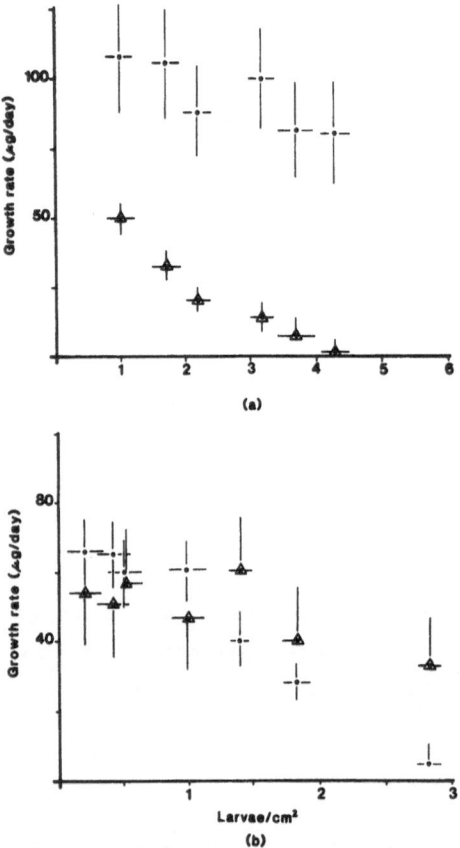

Figure 1   Mean larval growth rate (wet wt.) for *C. riparius*   (Δ) and *G. paripes* (·)
final instar larvae vs. density for   (a) 9:1 ratio   of *C.r.* to *G. p.*   and
(b) 1:9   ratio.   Results were determined from three-week enclosure experiments carried out in the field.

allowing the assimilation input to be limited by available feeding time (rather than
food availability) in such a way that density effects reduce the time available
for feeding, thus making the A term time-limited.   Translating Schoener's Equation
1 into the format employed here, we have

$$G = (1-\lambda(N/a))Fa - C - \gamma N/a \qquad N/a \geq 1 \qquad\qquad (6)$$

where   λ = the effect of density on time available for feeding.   Figure 3 shows
G(N) curves generated from equations 4-6 based as far as possible on plausible
parameter values.   Although both equations 4 and 5 produce hyperbolic relationships,
the shape of the curves in Figure 1 as well as the demonstrated presence of strong
interference effects, suggest equation 5 as best describing the mechanisms involved
in intraspecies competition in these larvae.   Equation 5 was approximated as a third
degree polynomial by expanding the exponential term into its series form, and the
curve was fitted to the data shown in Figure 1.   However, the parameter estimates

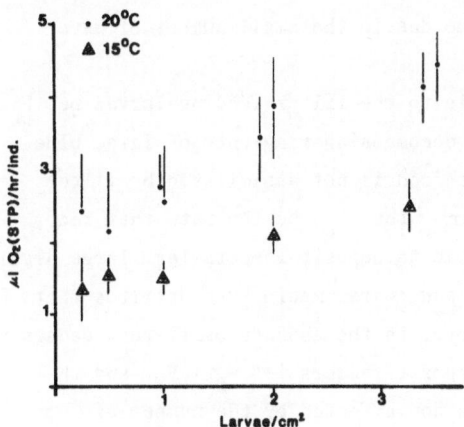

Figure 2   Oxygen uptake rate of final instar *G. paripes* larvae vs. larval density.
Respirometers contained sediment within which larvae built tubes; O_2
uptake of sediment and tubes was subtracted to arrive at estimates of O_2
uptake per larva.

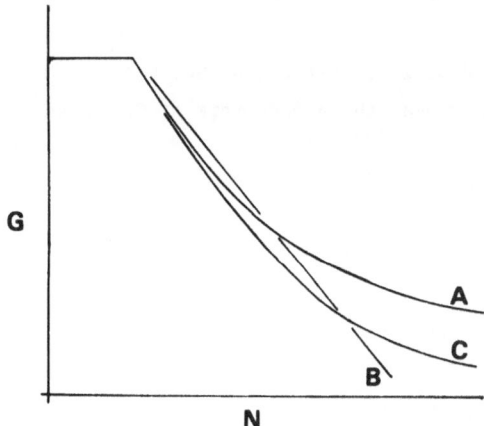

Figure 3   G(N)   curves for Equation 4 (curve A), Equation 5 (curve C), and
Equation 6 (curve B)   based on parameter values estimated from experiment
and from the literature.

lacked sufficient robustness to be of any value due to the small number of data points and their variance.

Let the rate that food becomes available to the filter-feeding larvae be F_F (this includes mainly small phytoplankton and decomposing fragments of large blue-green algal colonies), and assume that if this food is not assimilated by filter feeders it becomes available for deposit-feeders. Let F_D be the rate that food, unavailable to filter-feeders, becomes available to deposit-feeders (eg. large algal colonies that settle out of the water column, and coarse macrophyte detritus that fragments and decomposes in the mud). Therefore, in the absence of filter-feeders the rate at which food becomes available to deposit-feeders is $F_D + F_F$ and the rate that food becomes available to filters is not affected by the number of deposit-feeders present. Since *G. paripes* has been shown experimentally not to respond to addition of deposit-based detrital food, it is probably reasonable to assume that food depletion by *C. riparius* does not affect the quantity available to *G. paripes*. Assume for simplicity in the following argument, that $a = 1$ for both species; and, we can dispense with the proviso that $N/a \geq 1$ in the two-species analysis, since the zero-growth isoclines will involve only high larval densities.

The growth rate for *G. paripes* in the presence of *C. riparius* is therefore

$$G_F = \frac{F_F}{N_F} (1 - e^{-\Psi_F N_F}) - C_F - \gamma_F N_F - \gamma_{FD} N_D \qquad (7)$$

where γ_{FD} is the effect of an individual deposit-feeder on the metabolic costs of filter feeders. In the presence of *G. paripes*, the growth rate of *C. reiparius* is

$$G_D = \frac{F_F(e^{-\Psi_F N_F}) + F_D(1 - e^{-\Psi_D N_D})}{N_D} - C_D - \gamma_D N_D - \gamma_{DF} N_F \qquad (8)$$

where γ_{DF} is the effect of an individual filter-feeder on the metabolic costs of deposit-feeders.

In order to develop equations for zero-growth isoclines, equations 7 and 8 must be cast into a form where the numbers of one species can be written as a function of the numbers of the other. This can be accomplished by expanding the exponential expressions to yield polynomials in N_F and N_D. Using a quadratic approximation, equation 7 becomes

$$G_F = \frac{1}{6} F_F \Psi_F^3 N_F^2 - (\frac{1}{2} F_F \Psi_F^2 + \gamma_F) N_F + F_F \Psi_F - C_F - \gamma_{FD} N_D \qquad (9)$$

Then for $G_F = 0$,

$$N_F = \frac{\gamma_F + \frac{1}{2} F_F \psi_F^2 - [\gamma_F^2 + \gamma_F F_F \psi_F^2 + \frac{1}{4} F_F^2 \psi_F^4 - \frac{2}{3} F_F \psi_F^3 (F_F \psi_F - C_F - \gamma_{FD} N_D)]^{\frac{1}{2}}}{\frac{1}{3} F_F \psi_F^3} \qquad (10)$$

This isocline intercepts the N_D axis at

$$X_D = \frac{F_F \psi_F - C_F}{\gamma_{FD}} \qquad (11)$$

and the N_F axis at

$$K_F = \frac{\gamma_F + \frac{1}{2} F_F \psi_F^2 - [\gamma_F^2 + F_F \psi_F^2 \gamma_F + \frac{1}{4} F_F^2 \psi_F^4 - \frac{2}{3} F_F \psi_F^3 (F_F \psi_F - C_F)]^{\frac{1}{2}}}{\frac{1}{3} F_F \psi_F^3} \qquad (12)$$

Similarly

$$G_D = \frac{1}{2} \psi_D \psi_F^2 N_F^2 + [\frac{1}{2} F_F \psi_D^2 \psi_F N_D - F_F \psi_F \psi_D - \gamma_{DF}] N_F + F[\psi_D - \frac{1}{2} \psi_D^2 N_D + \frac{1}{6} \psi_D^3 N_D^2 - C_D - \gamma_D N_D] \qquad (13)$$

where $F = F_D + F_F$.

Then setting $G_D = 0$ we obtain the other isocline

$$N_F = \frac{F_F \psi_F \psi_D + \gamma_{DF} - \frac{1}{2} F_F \psi_D^2 \psi_F N_D}{\psi_D \psi_F^2}$$

$$\frac{- [(F_F \psi_F \psi_D + \gamma_{DF} - \frac{1}{2} \psi_D^2 \psi_F F_F N_D)^2 - 2\psi_D \psi_F^2 [F(\psi_D - \frac{1}{2} \psi_D^2 N_D + \frac{1}{6} \psi_D^3 N_D^2) - C_D - \gamma_D N_D]]^{\frac{1}{2}}}{\psi_D \psi_F^2} \qquad (14)$$

which intercepts the N_F axis at

$$X_F = \frac{F_F \psi_F \psi_D + \gamma_{DF} - [F_F^2 \psi_F^2 \psi_D^2 + 2\gamma_{DF} F_F \psi_F \psi_D + \gamma_{DF}^2 - 2\psi_D \psi_F^2 F + 2\psi_D \psi_F^2 C_D]^{\frac{1}{2}}}{\psi_D \psi_F^2} \qquad (15)$$

and the N_D axis at

$$K_D = \frac{\gamma_D + \frac{1}{2} F \psi_D^2 - [\gamma_D^2 + F \psi_D^2 \gamma_D + \frac{1}{4} F^2 \psi_D^4 - \frac{2}{3} F \psi_D^3 (F \psi_D - C_D)]^{\frac{1}{2}}}{\frac{1}{3} F \psi_D^3} \qquad (16)$$

When interference effects are low, equations 10 and 14 will yield strongly curved zero-growth isoclines, for which the quadratic approximation is not accurate. In fact, equations 12, 15, and 16, used to generate intercepts will yield complex numbers when γ values are low. However, the experiment depicted in Figure 2 indicates that metabolic costs account for at least half (low density) to four-fifths (high density) of total assimilation, and that therefore the values of C and γ should be quite high. With reasonably high values of C and γ, equation 10 and 14 yield isoclines that are nearly straight, and that intercept the N_F and N_D axes. Under these conditions then, the quadratic approximation used here seems sufficient for an accurate representation. Furthermore, the model is of no interest when values are low since coexistence is axiomatic (if $F_D > 0$) due to difference in the feeding mechanism of the two species. For these conditions then, the weak curvature of the isoclines allows the outcomes of competition to be approximately determined from the relationships of the intercepts on the N_F and N_D axes. The following criteria will then be accurate except when $X_D \simeq K_D$ and $X_F \simeq K_F$. When $X_F > K_F$ and $X_D > K_D$, only one intersection will occur and it will be stable; and, when $X_F < K_F$ and $X_D < K_D$ the intersection will represent a point of unstable equilibrium. Obviously then $X_D > K_D$ and $X_F > K_F$ results in extinction of the deposit-feeder, and $X_D < K_D$ and $X_F < K_F$ excludes the filter-feeder.

The outcomes predicted by this model are strongly affected by the feeding regime. When $F_D = 0$, and Ψ values are moderate to high, the deposit-feeder will tend to be eliminated unless metabolic costs for the filter-feeder are too high to permit growth at the levels of F_F present. A low value of γ_{DF} ($\gamma_{DF} < \gamma_F$) can prevent extinction of the deposit-feeder in this situation, but to make the co-existence stable $\gamma_{FD} < \gamma_D$ is also necessary. When Ψ values are very low, the two species can coexist even if $F_D = 0$ since exploitative competition is absent or nearly so. This requires however that $\gamma_{DF} < \gamma_F$ and $\gamma_{FD} < \gamma_D$.

When $F_D > 0$, but low in relation to F_F unstable equilibrium occurs for most parameter values, but when F_D becomes a substantial fraction of F, extinction of the filter-feeder occurs (unless of course $\Psi_D << \Psi_F$) when metabolic costs are all reasonably similar for the two species. Thus the fact that deposit-feeders have access to more food than filter-feeders leads to a tendency for extinction of the latter. If food particles deteriorate in quality rapidly as they settle to the substrate, this advantage to deposit-feeders would be reduced or even disappear; however, this factor has been ignored in this model. If metabolic costs for deposit-feeders are higher ($C_D > C_F$ and $\gamma_D > \gamma_F$) than those of filter-feeders, extinction of the latter is less likely. The most effective counterforce against extinction of filter-feeders here is $\gamma_{FD} < \gamma_F$. $\gamma_{DF} > \gamma_F$ will push toward unstable equilibrium.

In my study pond, sedimenting large *Aphanizomenon* (Cyanophyta) colonies and *Potamogeton* detritus are two abundant food sources not available to *G. paripes* but available to *C. riparius* (upon decomposition within the mud). *C. riparius* will tolerate higher densities than *G. paripes* (Figure 1b) within the enclosures, and this may in part reflect the greater range of food materials available to

C. riparius. If the carrying capacity for *C. riparius* is indeed higher than that for *G. paripes,* the analysis carried out in this paper suggests that, for stable coexistence between these larvae to be possible, the effect of *C. riparius* individuals on metabolic costs of *G. paripes* (γ_{FD}) should be less than the effect of *C. riparius* larvae on themselves (γ_D).

REFERENCES

Edgar, W.D., and P.S. Meadows (1969): Case construction, movement, spatial distribution and substrate selection in the larva of *Chironomus riparius* Meigen, *J. Exp. Biol.* 50:247-253.

Hassell, M.P. (1978): *The Dynamics of Arthropod Predator-Prey Systems,* Princeton University Press, Princeton, New Jersey.

McLachlan, A.J. (1977): Density and distribution in laboratory populations of midge larvae (Chironomidae: Diptera), *Hydrobiologia* 55:195-199.

Schoener, T.W. (1973): Population growth regulated by intraspecific competition for energy or time: some simple representations, *Theor. Pop. Biol.* 4:56-84.

_____ (1974): Competition and the form of the habitat shift, *Theor. Pop. Biol.* 6:265-307.

_____ (1978): Effects of density-restricted food encounter on some single-level competition models, *Theor. Pop. Biol.* 13:365-381.

Wiley, M.J. (1980): The components of substrate distribution, Ph.D. Dissertation, Univ. of Michigan, Ann Arbor, Michigan.

PART V:

PREDATOR-PREY SYSTEMS

THE DYNAMICS OF HOST-PARASITOID INTERACTIONS WITH HETEROGENEITY
HOST DENSITY DEPENDENCE AND VARIABLE SEX RATIOS

Michael P. Hassell

1. INTRODUCTION

Ever since the early work of Thompson (1924), Nicholson (1933) and Nicholson and Bailey (1935), there has been a tradition of modelling predator-prey interactions in discrete time that has run parallel with the considerable literature using differential equations following Lotka (1925) and Volterra (1926). The majority of these discrete models fall within the following general framework for a coupled predator and prey population:

$$N_{t+1} = \lambda N_t f(N_t, P_t) \tag{1a}$$

$$P_{t+1} = c N_t [1-f(N_t, P_t)] \tag{1b}$$

where N and P are, respectively, the prey and predator populations in successive generations, t and $t+1$, λ is the finite rate of increase of the prey population ($\lambda = e^r$, where r is the intrinsic rate of increase), f is a function defining the fraction of prey surviving predation by P_t searching predators, and c relates the total number of prey eaten to the numbers of predators of the next generation. Thus, in λ is contained all features affecting the prey's dynamics other than predation, and in f is contained all assumptions about the searching ability of the predators.

Equations (1a,b) suffer several weaknesses as a general predator-prey model. For instance, (1) most predators are polyphagous and have their reproduction little if at all dependent on just one of their prey species, making such coupled equations a very special case, (2) even for a specialist predator, reproduction will usually only be loosely coupled to the total number of prey eaten, making the structure of equation (1b) inappropriate, and (3) predators at different stages of development will search with quite different efficiencies, often for different sized prey, making the lack of age-structure in the model an important omission.

There exists, however, one class of predator whose life cycle is well suited to the simplifications of this model. These are the insect parasitoids that Thompson and Nicholson had primarily in mind as subjects for their models. Parasitoids make up about 10% of all Metazoan species, and play a most important part in the natural regulation of many other insect populations, as well as forming the basis of most biological pest control practices. Four features of their life cycle make them well suited to models of the form of equations (1a,b). (1) Several species are effectively monophagous or oligophagous. (2) Many species, especially in temperate regions, have fairly discrete generations and are well synchronized with their host populations. (3) It is only the adult female that searches for hosts, which somewhat reduces the need for a more complex age-structured model. (4) Instead of eating the prey they locate, parasitoids oviposit in or on them, so that the number of hosts attacked is a good indicator of parasitoid reproduction. Thus c in equation (1b) becomes the average number of parasitoid eggs laid per host attacked, tempered by any mortality of the parasitoid progeny occurring before adult emergence in the next generation.

One of the simplest versions of equations (1a,b) is the well-known model of Nicholson and Bailey (1935) in which f is given by the zero term of the Poisson distribution:

$$f = \exp(-aP_t) \tag{2}$$

and c = 1. The parameter a, the searching efficiency, is important as it recurs, in one guise or another, in almost all predator-prey or parasitoid host models. Simply stated, it is the proportion of hosts encountered per parasitoid throughout its searching life time. For each combination of λ and a the model has an unstable equilibrium, the slightest movement from which leads to rapidly expanding oscillations. This is quite at variance with the frequent observations of parasitoids and their hosts persisting over many generations, and also conflicts with what is seen from some of the spectacularly successful biological control programmes where the introduction of parasitoids has led to a marked reduction in pest numbers which are then maintained at these low levels in a relatively stable interaction.

It is in an effort to reconcile theory with such observations that parasitoid-host models have been considerably elaborated over the past few years. This paper examines several aspects of this development, concentrating on (1) some different forms for the function f, paying particular attention to spatial and other forms of heterogeneity, (2) the effects of including additional density dependence acting in the host life cycle and (3) the effects of biased parasitoid sex ratios.

2. SURVIVAL FROM PARASITISM

Ever since the papers of Holling (1959) and Watt (1959) there has been quite an industry putting forward different expressions for f in equations (1a,b). These in general fall into three categories: (1) elaboration of the functional response to host density, (2) provision for mutual interference between the parasitoids and (3) inclusion of some form of heterogeneity.

2.1 *Functional Responses*

Nicholson (1933) and Nicholson and Bailey (1935) implicitly assumed that the functional responses to host or prey density are linear with slope a. This assumes predators of unlimited appetite and parasitoids never short of eggs to lay. Much more realistic is the use of one of the equations describing type 2 or type 3 responses (see Hassell (1978) for a review). For instance, type 2 responses are often described by:

$$f = \exp\left[-\frac{a'P_t}{1+a'T_hN_t}\right] \tag{3}$$

where a' is the instantaneous version of a in equation (2), and T_h is the handling time expressed as a fraction of the total time available. These responses rise at a decreasing rate towards an upper asymptote representing the maximum attack rate per parasitoid. They are thus inversely density dependent and cannot contribute to population stability. Only the sigmoid type responses are density dependent, and then only over a range of host densities. While they must add to population stability to some extent, their effect on coupled interactions of the form of equations (1a,b) will in general be slight compared to some of the other factors discussed below.

2.2 *Mutual Interference*

Several expressions also exist for mutual interference between searching parasitoids (Hassell and Varley 1969, Rogers and Hassell 1974, Beddington 1975). For example, Beddington (1975) proposed:

$$f = \exp\left[-\frac{a'P_t}{1+T_hN+\gamma(P_t-1)}\right] \tag{4}$$

where γ is the 'time wasted' per encounter with another searching parasitoid per unit of searching time. This expression generates curvilinear decreases in searching efficiency (a) with increasing parasitoid density as shown in Figure 1. Within equations (1a,b) stability now hinges only upon the host rate of increase λ, the magnitude of the slope m^* evaluated at the equilibrium parasitoid density P^* and the value of T_h, as described by Hassell and May (1973) and further discussed below.

2.3 *Heterogeneity*

The introduction of heterogeneity in parasitoid-host and predator-prey models has been an important step removing the assumption of random parasitism that has afflicted so much past work. Heterogeneity can be manifest in many ways: for example, by the spatial distribution of parasitism from patch to patch, by temporal synchrony between host and parasitoid, or by differential susceptibility of the hosts to parasitism. All such mechanisms can have a marked impact on population

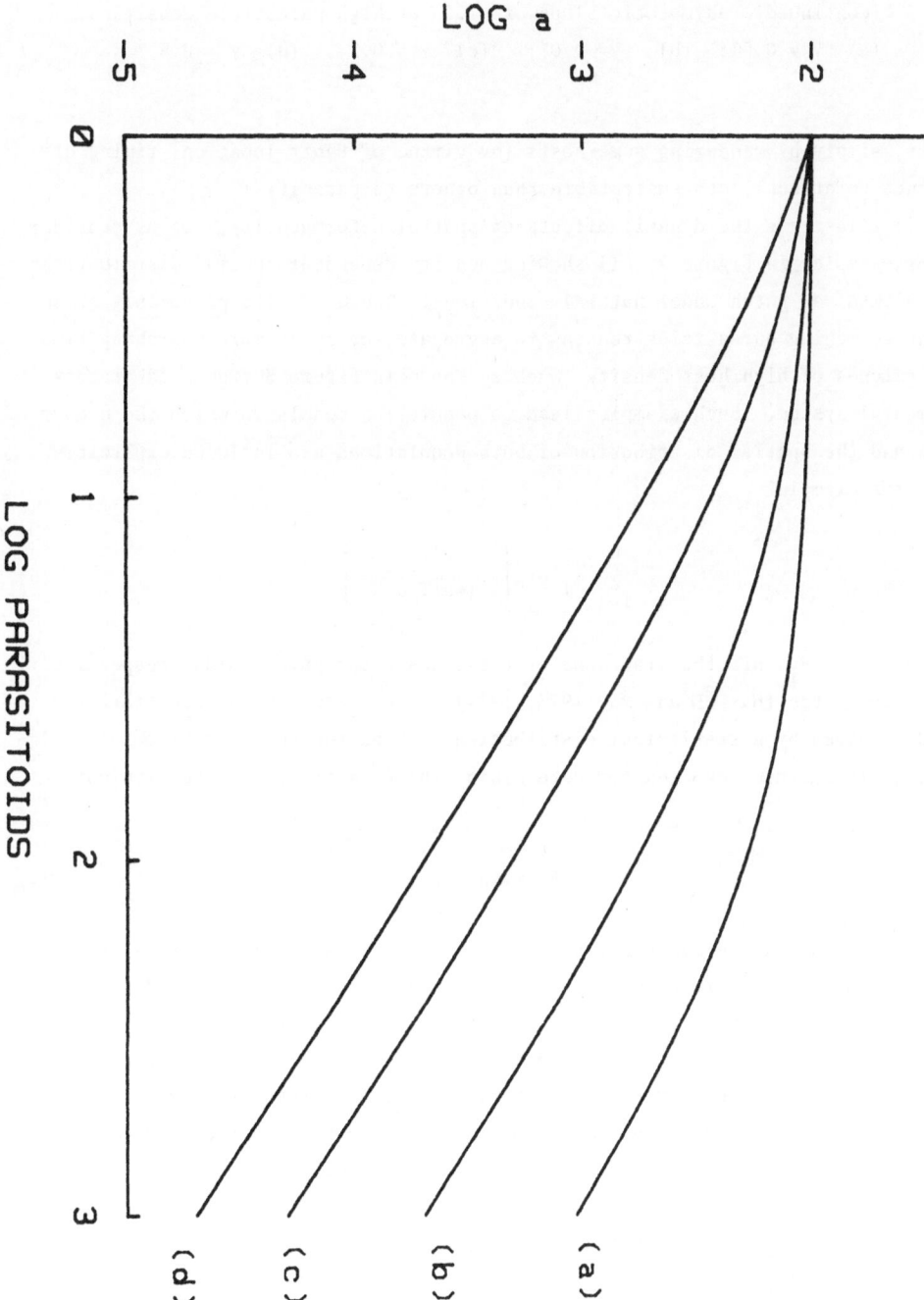

Figure 1 Interference relationships obtained from equation (4) for different values
of γ given that a' = 0.01 and T_h = 0. The searching efficiency term
a is derived from equation (2) and calculated from $a = \frac{1}{p} \ell n \frac{N}{N-N_a}$ where
N_a is the total number of hosts parasitized. All the curves have an

Figure 1 (continued) asymptotic slope of 1.0 at high parasitoid densities.
(a) $\gamma = 0.01$; (b) $\gamma = 0.05$; (c) $\gamma = 0.2$; (d) $\gamma = 0.5$

dynamics, simply by rendering some hosts (by virtue of their location, timing or resistance to attack) less susceptible than others to parasitism.

To illustrate the dynamic effects of spatial heterogeneity, let us consider the four examples in Figure 2, all showing density dependent spatial distributions of parasitism per patch under natural conditions. These results presumably arise from the searching parasitoids tending to aggregate, or spend more searching time, in the patches of high host density, much as shown in Figure 3 from a laboratory experimental system. Such examples lead to population models in which the number of patches and the spatial distribution of both populations are included explicitly giving, for example,

$$f = \sum_{i=1}^{n} \alpha_1 \exp\left[-\frac{a'\beta_i P_t}{1+a'T_h\alpha_i N_t}\right] \tag{5}$$

where α_i and β_i are the fractions of total hosts and parasitoids, respectively, in the ith patch (Hassell and May 1973, 1974). The α_i-set may be arbitrarily defined or given by a statistical distribution such as the negative binomial. Similarly, there are many choices for determining the β_i-set, one of the simplest being:

$$\beta_i = c\alpha_i\mu \tag{6}$$

where c is a normalisation constant and μ is the 'aggregation index'. With $\mu = 0$ the parasitoids are evenly distributed, independent of the host distribution; $\mu = 1$ corresponds to the fine-grained situation where the two distributions are perfectly correlated, and further increases in μ lead to increasing aggregation in the highest host density patches. Alternatively, more sophisticated expressions can be adopted as done by Comins and Hassell (1979) who assumed that the parasitoids forage optimally in each generation to maximise their rate of encounter with healthy hosts.

To consider just one level of patchiness will often be quite unrealistic, and a heirarchy of patches be more appropriate. Thus in the case of host insects feeding on herbaceous plants, there will be a distribution from leaf to leaf within plants, as well as from plant to plant, both of which a searching parasitoid may respond to in different ways. A model is now needed in which parasitism is summed over both levels, as in the expression:

$$f = \sum_{i=1}^{n} \alpha_i \sum_{x=1}^{x} \alpha_j \exp\left[-\frac{a'\beta_i\beta_j P_t}{1+a'T_h\alpha_i\alpha_j N_t}\right] \tag{7}$$

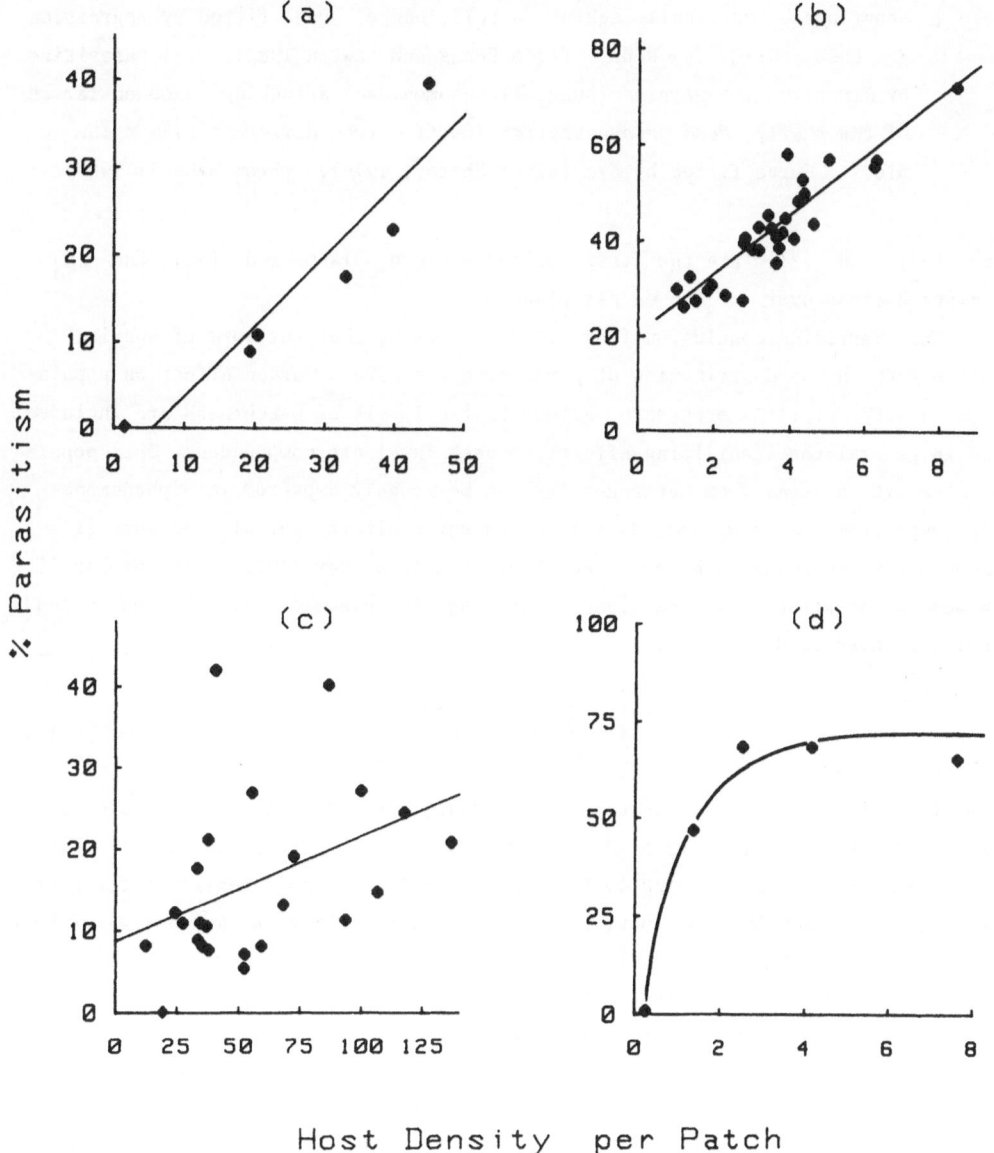

Host Density per Patch

Figure 2 Four examples showing density dependent parasitism per patch in the field.
(a) Parasitism by *Cyzenis albicans* (Fall.) Tachinidae) attacking winter
moth (*Operophtera brumata* (L.)) larvae per m^2 canopy area of different
trees. Data fitted by regression: Y = -4.51 + 0.82X (p < 0.05) (after
Hassell 1968). (b) Parasitism by *Aspidiotiphagus citrinus* (Craw.)
(Eulophidae) attacking the scale insect, *Fiorinia externa* Ferris on the
lower crown of 30 hemlock trees. Data fitted by regression: Y =
20.45 + 0.06X (p < 0.001) (from McClure 1977). (c) Parasitism by
Chrysocharis gemma (Walk.) (Eulophidae) attacking the holly leafminer
(*Phytomyza ilicis* Curtis) (Agromyzidae) per 0.25m^2 quadrats placed

Figure 2 (continued) vertically against a holly hedge. Data fitted by regression:
Y = 8.58 + 0.13X (p < 0.05) (from Heads and Lawton 1982). (d) Parasitism
by *Exenterus abruptorius* Thunb. (Ichneumonidae) attacking cocooned larvae
of the sawfly, *Neodiprion sertifer* (Geoff.) from different 15m × 30m
plots. Curve fitted by eye (after Sharov, 1979). (From Hassell 1982).

where $\{\alpha_i\}$ and $\{\beta_i\}$ are the distributions over n plants and $\{\alpha_j\}$ and $\{\beta\}_j$
the distributions over x leaves per plant.

The overriding conclusion from all such work is that any form of spatial
heterogeneity in the distribution of parasitism can have a marked effect on population stability. This is even more obvious if two levels of patchiness are included,
since in general the stabilizing effects at each level are compounded. This population stability arising from heterogeneity can be roughly captured in a phenomeno-
logical way from much simpler models in which any explicit spatial structure is
abandoned for something much more tractable, as done by May (1978). The probability
of a host being attacked is now given by the negative binomial distribution instead
of being assumed random. Thus,

$$f = \left[1 + \frac{aP_t}{k} \right]^{-k} \tag{8}$$

where k is the exponent of the negative binomial distribution. On inclusion in
equations (1a,b), the model is stable for all k < 1. As with equation (2), this
submodel for parasitism may readily be expanded to include more realistic features,
such as type 2 and type 3 functional responses, or making k a function of host
density (Hassell 1980).

The attractiveness of the idea that spatial heterogeneity is important to
the stability of natural interactions, lies in that its effects can operate at even
the lowest population levels (unlike host resource limitation or mutual interference). It therefore provides one of the most plausible explanations for the observed persistence of host-parasitoid interactions arising from successful biological control (Beddington, Free and Lawton 1978).

Spatial heterogeneity is, however, by no means the only kind of heterogeneity
that is important. Other forms can be just as important and have much the same
dynamic effects. Let us consider the case where there is some variability between
host individuals in their defences against parasitism (Hassell and Anderson 1983).
The almond moth *Ephestia cautella* (Walk.), for instance, can mount a haemocytic
defence against eggs and young larvae of the ichneumonid parasitoid, *Venturia
canescens* (Grav.). This effect is most marked in the older hosts, but is reduced at
the time of moulting when the haemocytes are elsewhere engaged (Rogers 1970). A
population of *E. cautella* will thus contain individuals of differing susceptibility
to parasitism by *V. canescens*. In other cases, some host individuals may be more

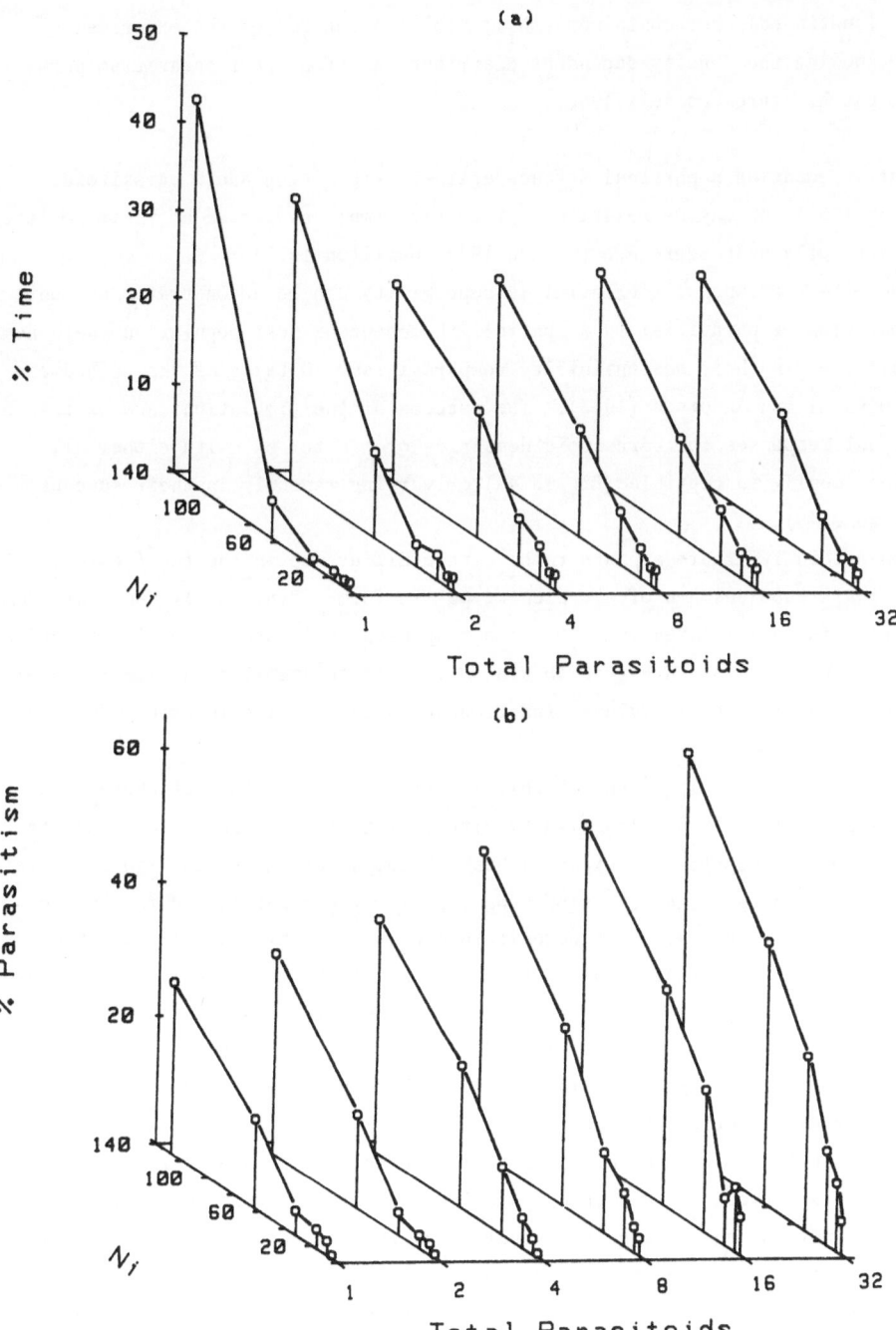

Figure 3 The aggregative responses of *Venturia canescens* (Grav.) and resulting
levels of parasitism per patch from a laboratory system in which the density of adult parasitoids per experiment if varied from 1 to 32 as shown.
(a) The aggregative response in terms of the percentage of the total observed time spent by the adult parasitoids on the different host densities

Figure 3 (continued) per container (N$_i$); (b) The outcome of the experiments
showing the density dependent distributions of percent parasitism per
patch. (From Hassell 1982).

successful at mounting a physical defence against ovipositing adult parasitoids
(Wellington 1957), or may be relatively protected simply by virtue of their position
in the centre of a host aggregate (Callen 1944, Hamilton 1971).

The effect of such differential susceptibility can be demonstrated by Monte
Carlo simulation of parasitism in a spatially homogeneous host population where host
individuals vary in their susceptibility to parasitism. Details of the method are
given in Hassell and Anderson (1983). The outcome of the simulations are expressed
as functional responses for parasitoid densities of 1 to 64, with either all
hosts being equally susceptible (Figure 4a) or varying randomly in their susceptibility (Figure 4b).

Superficially, there appears to be little difference in the two figures; all
the functional responses are of a typical type 2 form. Dynamically, however, there
is a great difference. Taken together, the responses in Figure 4a contribute nothing
to stability, while those in Figure 4b are sufficient to stabilize a wide range of
interactions, despite the individual functional responses being inversely density
dependent.

The underlying explanation for this is similar in kind to that where some
hosts are less at risk from parasitism by virtue of their location. The exploitation of the more susceptible (≡ exposed) hosts becomes very heavy at high parasitoid
densities, to the extent that the searching efficiency per parasitoid (a in equation (2)) is smaller than at lower parasitoid densities. This is well seen from
Figure 5 in which the estimated values of a from all the functional responses are
plotted against the corresponding parasitoid densities (P). Where all hosts are
susceptible the searching efficiency is independent of P (Figure 5a), but with
differential susceptibility it plumits at the higher values of P due to the very
high levels of exploitation (Figure 5b).

Figure 5b is reminiscent of the mutual interference relationships in Figure 1
where encounters between the searching parasitoids led to a reduction in searching
efficiency. Only now there is no behavioural interaction between the searching individuals, and the reduction in a with increasing P is solely due to the host
population being unevenly exploited. Similar apparent interference relationships
will be observed whatever the cause of this uneven exploitation and have been generally dubbed as 'pseudointerference' by Free, Beddington and Lawton (1977). As
with 'real interference', the contribution of pseudointerference to population
stability depends upon the slope, m*, of the relationship evaluated at the equilibrium parasitoid population density, P*. Within the framework of equations (1a,b),
and assuming type II functional responses as observed from the simulations, the

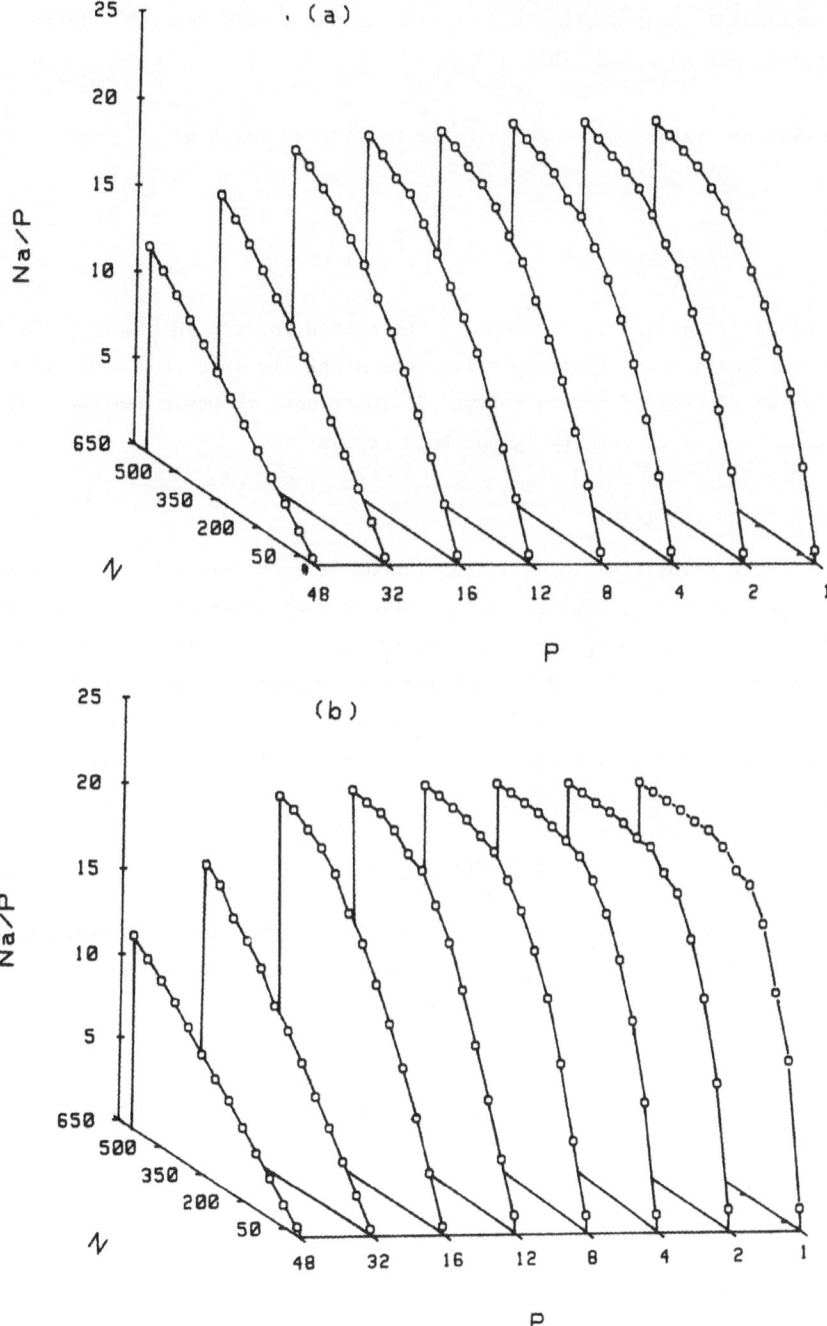

Figure 4 Functional responses for a range of parasitoid densities, generated by
Monte Carlo simulation. (a) All hosts equally susceptible. (b) Random
variability between host individuals in their susceptibility to parasitism.
Symbols: N_a/P is the number of hosts attacked per parsitoid, N is the

Figure 4 (continued) host density and P is the parasitoid density. (From Hassell and Anderson 1983).

populations will be stable if and only if the particular value of m* falls within the limits:

$$1 - \Theta > m > 1 - \left[\frac{\lambda - 1}{\lambda \ \ell n \ \lambda}\right] \tag{9}$$

where $\Theta = (a'T_h N^*)/(1+a'T_h N^*)$, as shown by the shaded regions in Figure 6 (Hassell and May 1973). Equation (9) thus emphasizes how stability is a trade-off between the destabilizing effects of having a type 2 functional response and the stabilizing effect of the heterogeneity in the host population.

3. THE HOST RATE OF INCREASE

However much parasitism in equations (1a,b) is made more realistic, there remains the important defect that the host's net rate of increase has so far been assumed constant and thus independent of population size. In this section, therefore, an additional density dependent factor acting on the host population is introduced.

Let us assume, for convenience, that this host density dependence takes the form of a discrete logistic (May 1976):

$$g = \exp(-rN_t/K) \tag{10}$$

where $r (= \ell n \ \lambda)$ is the intrinsic rate of increase and K is the 'carrying capacity' in the absence of the parasitoid. The problem now arises over where to position this density dependence relative to parasitism in the host's life cycle (May, Hassell, Anderson and Tonkyn 1981). The most obvious cases are:

Model 1 - density dependence acts first followed by parasitism of a later host stage:

$$N_{t+1} = \lambda N_t g(N_t) f \tag{11a}$$

$$P_{t+1} = N_t g(N_t)(1-f) \tag{11b}$$

Model 2 - parasitism acts first followed by the density dependence acting on the survivors from parasitism:

$$N_{t+1} = \lambda N_t g(N_t f) f \tag{12a}$$

$$P_{t+1} = N_t (1-f) \tag{12b}$$

The major effect of including such host density dependence is to add to the stability of the interactions wherever density dependence is sighted in the host's

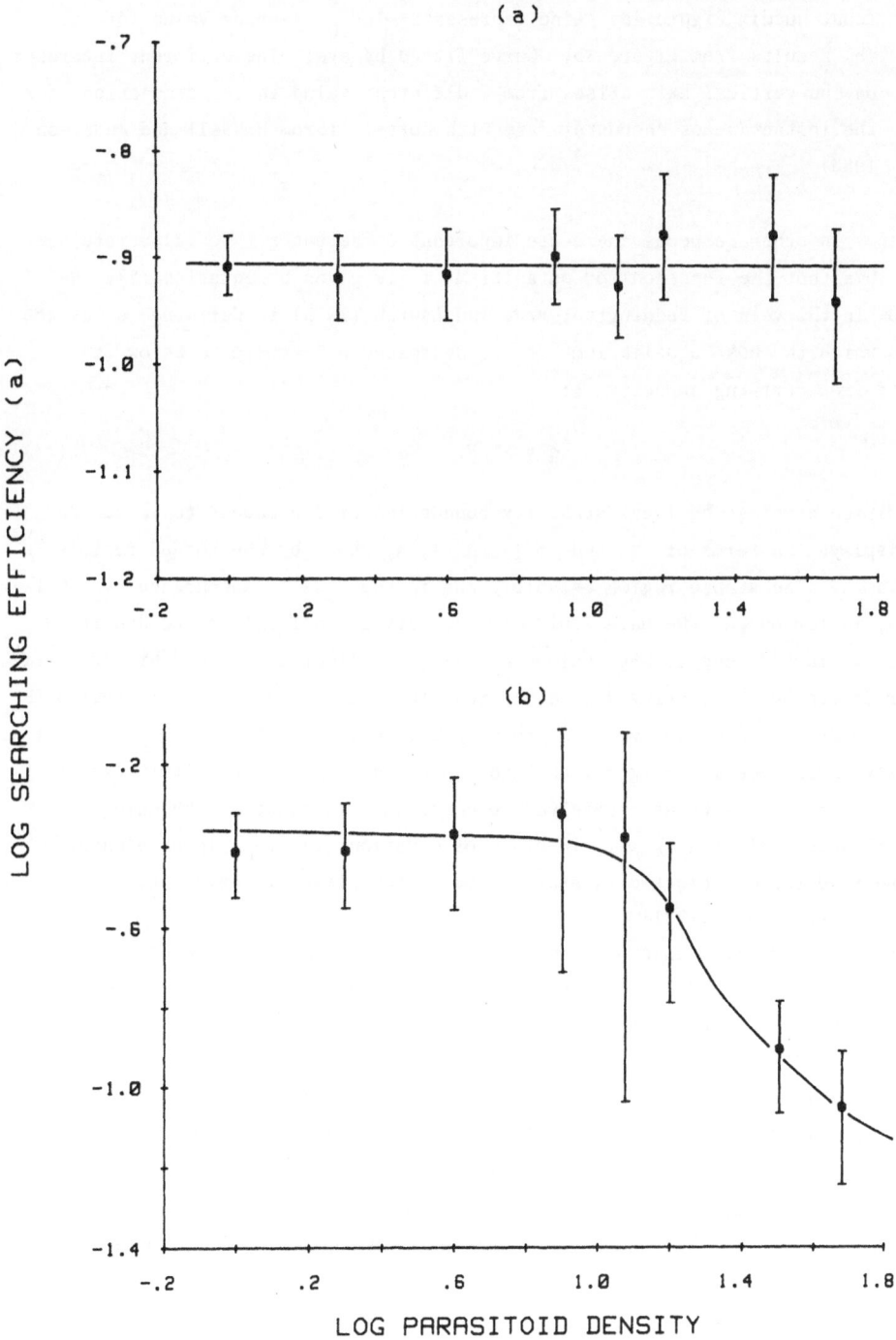

Figure 5 The relationship between the searching efficiency (a) (estimated from the
functional responses in Figure 4 using a non-linear least squares technique), and the corresponding parasitoid density. (a) Results from

Figure 5   (continued)   Figure 4a.   Line represents $\log_{10}$ average value for   a.
(b) Results from Figure 4b.   Curve fitted by eye.   The different intercept
on the vertical axis arises from a different value in the simulations for
the instantaneous encounter rate with hosts.   (From Hassell and Anderson
1983)

life cycle.   In other respects there are important differences.   To illustrate these,
let us assume that the function for parasitism   f   is given by equation (2).   We
now follow in the vein of Beddington, Free and Lawton (1975) in defining   q   as the
extend to which the host equilibrium   N*   is depressed by parasitism below its
parasitoid-free carrying capacity   K:

$$q = N^*/K. \tag{13}$$

The definition enables the local stability boundaries of the models to be conveniently displayed in terms of   q   and   r   (= $\ln \lambda$), as shown by the shaded regions in
Figures 7a, b.   The stable region is solely due to the density dependence   g.   Outside this, in region   A, the parasitoids are relatively ineffective and density
dependence so marked (due to high values of λ)   that limit cycles and higher order
behaviour driven by the density dependence tend to occur.   Below this, in region   B,
the parasitoids are so effective in depressing the host equilibrium (N* << K) that
the density dependence is insignificant and typical host-parasitoid limit cycles
result.   Were the parasitoids themselves to contribute to stability by using, for
example, equation (8) with   k < 1   in place of equation (2), this lower boundary
would tend to disappear creating a stable area in the place of region   B, as
discussed by May et al. (1981).

The striking difference between the two figures is that for Model 1 the introduction of a parasitoid can only depress the host equilibrium, while for Model 2
(Figure 7b) there can, for certain parameter combinations, be a stable host-
parasitoid equilibrium with   q > 1.   In other words, the introduction of a parasitoid acting before the density dependence can result in a stable host equilibrium
above the parasitoid-free level   K; a disturbing result for biological control!
The examples in Figure 8 illustrate how this seemingly strange result can arise.
The equilibrium population of eggs and young larvae in the absence of parasitoids
is   50   (Figure 8a).   Density dependence then acts on the pupal stage to give   6.77
adults which, after reproduction (λ = 7.39), recover the equilibrium population of
eggs for the next generation.   Let us now suppose that a parasitoid species attacking the third stage larvae is introduced in the manner of Model 2.   Depending upon
the choice of parameters, the equilibrium eggs and early larvae may either be depressed, or raised above   K   as in Figure 8b (q = 1.36).   This increase in host
levels will occur whenever the density dependence at equilibrium is overcompensating
(i.e. survivors decrease as initial population density rises).   Such 'scramble'

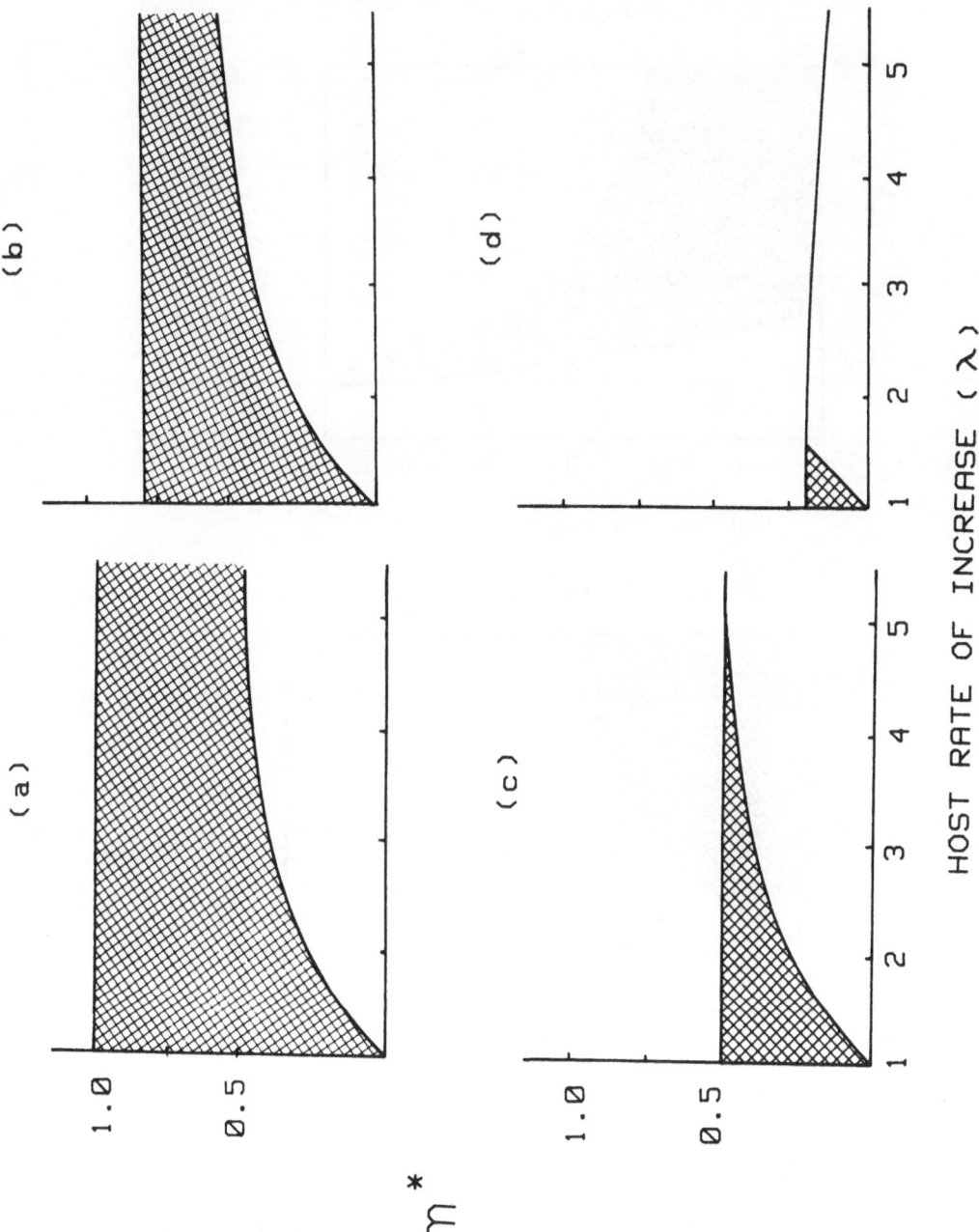

Figure 6 Local stability boundaries between the apparent interference constant m*
and the host rate of increase λ, for different values of Θ. Full details
are given in Hassell and May (1973). (a) Θ = 0; (b) Θ = 0.2;
(c) Θ = 0.5; (d) Θ = 0.8.

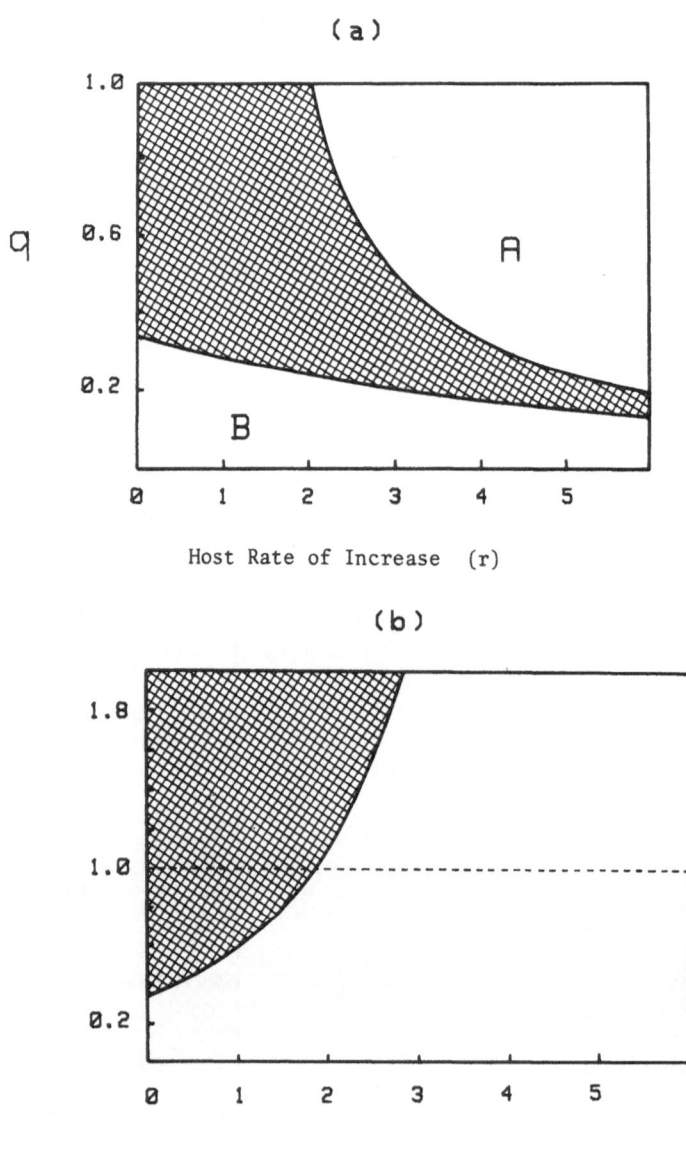

Figure 7 (a) Local stability boundaries for Model 1 (equations 11a,b) (with f
defined in equation (2) and g in equation (10)) in terms of the depression of the host equilibrium (q = N*/K) and the host rate of increase
(r = ℓn λ). (b) As for (a), but now using Model 2 (equations 12a,b).
(After May *et al.* 1981).

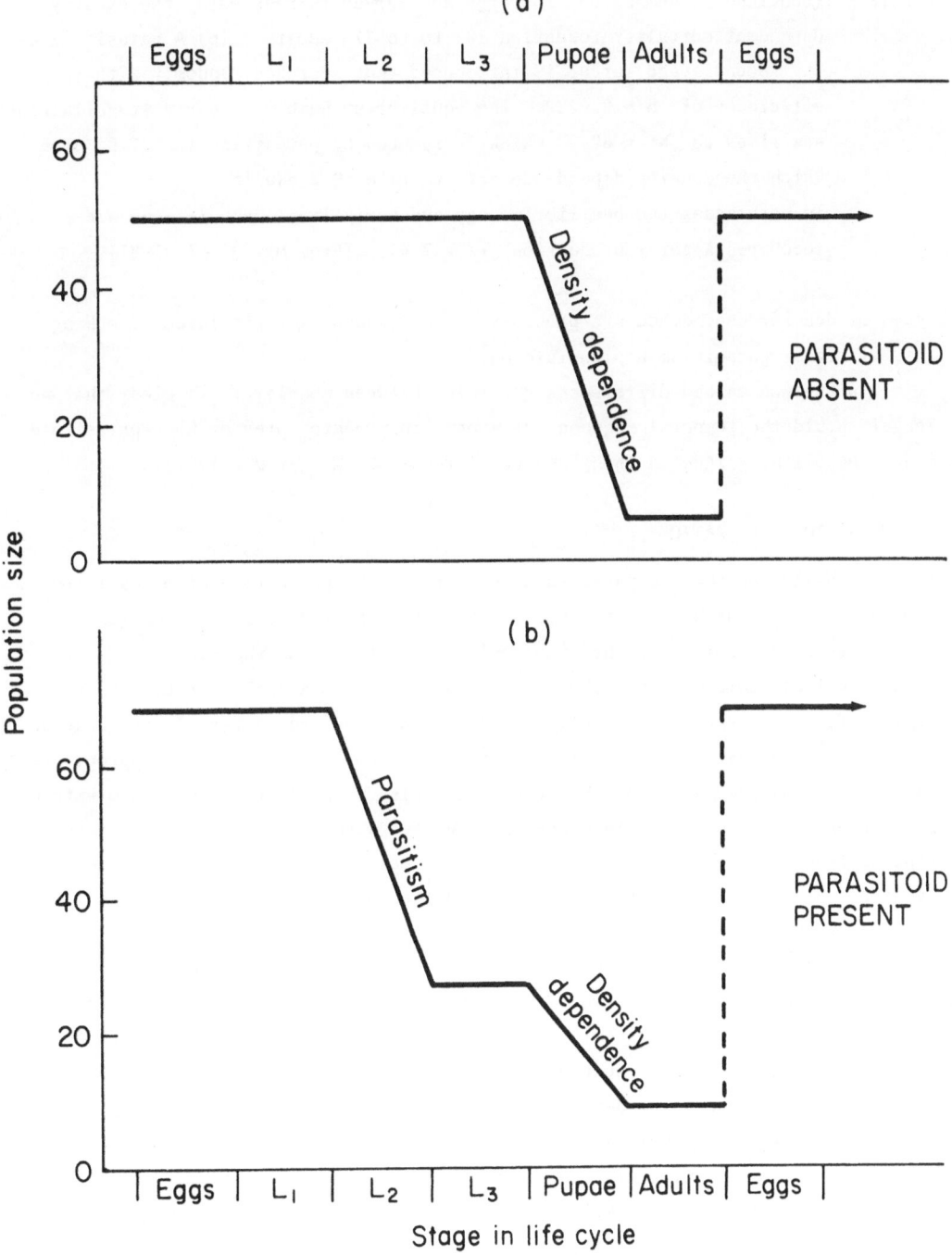

Figure 8 An example based on Model 2 illustrating how the introduction of a
         parasitoid can increase host population density. In each case r = 2
         and K = 50 and the host life cycle is divided into an egg stage, three
         larval stages, pupae and adults. (a) Parasitoids absent. The equilibrium

Figure 8 (continued) number of 50 eggs and larvae is reduced by the density
dependent mortality (equation 10) to 6.77 adults. (b) A parasitoid of
the second stage larvae is introduced that searches randomly with an
efficiency of a = 0.0222. The equilibrium number of early stage larvae
now rises to N* = 68.2 which is reduced by parasitism to 27.9 upon
which the density dependence acts to give 9.2 adults.
In both cases the equilibrium egg and early larval populations are
recovered after reproduction (λ = 7.4). (From May *et al.* 1981).

types of density dependence are probably quite frequent amongst insects, making
this effect of parasitism a plausible one.

With such marked differences occurring between models, it is clear that no
longer should the temporal sequence of events in predator-prey or host-parasitoid
models be blithely ignored (Wang and Gutierrez 1980, May *et al.* 1981).

4. PARASITOID SEX RATIOS

Finally, we turn to parasitoid sex ratios (the proportion of males in the
population) as a further factor that may influence host-parasitoid dynamics. Most
host-parasitoid models make the simplifying assumption that the parasitoids are
either entirely female (thelytoky, a relatively rare phenomenon) or that they exhibit a fixed sex ratio (e.g. Bellows, 1979). However, for a very large group of
parasitoids, the parasitic Hymenoptera, these assumptions are often inappropriate.
Parasitic wasps are known to exhibit considerable variability in sex ratio within
populations. This property is closely associated with their haplodiploid mode of
reproduction in which unfertilized eggs become males, fertilized eggs become
females, and the act of fertilization is mediated by the ovipositing female in
response to varying internal and external stimuli.

A largely anecdotal literature indicates that individual hymenopterous parasitoids will alter the sex ratio of their progeny in response to host species, host
size, host density and parasitoid density (Flanders 1939, Viktorov 1976, Kochetova
1978, Waage 1982). In this section, the dynamical consequences are considered of
parasitoid sex ratios being dependent on the density of adult females (P_t), as
observed for example by Wilkes (1963), Wylie (1965) and Walker (1967). A fuller
treatment, with sex ratios also depending on the ratio of parasitoids to hosts
(P_t/N_t), and including an example of a deviant male ontogeny found in certain
Aphelinidae, is given in Hassell, Waage and May (1983).

We commence with the general model in equations (1a,b) extended to give:

$$N_{t+1} = \lambda N_t f(N_t, P_t) \tag{14a}$$

$$P_{t+1} = s(P_t) c N_t [1 - f(N_t, P_t)] \tag{14b}$$

where $s(P_t)$ is the proportion of females in the total parasitoid progeny. Defining f from equation (2) and setting $N_{t+1} = N = N^*$ and $P_{t+1} = P_t = P^*$, we have the equilibrium populations:

$$aP^* = \ln \lambda \tag{15a}$$

$$acN^* = \frac{\lambda \ln \lambda}{\lambda - 1} \cdot \frac{1}{s(aP^*)} . \tag{15b}$$

Not surprisingly, the host equilibrium is raised as the male bias in the equilibrium parasitoid population increases.

Following the recipe in Hassell and May (1973), the model proves stable if and only if the following criterion is satisfied:

$$\frac{2 + (\frac{\lambda + 1}{\lambda - 1}) \ln \lambda}{2 \ln \lambda} > \Theta > \frac{\lambda}{\lambda - 1} - \frac{1}{\ln \lambda} \tag{16}$$

where $\Theta \equiv - (\frac{1}{s} \frac{ds}{d(aP)})_{P=P^*}$. As a specific example for $s(P_t)$ that adequately describes several data sets, Hassell *et al.* (1983) chose the expression:

$$s(P_t) = \frac{\alpha \beta}{\beta + P_t} \tag{17}$$

where α is a constant representing the proportion of females as $P_t \rightarrow 0$ and β is another constant where $1/\beta$ indicates the rate of increase in sex ratio $(1-s)$ as P_t increases from very low levels. Θ in equation (16) is now given by:

$$\Theta = \frac{1}{\ln \lambda + \alpha \beta} . \tag{18}$$

Stability is thus promoted by small host rates of increase (λ) and pronounced sex ratio shifts at relatively low parasitoid densities, represented by the dimensionless combination $\alpha \beta$.

Clearly sex ratios can have a marked impact on dynamics, and should not be blithely ignored in the development of models as has largely been the case in the past.

5. CONCLUSION

There are many factors that can influence the dynamics of host-parasitoid interactions, without even considering the impact of environmental vagaries. Equilibrium levels are determined by a straightforward trade off between two opposing sets of factors. On the one hand, there are all factors tending to increase the average level of parasitism, such as (1) higher overall searching efficiencies, (2) female-biased parasitoid sex ratios and (3) high survival of parasitoid progeny. Acting against these are all factors enhancing the host's net rate of increase (λ), such as (1) high fecundity per adult host and (2) few additional mortalities to offset this.

Stability is also affected by many factors. It is enhanced (1) by sigmoid functional responses, (2) by most forms of heterogeneity, spatial or otherwise, (3) by any mutual interference, (4) by density dependent parasitoid sex ratios and (5) by any additional density dependence that occurs in the host or parasitoid life cycles. With many of these acting in concert, we are left with the picture of insect predation and parasitism being almost inevitably markedly stabilizing processes that are often likely to be playing a major part in the persistence of complex assemblies of insect species in food webs.

REFERENCES

Beddington, J.R. (1975): Mutual interference between parasites or predators and its effect on searching efficiency, *Journal of Animal Ecology* 44:331-40.

Beddington, J.R., C.A. Free, and J.H. Lawton (1975): Dynamic complexity in predator-prey models framed in difference equations, *Nature, Lond.*, 225:58-60.

_____(1978): Modelling biological control: on the characteristics of successful natural enemies, *Nature, Lond.*, 273:513-519.

Bellows, T.S. (1979): The modelling of competition and parasitism in laboratory insect populations, *Unpublished Ph.D. thesis, University of London*.

Callan, E.Mc.C. (1944): A note on *Phanuropis semiflaviventris* Girault (Hym., Scelionidae), an egg-parasite of cacao stink-bugs, *Proceedings of the Royal Entomological Society of London* (A), 19:48-49.

Comins, H.N., and M.P. Hassell (1979): The dynamics of optimally foraging predators and parasitoids, *Journal of Animal Ecology*, 48:335-351.

Flanders, S.E. (1939): Environmental control of sex in hymenopterous insects, *Annals of the Entomological Society of America*, 32:11-26.

Free, C.A., J.R. Beddington, and J.H. Lawton (1977): On the inadequacy of simple models of mutual interference for parasitism and predation, *Journal of Animal Ecology*, 46:543-554.

Hamilton, W.D. (1971): Geometry for the selfish herd, *Journal of Theoretical Biology*, 31:295-311.

Hassell, M.P. (1968): The behavioural response of a tachinid fly (*Cyzenis albicans* (Fall.)) to its host, the winter moth, (*Operophtera brumata* (L.)). *Journal of Animal Ecology*, 37:627-639.

_____(1978): *The Dynamics of Arthropod Predator-Prey Systems*. Princeton University Press, Princeton.

_____(1980): Foraging strategies, population models and biological control: a case study. *Journal of Animal Ecology*, 49:603-628.

_____(1982): Patterns of parasitism in patchy environments, *Ecological Entomology*, 7: (in press).

Hassell, M.P., and R.M. Anderson (1983): Host susceptibility: a component of heterogeneity in insect host-parasitoid models, (manuscript).

Hassell, M.P., and R.M. May (1973): Stability in insect host-parasite models, *Journal of Animal Ecology*, 42:693-726.

_____(1974): Aggregation in predators and insect parasites and its effect on stability, *Journal of Animal Ecology*, 43:567-594.

Hassell, M.P., and G.C. Varley (1969): A new inductive population model for insect parasites and its bearing on biological control, *Nature, Lond.*, 223:1133-1136.

Hassell, M.P., J.K. Waage, and R.M. May (1983): Variable parasitoid sex ratios and their effect on host-parasitoid dynamics (manuscript).

Heads, P.A., and J.H. Lawton (1982): Studies on the natural enemy complex of the holly leaf miner: the effects of scale on aggregation responses and the implications for biological control, *Oikos* (in press).

Holling, C.S. (1959): Some characteristics of simple types of predation and parasitism, *Canadian Entomologist*, 91:385-398.

Kochetova, N.I. (1978): Factors determining the sex ratio in some entomophagous Hymenoptera, *Entomological Reviews*, 60:1-5.

Lotka, A.J. (1925): *Elements of Physical Biology*. Williams and Wilkins, Baltimore. (Reissued as *Elements of Mathematical Biology* by Dover, 1956).

May, R.M. (1976): Simple mathematical models with very complicated dynamics, *Nature, Lond.*, 261:459-467.

_____(1978): Host-parasitoid systems in patch environments: a phenomeno-logical model, *Journal of Animal Ecology*, 47:833-844.

May, R.M., M.P. Hassell, R.M. Anderson, and D.W. Tonkyn (1981): Density dependence in host-parasitoid models, *Journal of Animal Ecology*, 50:855-865.

McClure, M.S. (1977): Parasitism of the scale insect, *Fiorinia externa* (Homoptera: Diaspididae), by *Aspidiotiphagus citrinus* (Hymenoptera: Eulophidae) in a hemlock forest: density dependence. *Environmental Entomology*, 6:551-555.

Nicholson, A.J. (1933): The balance of animal populations, *Journal of Animal Ecology*, 2:132-178.

Nicholson, A.J., and V.A. Bailey (1935): The balance of animal populations, *Part I, Proceedings of the Zoological Society of London*, 1935:551-598.

Rogers, D.J. (1970): Aspects of host-parasite interactions in laboratory popula-tions of insects, Unpublished D. Phil. Thesis, Oxford.

Rogers, D.J., and M.P. Hassell (1974): General models for insect parasite and predator searching behaviour: interference, *Journal of Animal Ecology*, 43: 239-253.

Sharov, A.A. (1979): Effect of spatial structure of the interacting populations of *Neodiprion sertifer* and its parasite *Exenterus abruptorius* on the dynamics of their numbers, *Zoologicheskii Zhurnal*, 58:356-365 (In Russian).

Thompson, W.R. (1924): La théorie mathématique de l'action des parasites entomophages et le facteur du hasard, *Annls. Fac. Sci. Marseille*, 2:69-89.

Viktorov, G.A. (1976): *The Ecology of Entomophagous Parasites*, Moscow: Izdatel'stvo Nanka. (Russian).

Volterra, V. (1926): Variazioni e fluttnazioni del numero d'individui in specie animali conviventir, *Mem. Acad. Lincei.*, 2:31-113. (Translation in: Chapman, R.N. 1931. *Animal Ecology*, McGraw-Hill, New York, pp. 409-448.)

Waage, J.K. (1982): Sib-mating and sex ratio strategies in scelionid wasps, *Ecological Entomology*, 7:103-112.

Walker, I. (1967): Effect of population density on the viability and fecundity of *Nasonia vitripennis* Walker (Hymenoptera, Pteromalidae), *Ecology*, 48:294-301.

Wang, Y.H., and A.P. Gutierrez (1980): An assessment of the use of stability analyses in population ecology, *Journal of Animal Ecology*, 49:435-452.

Watt, K.E.F. (1959): A mathematical model for the effect of densities of attacked and attacking species on the number attacked, *Canadian Entomologist*, 91:129-144.

Wellington, W.G. (1957): Individual difference as a factor in population dynamics: The development of a problem, *Canadian Journal of Zoology*, 35:293-323.

Wilkes, A. (1963): Environmental causes of variation in the sex ratio of an arrhenotokous insect, *Dahlbominus fuliginosus* (Nees) (Hymenoptera: Eulophidae), *Canadian Entomologist*, 95:183-202.

Wylie, H.J. (1965): Some factors that reduce the reproductive rate of *Nasonia vitripennis* at high adult population densities, *Canadian Entomologist*, 97:970-977.

SOME APPLICATIONS OF COHERENT STATE REPRESENTATION TO NONLINEAR BIO-OSCILLATORS

Ranabir Dutt

ABSTRACT

It is shown that the 'coherent state representation' which is commonly used for physical problems, can be used to obtain perturbative solutions for the amplitude and the period of nonlinear oscillation in a biological system. The method is illustrated by treating the well-known Lotka-Volterra predation model of population dynamics which describes oscillation in a predator-prey ecosystem. It is observed that the method may be useful to estimate statistical averages of stochastic dynamical variables associated with biological models.

Recently there have been several interesting applications of the coherent state representation to obtain perturbative solutions to both classical and quantum mechanical nonlinear oscillators, Bhaumik and Dutta Roy (1975), Dutt and Lakshmanan (1976), Malkin et al. (1973). In this note, we propose to extend the use of this representation to study the basic features of nonlinear oscillations in biological systems. This work is motivated by the recent proposition of Fröhlich (1968, 1975) about the coherent excitation of a single oscillatory mode of macromolecules when stimulated by millimeter band electromagnetic radiation. Fröhlich's proposition for Bose -- condensation like excitation in biological systems, has been approached by Bhaumik et. al. (1976) by using the coherent state formalism.

For the illustration of our method, we consider the nonlinear Lotka-Volterra (LV) population model Goel et. al. (1971) which exhibits the characteristic oscillation of populations in a prey-predator ecosystem. This model sets the guidelines for the formulation of realistic biological models such as host-parasite (Leslie and Gower, 1960) antigen-antibody (Bell, 1973) models. The LV rate equations for two conflicting populations are given by

$$dN_1/dt = \alpha_1 N_1 - \lambda_1 N_1 N_2$$
$$dN_2/dt = -\alpha_2 N_2 + \lambda_2 N_1 N_2 \tag{1}$$

where N_i ($i = 1,2$) is the number of individuals of species i at a given time, α_i is the innate capacity for increase per individual (intraspecific coefficient) and λ_i is the interspecific coefficient (niche overlap parameter). The steady state populations $\{q_i\}$ which are obtained by setting $dN_{1,2}/dt = 0$ in Eqs. (1) are

$$q_1 = \alpha_2/\lambda_2, \quad q_2 = \alpha_1/\lambda_1 . \tag{2}$$

In the $N_1 N_2$ phase plane, the Equations (1) describe a closed trajectory around the equilibrium point (q_1, q_2) indicating that both the population exhibit periodicity, Andronov and Chaiken (1953), Davis (1962). However, exact solutions of the problem are not yet obtained because of the complexity of these nonlinear equations.

When exact solutions are not available, the standard procedure (Rosen, 1970) generally followed, is to linearize the rate equations in the neighbourhood of the equilibrium point in the phase plane, assuming a priori that the effect of the nonlinear terms would be small. Essentially, this assumption motivates us to do a perturbative calculation to determine the effects of nonlinearity in the frequency and amplitude of oscillation. Choosing a new set of variables

$$p(t) = \log_e (N_1(t)/q_1), \quad x(t) = \log_e (N_2(t)/q_2), \tag{3}$$

Equations (1) are reduced to the form

$$\dot{p} = \alpha_1 (1-e^x), \quad \dot{x} = -\alpha_2 (1-e^p), \tag{4}$$

a form which is seen to be canonical with the Hamiltonian

$$H(x,p) = \alpha_1 (e^x - x - 1) + \alpha_2 (e^p - p - 1). \tag{5}$$

If we now assume that the amplitudes x and p are small, the Hamiltonian can be approximated by a few terms in the expansion

$$H(x,p) = H_2(x,p) + H_3(x,p) + H_4(x,p) + 0(x^2+p^2)^{5/2} \tag{6}$$

where

$$H_n = (\alpha_1 x^n + \alpha_2 p^n)/n!, \quad n = 2,3,4. \tag{7}$$

We retain terms up to the fourth powers of x and p because we find that the first order correction to the linear time period associated with H_2 comes from H_4 whereas that due to H_3 vanishes identically. The lowest order term $H_2 = 1/2(\alpha_1 x^2 + \alpha_2 p^2)$ corresponds to the linear harmonic oscillator and is the dominant part of H. The linearized equations of motion which follow from H_2 are

$$\ddot{x} = \omega_0^2 x = 0, \quad \ddot{p} + \omega_0^2 p = 0, \tag{8}$$

where $\omega_0 = \sqrt{\alpha_1 \alpha_2}$. The linear time period is $T_0 = 2\pi/\sqrt{\alpha_1 \alpha_2}$. This depends only on the intraspecific parameters α_1 and α_2.

The approximate form of the Hamiltonian in (6) indicates that LV equations correspond to an anharmonic oscillator and can be solved perturbatively by using well-known coherent states represented by (Glauber 1964, Klauder and Sudarshan 1968)

$$|\alpha> = e^{-|\lambda|^2/2} \sum_{n=0}^{\infty} \frac{(-i\lambda)^n}{\sqrt{n!}} e^{iE_n t/\hbar} |n> \tag{9}$$

where $|n>$ represents a harmonic oscillator state of "n quanta" and is the eigenstate of H_2 corresponding to the eigenenergy

$$E_n = \frac{1}{2} \hbar \omega_0 (n + \frac{1}{2}) . \tag{10}$$

The coherent state $|\alpha>$, being an eigenstate of the annihilation operator, corresponds to the lowest uncertainty state and is suitable to reproduce classical results from the quantum description in the appropriate classical limit. This procedure is clearly illustrated in Bhaumik and Dutta Roy (1975) and Dutt and Lakshmanan (1976).

Considering the cubic and quartic terms, H_3 and H_4 in (6) to be the perturbations to the linear Hamiltonian H_2, we may obtain the perturbed eigenstates and eigenenergies. Expressing x and p in terms of the creation and the annihilation operators a^+ and a as

$$x = i \left(\frac{\hbar \alpha_2}{2\omega_0} \right)^{1/2} (a-a^+)$$

$$p = \left(\frac{\omega_0 \hbar}{2 \alpha_2} \right)^{1/2} (a+a^+) \tag{11}$$

we obtain the following perturbed eigenenergy and eigenstate of "n quanta"

$$E_n = \hbar \omega_0 (n + \frac{1}{2}) + \frac{\hbar^2}{32} (\alpha_1 + \alpha_2)(2n^2 + 2n + 1) + 0(\hbar^3) \tag{12}$$

$$|n>' = |n> - \left(\frac{1}{4\hbar \omega_0} \right) \frac{\hbar^2}{96} [(\alpha_1 + \alpha_2)\sqrt{(n+1)(n+2)(n+3)(n+4)} |n+4>$$

$$+ 4(\alpha_1 - \alpha_2)(2n+3)\sqrt{(n+1)(n+2)} |n+2> - 4(\alpha_1 - \alpha_2)(2n-1) \times$$

$$\times \sqrt{n(n-1)} |n-2> - (\alpha_1 + \alpha_2)\sqrt{n(n-1)(n-2)(n-3)} |n-4>] + 0(\hbar^2) \tag{13}$$

Clearly, the cubic terms in H_3 do not contribute to (12) and (13). The corresponding perturbed coherent state may be given in the normalized form

$$|\alpha>' = e^{-|\lambda|^2/2} \sum_{n=0}^{\infty} \frac{(-i\lambda)^n}{\sqrt{n!}} e^{iE_n' t/\hbar} |n>' . \tag{14}$$

It is now straightforward to obtain perturbative solutions for x and p by computing the expection values of these quantities with respect to $|\alpha>'$ in the appropriate classical limit. We give the first order result for $x(t)$ and the period of oscillation:

$$x(t) = \underset{\substack{\hbar \to 0 \\ \lambda \to \infty}}{\alpha_1} \,\, '<\alpha|x|\alpha>' = x(0) \left[\cos \Omega t \left(1 + \frac{\alpha_1(\alpha_1 - \alpha_2)}{32} \cdot \frac{x^2(0)}{\omega_0^2} \right) \right. $$
$$\left. - \alpha_1 \frac{(3\alpha_1 - \alpha_2)}{192} \cdot \frac{x^2(0)}{\omega_0^2} \cos 3\Omega t \right] \tag{15}$$

$$T = \frac{2\pi}{\Omega} = T_0 \left[1 + \frac{(\alpha_1 + \alpha_2)}{16\omega_0^2} (\alpha_1 x^2(0) + \alpha_2 p^2(0)) \right] \tag{16}$$

In principle, the same procedure can be continued with more computational labour to higher order terms in (6) so as to get hgiher order corrections to the amplitudes and the period of oscillation of the species populations. However, we shall not do this because our first order results have the basic features of the nonlinearity such as dependence of the period on the amplitudes and vice versa. Further, since x and p depend on λ_1 and λ_2 because of (2) and (3), the corrected period depends on both intra- and interspecific coefficients, unlike the period T_0 which depends on the intraspecific parameters only.

The size of the correction term in (16) can be estimated numerically. In their review work, Goel et. al. (1971) gave the time variations of the two populations by numerical analysis with the help of a computer for nine sets of parameters α_1 and α_2 with different initial amplitudes $x(0)$ and $p(0)$. (Our definition for x and p correspond to their f_1 and f_2 through the relations $x = \log_e f_2$ and $p = \log_e f_1$). From the graphs (see Figure 3 of Goel, Maitra and Montroll, 1971) we measure the time periods which we consider to be nearly exact ones for this problem. We utilize the same set of initial conditions to evaluate our corrected period T in (16). For comparison, we tabulate our results in Table 1 along with the linear period T_0 as well as the period measured from the graphs. The data reveal that the first order correction not only gives an appreciable contribution ($\sim 20\%$) to the period of the linearized system, but also agrees with that given by computer analysis.

A few general remarks may be made here about other possible applications of the coherent state representation to biological problems. It is pointed out by Glauber (1964) that the 'P-representation' in the coherent state formalism might play a role analogous to probability distribution in the appropriate classical limit and hence may be used to compute statistical averages. In a stochastic approach to deal with biological problems, one essentially needs to obtain the probability distributions of various species populations and then to compute statistical averages of correlation functions which may be experimentally measured. In this regard, the coherent state method will have access to the experimental aspects of the biological problem. These points are presently under investigation.

Table 1 Period of Oscillation for Different Input Values of the Intraspecific Parameters and Initial Amplitudes

Graph No. in Fig. 3 of Goel, Maitra & Montroll (1971)	α_1	α_2	x(0)	p(0)	Linear $T_0=2\pi/\omega_0$	Period of Oscillation	
						Corrected Up to First Order T = $2\pi/\Omega$	Computer Simulated (Measured From Graphs)
1	1.0	2.0	-0.70	0.70	4.44	5.1	5.5
2	1.0	2.0	0.70	-0.70	4.4	5.1	5.2
3	1.0	2.0	-0.36	-1.20	4.4	5.7	6.0
4	1.0	1.0	-0.22	0.70	6.3	6.7	6.8
5	1.0	1.0	0.70	-0.70	6.3	7.1	7.2
6	1.0	1.0	-0.36	-1.20	6.3	7.5	7.7
7	2.0	1.0	0.70	-0.70	4.4	5.1	5.5
8	2.0	1.0	-0.70	0.70	4.4	5.1	5.2
9	2.0	1.0	-1.20	-0.36	4.4	5.7	6.0

REFERENCES

Andronov, A.A., and C.E. Chaikin (1953): *Theory of Oscillations*, Princeton Univ. Press, New Jersey.

Bell, G.I. (1973): Predator prey equations simulating an immune response, *Math. Bioscience*, Vol. 16, p. 291.

Bhaumik, K., and B. Dutta Roy (1975): The classical nonlinear oscillator and the coherent state, *Jour. Math. Phys.*, Vol. 16, p. 1131.

Bhaumik, D., K. Bhaumik and B. Dutta Roy (1976): On the possibility of Bose-condensation in the excitation of coherent modes in biological systems, *Phys. Lett.*, Vol. 56A, p. 145.

Davis, H.T. (1962): *Introduction to Nonlinear Differential and Integral Equation*, Dover publication, New York.

Dutt, R., and M. Lakshmanan (1976): Application of coherent state representation to classical x^6 and coupled anharmonic oscillators, *Jour. Math. Phys.*, Vol. 17, p. 482.

Frölich, H. (1968): Bose condensation of strongly excited longitudinal electric modes, *Phys. Lett.*, Vol. 26A, p. 402.

_____ (1975): Evidence for Bose condensation-like excitation of coherent modes in biological systems, *Phys. Lett.*, Vol. 51A, p. 21.

Glauber, R.J. (1964): *Quantum Optics and Electronics*, Gordon Breach, New York.

Goel, N.S., S.C. Maitra, and E. Montroll (1971): On the Volterra and other non-linear models of interacting populations, *Rev. Mod. Phys.*, Vol. 43, p. 231.

Klauder, J.R., and E.C.G. Sudarshan (1968): *Foundations of Quantum Optics*, W.A. Benjamin, Inc., New York.

Leslie, P.H., and J.C. Gower (1960): The properties of a stochastic model for the predator-prey type of interaction between two species, *Biometrica*, Vol. 47, p. 219.

Malkin, I.A., V.I. Manko, and D.A. Trifonov (1973): Linear adiabatic invariants and coherent states, *Jour. Math. Phys.*, Vol. 14, p. 576.

Rosen, R. (1970): *Dynamical System Theory in Biology*, Vol. I, Willey Interscience, New York.

MATHEMATICAL ANALYSIS OF SOME RESOURCE-PREY-PREDATOR MODELS:
APPLICATION TO A NPZ MICROCOSM MODEL

Thomas C. Gard*

1. INTRODUCTION

It has been proposed (Lassiter, 1978) that nutrient-phytoplankton-zooplankton (NPZ) submodels of ecosystem microcosm models exhibit the dynamics of a three-species food chain with Monod (Michaelis-Menton, Holling) functional response, except that the resource growth rate is a decreasing function of resource density. In this case, the basal prey species of the food chain is replaced by an abiotic resource which is supplied externally and is subject to density dependent dissipation as in a chemostat. Such a chemostat-chain, representing a sugar-bacteria-protozoa (SBP) system, has been studied qualitatively (local stability of equilibria) by Canale (1969,1970) and numerically by Jost et al. (1973).

This paper gives a qualitative analysis of a general class of such autonomous differential equation models, along the lines of Freedman and Waltman (1977). As such, the analysis focuses on persistence - the global stability type property that asserts that no component of a solution having positive initial values can tend to zero. In chemostat-chain models, persistence means that neither predator nor prey washes out no matter what their (positive) starting densities. The decreasing resource growth rate assumption allows the computation of a sharp persistence criterion in terms of model parameters, as a corollary of a result in Freedman and Waltman (1977).

For the open NPZ model alluded to above, the criterion takes the form of a threshold level for nutrient input required for persistence. Hallam (1977) has obtained such results for similar closed NPZ models. A mathematical consequence of persistence for autonomous models is the existence of a global attractor and the possibility of undamped oscillatory type solutions in the positive cone or feasible region. Freedman (1980) gives a detailed discussion of such qualitative properties for basic deterministic mathematical models in ecology, including simple food chain models. A persistence criterion for food chains of arbitrary length and general functional response can be found in Gard (1980). A main point of this article is that a simpler and sharper persistence criterion exists for chemostat-chains than food chains. Also it is indicated here how persistence criteria can be determined for nonautonomous type models arising, say, when the nutrient input rate may be time-varying. The latter can be applied to obtain predator persistence conditions in certain multicomponent microcosm models. Finally, note that the question of

---

*Research for this paper was supported by the U.S. Environmental Protection Agency under Cooperative Agreement No. CR807830.

competition among prey in chemostats and among predators in chemostat-chains is discussed in the paper by Waltman (this Proceedings); the references contained therein include some basic survey articles on chemostats.

2. THE GENERAL AUTONOMOUS MODEL

The general model discussed in this section has the form

$$x' = g(x) - p(x)y, \qquad y' = y(ap(x)-b) - q(y)z$$
$$z' = z(cq(y)-d), \qquad ' = d/dt \tag{1}$$

Here a, b, c, and d denote positive constants, and g, p, and q represent continuous functions defined on $R_+ = [0, \infty)$. Also, the following assumptions will be invoked.

The function g is decreasing on $R_+^O = (0, \infty)$, and there exists $K > 0$ such that $g(K) = 0$. (H1)

The functions p and q are increasing on R_+^O, and $p(0) = 0 = q(0)$. (H2)

Hypothesis (H2) is the same as in Freedman and Waltman (1977) and, as noted by them, includes the usual predation curves found in the literature.

Basic facts, definitions, and notation to be used in the main results are now stated. Corresponding to each point $(x_0, y_0, z_0) \in R_+^3 = \{(x,y,z): x \geq 0, y \geq 0, z \geq 0\}$ there exists a unique solution $\phi(t) = (x(t), y(t), z(t))$ of (1) with initial value $\phi(0) = (x_0, y_0, z_0)$ and defined on some maximal interval $[0, T]$. Furthermore $\phi(t) \in R_+^3$, all $t \in [0, T]$. (i.e. R_+^3 is an invariant set for (1).) Theorem 1 establishes that $T = \infty$. A solution $\phi(t)$ is recurrent if it satisfies the following property: for each $\epsilon > 0$, there is a $\tau > 0$, such that the ϵ-neighborhood of any trajectory segment $\{\phi(t): t_0 \leq t \leq t_0 + \tau\}$ contains the entire trajectory. Bounded solutions approach sets containing trajectories of recurrent solutions. In the plane trajectories of recurrent solutions are equilibria or periodic orbits. Hale (1969), Nemytskii and Stepanov (1960), and Andronov et al. (1973) are good sources for details. Persistence of the z component means that for any solution $\phi(t) = (x(t), y(t), z(t))$ with initial value $\phi(0) \in R_+^{3,O} = \{(x,y,z): x > 0, y > 0, z > 0\}$, $\lim\sup_{t \to \tau} z(t) > 0$, for all $\tau \in [0, \infty]$. Persistence of the z component is equivalent to entire system persistence (Gard and Hallam, 1979) for models of the form (1).

THEOREM 1. *Assume* (H1). *There exist positive constants* s_1 *and* s_2, *depending on* a, b, c, d, *and* g, *such that any solution* $\phi(t) = (x(t), y(t), z(t))$ *with* $\phi(0) \in R_+^3$ *satisfies*

$$\phi(t) \to S = \{(x,y,z) \in R_+^3 : s_1 \leq acx + cy + z \leq s_2\}$$

as $t \to \infty$.

THEOREM 2. *Assume* (H1), (H2). *If, in addition*

$$cq(ag(p^{-1}(b/a))/b) - d > 0 \qquad\qquad (H3)$$

then the system (1) *is persistent. Nonnegativity of the expression in* (H3) *is necessary for persistence in* (1). *Furthermore, under* (H3), *system* (1) *has a unique equilibrium in* $R_+^{3,0}$, *and if the equilibrium is unstable, there exists a non-constant recurrent solution of* (1) *with trajectory in* $S \cap R_+^{3,0}$.

For the case $a = c = 1$, and $g(x) = a_0 - d_0 x$, Saunders and Bazin (1975) have determined that (H3) implies instability of the equilibria of (1) corresponding to prey or predator washout.

3. THE CONSTANT INPUT NPZ MODEL

The NPZ microcosm model (Lassiter, 1978) can be written in the form

$$x' = a_0 - x\left(d_0 + \frac{b_1 y}{c_1 + x}\right), \qquad y' = y\left(\frac{b_1 x}{c_1 + x} - d_1 - \frac{b_2 z}{k(c_2 + y)}\right)$$

$$z' = z\left(\frac{b_2 y}{c_2 + y} - d_2\right) \qquad\qquad (2)$$

where a_0 (nutrient input), k, b_i, c_i, d_i, are positive constants.

With $d_0 = d_1 = d_2$ (= 1/retention time), (2) was first postulated as a model for the SBP chemostat-chain by Bungay and Bungay (1968), and Drake et al. (1966); local stability of equilibria analysis was carried out by Canale (1969, 1970); Tsuchiya et al. (1972) observed predicted sustained oscillations with all species present; and Jost et al. (1973) determined five stability regimes arising from increasing input and retention times, for (2) as well as a so-called multiple saturation (replace $\frac{y}{c_2 + y}$ in (2) by $\frac{y^2}{(c_{21} + y)(c_{22} + y)}$) model. For the case $a_0 = d_0 = d_1 = d_2$, $k = 1$, Sell (1977) has mathematically determined ranges of values for a_0 (in terms of the other model parameters) for which three (corresponding to, respectively, predator washout, prey and predator washout, or sustained oscillations with all species present) of those five regimes occur.

The invariant set S in Theorem 1, in this case, can be shown to be $S = \{(x,y,z) \in R_+^3 : a_0/d_M \leq x + y + z/k \leq a_0/d_m\}$ where $d_m = \min\{1, d_1, d_2\}$, $d_M = \max\{1, d_1, d_2\}$, and without loss of generality, $d_0 = 1$. (H3) in Theorem 2 which guarantees persistence can be written in the form

$$a_0 > \frac{c_1 d_1}{b_1 - d_1} + d_1\left(\frac{c_2 d_2}{b_2 - d_2}\right)$$

with $b_i - d_i > 0$, $i = 1,2$. The unique equilibrium $E(x^*,y^*,z^*)$ in $R_+^{3,0}$ is given by

$$y^* = \frac{c_2 d_2}{b_2 - d_2} , \qquad x^* = \frac{1}{2} \left(a_0 - c_1 - b_1 y^* + \sqrt{(a_0 - c_1 - b_1 y^*)^2 + 4a_0 c_1} \right) ,$$

$$z^* = \frac{k(a_0 - x^* - d_1 y^*)}{d_2} .$$

The existence of nontrivial recurrent solutions in $S \cap R_+^{3,0}$ follows if the matrix corresponding to linearization of (2) at the equilibrium E has an eigenvalue with positive real part.

4. THE TIME-VARYING INPUT MODELS

Denote by (2') the model obtained when the constant input a_0 is replaced in (2) by a positive continuous function $a(t)$ defined on R_+. The system is no longer autonomous and the results in section 2 do not apply. However, the following results can be established if $a(t)$ is bounded. Indeed, suppose there exist positive constants a_0 and a_1 such that $a_0 \leq a(t) \leq a_1$, for all $t \in R_+$. Let $A_0 = a_0/d_M$ and $A_1 = a_1/d_m$.

THEOREM 3. *The conclusion of Theorem 1 holds for* (2') *with*

$$S = \{(x,y,z) \in R_+^3 : A_0 \leq x + y + z/k \leq A_1\}.$$

THEOREM 4. *Persistence holds for* (2') *if*

$$a_0 > \frac{d_1}{b_1 - d_1} \left[c_1 + \frac{(A_1 + c_2)b_1^2 d_2}{(b_1 - d_1)b_2} \right] . \tag{H4}$$

5. DISCUSSION

The results given in section 2 indicate that solutions of (1) having positive initial values are asymptotically uniformly bounded and have components which do not tend to zero under a certain condition (H3) which is equivalent to the existence of a positive equilibrium. Furthermore oscillatory type solutions exist when that equilibrium is unstable. Condition (H3) means that near the positive equilibrium in the xy plane, the growth rate of z is positive.

For the NPZ example (2), (H3) is expressed as the constant nutrient input a_0 exceeding a certain algebraic expression in the other model parameters. In fact, it can be seen that there are three regions for (2) which can be characterized in terms of a_0. (Writing $K_i = c_i d_i / (b_i - d_i)$, $i = 1,2$), positive initial value solution trajectories approach the equilibrium $E_0(a_0,0,0)$ if $a_0 < K_1$, no such trajectories approach E_0, and at least some approach $E_1(K_1,(a_0-K_1)/d_1,0)$ if $K_1 < a_0 < K_1 + d_1 K_2$; and if $a_0 > K_1 + d_1 K_2$, none approach E_0 or E_1, at least some will approach E, with the possibility of oscillations.

That (H4), the persistence criterion for the time-varying NPZ model (2'), is more conservative than (H3) can be seen by applying it to the constant input case. Here, $a(t) = a_0$, so that $A_1 = a_0/d_m \geq a_0/d_1$ and so (H4) implies that

$$b_1 - d_1 - \frac{b_1^2 d_2}{(b_1-d_1)b_2} > 0 \quad \text{and} \quad a_0 > \frac{c_1 d_1}{b_1-d_1 - \frac{b_1^2 d_2}{(b_1-d_1)b_2}} + d_1 \left(\frac{c_2 d_2}{b_2 \left(\frac{b_1-d_1}{b_1}\right)^2 - d_2} \right)$$

This is not surprising, since the criterion applies to models which can exhibit more diverse dynamical behaviors than (2). For example, it is no longer possible to rule out periodic solutions in the xy plane, as in the proof of Theorem 2.

6. PROOFS OF THE MAIN RESULTS.

PROOF OF THEOREM 1. Let $u(t) = acx(t) + cy(t) + z(t)$. From (2) one obtains (suppressing t)

$$u' = acg(x) - bcy - dz \qquad (3)$$

Hypothesis (H1) implies that there exist positive constants k_i, $i = 1,2,3,4$, such that

$$k_1 - k_2 x \leq g(x) \leq k_3 - k_4 x \qquad (4)$$

at least for $0 \leq x \leq K$. Since $g(x) < 0$, for $x > K$, it follows, from (3) and (4), that for sufficiently large t

$$ack_1 - Mu \leq u' \leq ack_3 - mu \qquad (5)$$

where $M = \max\{k_2,b,d\}$ and $m = \min\{k_4,b,d\}$. Taking $s_1 = ack_1/M$ and $s_2 = ack_3/m$, integrating (5) leads to

$$s_1 + (u(0)-s_1)e^{-Mt} \leq u(t) \leq s_2 + (u(0)-s_2)e^{-mt} \qquad (6)$$

for sufficiently large t. The conclusion of the theorem follows from (6).

PROOF OF THEOREM 2. The first part of this result follows from Theorem 3.1 (Freedman and Waltman, 1977) if periodic orbits can be ruled out in the xy plane. The latter is accomplished by application of the Bendixon-Dulac criterion (see

Andronov et al. 1973, for example) to the xy subsystem of (1)

$$x' = g(x) - p(x)y$$
$$y' = y(ap(x)-b)$$

(7)

This analysis has been carried out for some particular models of this form (Schoener, 1973).

PROOF OF THEOREM 3. The proof follows as in Theorem 1. With $u = x + y + z/k$, the inequality

$$a_0 - d_M u \leq u' \leq a_1 - d_m u$$

is established which leads to the result.

PROOF OF THEOREM 4. One constructs a nonnegative continuous function $\rho(x,y,z)$ on R_+^3 which has continuous first order partial derivatives on $R_+^{3,o}$, and satisfies the condition

$$\rho \to 0 \quad \text{if either} \quad x,y, \quad \text{or} \quad z \to 0.$$

(8)

Supposing, by way of contradiction, that persistence fails, one assumes a solution $(x(t),y(t),z(t))$ with $(x(0),y(0),z(0)) \in R_+^{3,o}$ such that $z(t) \to 0$ as $t \to \infty$, and considers the function $\rho(t) = \rho(x(t),y(t),z(t))$. If, for sufficiently large t, a differential inequality of the form

$$\dot{\rho}(t) \equiv \nabla\rho \cdot (x',y',z') \geq \lambda\rho(t)$$

(9)

where λ is a positive constant, can be established, the desired contradiction is obtained. This is so because of (8) and the fact that (9) implies that

$$\rho(t) \geq \rho(0)e^{\lambda t} \neq 0 \quad \text{as} \quad t \to \infty.$$

The resulting persistence criterion is a condition on the model parameters that allows the existence of such a λ .

To apply this method to system (2') one chooses the function

$$\rho = \left(\frac{x}{c_1+x}\right) y^r z^s$$

(10)

where r and s are positive constants to be determined. Then one has, by making use of (2'),

$$\dot{\rho} = \frac{c_1}{(c_1+x)^2} y^r z^s \left(a(t)-x \left(1 + \frac{b_1 y}{c_1+x} \right) \right)$$

$$+ \left(\frac{x}{c_1+x} \right) y^r z^s r \left(\frac{b_1 x}{c_1+x} - d_1 - \frac{b_2 z}{k(c_2+y)} \right) + \left(\frac{x}{c_1+x} \right) y^r z^s s \left(\frac{b_2 y}{c_2+y} - d_2 \right).$$

(11)

By using the lower bound for $a(t)$ and rearranging terms in (11), one obtains the inequality

$$\dot{\rho} \geq \frac{c_1}{c_1+x} \rho \left\{ \frac{a_0}{x} - 1 - rd_1 - sd_2 + \frac{x}{c_1}[r(b_1-d_1)-sd_2] \right.$$

$$\left. + y \left[s \left(\frac{c_1+x}{c_1} \right) \left(\frac{b_2}{c_2+y} \right) - \frac{b_1}{c_1+x} \right] - z \left[r \left(\frac{c_1+x}{c_1} \right) \frac{b_2}{k(c_2+y)} \right] \right\}.$$

(12)

Now, for any solution $y(t) \leq A_1 + \epsilon_1$ for sufficiently large t where ϵ_1 is an arbitrarily small positive number, by Theorem 3. Therefore, for arbitrary small $\epsilon_2 > 0$, it can be seen that

$$y \left[s \left(\frac{c_1+x}{c_1} \right) \left(\frac{b_2}{c_2+y} \right) - \frac{b_1}{c_1+x} \right] \geq y \left[s \frac{b_2}{c_2+A_1+\epsilon_1} - \frac{b_1}{c_1} \right] \geq -\epsilon_2$$

(13)

for sufficiently large t, if $s = b_1(c_2+A_1)/b_2 c_1$. For any solution which exhibits failure of persistence

$$z(t) \to 0 \quad \text{as} \quad t \to \infty.$$

This together with (12) and (13) imply that the simplified inequality

$$\dot{\rho}(t) \geq k_1 \rho(t) \{\mu_r(x(t)) - \epsilon\}$$

(14)

holds for sufficiently large t, where $\epsilon > 0$ is arbitrary, $k_1 = c_1/(c_1 + A_1)$, and

$$\mu_r(x) = a_0/x - 1 - rd_1 - sd_2 - x[r(b_1-d_1)-sd_2]/c_1.$$

In view of (9) and (14), it suffices to show that a positive r exists such that for some positive number k_2

$$\mu_r(x) \geq k_2 \quad \text{for all} \quad x > 0.$$

(15)

A calculus argument verifies (15) provided

$$a_0 \left(\frac{b_1-d_1}{c_1} \right) - d_1(1+sd_2) - \frac{sd_1^2 d_2}{b_1-d_1} > 0.$$

Observing that this inequality, with s chosen as in (13), is equivalent to (H4) completes the proof.

REFERENCES

Andronov, A.A., E.A. Leontovich, I.I. Gordon, and A.G. Maier (1973): *Qualitative Theory of Second Order Dynamical Systems*, Wiley, New York.

Bungay, H.R. III, and M.L. Bungay (1968): Microbial interactions in continuous cultures, *Adv. Appl. Microbiol.* 10:269-290.

Canale, R.P. (1969): Prey-predator relationships in a model for the activated process, *Biotech. Bioengng.* 11:887-907.

_____ (1970): An analysis of models describing predator-prey interaction, *Biotech. Bioengng.* 12:353-378

Drake, J.F., J.L. Jost, A.G. Fredrickson, and H.M. Tsuchiya (1966): The food chain in bio-regenitive systems, *NASA SP-165*, Washington, D.C.

Freedman, H.I. (1980): *Deterministic Mathematical Models in Population Ecology*, Marcel Dekker, New York.

Freedman, H.I., and P. Waltman (1977): Mathematical analysis of some three-species food-chain models, *Math. Biosci.* 33:257-276.

Gard, T.C. (1980): Persistence in food chains with general interactions, *Math. Biosci.* 51:165-174.

Gard, T.C., and T.G. Hallam (1979): Persistence in food webs: I. Lotka-Volterra food chains, *Bull. Math. Biol.* 41:877-891.

Hale, J.K. (1969): *Ordinary Differential Equations*, Wiley-Interscience, New York.

Hallam, T.G. (1977): Controlled persistence in rudimentary plankton models, *Proc. 1st. Int. Conf. Math. Modelling*, St. Louis.

Jost, J.L., J.F. Drake, H.M. Tsuchiya, and A.G. Fredrickson (1973): Microbial food chains and food webs, *J. Theor. Biol.* 41:461-484.

Lassiter, R.R. (1978): Microcosms as ecosystems for testing ecological models, State-of-the-Art in Ecological Modelling, Vol. 7, *Proc. 1st Int. Conf. Ecol. Modelling (ISEM)* (S.E. Jorgensen, editor), Copenhagen, pp. 127-161.

Nemytskii, V.V. and V.V. Stepanov (1960): *Qualitative Theory of Differential Equations*, Princeton University Press, Princeton.

Saunders, P.T. and M.J. Bazin (1975): On the statbility of food chains, *J. Theor. Biol.* 52:121-142.

Schoener, T.W. (1973): Population growth regulated by intraspecific competition for energy and time: Some simple representations, *Theor. Pop. Biol.* 4:56-84.

Sell, G.R. (1977): What is a dynamical system? *Studies in Ordinary Differential Equations* (J. Hale, editor), Math. Assoc. Amer.

Tsuchiya, H.M., J.F. Drake, J.L. Jost, and A.G. Fredrickson (1972): Predator-prey interactions of dictyostelium-discoidium and escherichia-coli in continuous culture, *J. Bact.* 110:1147-1153.

Waltman, P.: Competition for a renewable resource, this *Proceedings*.

RESONANCE IN PREY-PREDATOR SYSTEMS

P.K. Ghosh

ABSTRACT

The role of periodically varying parameters in prey-predator systems is studied. In Lotka-Volterra (LV) and Volterra-Gause-Witt models, it is found that under certain conditions populations can resonate with the periodicities of the parameters. Parametric instability of LV oscillations is studied. Biological implications of the results are discussed.

1. INTRODUCTION

The usual procedure of studying population models has been on the basis of constant values of parameters. In single species models and in interacting species models, the biological and environmental parameters are generally assumed to be constants. Few populations, however, live in a constant environment. Real environments are uncertain, stochastic. The birth and death rates, carrying capacities and other "rate constants" which characterize populations are expected to be affected by changes in temperature, humidity, seasons and other ecological factors. In most realistic ecological systems, the parameters are time-dependent either in periodic or random manner (Nisbet and Gurney, 1982). Detailed studies have been made on single species and interacting species models with the parameters considered as stochastic variables (May, 1974, 1981; Goel et al., 1971).

Several rodent and plant populations exhibit well-defined oscillations. Meyers and Krebs (1974) have discussed population cycles in rodents and have shown that the rodent populations possess some periodically varying parameters. Wiegart et al. (1975) have modeled coastal Georgia Spartina Marsh ecosystem with certain periodic parameters. It is known that to obtain cyclic equilibrium in populations one has to use either functional terms (e.g. age structure or delay term) or parameters with periodic time-dependence in population models. In fact, Gilpin (1973) modeled the lynx-hare oscillations using time-varying coefficients. Kannan (1979) has discussed the Volterra-Verhulst prey-predator systems with time-dependent coefficients and established the periodic behavior under periodicity conditions on the coefficients and perturbing random forces.

Recently some work has been done for interacting species models with periodic parameters. Cushing (1977) has studied the Lotka-Volterra (LV) prey-predator model with periodic coefficients and obtained conditions for the existence of periodic solutions. Rosenblatt (1980) has considered LV competition model for two species with periodic coefficients, while Freedman and Manetto (1981) and Butler and

Freedman (1981) have discussed Gause-type and Kolmogorov-type prey-predator system with periodic coefficients.

In this work we study the role of periodically varying parameters in two-species predation models, viz., in the LV and in Volterra-Gause-Witt (VGW) models. It is assumed that the parametric variations are small enough so that perturbative techniques may be used.

2. THE LOTKA-VOLTERRA MODEL

The prototype of all mathematical models for interacting species is the LV model (Goel et al., 1971). In the LV model for a deterministic one-prey-one-predator system, it is assumed that the prey (species 1) would grow exponentially in the absence of the predator (species 2), while the predator dies out exponentially in the absence of its prey. The LV rate equations are

$$
\begin{aligned}
dN_1/dt &= \alpha_1 N_1 - \beta_1 N_1 N_2 \\
dN_2/dt &= -\alpha_2 N_2 + \beta_2 N_1 N_2
\end{aligned}
\tag{1}
$$

where the quadratic nonlinear terms describe the interaction between the prey and the predator. Here N_i $(i = 1,2)$ is the size of individuals of species i at a given time. The parameter α_1 relates to the birth rate of the prey, α_2 to the death rate of the predator, and β_1, β_2 to the interaction between the species: all are positive numbers. Let us now set

$$
\begin{aligned}
\alpha_1 &= \alpha_0(1 + \rho \sin \omega_1 t) \\
\alpha_2 &= \alpha_0'(1 + \rho \sin \omega_2 t)
\end{aligned}
\tag{2}
$$

where α_0, $\alpha_0' > 0$ and ρ is a small positive quantity ($\rho \ll 1$). This incorpo-ates the possibility of temporal changes in the birth and death rates. To study LV equations (1) in conjunction with equations (2), we proceed à la Poincaré (Minorsky, 1962; Hyver, 1973). We set

$$
\begin{aligned}
N_1(t) &= N_{10} + N_{11}(t) + \rho^2 N_{12}(t) + \ldots \\
N_2(t) &= N_{20} + N_{21}(t) + \rho^2 N_{22}(t) + \ldots
\end{aligned}
\tag{3}
$$

in equations (1) and collect terms to each order in ρ. To order ρ^0, we get

$$
\begin{aligned}
0 &= \alpha_0 N_{10} - \beta_1 N_{10} N_{20} \\
0 &= -\alpha_0' N_{20} + \beta_2 N_{10} N_{20}
\end{aligned}
\tag{4}
$$

giving two pairs of solutions: (a) $N_{10} = N_{20} = 0$, and (b) $N_{10} = \alpha_0'/\beta_2$, $N_{20} = \alpha_0/\beta_1$.

To order ρ^1, we get

$$dN_{11}/dt = \alpha_0 N_{11} + \alpha_0 N_{10} \sin \omega_1 t - \beta_1 N_{10} N_{21} - \beta_1 N_{11} N_{20}$$

$$dN_{21}/dt = -\alpha_0' N_{21} - \alpha_0' N_{20} \sin \omega_2 t + \beta_2 N_{10} N_{21} + \beta_2 N_{11} N_{20} \ .$$

(5)

If we put $N_{10} = N_{20} = 0$ in equations (5), we get $dN_{11}/dt = \alpha_0 N_{11}$ and $dN_{21}/dt = -\alpha_0' N_{21}$, indicating unlimited growth of prey and extinction of predator species. If, however, we set $N_{10} = \alpha_0'/\beta_2$, $N_{20} = \alpha_0/\beta_1$ in equations (5), then we obtain

$$dN_{11}/dt = (\alpha_0 \alpha_0'/\beta_2)\sin \omega_1 t - (\alpha_0' \beta_1/\beta_2)N_{21}$$

$$dN_{21}/dt = (-\alpha_0 \alpha_0'/\beta_1)\sin \omega_2 t + (\alpha_0 \beta_2/\beta_1)N_{11}$$

(6)

or

$$d^2 N_{11}/dt^2 + \omega_0^2 N_{11} = \frac{\omega_0^2}{\beta_2} (\omega_1 \cos \omega_1 t + \alpha_0' \sin \omega_2 t)$$

$$d^2 N_{21}/dt^2 + \omega_0^2 N_{21} = - \frac{\omega_0^2}{\beta_1} (\omega_2 \cos \omega_2 t - \alpha_0 \sin \omega_1 t)$$

(7)

where $\omega_0^2 = \alpha_0 \alpha_0'$. Equations (7) represent the equations of two harmonic oscillators, both with natural frequency ω_0, executing forced oscillations. The general solutions of equations (7) may be written as

$$N_{11} = a_{11} \cos(\omega_0 t + \eta_{11}) - \frac{\omega_0^2 \omega_1}{\beta_2(\omega_1^2 - \omega_0^2)} \cos \omega_1 t - \frac{\omega_0^2 \alpha_0'}{\beta_2(\omega_2^2 - \omega_0^2)} \sin \omega_2 t \ ,$$

$$N_{21} = a_{21} \cos(\omega_0 t + \eta_{21}) + \frac{\omega_0^2 \omega_2}{\beta_1(\omega_2^2 - \omega_0^2)} \cos \omega_2 t - \frac{\omega_0^2 \alpha_0}{\beta_1(\omega_1^2 - \omega_0^2)} \sin \omega_1 t \ ,$$

(8)

where a's and η's are found from the initial conditions. Thus, in general, both the prey and the predator species execute motions about their respective steady-spates and the motions are a combination of three oscillations. The solutions (8) are not valid when resonance occurs, i.e. when either ω_1 or ω_2 is equal to ω_0. At resonance $(\omega_1 = \omega_0)$,

$$N_{11} = a_{11} \cos(\omega_0 t + \eta_{11}) + \frac{\omega_0}{2\beta_2} (\omega_0 t \sin \omega_0 t - \cos \omega_0 t) - \frac{\omega_0^2 \alpha_0'}{\beta_2(\omega_0^2 - \omega_2^2)} \sin \omega_2 t \ ,$$

$$N_{21} = a_{21} \cos(\omega_0 t + \eta_{21}) - \frac{\omega_0^2 \omega_2}{\beta_1(\omega_0^2 - \omega_2^2)} \cos \omega_2 t - \frac{\omega_0^2 \alpha_0 t}{2\beta_1} \cos \omega_0 t .$$

(9)

Thus the amplitude of oscillations in resonance increases linearly with time (until the oscillations are no longer small). The behavior of small oscillations near resonance can easily be ascertained. Because of the coupled equations, resonance in prey (predator) amplitude implies ultimately resonance in predator (prey) amplitude. If, in equations (1) we keep α_1, α_2 constants and let β_1, β_2 vary periodically, allowing for changes in alternative food supplies, then we can study the equations in the above manner. The conclusions are essentially the same. Next, we can analyse the most general case of equations (1) where all the parameters are periodically varying. Here resonance occurs whenever the frequencies associated with them become equal to ω_0, and the amplitudes of oscillations in resonance increase linearly with time.

3. THE VOLTERRA-GAUSE-WITT MODEL

The Volterra-Gause-Witt (VGW) model (Dutt et al., 1975) is described by the equations

$$dN_1/dt = \alpha_1 N_1 (1-N_1/\theta) - \beta_1 N_1 N_2 = \alpha_1 N_1 - \alpha_1' N_1^2 - \beta_1 N_1 N_2$$
$$dN_2/dt = -\alpha_2 N_2 + \beta_2 N_1 N_2 \tag{10}$$

where $\alpha_1' = \alpha_1/\theta$, and θ is the carrying capacity of the environment for the prey. Let

$$\alpha_1 = \alpha_0 (1 + \rho \sin \omega_1 t)$$
$$\alpha_1' = \alpha_0' (1 + \rho \sin \omega_1 t) \tag{11}$$
$$\alpha_2 = \alpha_0'' (1 + \rho \sin \omega_2 t) \quad \text{where} \quad \alpha_0' = \alpha_0/\theta$$

In this case, the equations for N_{11} and N_{21} are the same as those for systems executing forced oscillations under friction. Damping in the VGW model is due to the "overcrowding" effect in the prey. It is well known that in case of forced oscillation with damping, the general solution consists of a transient part which decreases exponentially with time and a steady oscillatory motion (Landau and Lifshitz, 1969). Clearly the expression for the amplitude of oscillation increases at resonance $[\omega_1 \to \omega_0'$, or, $\omega_2 \to \omega_0'$ where $\omega_0'^2 = \alpha_0''(\alpha_0 - \alpha_0' \alpha_0''/\beta_2)]$ but does not become infinite! The resonance curve becomes more peaked as θ becomes large, the half-width being equal to $\lambda = (\alpha_0' \alpha_0'')/2\beta_2$, the damping coefficient. Thus, we see that the larger the carrying capacity of the environment, the more peaked is the resonance curve i.e. the sharpness of resonance is proportional to the value of θ.

4. PARAMETRIC RESONANCE

Let us now return to the LV equations (1). We see that the trivial stationary state is a saddle point (and thus always unstable), whereas the non-trivial one is a center. If we write $N_1 = (\alpha_2)/(\beta_2) + n_1$ and $N_2 = (\alpha_1)/(\beta_1) + n_2$, then for small oscillations about the center, we get

$$dn_1/dt = -k_1 n_2,$$
$$dn_2/dt = k_0' n_1, \tag{12}$$

where $k_1 = (\alpha_2 \beta_1)/(\beta_2)$ and $k_0' = (\alpha_1 \beta_2)/(\beta_1)$. Thus, small perturbation about the center are periodic with a universal frequency $\sqrt{\alpha_1 \alpha_2}$.

We now inquire into the possibility of parametric instability of a periodic orbit about the center. We suppose that $k_1 = k_0(1 + \rho \cos \gamma t)$. This may happen when the death rate (α_2) of the predator is periodic. Then equations (12) become

$$d^2 n_2/dt^2 + \omega^2(t) n_2 = 0 \tag{13}$$

where $\omega^2(t) = \omega_0^2(1 + \rho \cos \gamma t)$ with $\omega_0^2 = k_0 k_0'$. Analysis of equation (13) shows that the system at rest in equilibrium $(n_2 = 0)$ is unstable; any deviation from this state, however small, is sufficient to lead to an exponentially increasing displacement. This is called "parametric resonance " (Landau and Lifshitz, 1969). Parametric resonance is strongest when the frequency of $\omega(t)$ is nearly twice ω_0. Hence we put $\gamma = 2\omega_0 + \epsilon$ where $\epsilon << \omega_0$. Equation (13) then becomes

$$d^2 n_2/dt^2 + \omega_0^2[1 + \rho \cos(2\omega_0 + \epsilon)t] n_2 = 0 \tag{14}$$

This is the well-known Mathieu equation. It can be shown that parametric resonance occurs in the range

$$-\frac{1}{2} \rho \omega_0 < \epsilon < \frac{1}{2} \rho \omega_0 \tag{15}$$

on either side of the frequency $2\omega_0$. The values of the amplification coefficient of the oscillations in the range are of the order of ρ. Similar considerations apply to n_1.

5. CONCLUSION

In this work we have studied the role of periodically varying parameters in prey-predator systems. It is seen that in general both the prey and the predator species execute small oscillations about their respective steady states. Under certain conditions, the populations can resonate with the periodicities in the parameters, e.g. with the periodicities in the environment.

In 1954 Hutchinson and Slobodkin suggested that populations can be considered as feedback systems and hence can resonate in response to environmental fluctuations

(Slobodkin, 1961). To study this, Oster and Takahasi (1974) and Auslander, Oster and Huffaker (1974) have considered several models for populations coupled by age-specific interactions in a periodic environment. Our study has shown that resonance effects can be observed in prey-predator models with periodic parameters. It is clear that such models can explain certain long-term periodicities and population irruptions [e.g. population irruption of an Australian psyllid-insect living on eucalyptus trees (Odum, 1971); cyclic irruptions of elephant populations in East Africa (May, 1974)] observed in both laboratory and natural ecosystems. We believe that this study may be found useful also in pest-control studies and in studies of prey-predator models of immune response.

Acknowledgements

The author is grateful to B. Dutta Roy and D. Bhaumik for useful discussions.

REFERENCES

Auslander, D.M., G.F. Oster and C.B. Huffaker (1974): Dynamics of interacting populations, *J. Franklin Inst.* 297:345-376.

Butler, G.J. and H.I. Freedman (1981): Periodic solutions of a predator-prey system with periodic coefficients, *Math. Biosc.* 55:27-38.

Cushing, J.M. (1977): Periodic time-dependent predator-prey systems, *SIAM J. Appl. Math.* 32:82-95.

Dutt, R., P.K. Ghosh and B.B. Karmakar (1975): Application of perturbation theory to the nonlinear Volterra-Gause-Witt model for prey-predator interaction, *Bull. Math. Biol.* 37:139-146.

Freedman, H.I., and L. Manetta (1981): Predator-prey systems with a perturbed periodic carrying capacity, *Utilitas Math.* 19:141-155.

Gilpin, M.E. (1973): Do hares eat lynx?, *Amer. Nat.* 107:727-730.

Goel, N.S., S.C. Maitra and E.W. Montroll (1971): On the Volterra and other non-linear models of interacting populations, *Rev. Mod. Phys.* 43:231-276.

Hyver, C. (1975): Existence de resonances parametriques dans des systemes de transformation d'interact biologique, *Bull. Math. Biol.* 37:1-9.

Kannan, D. (1979): Volterra-Verhulst prey-predator systems with time dependent coefficients. Diffusion type approximation and periodic solutions, *Bull. Math. Biol.* 41:229-251.

Landau, L.D. and E.M. Lifshitz (1969): *Mechanics*, Pergamon Press, Oxford.

May, R.M. (1974): *Stability and Complexity in Model Ecosystems*, Princeton Univ. Press, Princeton.

_____(1981): *Theoretical Ecology: Principles and Applications*, Blackwell, Oxford.

Minorsky, N. (1962); *Nonlinear Oscillations*, Van Nostrand, Princeton, N.J.

Myers, J.H., and C.J. Krebs (1974): Populations cycles in rodents, *Scientific Amer.* 230-38-46.

Nisbet, R.M., and W.S.C. Gurney (1982): *Modelling Fluctuating Populations,* John Wiley and Sons, New York.

Odum, E.P. (1971): *Fundamentals of Ecology,* W.B. Saunders Co., Philadelphia.

Oster, G.F., and Y. Takahasi (1974): Models for age-specific interactions in a periodic environment, *Ecol. Monographs* 44:483-501.

Rosenblatt, S. (1980): Population models in a periodically fluctuating environment, *J. Math. Biol.* 9:23-36.

Slobodkin, L.P. (1961): *Growth and Regulation of Animal Populations,* Holt, Rinehart and Winston, New York.

Wiegart, R.G., et al. (1975): A preliminary ecosystem model of coastal Georgia Spartina marsh, *Estuarine Research* 1:563-601.

A MATHEMATICAL MODEL OF POPULATION REGULATION IN CYCLIC MAMMALS

Jay B. Hestbeck

1. INTRODUCTION

The cyclic fluctuations of animal abundance has puzzled researchers for many years. A wide diversity of mathematical models has been developed to aid the study of cycles. The earliest model of cycles was the predator-prey model developed by Lotka (1925) and Volterra (1926).

$$dN_1/dt = N_1(b_1-p_1N_2)$$

$$dN_2/dt = N_2(-d_2+p_2N_1)$$

where N_1 is prey density, N_2 is predator density, b_1 is the birth rate of prey, d_2 is the death rate of predator, p_1 is the consumption rate of prey, and p_2 represents the conversion rate of prey into predator. This model is neutrally stable and has either a stable point or a stable limit cycle (May 1876b).

If a time lag is intrinsic to a population, a Verhulst-Pearl logistic equation in which the rate dN/dt is determined by $N(t-T)$ produces density oscillations (Hutchinson 1948).

$$dN/dt = Nr(1-N(t-T)/K)$$

where r is the intrinsic growth rate, K is the carrying capacity, and T is the time lag. This equation produces damped oscillations when $e^{-1} < rT < \pi/2$ and exhibits stable limit cycles when $rT > \pi/2$ (May 1976a).

The random series model for cycles was proposed by Palmgren (1949) and Cole (1958) in response to a prevailing trend among biologists to find cycles in a very diverse array of species. By taking a two-point moving average of a series of random numbers, Palmgren (1949) produced cycles similar to those observed in natural populations. This led Palmgren to conclude that populations may actually fluctuate at random. Cole (1958) strengthened the random model position and concluded that unless non-random components could be indentified in population data, cosmic influences, intra-specific or inter-specific causes for population cycles are unnecessary.

Lack (1954) suggested that cycling lemmings may result from a vegetation-herbivore interaction. May (1973), Noy-Meir (1975) and Caughley (1976) developed several vegetation-herbivore models by generalizing the predator-prey equations. Although conditions necessary to produce oscillations were noted, these analyses were directed at large grazing ungulates.

Recent models have been directed at the Chitty hypothesis. Chitty (1960,1967) postulated that as density increased, selection for aggression changed the genetic composition of the population such that individuals become more susceptible to normal mortality sources. Although Schaffer and Tamarin (1973) and Thue Poulsen (1979) have presented models, Stenseth (1981) concluded that no model has demonstrated Chitty's predictions.

The most complex models of cycling were created by the International Biological Program. Bunnell (1972) concluded that no single hypothesis was sufficient to explain lemming cycles. Timin and Collier (1972) reported that cycling primarily resulted from a lemming-weasel interaction. Liestøl et al. (1975) found that predation, lemming-vegetation and density dependent behavioral changes acting singly or in combination sufficiently produced cycles. Liestøl et al. could not, however, give a conclusive answer as to which factor(s) caused the cycles. Miller et al. (1975) concluded that cycling primarily resulted from the lemming-weasel interaction.

Although this great diversity of models has aided the study of cycles, the explanation of cycles presently is limited by a lack of understanding about the underlying biological processes. To provide a better understanding of population cycles, an experiment was designed to explore the effects of spacing behavior and initial breeding density (Hestbeck in preparation). Two main conclusions were suggested by this experiment: (1) Regardless of initial breeding density, if a dispersal sink exists adjacent to a population, spacing behavior limits that population to low densities. (2) If spacing behavior is prevented, the enclosed population increases to very high densities. These higher densities reached by the enclosed populations, however, correspond to peak densities recorded by Batzli and Pitelka (1971) and Kishler (1972) for natural populations. Naturally occurring populations of *Microtus* appear to be regulated by two separate mechanisms. At low to moderate densities, spacing behavior limits densities. Once spacing is prevented, densities rise until resource exhaustion limits the population.

The social fence hypothesis provides a mechanism which explains how population regulation shifts from spacing behavior to resource exhaustion (Hestbeck 1982). Spacing behavior limits densities at low densities. As neighboring group densities rise, the effectiveness of spacing behavior to limit a more central group is impaired. Once emigration is blocked, densities in the central area rise until resource limitation occurs.

In order to clarify the social fence mechanism and to examine its role in population regulation and in the generation of cycles, a mathematical model of the social fence hypothesis is presented.

2. MODEL

The population is modeled with four differential equations. One represents the central group dynamics. One describes the neighbor group. One represents the vegetation in the central area. The last one describes the neighbor area vegetation.

The relationship between vole densities and grassland carrying capacities are modeled as a herbivore-vegetation interaction. The dynamics between vole groups occurred by emigration and immigration. Following Caughley's (1976) interactive herbivore-vegetation model, vegetative production was modeled with a logistic growth rate

$$dK_i/dt = r_v(1-K_i/K_{max})K_i - p_N N_i$$

where K_i is the carrying capacity for the ith area, r_v is the growth rate of the vegetation, K_{max} is the maximum carrying capacity, p_N is the grazing rate and N_i is the vole density of the ith area.

The central-group dynamics were modeled as

$$\frac{dN_1}{dt} = b\left[e^{-2N_1/K_1}\right]N_1 - d\left[\frac{N_1}{K_1}\right]N_1 - em\left[\frac{(1-e^{-\alpha N_1})}{e^{\beta N_2}}\right]N_1 + im\left[\frac{(1-e^{-\alpha N_2})}{e^{\beta N_1}}\right]\left[\frac{1}{e^{\beta N_1}}\right]N_2$$

where b is the birth rate, d the death rate, em the emigration rate and im the immigration rate. Also, αN_i is the aggression intensity experienced by an individual within its group and βN_i is the aggression intensity experienced by an individual entering the neighboring group. The birth rate ranges from b to 0. The death rate linearly from 0 as N/K. Emigration is a function of within group aggression level and between group aggression level. Immigration is the neighbor group emigration multiplied by the colonization success of the central group $(1/e^{\beta N})$.

The neighbor-group dynamics were modeled similarly to the central group

$$\frac{dN_2}{dt} = b\left[e^{-2N_2/K_2}\right]N_2 - d\left[\frac{N_2}{K_2}\right]N_2 - em\left[(1-e^{-\alpha N_2})\right]N_2 + im\left[\frac{(1-e^{-\alpha N_1})}{e^{\beta N_2}}\right]\left[\frac{1}{e^{\beta N_2}}\right]N_1$$

except that emigration is only a function of within group aggression.

This model, using the parameter values listed in Figure 1 and a K_{max} of 2500, results in a stable solution with the central group reaching a density three time higher than the neighbor group. The central group dynamics are determined solely by the birth and death rates. The neighbor group dynamics, however, are determined by the birth and emigration rates. Mortality occurs in the neighbor group but at an insignificant level since individuals can readily emigrate. The social fence thus provides a mechanism by which populations can be limited by different mechanisms. Also, the social fence provides an explanation of how a population can first be regulated by spacing behavior, become socially fenced, and rise to the level dictated by resources. A social fence by itself cannot produce oscillations.

Multi-year cycles can be produced from the herbivore-vegetation interaction

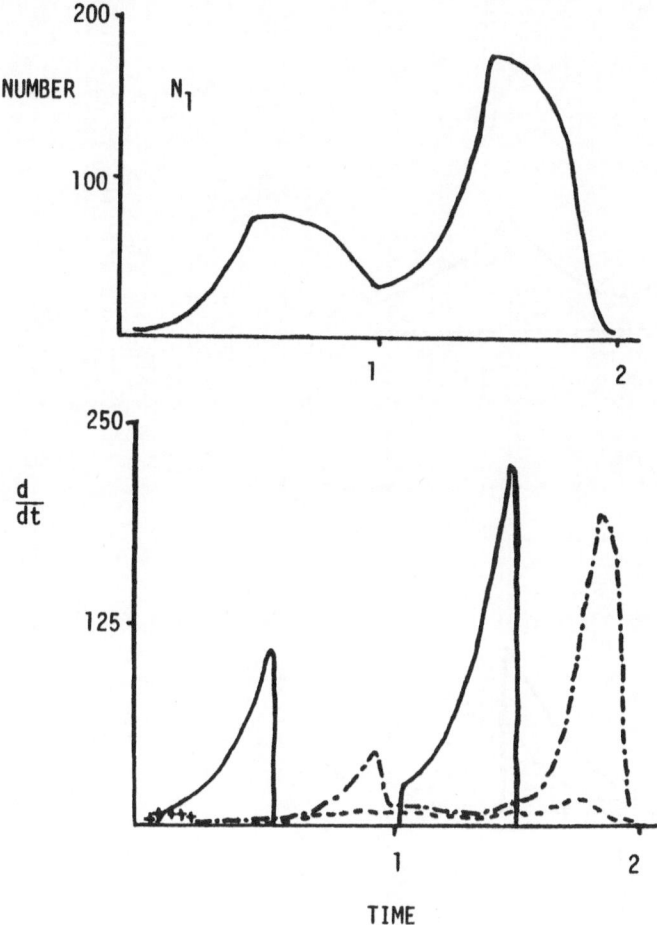

Figure 1   (a) Multi-year population cycle for central group.

(b) Birth   rate (——), death rate (-·-), emigration rate (---), and
immigration rate (+++) for the central group.

when seasonality is introduced. To model seasonality the logistic growth term for
vegetative production was replaced by a seasonal oscillator (sin wt). For every
t = 2πj where j = 0,1,2,..., the carrying capacity was reinitialized to a constant
K_0. This reflects annual germination and prevents the cycle from resulting from
long term habitat damage.

When the seasonal productivity model was simulated over time, the central
group exhibits a two year density cycle (Figure 1a) and the neighbor group undergoes
annual density fluctuations (Figure 2a). As before, the central group dynamics are
explained by the birth and death rates (Figure 1b). Due to the social fence, emigration is insignificant, and immigration occurs only after a crash. Again, the
neighbor group dynamics are determined by birth and emigration rates. Mortality

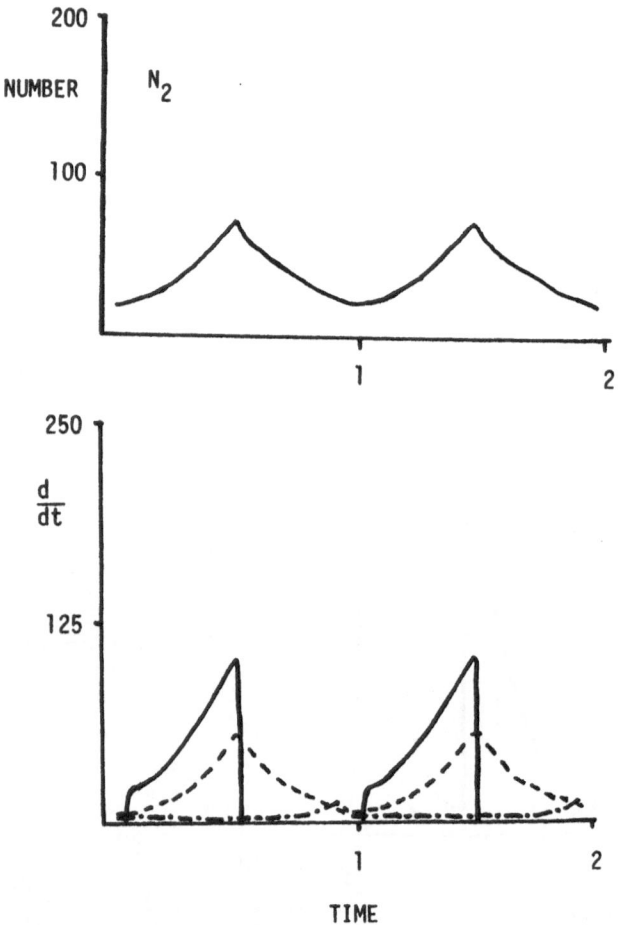

Figure 2 (a) Annual population fluctuation for neighbor group.
(b) Birth rate (——), death rate (-·-), and emigration rate (---)
for the neighbor group.

rates are very low since most individuals emigrate into a dispersal sink. When
vegetative production was modeled with a logistic term, the vole-vegetation phase
diagram displays a stable focus. When seasonal production was used the vole-
vegetation model undergoes a stable limit cycle with a period of two years. The
social fence results in regulation occurring by two different mechanisms, and
seasonal variation drives the multi-year cycles.

The nature of the density fluctuation depends on the amount of grazing which
takes place during the early, slow growth germination phase. This relationship
between habitat destruction and multi-year cycles was demonstrated by varying
grazing pressure. To concentrate attention on the multi-year cycle, a low pass
filter was used to remove oscillations with a frquency corresponding to a year or

less. The amplitude of the remaining multi-year cycle was then plotted against grazing intensity (Figure 3). The graph can be broken into three parts. The first part depicts a herbivore which has little impact on the vegetation resource. The second part describes a herbivore or small mammal grazer which can damage a resource once it reaches a sufficiently high density. The third part represents a larger herbivore or ungulate which can greatly damage the vegetation if the individual remains in a small local area. These individuals must either remain at low densities or migrate to new resource areas.

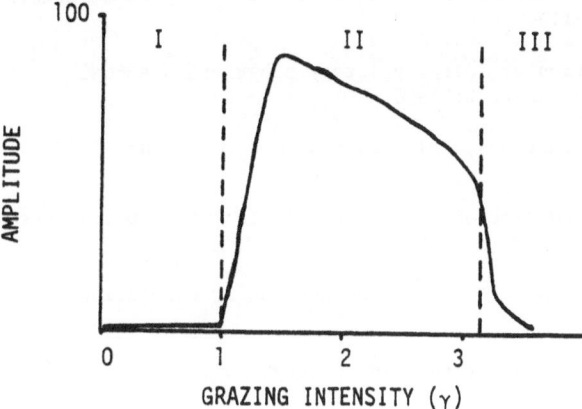

Figure 3   Amplitude of oscillations with a frequency greater than one season versus grazing intensity.

3. CONCLUSION

The social fence hypothesis proposes a mechanism in which population regulation can occur from both spacing behavior and resource exhaustion. The social fence also provides a mechanism which causes regulation to shift from spacing behavior to resource exhaustion. The social fence was mathematically modelled to clarify the mechanism and explore its relationship to population cycles in small mammals.

The results from the model illustrate that cycling is not caused by the development of the social fence. The social fence mechanism inhibits emigration and causes densities to approach the resource limitation level. The groups limited by spacing behavior undergo annual fluctuations at a moderate density. The density fluctuation for the resource limited groups is determined by the degree of habitat destruction occuring through the herbivore-vegetation interaction. This relationship is general and matches the observations of Jameson (1955) concerning cyclic and acyclic rodents, the experiment by Hestbeck (in preparation) and Caughley's (1970) description of ungulate population eruptions.

REFERENCES

Batzli, G., and F.A. Pitelka (1971): Condition and diet of cycling populations of the California vole, *Microtus californicus*, *J. Mammal.* 52:141-163.

Bunnell, F.L. (1972): Lemmings-models and the real world, *Proc. Summer Comp. Sim. Conf.* 2:1183-1197.

Caughley, G. (1970): Eruption of ungulate populations, with emphasis on Himalayan thar in New Zealand, *Ecology* 51:53-72.

_____(1970): Plant-Herbivore systems. In R.M. May (ed.) *Theoretical Ecology*, Saunders, Philadelphia.

Chitty, D. (1960): Population processes in the vole and their relevance to general theory, *Can. J. Zool.* 38:99-113.

_____ (1967): The natural selection of self-regulatory behavior in Animal Populations, *Proc. Ecol. Soc. Aust.* 2:51-78.

Cole, L.C. (1958): Population fluctuations, *Proc. Intern. Congr. of Ent.* 10(2): 639-647.

Hestbeck, J.B. (1982): Population regulation of cyclic mammals: the social fence hypothesis, *Oikos* 39:157-163.

_____(in preparation): Demography of spacing and nonspacing populations of the California vole.

Hutchinson, G.E. (1948): Circular causal systems in ecology, *Ann. N.Y. Acad. Sci.* 50:221-246.

Jameson, E.W., Jr. (1955): Some factors affecting fluctuations of *Microtus* and *Peromyscus*, *J. Mammal.* 36:206-209.

Kishler, C.L., Jr. (1972): Some behavioral and metabolic comparisons of *Microtus californicus* from different phases of the population cycle, *M.A. Thesis*, Univ. of California, Berkeley.

Lack, D. (1954): *The Natural Regulation of Animal Numbers*, Oxford Univ. Press, New York.

Liestøl, K., E. Østbye, H.-J. Skar, G. Swartzman (1975): A simulation model of a small rodent population. In F.E. Wielolaski (ed.) *Fennoscandian Tundra Ecosystems Part 2. Animals and Systems Analysis*.

Lotka, A.J. (1925): *Elements of Physical Biology*, Williams and Wilkins, Baltimore.

May, R.M. (1973): Time-delay versus stability in population models with two and three trophic levels, *Ecology* 54:315-325.

_____(1976a): Models for single populations. In R.M. May (ed.) *Theoretical Ecology*, Saunders, Philadelphia, pp. 4-25.

_____(1976b): Models for two interacting populations. In R.M. May (ed.) *Theoretical Ecology*, Saunders, Philadelphia, pp. 49-70.

Miller, P.C., B.D. Collier, and F.L. Bunnell (1975): Development of ecosystem modeling in the tundra biome. In B.C. Patten (ed.) *Systems Analysis and Simulations in Ecology*, Vol. 3, pp. 95-115.

Noy-Meir, I. (1975): Stability of grazing systems: an application of predator-prey graphs, *J. Ecol.* 63:459-481.

Palmgren, P. (1949): Some remarks on the short-term fluctuations in the numbers of northern birds and mammals, *Oikos* 1:114-121.

Schaffer, W.M., and R.H. Tamarin (1973): Changing reproductive rates and population cycles in lemmings and voles, *Evolution* 27:111-124.

Stenseth, N.C. (1981): On Chitty's theory for fluctuating populations: the importance of genetic polymorphism in the generation of regular density cycles, *J. Theor. Biol.* 90:9-36.

Thue Poulsen, E. (1979): A model for population regulation with density- and frequency-dependent selection, *J. Math. Biol.* 8:325-343.

Timin, M.E., and B.O. Collier (1972): Simulating the arctic tundra ecosystem near Barrow, Alaska, *Proc. Summer Comp. Sim. Conf.* 2:1198-1204.

Volterra, V. (1926): Variations and fluctuations of the number of individuals in animal species living together, *J. Cons. Perm. Int. Ent. Mer.* 3:3-51.

MONOTONE SCHEMES FOR THREE SPECIES
PREY-PREDATOR REACTION-DIFFUSION

Anthony W. Leung*

1. INTRODUCTION

This is a shortened version of Leung (to appear), we consider the boundary value problem:

$$\Delta u + u[a+f_1(u,v,w)] = 0$$

$$\Delta v + v[b+f_2(u,v,w)] = 0 \quad \text{in } \mathcal{D}$$

$$\Delta w + w[c+f_3(u,v,w)] = 0 \tag{1.1}$$

$$u = g_1, \ v = g_2, \ w = g_3 \quad \text{on } \delta\mathcal{D}$$

where $\Delta = \sum_{i=1}^{n} (\partial^2/\partial x_i^2)$, a, b, c are constants. $\mathcal{D}$ is a bounded domain in R^n, $n \geq 2$, with boundary $\delta\mathcal{D}$. The system studies spatial equilibrium (or steady-state) for three interacting species under diffusion, with space variables $(x_1,\ldots,x_n)$. We will assume conditions on the signs of the first partial derivatives of the interaction functions f_i, so that (1.1) describes three interacting prey-predator species (with concentrations u, v, w, and corresponding intrinsic growth or death rates a, b, c). To simplify results, we assume $f_i(u,v,w)$, i = 1,2,3, have uniformly Hölder continuous partial derivatives up to second order in compact sets of the first closed octant in R^3. $H^{2+\ell}(\mathcal{D})$, $0 < \ell < 1$ denotes the Banach space of all real-valued functions u continuous in $\mathcal{D}$ (closure of $\mathcal{D}$) with all first and second derivatives also continuous in $\bar{\mathcal{D}}$, and with finite value for the norm $|u|_{\mathcal{D}}^{2+\ell}$ (described in Williams and Chow, 1978, p. 159). We assume that $\delta\mathcal{D} \in H^{2+\ell}$ (see Williams and Chow, 1978), and the functions g_i, i = 1,2,3, have extensions $\hat{g}_i \in H^{2+\ell}(\bar{\mathcal{D}})$, and $g_i(x) \geq 0$, $\not\equiv 0$ for $x \in \delta\mathcal{D}$.

We will classify the interactions into four cases: (I) food chain, (II) two predators with one prey, (III) one predator with two preys, and (IV) mutualist loop. (The situation when all species compete with each other is not strictly prey-predator interaction, and is thus excluded here). In the four cases, monotone sequences of functions are constructed which converge to $u^*, u_*, v^*, v_*, w^*, w_*$ so that all positive solutions of (1.1) in $\mathcal{D}$ will satisfy $u_* \leq u \leq u^*$, $v_* \leq v \leq v^*$, $w_* \leq w \leq w^*$. The functions in those monotone sequences satisfy scalar equations constructed according to different elaborate schemes depending on the

─────────────
*This research was partially supported by a grant from the National Science Foundation, MCS 80-01851.

different cases considered. The schemes become complicated because the partial derivatives of f_i do not all have the same signs. The scheme also give numerical (see e.g. Lazer, et al., to appear, for two-species case) as well as analytical contribution to the study of our problem. Applications to uniqueness and stability of the time-dependent model, as well as detailed proofs will appear elsewhere (Leung, to appear).

In Leung (1982) the case of two interacting prey-predators is similarly studied. However, in Leung (1982) (and Lazer et al., to appear) the interactions are restricted to Volterra-Lotka type and the boundary conditions are identically homogeneous. With the present boundary conditions, many restrictions on the intrinsic and interaction rates in Leung (to appear) can be removed. Furthermore, the situation in case IV describes a relationship which does not have any analogy in two species prey-predator interaction. The recent interests in the study of many species interactions can be seen in, for example, Krikorian (1979), May (1973), Redheffer and Zhou (1981), Williams and Chow (1978).

The following self-crowding effects are always assumed in this article:

$$\frac{\partial f_1}{\partial u} < 0, \qquad \frac{\partial f_2}{\partial v} < 0, \qquad \frac{\partial f_3}{\partial w} < 0 \qquad (1.2)$$

in the first closed octant in R^3;

$$\lim_{u \to +\infty} f_1(u,v,w) = -\infty, \qquad \lim_{v \to +\infty} f_2(u,v,w) = -\infty, \qquad \lim_{w \to +\infty} f_3(u,v,w) = -\infty \qquad (1.3)$$

In (1.3), the limits are uniform when the independent variables not tending to $+\infty$ are to remain in compact sets.

2. SCHEMES FOR CASES (I) TO (III)

2.1 *Food chain*. We consider species A, B, C (with corresponding concentrations u, v, w) where C preys on B, B preys on A, and C preys on or has no direct relation with A. Mathematically speaking, the specific assumptions are

(a) $\partial f_1/\partial v \leq 0$, $\partial f_1/\partial w \leq 0$;

(b) $\partial f_2/\partial u \geq 0$, $\partial f_2/\partial w \leq 0$;

(c) $\partial f_3/\partial u \geq 0$, $\partial f_3/\partial v \geq 0$;

(d) $\partial f_1/\partial v$ and $\partial f_2/\partial u$ cannot be both identically zero;

(e) $\partial f_2/\partial w$ and $\partial f_3/\partial v$ cannot be both identically zero.

Relations (a) to (e) are all considered in the first closed octant in R^3. We construct sequences of functions u_i, v_i, w_i, $i = 1,2,\ldots$ to be strictly positive functions in $\mathcal{D}$ as follows: Let $u_0 \equiv v_0 \equiv w_0 \equiv 0$ in $\bar{\mathcal{D}}$, and let u_i, v_i, w_i, i = 1,2,... be the uniquely defined positive functions in $\mathcal{D}$ satisfying

$$\Delta u_i + u_i[a+f_1(u_i,v_{i-1},w_{i-1})] = 0 \quad \text{in } \mathcal{D}, \quad u_i = g_1 \quad \text{on } \delta\mathcal{D}; \qquad (2.1)$$

$$\Delta v_i + v_i[b+f_2(u_i,v_i,w_{i-1})] = 0 \quad \text{in } \mathcal{D}, \quad v_i = g_2 \quad \text{on } \delta\mathcal{D}; \qquad (2.2)$$

$$\Delta w_i + w_i[c+f_3(u_i,v_i,w_i)] = 0 \quad \text{in } \mathcal{D}, \quad w_i = g_3 \quad \text{on } \delta\mathcal{D}. \qquad (2.3)$$

The existence and uniqueness of such functions follow from the method of upper and lower solutions and Lemma 2.1 below.

2.2 *Two predators with one prey.* We consider species A, B, C (with corresponding concentrations u, v, w) where B preys on A and/or C preys on A, B competes or has no direct relation with C. More precisely, we assume mathematically:

(a) $\partial f_1/\partial v \le 0, \quad \partial f_1/\partial w \le 0;$

(b) $\partial f_2/\partial u \ge 0, \quad \partial f_2/\partial w \le 0;$

(c) $\partial f_3/\partial u \ge 0, \quad \partial f_3/\partial v \le 0;$

(d) if (1) $\partial f_1/\partial v \equiv \partial f_2/\partial u \equiv 0$ then one cannot have (2) $\partial f_1/\partial w \equiv \partial f_3/\partial u \equiv 0$, conversely if (2) holds then (1) cannot hold;

(e) if $\partial f_2/\partial w \equiv \partial f_3/\partial v \equiv 0$ then both (1) and (2) above cannot hold.

Relations (a) to (e) are all considered in the first closed octant in R^3. Let $u_0 \equiv v_0 \equiv w_0 \equiv 0$ in $\bar{\mathcal{D}}$. For i = 1,2,... recursively define u_i, v_i, w_i as the unique positive functions in $\mathcal{D}$ satisfying:

$$\Delta u_i + u_i[a+f_1(u_i,v_{i-1},w_{i-1})] = 0 \quad \text{in } \mathcal{D}, \quad u_i = g_1 \quad \text{on } \delta\mathcal{D}; \qquad (2.4)$$

$$\Delta v_i + v_i[b+f_2(u_i,v_i,w_{i-1})] = 0 \quad \text{in } \mathcal{D}, \quad v_i = g_2 \quad \text{on } \delta\mathcal{D}; \qquad (2.5)$$

$$\Delta w_i + w_i[c+f_3(u_i,v_{i-1},w_i)] = 0 \quad \text{in } \mathcal{D}, \quad w_i = g_3 \quad \text{on } \delta\mathcal{D}. \qquad (2.6)$$

The existence and uniqueness of such functions follow from reasons as before.

2.3 *One predator with two preys.* We consider species A, B, C (with concentrations u, v, w) where A competes or has no direct relation with B, C preys on A and/or C preys on B. Specifically, we assume in the first closed octant in R^3 that

(a) $\partial f_1/\partial v \le 0, \quad \partial f_1/\partial w \le 0;$

(b) $\partial f_2/\partial u \le 0, \quad \partial f_2/\partial w \le 0;$

(c) $\partial f_3/\partial u \ge 0, \quad \partial f_3/\partial v \ge 0;$

(d) if (1) $\partial f_1/\partial w \equiv \partial f_3/\partial u \equiv 0$ then one cannot have (2) $\partial f_2/\partial w \equiv \partial f_3/\partial v \equiv 0$, conversely if (2) holds then (1) cannot hold;

(e) if $\partial f_1/\partial v \equiv \partial f_2/\partial u \equiv 0$ then (1) and (2) above cannot hold.

Let $u_0 \equiv v_0 \equiv w_0 \equiv 0$ in $\bar{D}$. For $i = 1,2,\ldots$ recursively define analogously u_i, v_i, w_i as the unique positive functions in D satisfying:

$$\Delta u_i + u_i[a+f_1(u_i,v_{i-1},w_{i-1})] = 0 \quad \text{in } D, \qquad u_i = g_1 \quad \text{on } \delta D; \tag{2.7}$$

$$\Delta v_i + v_i[b+f_2(u_{i-1},v_i,w_{i-1})] = 0 \quad \text{in } D, \qquad v_i = g_2 \quad \text{on } \delta D; \tag{2.8}$$

$$\Delta w_i + w_i[c+f_3(u_i,v_i,w_i)] = 0 \quad \text{in } D, \qquad w_i = g_3 \quad \text{on } \delta D. \tag{2.9}$$

MAIN THEOREM *In each case from (2.1) to (2.3), any solution (u,v,w) of the boundary value problem (1.1) with u, v, w in $H^{2+\ell}(\bar{D}) \geq 0$, $\neq 0$ must satisfy:*

$$u_* \leq u \leq u^*, \qquad v_* \leq v \leq v^*, \qquad w_* \leq w \leq w^* \tag{2.10}$$

for all $x \in \bar{D}$. Here $u^* = \lim_{n\to\infty} u_{2n+1}$, $u_* = \lim_{n\to\infty} u_{2n}$, $v^* = \lim_{n\to\infty} v_{2n+1}$, $v_* = \lim_{n\to\infty} v_{2n}$, $w^* = \lim_{n\to\infty} w_{2n+1}$, $w_* = \lim_{n\to\infty} w_{2n}$, *for* $x \in \bar{D}$. *In each case, we have the following order relations for* $x \in \bar{D}$:

$$0 \leq u_2 \leq u_4 \leq u_6 \leq \cdots \leq u_5 \leq u_3 \leq u_1$$

$$0 \leq v_2 \leq v_4 \leq v_6 \leq \cdots \leq v_5 \leq v_3 \leq v_1 \tag{2.11}$$

$$0 \leq w_2 \leq w_4 \leq w_6 \leq \cdots \leq w_5 \leq w_3 \leq w_1$$

2.4 *Remarks.* The theorem gives bounds to the solution of the boundary value problem. The scheme can be adapted to finite difference method for numerical calculations. This had been done for two species case in Lazer, et al. (to appear). Application to the problem of uniqueness can readily be made. If $u_* = u^*$, $v_* = v^*$ and $w_* = w^*$, we clearly have uniqueness of solution with properties described in the Main Theorem.

The detailed proof of the Main Theorem can be found in Leung (to appear). It essentially involved repeated application of the two following lemmas:

LEMMA 2.1 *(Uniqueness). Let $h(x,z)$ be a real function defined for (x,z) in $\bar{D}\times[0,\infty)$ with Hölder continuous first partial derivatives in compact sets of $\bar{D}\times[0,\infty)$. Suppose that $\partial h/\partial z < 0$ at each point in $D\times(0,\infty)$ and there exists a constant $C > 0$ such that $h(x,z) \leq 0$ for all $x \in \bar{D}$, $z \geq C$. Let $z_i \in H^{2+\ell}(\bar{D})$, $i = 1,2$ be solutions of*

$$\Delta z + zh(x,z) = 0, \qquad x \in D$$

with the property $z_i(x) > 0$ for each $x \in D$, $i = 1,2$ and $z_1(x) = z_2(x)$ for

$x \in \delta D$. *Then* $z_1(x) \equiv z_2(x)$ *for all* $x \in \bar{D}$.

LEMMA 2.2 *(Comparison)*. *Let* $h_i(x,z)$, $i = 1,2$ *be functions defined in* $\bar{D} \times [0,\infty)$ *satisfying assumptions concerning* $h(x,z)$ *in Lemma 2.1. Further, suppose* $h_1(x,z) \geq h_2(x,z)$ *for all* $(x,z) \in \bar{D} \times [0,\infty)$. *For each* $i = 1,2$, *let* $z_i \in H^{2+\ell}(\bar{D})$, $z_i(x) > 0$ *for each* $x \in D$, *satisfies*

$$\Delta z_i + z_i h_i(x,z_i) = 0 \quad in \ D, \quad z_i = g \quad on \ \delta D$$

where $g \geq 0$, $\ddagger 0$ *on* δD *and has extension* $\hat{g} \in H^{2+\ell}(\bar{D})$. *Then* $z_1(x) \geq z_2(x)$ *for all* $x \in D$.

3. SCHEME FOR MUTUALISTIC LOOP (Volterra-Lotka type interactions)

We consider species A, B, C (with corresponding concentrations u, v, w) where A preys on C, C preys on B, and B preys on A. In order to obtain convergent sequences in this case, we need more restrictive conditions. We therefore consider only Volterra-Lotka type reactions to simplify these conditions. We consider problem (1.1) with $f_i(u,v,w) = \lambda_{i1}u + \lambda_{i2}v + \lambda_{i3}w$, $i = 1,2,3$, where λ_{ij} are constants with λ_{11}, λ_{22} and λ_{33} being negative, $\lambda_{12} \leq 0$, $\lambda_{13} \geq 0$, $\lambda_{21} \geq 0$, $\lambda_{23} \leq 0$, $\lambda_{31} \leq 0$, $\lambda_{32} \geq 0$. We assume that for each pair (i,j), λ_{ij} and λ_{ji} cannot be both zero. The following conditions will insure that the sequences we construct will be monotonic

$$\frac{\lambda_{32}\lambda_{21}\lambda_{13}}{|\lambda_{11}\lambda_{22}\lambda_{33}|} < 1, \quad \lambda_{13} > 0, \quad \lambda_{21} > 0, \quad \lambda_{32} > 0. \tag{3.1}$$

Let Ω be a large enough positive constant depending on a, b, c, λ_{ij} and g_i, i, j = 1,2,3 (see (6.4) to (6.7) in Leung (to appear)). Let $u_0 \equiv v_0 \equiv w_0 \equiv 0$, $w_{-1} \equiv \Omega$ in $\bar{D}$. Define u_i, v_i, w_i, i = 1,2,... to be unique strictly positive functions in D as follows:

$$\Delta u_i + u_i[a + \lambda_{11}u_i + \lambda_{12}v_{i-1} + \lambda_{13}w_{i-2}] = 0 \quad in \ D, \quad u_i = g_1 \quad on \ \delta D; \tag{3.2}$$

$$\Delta v_i + v_i[b + \lambda_{21}u_i + \lambda_{22}v_i + \lambda_{23}w_{i-1}] = 0 \quad in \ D, \quad v_i = g_2 \quad on \ \delta D; \tag{3.3}$$

$$\Delta w_i + w_i[c + \lambda_{31}u_{i-1} + \lambda_{32}v_i + \lambda_{33}w_i] = 0 \quad in \ D, \quad w_i = g_3 \quad on \ \delta D. \tag{3.4}$$

A nonnegative solution (u,v,w) of the boundary value problem in this case, with $w(x) \leq \Omega$ in $\bar{D}$, will satisfy properties (2.10) in the Main Theorem. (2.11) also holds.

REFERENCES

Conway, E., and J. Smoller (1977): Diffusion and the predator-prey interaction, *SIAM J. Appl. Math.* 33:673-686.

Krikorian, N. (1979): The Volterra model for three species predator-prey systems: boundedness and stability, *J. Math. Biol.* 7:117-132.

Lazer, A., A. Leung, and D. Murio (to appear): Monotone scheme for finite difference equations concerning steady-state prey-predator interaction, *J. Comp. Appl. Math.*

Leung, A. (1982): Monotone schemes for semilinear elliptic systems related to ecology, *Math. Meth. in Appl. Sci.* 4:272-285.

_____(to appear): A study of three species prey-predator reaction-diffusions by monotone schemes, *J. Math. Anal. Appl.*

May, R.M. (1973): *Stability and Complexity in Model Ecosystems*, Princeton Univ. Press, Princeton, New Jersey.

Redheffer, R., and Z. Zhou (1981): Global asymptotic stability for a class of many variable Volterra prey-predator systems, *Nonlin. Anal.* 5:1309-1329.

Williams, S., and P.L. Chow (1978): Nonlinear reaction-diffusion models for interacting populations, *J. Math. Anal. Appl.* 62:157-169.

SOME AGE-STRUCTURE EFFECTS
IN PREDATOR-PREY MODELS

Daniel S. Levine

ABSTRACT

First a model for a single species cannibalizing its own young is studied. If the birth rate is high enough to prevent extinction, a stable positive equilibrium or a periodic solution exists. Then three cases of a two-species model are studied. If predators eat all ages of prey equally, the system behaves asymptotically like a predator-prey system without age structure. If predators eat only newborn prey, unbounded oscillations can arise, but realistic refinements stabilize the model. If predators eat all ages but more of the very young and very old, bifurcating periodic solutions exist if the age bias is small enough.

1. INTRODUCTION

McKendrick (1926) modeled a single population with age structure. The density $\rho(a,t)$ of individuals of age a at time t is described by

$$\frac{\partial \rho}{\partial a} + \frac{\partial \rho}{\partial t} + \mu\rho = 0 \qquad (1)$$

the death rate μ may be a function of a, t, total population, or amount of another population (such as predators or competitors). The total population is

$$P(t) = \int_0^\infty \rho(a,t)da. \qquad (2)$$

The models herein combine (1) with the predator-prey equations of Lotka-Volterra or of Kolmogorov (1936). In one case, predators and prey are the same species which cannibalizes its own young. In the other three cases, predator and prey are separate species. Predation is either equally on all prey ages, only on newborn prey, or on all ages but weighted toward very young and very old prey. This article summarizes results from Gurtin and Levine (1979, 1982) and Levine (1981, to appear).

As in Gurtin and MacCamy (1979), the prey birth rate is given by

$$B(t) = \int_0^\infty \beta(a)\rho(a,t)da \qquad (3)$$

$\beta(a)$ is the *fecundity function* (expected rate at which offspring are born to an individual of age a). Many, but not all, of our examples use

$$\beta(a) = \beta_0 a \exp(-\alpha a) \quad (\beta_0 \geq 0, \ \alpha \geq 0) \tag{4}$$

If $\alpha > 0$, (4) describes behavior like that of many mammals whose fecundity is greatest at fairly young ages. If $\alpha = 0$, (4) could be appropriate for many fish species whose fecundity increases with size (thus with age).

The equations used for age-dependent predation are derived from (1)-(3). For each case there is variation in the functional dependence of the death function, the survival rate of newborns, and the dynamics of the predator population.

2. ONE SPECIES: CANNIBALISM

Cannibalism of the young has been observed in many fish (for example, Chadwick et. al., 1977) and in Dungeness crabs (Botsford and Wickham, 1978). This model, developed in Levine (1981), was designed to test the common belief that cannibalism is a means of population control.

Assume, for simplicity, that only newborns are eaten. $(0,t)$ represents the number of newborns surviving cannibalism per unit time, thus

$$\rho(0,t) = B(t)g(B(t),P(t)) \tag{5}$$

where $g(B,P)$ is the fraction surviving. Conversion of newborns into food is modeled by a death function

$$\mu = f(B(t),P(t)) \tag{6}$$

The system (1)-(6) reduces to ordinary differential equations via a technique from Gurtin and MacCamy (1979). Assuming $\rho(a,t) \to 0$ as $a \to \infty$, an equation for P is obtained by integrating (1) with respect to a. Similarly, if we assume (3) and (4) and set

$$A(t) = \int_0^\infty \exp(-\alpha a)\rho(a,t)da \tag{7}$$

we obtain equations for B and A. The resulting system is

$$\dot{P} = -Pf(B,P) + Bg(B,P)$$
$$\dot{B} = -Bf(B,P) - \alpha B + \beta_0 A \tag{8}$$
$$\dot{A} = -Af(B,P) - \alpha A + Bg(B,P)$$

In Gurtin and Levine (1982) it was shown that $r(B,P) = \beta_0 g(B,P)/(\alpha+f(B,P))^2$ is the expected number of offspring born to an individual in its lifetime. Eigenvalue analysis shows that the equilibrium $(0,0,0)$ of (8) is stable if and only if $r(0,0) > 1$. An analogous assumption will be imposed to prevent *prey* extinction in the two-species case.

Let $X = B/P$, the food supply per individual. Effects of total population and food supply on the death and newborn survival rates are modeled by setting $F(X,P) = f(B,P)$, $b(X,P) = G(B,P)$ and assuming that

$$\frac{\partial F}{\partial X} \leq 0, \quad \frac{\partial F}{\partial P} \geq 0, \quad \frac{\partial G}{\partial X} > 0, \quad \frac{\partial G}{\partial P} < 0 \tag{9}$$

Assume also that

(a) $G(X,0) = g(0,0) = 1$

(b) $\lim_{P \to \infty} [\sup_{X \to 0} G(X,P)] = 0$ $\tag{10}$

(c) $\beta_0 > (\alpha + F(X,0))^2 = (\alpha + f(0,0))^2$

(10a) means that all newborn survive as $P \to 0$, and (10b) that no newborns survive as $P \to \infty$. (10c) means that the net reproduction rate $r(0,0) > 1$.

The above assumptions lead to stable equilibria or periodic solutions. Using the variables $X = B/P$, $Z = A/B$, (8) is transformed to

$$\dot{P} = P[-F(X,P)+XG(X,P)]$$
$$\dot{X} = X[-\alpha+\beta_0 Z-XG(X,P)] \tag{11}$$
$$\dot{Z} = G(X,P) - \beta_0 Z^2$$

THEOREM 1 Assume (9) and (10), and assume that (11) has exactly one equilibrium (P_0,X_0,Z_0) in the positive octant, which is unstable. If the Jacobian matrix of (11) at that point has non-zero determinant and no pure imaginary eigenvalues, then (11) has a nonconstant periodic solution.

The proof, from Gurtin and Levine (1982), will be sketched. First a positively invariant set for (11) is found, namely, $S = \{(P,X,Z) \mid 0 \leq X \leq \beta_0/\alpha,\ 0 \leq Z \leq 1/\sqrt{\beta_0},\ 0 \leq XZ \leq 1\}$, and (P_0,X_0,Z_0) is in the interior of S. We need thus consider only trajectories contained in S.

The planes $P = P_0$, $X = X_0$, $Z = Z_0$ cut S into eight regions, I, II, III, IV, V, VI, VII and VIII. It is shown that as time increases, only the following transitions can occur:

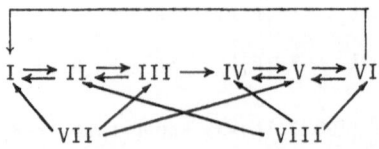

The eigenvalues of the Jacobian matrix of (11) at (P_0,X_0,Z_0) are of the form $-\alpha$, $\nu + i\eta$, $\psi - i\eta$ ($\alpha > 0$, $\nu > 0$, $\eta \geq 0$, $\nu = \psi$ unless $\eta = 0$). The eigenvectors associated with $-\alpha$ point into VII $\cup$ VIII. Thus, since no trajectory can enter $VII \cup VIII$, $K = S - (VII \cup VIII) - \{(P,X,Z) | P = 0$ or $X = 0\}'$ cannot intersect the stable manifold of (P_0, X_0, Z_0). The only other equilibria in the closure of K correspond to the equilibrium $(0,0,0)$ of (8); eigenvector analysis of (8) shows that K cannot intersect the stable manifold of either equilibrium. Hence a trajectory starting in K must stay in $S - (VII \cup VIII)$ for $t > 0$.

A trajectory in K must proceed in the order $I \rightleftarrows II \rightleftarrows III \rightarrow IV \rightleftarrows V \rightleftarrows VI$

Let $R_1 = I \cup II \cup III = K \cap \{(P,X,Z) | P \geq P_0\}$, $R_2 = IV \cup V \cup VI = K \cap \{(P,X,Z) | P \leq P_0\}$. $P > P_0$ at every maximum of P past the first, and $P < P_0$ at every minimum of P past the first. Thus if the solution stays in R_1 or R_2, P is eventually monotone. By asymptotic arguments, $P \to \infty$ is shown to lead to a contradiction. $P \to P_0 \to \infty$ leads to (P,X,Z) converging to some equilibrium, which was already ruled out. A trajectory in K cannot then remain in either R_i for large t, so must pass infinitely often from VI to I.

The fact F between VI and I is bounded, and $F_0 = F - \{(P_0, X_0, Z_0)\} \subset K$. Any trajectory starting at $u \in F_0$ must return to F_0. If the return point is called $p(u)$, this defines the Poincare map p. A small "cylinder" can be drawn around (P_0, X_0, Z_0) such that solutions cross the cylinder from inside out. If $C = F_0 - $ (interior of cylinder), then C is compact and $p(C) \subset C$. By the Brouwer fixed point theorem, a point $u \in C$ exists such that $p(u) = u$. Since u is not an equilibrium for (11), the trajectory through u is non-constant and periodic, completing the proof.

Theorem 1 does not say whether the periodic solution is stable, nor what happens if there are multiple positive equilibria. A numerical study by Frauenthal (to appear) shows that for some subcases with multiple equilibria, parameter changes can cause a sudden switch from a stable equilibrium to a stable limit cycle.

These results indicate that a cannibalistic population can survive if its reproduction rate is high enough, supporting the idea of cannibalism as population control. Variations of our model could be used for other density-dependent effects, such as compensation for increased mortality (due to hunting or pollution) by higher egg survival rate

3. TWO SPECIES: INDISCRIMINATE EATING VERSUS EGG-EATING

Two cases of age-dependent predation were discussed in Gurin and Levine (1979). The equations for the prey population P_1 are based on (1)-(4) and (7).

For the case of age-indiscriminate predation, the death function depends linearly on predator population P_2, thus $\mu = \mu_0 + rP_2$. Since newborn prey are no more vulnerable than other ages, $\rho(0,t) = B(t)$. The predator population obeys a Volterra-Lotka equation. Thus we derive the system

$$\dot{P}_1 = -\mu_0 P_1 - r P_1 P_2 + B$$
$$\dot{B} = -\gamma B - r B P_2 + \beta_0 A$$
$$\dot{A} = -\gamma A - r A P_2 + B \tag{12}$$
$$\dot{P}_2 = -b P_2 + c P_1 P_2$$

where all small-letter parameters are positive constants and $\gamma = \mu_0 + \alpha$. The assumption that expected offspring be at least 1 translates to $\beta_0 > \gamma^2$.

The results of (12) show that the system behaves like a predator-prey system with age ignored. As $t \to \infty$, A/B approaches $1/\sqrt{\beta_0}$ and B/P approaches $-\alpha + \sqrt{\beta_0}$, both at exponential rates. This reduces (12) asymptotically to a Volterra-Lotka system in P_1 and P_2. Hence the P_1, P_2 coordinates of (12) approach one of a continuum of neutrally stable solutions. Asymptotic equivalence to a system without age structure still holds if the Volterra-Lotka equations are replaced by the more realistic equations of Kolmogorov (1936).

For the case of predation on newborns (the "egg-eating" model), is a constant, since predators do not affect survival of prey past birth. $\rho(0,t)$ equals B(t) minus number of newborns eaten per unit time. In Gurtin and Levine (1982), the number eaten was a constant multiple exceeded product (birth rate times number of predators), until that multiple exceeded B, hence $\rho(0,t) = \max(B - K P_2 B, 0)$. In Levine (1981), this function was replaced by the continuously differentiable approximation $\rho(0,t) = B/(1+\kappa P_2)$, with no change in qualitative behavior. In (12d), P_1 is replaced by B, thus

$$P_1 = -\mu_0 P_1 + (B/(1+\kappa P_2))$$
$$B = -\gamma B + \beta_0 A$$
$$A = -\gamma A + (B/(1+\kappa P_2)) \tag{13}$$
$$P_2 = -b P_2 + c B P_2$$

Again it is assumed that $\beta_0 > \gamma^2$, making $(0,0,0,0)$ an unstable equilibrium.

Figure 1, reproduced from Levine (to appear), shows a typical numerical simulation of (13). The predator and prey populations oscillate with maxima increasing to infinity and minima decreasing to zero. The result corresponds to extinction, first of the prey and then of the predators. A proof of this unboundedness is in progress. (13b)-(13d) do not depend on P_1, and the substitutions $X = A/B$, $Y = \ln B$, $Z = \ln P_2$ reduce those equations to the form

$$X = k(Z) - f(X)$$
$$Y = g(X) \qquad k' < 0, \; f' > 0, \; g' > 0, \; h' > 0 \tag{14}$$
$$Z = h(Y)$$

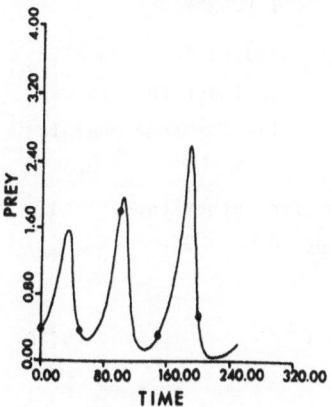

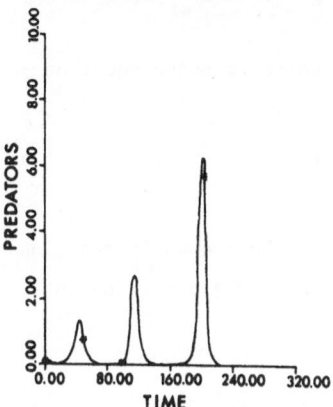

Figure 1   Prey and predator populations as functions of time for equations (15)

      when $\beta_0 = 5$, $\gamma = 2$, $b = c = \mu = \kappa = 1$.

If $g^{-1}(0)$, $h^{-1}(0)$, and $k^{-1}(f(g^{-1}(0)))$ all exist, (14) has a unique equilibrium, and its unstable manifold is 2-dimensional. All trajectories not tending to the equilibrium tend to the unstable manifold, and on the unstable manifold they spiral out from the equilibrium.

     In Levine (1981) it was shown that the egg-eating system is stabilized by either of two modifications which make the model more realistic. These modifications were making the predator birth rate depend on numbers of newborn prey ingested rather than total contacts with newborn prey (that is, incorporating $1/(1+\kappa P_2)$ into the equation for P_2) introducing a carrying capacity for prey (that is, making μ depend linearly on P_1).

4. TWO SPECIES: PREDATION ON ALL AGES WITH EXTREME AGES MOST VULNERABLE

In Chadwick et. al. (1977), the two-species case was studied in which all ages of prey are eaten, but the very young and the very old are least able to escape predation and so are eaten most. This pattern is realistic for some mammals (cf. Mech (1970)).

The death function μ has an age-independent term (from other causes) and one that is proportional to the predation rate $D(a)$. Thus

$$\mu = \mu_0 + P_2(t)D(a)$$

$$D(a) = \gamma + \lambda_1 S_1(a) + \lambda_2 S_2(a) \tag{15}$$

where γ, λ_1, and λ_2 are positive constants. $S_1(a)$ is greatest for young ages and $S_2(a)$ is greatest for old ages.

The predators are described by the age-weighted Volterra-Lotka equation

$$\dot{P}_2 = -bP_2 + cP_2 \int_{a=0}^{\infty} D(a)\rho(a,t)dt \tag{16}$$

For the birth law (3), it is assumed that $\rho(0,t) = B(t)$. The fecundity function $\beta(a)$ need not be of the form $\beta_0 a \exp(-\alpha a)$, but must again obey a condition that prevents extinction, here

$$\int_{a=0}^{\infty} \beta(a) \exp(-\mu_0 a) > 1 \tag{17}$$

From (1), (3), (15), and (16) are derived the system

$$\frac{\partial \rho}{\partial a} + \frac{\partial \rho}{\partial t} + (\mu_0 + P_2(t)D(a))\rho = 0$$

$$\dot{P}_2 = -bP_2 + cP_2 \int_{a=0}^{\infty} D(a)\rho(a,t)da$$

$$\rho(0,t) = \int_{a=0}^{\infty} \beta(a)\rho(a,t)da \tag{18}$$

$$D(a) = \gamma + \lambda_1 S_1(a) + \lambda_2 S_2(a)$$

(18) includes as special cases both classes of systems discussed in Section 3.

The system (18) was studied in Levine (to appear) using a bifurcation theory for integrodifferential equations developed in Cushing (1982). This theory is analogous to Hopf bifurcation theory for ordinary differential equations. The system for $\lambda_1 = \lambda_2 = 0$ has two pure imaginary eigenvalues $\pm i\sigma$. It was shown that for small λ_i, there exists a time-periodic solution of (18) with period $T = 2\pi/\sigma$.

First, (18) has a unique positive steady state for any given λ_1 and λ_2. For let $\rho = \bar{\rho}(a)$, $P_2 = P_0$ be a steady state. Then, for (18a), $\bar{\rho}'(a) + (\mu_0 + P_0 D(a))\bar{\rho}(a) = 0$. So if $E(a) = \int_{\alpha=0}^{a} D(\alpha)d\alpha$, then $\bar{\rho}(a) = \rho_0 \exp(-\mu_0 a - P_0 E(a))$ for some constant ρ_0. (18c) yields

$$1 = \int_{a=0}^{\infty} \beta(a) \exp(-\mu_0 a - P_0 E(a))da \tag{19}$$

The right-hand side of (19) is a decreasing function of P_0; it tends to 0 as $P_0 \to \infty$, and at $P_0 = 0$ its value is > 1 by (17). By continuity, then, (19) is satisfied for a unique positive value P_0. From (18b), $b/c = \int_{a=0}^{\infty} D(a)\bar{\rho}(a)da$, yielding a unique ρ_0 and thus a unique $\bar{\rho}(a)$.

Now we can state the bifurcation result.

THEOREM 2 Define $P_0^0 = P_0|_{\lambda_1=\lambda_2=0}$, $\bar{\rho}^0(a) = \bar{\rho}(a)|_{\lambda_1=\lambda_2=0}$. Then (18) has a nontrivial solution of the form $\rho(t,a) = \bar{\rho}(a) + \epsilon z_1(t,a) + \epsilon z_2(t,a,\epsilon)$, $P_2(t) = P_0^0 + \epsilon p_1(t,a) + \epsilon p_2(t,a,\epsilon)$, z_i and p_i T-periodic in t_0 $z_i = O(|\epsilon|)$, $p_i = O(|\epsilon|)$ near $\epsilon = 0$, if a certain Jacobian is non-zero. (The exact form of the Jacobian is too long to be included here.)

The proof, from Levine (to appear), will be sketched. Set $z(a,t) = \rho(a,t) - \bar{\rho}(a)$, $p(t) = P_2(t) - P_0$. For $\lambda_1 = \lambda_2 = 0$, the linearization of (19) is $L_i(z,p) = 0$, $i = 1,2,3$, where:

$$L_1(z,p) = z_a + z_t + \nu z + \phi(a)p$$

$$L_2(z,p) = p - cP_0^0 \int_{a=0}^{\infty} z(a,t)da \tag{20}$$

$$L_3(z,p) = z(0,t) - \int_{a=0}^{\infty} \beta(a)z(a,t)da,$$

where $\phi(a) = r\rho_0^0(a)$.

Then (18) can be written in the form

$$L_i(z,p) = K_i(\lambda_1,\lambda_2,a,z,p) + T_i(\lambda_1,\lambda_2,a,z,p), \quad i = 1,2$$

$$L_3(z,p) = 0 \tag{21}$$

where $T_i = O(z,p)$, $K_i = O(z,p)$, $K_i = O(\lambda_j)$, $i,j = 1,2$.

Now make the change of variables $\alpha = a$, $\tau = t-a$, thus replacing the terms $z_a + z_t$ in (20a) by z_τ. Consider the general linear system

$$L_1(z,p) = f(\alpha,\tau)$$
$$L_2(z,p) = g(\tau) \tag{22}$$
$$L_3(z,p) = h(\tau)$$

f,g,h all T-periodic in τ. Into (22), substitute $z(\alpha,\tau) = z(\alpha) \exp(i\sigma\tau)$,
$p(\tau) = \exp(i\sigma(\tau+\alpha))$, $f(\alpha,\tau) = f(\alpha) \exp(i\sigma\tau)$, $g(\tau) = g \exp(i\sigma\tau)$, $h(\tau) = h \exp(i\sigma\tau)$,
with p,g,h complex constants. Then eliminate p to obtain

$$z'(\alpha) + \nu z(\alpha) + \frac{cP_0^0\phi(\sigma)}{i\sigma} \int_{a=0}^{\infty} \exp(-i\sigma a)z(a)d(a) = f(\alpha) - \frac{g}{i\sigma}\phi(\alpha)$$

(23)

$$z(0) - \int_{a=0}^{\infty} \beta(a) \exp(-i\sigma a)z(a)da = h$$

By the choice of σ, the homogeneous equations corresponding to (23) have
a non-trivial solution. Any solution of (23) can be written $z(\alpha) = z(0)y(\alpha) + \Omega_{f,g}(\alpha)$, where $y(\alpha)$ is the solution with $y(0) = 1$ of the homogeneous equation
corresponding to (23a), and $\Omega_{f,g}(\alpha)$ is the particular solution of (23a)
vanishing at $\alpha = 0$. Define the complex linear functional Q on $C^1(\mathbb{C}) \times \mathbb{C} \times \mathbb{C}$ by

$$Q[f,g,h] = (\int_0^{\infty} \beta(a) \exp(-i\sigma a)\Omega_{f,g}(a)da) + h$$

(24)

Then $z(\alpha)$ satisfies (23b) if and only if $Q[f,g,h] = 0$. The right-hand side of
(22) is T-periodic in τ if z and p are, so if (22) has a T-periodic solution,

$$Q[K_1+T_1,K_2+T_2,0] = 0$$

(25)

Define the linear operator $V: N(Q) \to C^1(\mathbb{C}) \times \mathbb{C}$, with $N(Q)$ the space of all
(f_1,f_2) in range V such that $Q[f_1,f_2,0] = 0$ and $V(f_1,f_2)$ the unique (g_1,g_2)
in $C^1(\mathbb{C}) \times \mathbb{C}$ such that $L_i(g_1,g_2) = f_i$ for $i = 1,2$, $L_3(g_1,g_2) = 0$, and
$g_1(0) = 0$. Then V is compact.
    Define $\bar{K}_i$, $\bar{T}_i$, $i = 1,2$, by $T_i(\lambda_1,\lambda_2,a,\epsilon z,\epsilon p) = \epsilon\bar{T}_i(\lambda_1,\lambda_2,a,z,p,\epsilon)$,
$K_i(\lambda_1,\lambda_2,a,\epsilon z,\epsilon p) = \epsilon\bar{K}_i(\lambda_1,\lambda_2,a,z,p,\epsilon)$. For given ε, $(z(\alpha) \exp(i\sigma\tau)$,
$p\exp(i\sigma(\tau+\alpha))$ is a T-periodic (in τ) solution of (22) if and only if $(z(\alpha),p)$
is a fixed point of the operator

$$(u_1,u_2) \to B(\epsilon,u_1,u_2) = \epsilon V(\bar{K}_1(\lambda_1,\lambda_2,\alpha,u_1,u_2,\epsilon) + \bar{T}_1(\lambda_1,\lambda_2,\alpha,u_1,u_2,\epsilon),$$
$$\bar{K}_2(\lambda_1,\lambda_2,\alpha,u_1,u_2,\epsilon) + \bar{T}_2(\lambda_1,\lambda_2,\alpha,u_1,u_2,\epsilon)).$$

(26)

Since V is compact and K_i and T_i bounded, B is completely continuous for
each ε. Express z,p as

$$z(\tau,\alpha,\epsilon) = \epsilon Y_1(\tau,\alpha) + W_1(\tau,\alpha,\epsilon)$$
$$p(\tau,\alpha,\epsilon) = \epsilon Y_2(\tau,\alpha) + W_2(\tau,\alpha,\epsilon)$$

(27)

where the Y_i are non-trivial periodic solutions of (18) for $\epsilon = 0$, and the W_i are $O(|\epsilon|)$. From (26) and (27), $B(\epsilon, u_1, u_2) = O(|\epsilon|)$ uniformly in ϵ.

If the Jacobian with respect to λ_1, λ_2 of $\text{Re}(B(\epsilon, u_1, u_2))$ and $\text{Im}(B(\epsilon, u_1, u_2))$ at $z = z_0(\alpha) \exp(i\sigma\tau)$, $p = p \exp(i\sigma(\tau+\alpha))$, $\lambda_i = 0$, $\epsilon = 0$ is non-zero, that is, if $\text{Im}(\bar{Q}[\partial\bar{K}_1/\partial\lambda_1, \ \partial\bar{K}_2/\partial\lambda_1, \ 0]Q[\partial\bar{K}_2/\partial\lambda_1, \ \partial\bar{K}_2/\partial\lambda_2, \ 0]) \neq 0$, then ϵ_1 can be chosen so that λ_1 and λ_2 can be defined implicitly from (26) for $|\epsilon| < \epsilon_1$. Since $B(\epsilon, u_1, u_2) = O(|\epsilon|)$, the range of B lies in $\beta_0(\epsilon_1)$, so $(u_1, u_2) \rightarrow B(\epsilon, u_1, u_2)$ maps $\beta_0(\epsilon_1)$ to itself. Thus by the Schauder fixed point theorem, B has a fixed point. Hence (22) has a periodic solution, completing the proof.

Theorem 2 does not say whether the periodic solutions are stable nor what the direction of bifurcation is Figures 2, 3, and 4, reproduced from Chadwick et. al. (1977), give a computer result with $\lambda_1 > 0$, $\lambda_2 > 0$, and birth and predation rates chosen proportional to known data for wolves and white-tailed deer (Mech, 1970; Cheatum and Severinghaus, 1950). These parameters yield a stable periodic solution, indicating stable predator-prey coexistence.

Deer killed by

Age (years)	Wolf Predation	Other Causes	Ratio Killed By Wolves
0 - 1	56	54	.509
1 - 2	19	77	.198
2 - 3	10	47	.175
3 - 4	24	39	.381
4 - 5	32	22	.593
5 - 6	39	16	.709
6 - 7	50	7	.877
7 - 8	29	8	.784
8 - 9	72	5	.935

Age (years)	Number of Animals	Average Numbers of Fawns Per Doe
0 - 1	944	0.32
1 - 2	915	1.54
2 - 3	433	1.57
3 - 4	245	1.65
4 - 5	115	2.00
5 - 6	75	2.00
6 - 7	39	2.00
7 - 8	30	2.00
8 - 9	20	1.22

Figure 2 (a) Comparison of age distributions of white-tailed deer killed by wolves and killed by other means (automobiles or humans) in Algonquin

Figure 2 (continued). Park, Ontario, Modified from Mech [16], p. 251.
(b) Predicted effect of age on fecundity of white-tailed deer based on examination of hunters' kills in western New York State. Modified from Cheatum and Severinghaus [19].

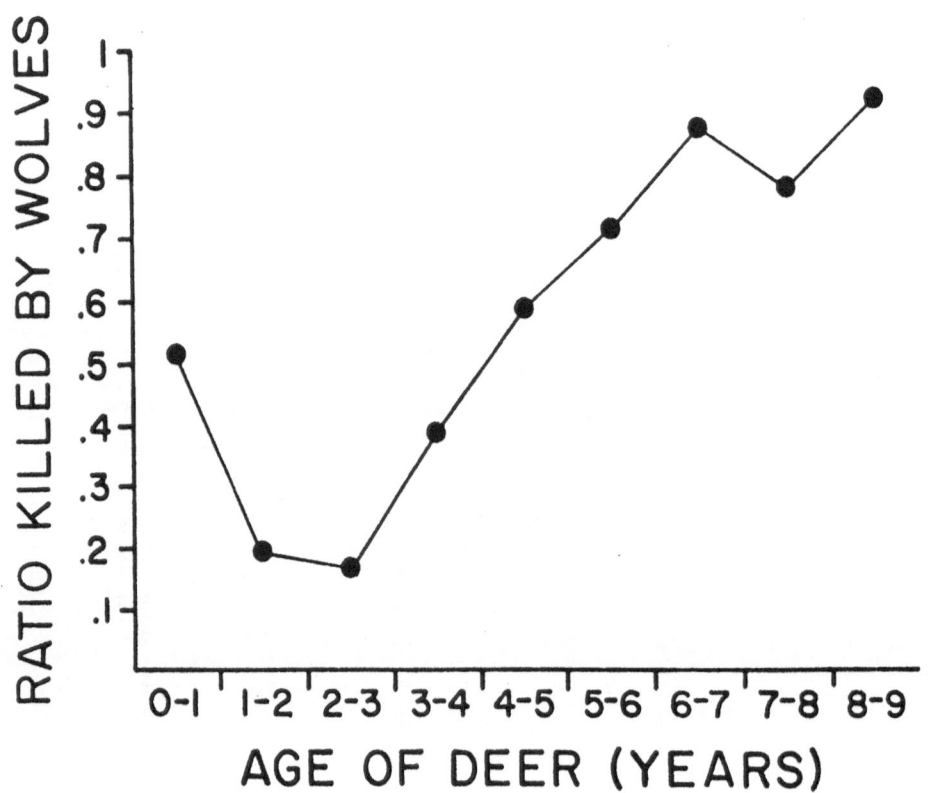

Figure 3 Plot of predation ratios in final column of Figure 3a.

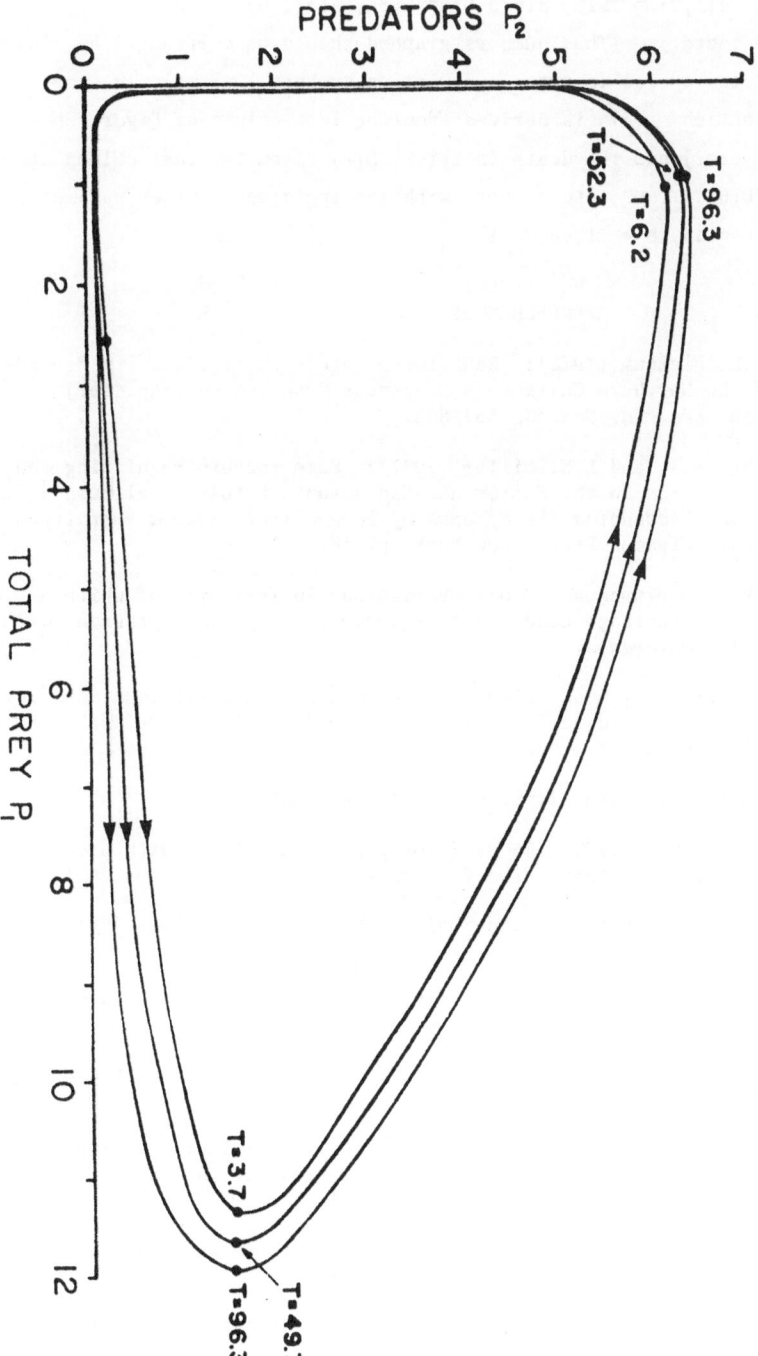

Figure 4   Plot of numerical data up to t = 100   for a solution of equations (1)-
(4).   Age is discretized into   25   classes, 1/3   year apart initial conditions are derived by dividing the middle column in Figure 2b by 1000 and
interpolating   (i.e. $\rho(0,0)$ = .944, $\rho(1/3,0)$ = .934, $\rho(2/3,0)$ = .924,

Figure 4 (continued). $\rho(1,0) = .915$, $\rho(4/3,0) = .754$, $\rho(5/3,0) = .593$, $\rho(2,0) = .433$, etc.). (Thus numbers graphed should be multiplied by 1000 to get predicted cycles in actual population numbers.) Similarly, the fecundity function (a) is derived from the last column of Figure 2b ($\beta(0) = .32$, etc.) and the death function $D(a)$ from the last column of Figure 2a ($D(0) = .509$, etc.), both with interpolation. Other parameter values are $\mu = .5$, $b = .1$, $c = .1$.

REFERENCES

Botsford, L.W., and D.E. Wickham (1978): Behavior of age specific, density dependent models and the Northern California Dungeness Crab (Cancer magister) fishery, *J. Fish. Res. Bd. Can.* 35:833-843.

Chadwick, H.K., D.E. Stevens, and L.W. Miller (1977): Some factors regulating the striped bass population in the Sacramento-San Joaquin Estuary, California, in: W. Van Winkle, ed., *Assessing the Effects of Power-Plant-Induced Mortality on Fish Populations*, Pergamon Press, New York, p. 18.

Cheatum, E.L., and C.W. Severinghaus (1950): Variations in fertility of white tailed deer related to range conditions, *Transactions of the Fifteenth North American Wildlife Conference*.

Cushing, J.M. (1982): Bifurcation of periodic solutions of nonlinear equations in age-structured population dynamics, *Proc. Int. Conf. on Nonlinear Math. and Applications*, Arlington, TX., Academic Press, New York.

Frauenthal, J.C. (to appear): Some simple models of cannibalism.

Gurtin, M.E., and D.S. Levine (1979): On predator-prey interactions with predation dependent on age of prey, *Math. Biosci.* 47:207-219.

_____ (1982): On populations that cannibalize their young, *SIAM J. Appl. Math.* 42:94-108.

Gurtin, M.E., and R.C. MacCamy (1979): Some simple models for non-linear age-dependent population dynamics, *Math. Biosci.* 43:199-211.

Kolmogorov, A.N. (1936): Sulla teoria di Volterra della lotta per l'esistenza, *Giorn. Inst. Ital. Attuari* 7:74-80.

Levin, S.A. (1979): The concept of compensatory mortality in relation to impacts of power plants on fish populations. Testimony prepared for the United States Environmental Protection Agency, Region II, April.

Levine, D.S. (1981): On the stability of a predator-prey system with egg-eating predators, *Math. Biosci.* 56:27-46.

_____ (to appear): Bifurcating periodic solutions for a class of age-structured predator-prey systems, *Bull. Math. Biol.*

McKendrick, A.G. (1926): Applications of mathematics to medical problems, *Proc. Edinburgh Math. Soc.* 44:98-130.

Mech, L.D. (1970): *The Wolf: Ecology and Behavior of an Endangered Species*, Garden City, N.Y.: Natural History Press.

A MODEL PREDATOR-PREY SYSTEM
WITH MUTUAL INTERFERENCE AND TIME DELAY

V.S.H. Rao and H.I. Freedman*

1. INTRODUCTION

The Gause predator-prey model (Freedman, 1980, Chapter 4) separates the specific growth rate of the prey from the predator functional response. Hassell (1971) and Rogers and Hassell (1974) introduced the notion of mutual interference of predators searching for prey. This idea was incorporated into the Gause predator-prey model in Freedman (1979). It was shown there that mutual interference is a "stabilizing" process.

May (1973) discussed time delays in predator-prey models. He pointed out that delays tend to "destabilize" the system. We are interested in this paper to look at the combined effect of mutual interference and time delays, and in partic-ular for given values of the parameters of the model stated in Section 3 to deter-mine whether the combined effect is "stabilizing" or "destabilizing."

Note that stability or instability is with regard to the interior equilibrium of our model.

2. THE MODEL

We propose as our model the system of ordinary-delay differential equations

$$x'(t) = x(t)g(x(t)) - y(t)^m p(x(t))$$
$$y'(t) = y(t)(-s+cy(t)^{m-1}p(x(t-\tau)) - q(y(t)), \qquad (' = \frac{d}{dt}) \qquad (1)$$

where $g(x)$ is the specific growth rate of the prey x, $g(0) > 0$, $dg/dx \leq 0$, and $\exists K > 0$ (the carrying capacity of the environment) $\ni g(K) = 0$; $p(x)$ is the pred-ator functional response, $p(0) = 0$, $dp/dx > 0$; $0 \leq m \leq 1$ is the mutual interfer-ence parameter; $-s + q(y)$ is the density dependent predator death rate, $q(0) = 0$, $dq/dy \geq 0$; and τ is the time delay due to gestation.

This model is studied in detail in Freedman and Rao (1983). Here we present some of the results of that analysis and refer the reader to the above paper for the proofs.

*This author wishes to acknowledge the Natural Sciences and Engineering Research Council of Canada, Grant No. NSERC A4823 for partially supporting the research in this paper.

3. THE EQUILIBRIUM

We suppose the algebraic system

$$xg(x) - y^m p(x) = 0$$

$$-s + cp(x)y^{m-1} - q(y) = 0$$

(2)

has one or more positive solutions (see Freedman, 1979). Let $E(x^*,y^*)$ denote such a solution. Then E is an interior equilibrium.

The stability of E is, except for critical cases equivalent to the signs of the real parts of the solutions of the characteristic equation

$$\lambda^2 - (H+R)\lambda + HR - NQe^{-\tau\lambda} = 0,$$

(3)

where $H = x^* dg(x^*)/dx + g(x^*) - y^{*m} dp(x^*)/dx$, $N = -my^{*m-1}p(x^*) < 0$, $Q = cy^{*m} dp(x^*)/dx > 0$, and $R = (m-1)cy^{*m-1}p(x^*) - y^* dq(y^*)/dy \leq 0$. We denote solutions of (3) by $\lambda(\tau)$.

4. RESULTS

The first property of the model we note is that if E is unstable for $\tau = 0$, then it remains unstable for all positive τ, i.e. delays cannot stabilize system (1).

The remaining results depend on HR and NQ (note $NQ < 0$).

Case (i): $NR < NQ$. In this case, since there exists a positive $\lambda(0)$, E is unstable at $\tau = 0$, and hence must remain so for all positive τ.

Case (ii): $NQ \leq HR < 0$. Even if all real parts of $\lambda(0)$ are negative, there exists a $\tau_0 > 0$ such that for $\tau > \tau_0$, one of the $\lambda(\tau)$ is positive. Hence for fixed mutual interference, sufficiently long delays can destabilize the system.

Case (iii): $0 \leq NR \leq -NQ$. In this case there are no real roots $\lambda(\tau)$ such that $\lambda(\tau) > 0$. However, even if the real parts of all roots $\lambda(0)$ are negative, there is a value of τ, τ^* such that $\text{Re }\lambda(\tau^*) = 0$ for one of the roots, and $\text{Re}(\lambda(\tau)) > 0$ for $\tau > \tau^*$. As τ passes through τ^*, a Hopf bifurcation occurs giving the existence of limit cycles. Again, sufficiently long delays are destabilizing.

Case (iv): $-NQ \leq HR$. In this case, $\text{Re }\lambda(\tau) < 0$ for all $\tau \geq 0$. Even infinite delays cannot destabilize the system.

5. DISCUSSION

For fixed mutual interference of predators, a sufficiently long delay in gestation will usually cause an otherwise stable equilibrium to become unstable. However, if the mutual interference is sufficiently strong, even an infinite delay cannot destabilize the system.

Details of the proofs of these results may be found in Freedman and Rao (1983).

REFERENCES

Freedman, H.I. (1979): Stability analysis of a predator-prey system with mutual interference and density-dependent death rates, *Bul. Math. Biol.* 41:67-78.

_____ (1980): *Deterministic Mathematical Models in Population Ecology,* Marcel Dekker, Inc., New York.

Freedman, H.I., and V.S.H. Rao (1983): The tradeoff between mutual interference and time lags in predator-prey systems, *Bul. Math. Biol.* (in press).

Hassell, M.P. (1971): Mutual interference between searching insect parasites, *J. Anim. Ecol.* 40:473-486.

May, R.M. (1973): Time-delay versus stability in population models with two and three trophic levels, *Ecology* 54:315-325.

Rogers, D.J., and M.P. Hassell (1974): General models for insect parasite and predator searching behaviour: interference, *J. Anim. Ecol.* 43:239-253.

OSCILLATIONS IN PREY-PREDATOR VOLTERRA MODELS

Yasuhiro Takeuchi and Norihiko Adachi

ABSTRACT

This paper gives some results on oscillatory behaviors of Lotka-Volterra models. We are concerned with dynamics of biological communities with two trophic levels, particularly with those of two-prey, one-predator systems and two-prey, two-predators systems. The possible behaviors of trajectories of the systems are discussed by perturbation methods and Hopf bifurcation theory. It is shown that the addition of one or two predators to two-species competing system can increase species diversity. Three patterns of the coexistence are possible mathematically;

  (i)   coexistence at the globally (or locally) stable equilibrium,
 (ii) coexistence in the stable periodic motion of Hopf type (a limit cycle),
(iii)   the coexistence in chaotic motions.

1. INTRODUCTION

It is well known that there exist closed orbits for classical one-prey, one-predator models of Lotka-Volterra type. These orbits, however, are not limit cycles and are structurally unstable. In recent years, some interesting classes of trajectories in Lotka-Volterra models have been discovered. May and Leonard (1975) have shown that two types of oscillatory trajectories exist in Lotka-Volterra models of three-competing species: (i) one parameter family of closed orbits on an invariant hyperplane, (ii) non-periodic oscillations of bounded amplitudes with increasing cycle times. Coste, Peyraud and Coullet (1979) have proved that, for more general competitive systems, stable limit cycles can be generated by Hopf bifurcation. They have shown that closed orbits of type (i) observed by May and Leonard are not stable limit cycles and are quite analogous to those obtained in classical one-prey, one-predator models of Volterra type. On the other hand, with respect to two-prey, one-predator Volterra models, Fujii (1977) suggested the existence of a limit cycle, but his proof is insufficient. Vance (1978) discovered a "quasi-cyclic" motion, which was named a spiral chaos by Gilpin (1978). Hsu (1981) gave a numerical example of a limit cycle.

These recent results imply that the simplest model of population biology such as Volterra type shows highly nonlinear phenomena. Hence, conventional perturbation techniques are insufficient to analyze the behaviors of population models.

In this paper, we give some results on oscillatory behaviors of Volterra models. Possible types of trajectories of the system are discussed by perturbation technique, by Lyapunov's direct method and by Hopf bifurcation theory. We show that the equilibrium state bifurcates to oscillatory motion, that is, to a stable limit cycle or to a chaotic motion.

2. MODELS AND DEFINITIONS

In this paper, we are particularly concerned with models described by the following system of differential equations:

Two-Species-Competing System

$$\frac{d}{dt}\begin{bmatrix} x_1(t) \\ x_2(t) \end{bmatrix} = \begin{bmatrix} x_1(t)(1 - x_1(t)-\alpha x_2(t)) \\ x_2(t)(1 - \beta x_1(t) - x_2(t)) \end{bmatrix}, \tag{1}$$

Two-Prey, One-Predator System

$$\frac{d}{dt}\begin{bmatrix} x_1(t) \\ x_2(t) \\ y(t) \end{bmatrix} = \begin{bmatrix} x_1(t)(1- x_1(t)- \alpha x_2(t)-\varepsilon y(t)) \\ x_2(t)(1- \beta x_1(t)- x_2(t)-\mu y(t)) \\ y(t)(-1+d\varepsilon x_1(t)+d\mu x_2(t)) \end{bmatrix}, \tag{2}$$

Two-Prey, Two-Predators System

$$\frac{d}{dt}\begin{bmatrix} x_1(t) \\ x_2(t) \\ y_1(t) \\ y_2(t) \end{bmatrix} = \begin{bmatrix} x_1(t)(1 - x_1(t) - \alpha x_2(t)-\varepsilon_1 y_1(t)-\varepsilon_2 y_2(t)) \\ x_2(t)(1 - \beta x_1(t) - x_2(t)-\mu_1 y_1(t)-\mu_2 y_2(t)) \\ y_1(t)(-1+d_1\varepsilon_1 x_1(t)+d_1\mu_1 x_2(t)) \\ y_2(t)(-1+d_2\varepsilon_2 x_1(t)+d_2\mu_2 x_2(t)) \end{bmatrix}. \tag{3}$$

Here $x_i(t)$ (or $y_i(t)$), $(i = 1,2)$ are population sizes of prey (or predators), positive parameters α and β represent competitive effects between two prey, positive parameters ε_i and μ_i are coefficients of decrease of prey due to predation and $d_i > 0$ are transformation rates of i-th predator. For the generalized Volterra system;

$$\frac{d}{dt} x_i(t) = x_i(t)[b_i - \sum_{j=1}^{n} a_{ij}x_j(t)], \quad i = 1,2,\ldots,n \tag{4}$$

the stability of equilibria is defined as follows:

DEFINITION A nonnegative equilibrium x^* of (4) is called globally stable if and only if

(i) x^* is locally stable, that is, for any $\varepsilon > 0$, there exists a $\delta(\varepsilon)$ such that $|x^0-x^*| < \delta(\varepsilon)$ and $x(t) \in R_I^n$, then $|x(t)-x^*| < \varepsilon$ for $t \geq 0$, and

(ii) every solution converges to x^* at $t \to +\infty$, if $x^0 \in R^n_I$.

Here x^0 is an initial state, $x(t)$ is the solution of (4) such that $x(0) = x^0$ and R^n_I is the set such as $\{x \mid x_i \geq 0$ for $i \in I$ and $x_j > 0$ for $j \in J\}$, where $x^*_i = 0$ for $i \in I$ and $x^*_j > 0$ for $j \in J$.

Let the possible equilibria of system (2) be denoted by (E_{+++}), (E_{++0}), (E_{+0+}), (E_{0++}), (E_{+00}) and (E_{0+0}). Here (E_{+0+}), for example, denotes the equilibrium in which prey x_1 and predator y remain positive, and prey x_2 is extinct. For system (1) and (3), equilibria are defined similarly.

3. PREY-PREDATOR MODELS

For system (1) of two competing species, the following theorem is known.

THEOREM 1 (Takeuchi and Adachi, manuscript) *Let us consider system (1) satisfying* $\alpha \neq 1$ *or* $\beta \neq 1$. *Then* (E_{++}) *is globally stable if and only if* $\alpha < 1$ *and* $\beta < 1$. (E_{+0}) *(or* (E_{0+})*) is globally stable if and only if* $\alpha \leq 1$ *and* $\beta \geq 1$ *(or* $\alpha \geq 1$ *and* $\beta \leq 1$*). When* $\alpha > 1$ *and* $\beta > 1$, (E_{+0}) *and* (E_{0+}) *are locally stable.*

Theorem 1 shows that no oscillatory motion exists and that two-species co-existence is possible if and only if $\alpha < 1$ and $\beta < 1$. The addition of one or two predators to (1) enlarges the possibility of coexistence of the species.

THEOREM 2 (Takeuchi and Adachi, manuscript) *Let us consider system* (2).
  (i) *Suppose that* $\alpha + \beta < 2$. *If* $(E_{+++}) = (x^*_1, x^*_2, y^*)$ *is nonnegative, then* (E_{+++}) *is globally stable. If* (E_{+++}) *is not nonnegative, then one of* (E_{0++}), (E_{+0+}), (E_{++0}), (E_{+00}) *or* (E_{0+0}) *is globally stable.*
  (ii) *Suppose that* $\alpha\beta \leq 1$ *but* $\alpha + \beta \geq 2$. *Then there exists at least one locally stable equilibrium for any* ϵ *and* μ.
  (iii) *Suppose that* $\alpha > 1$ *and* $\beta > 1$. *In the parametric region* (A) *(or* (B)*) of Figure 1, there exists at least one Hopf bifurcation value* ϵ^* *(or* μ^**) for any* μ *(or* ϵ*) fixed if*

$$\frac{d}{d\epsilon}(a_0 a_1 - a_2)\Big|_{\epsilon=\epsilon^*} \neq 0, \quad (or \ \frac{d}{d\mu}(a_0 a_1 - a_2)\Big|_{\mu=\mu^*} \neq 0) \tag{5}$$

where $a_0 = x^*_1 + x^*_2$, $a_1 = (1-\alpha\beta)x^*_1 x^*_2 + d\mu^2 x^*_2 y^* + d\epsilon^2 y^* x^*_1$ *and* $a_2 = d(\epsilon^2 + \mu^2 - (\alpha+\beta)\epsilon\mu)x^*_1 x^*_2 y^*$.

  (iv) *Suppose that* $\alpha \leq 1$, $\beta \geq 1$ *(or* $\alpha \geq 1$, $\beta \leq 1$*) and* $\alpha\beta \geq 1$. *There exists also at least one Hopf bifurcation value if* (5) *is satisfied.*

Numerical calculations show that a stable equilibrium bifurcates super-critically to a periodic motion (Figure 2(a)). Further, when ϵ increases $(\epsilon > \epsilon^*)$,

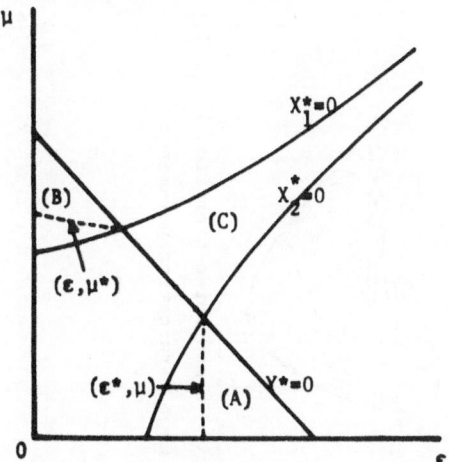

Figure 1 In the regions (A), (B) and (C) of the figure, a positive equilibrium
for system (2) (E_{+++}) exists. In (A) (or (B)), there exists at
least one Hopf bifurcation parameter ε^* (or μ^*) for any straight line
μ = constant (or ε = constant), if (5) is satisfied. See (iii) of
Theorem 2.

a periodic motion in case (iii) degenerates into one of (E_{+00}) or (E_{0+0}) and,
in case (iv), a spiral chaos of Vance emerges from a limit cycle (Figures 2(b),(c)).

Next, let us consider (3). We simplify the model by reducing the number of
parameters as follows; $\varepsilon_1 = \mu_2 = \varepsilon$, $\varepsilon_2 = \mu_1 = \mu$, $d_1 = d_2 = d$ and $\alpha = \beta$.

By Lyapunov's direct method, the global stability of equilibria is estab-
lished for $\alpha < 1$. When $\alpha = 1$, the linearized matrix of (3) at (E_{++++}) has
two pure imaginary eigenvalues $(\lambda, \bar{\lambda})$ and two eigenvalues with negative real
parts. Further, at $\alpha = 1$, λ and $\bar{\lambda}$ cross the imaginary axis with non-zero speed.
Hence, a Hopf bifurcation occurs at $\alpha = \alpha^* = 1$. A very lengthy computation of the
stability criterion (Marsden and McCraken (1976)) shows that the bifurcation is
supercritical, that is, the closed orbits emerge for $\alpha > \alpha^* = 1$ and are attracting.

THEOREM 3 *Let us consider system (3) with* $\varepsilon_1 = \mu_2 = \varepsilon$, $\varepsilon_2 = \mu_1 = \mu$, $d_1 = d_2 = d$
and $\alpha = \beta$.

(i) *Suppose that* $\alpha < 1$. *Then* (E_{++++}) *is globablly stable if and only if*
$\varepsilon + \mu > (\alpha+1)/d$. *When* $\varepsilon + \mu \leqq (\alpha+1)/d$, (E_{++00}) *is globally stable.*

(ii) (E_{++++}) *bifurcates for* $\alpha > 1$ *to a stable limit cycle of a period about;*

$$2\pi [d(\varepsilon+\mu)/(d(\varepsilon+\mu)-2)]^{\frac{1}{2}} |\varepsilon+\mu|/|\varepsilon-\mu|, \tag{6}$$

if $\alpha < d(\varepsilon+\mu) - 1$ *and* $\varepsilon \neq \mu$.

Figures 3 are examples of computer simulations. A stable limit cycle of

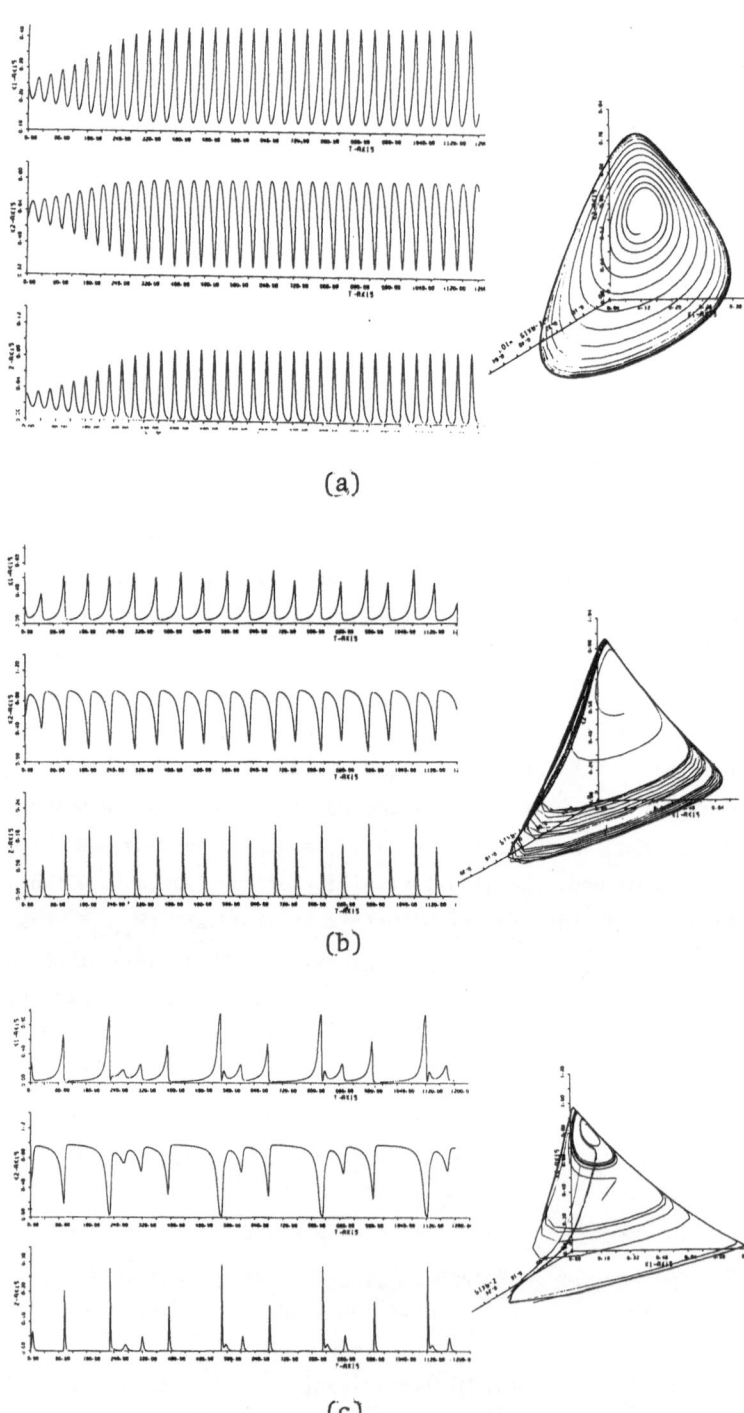

Figure 2 Each figure depicts one-dimensional x_1, x_2, y population changes with respect to time t and three-dimensional x_1, x_2, y population space for system (2) with $\alpha = 1.0$, $\beta = 1.5$, $d = 0.5$, $\mu = 1.0$ The Hopf bifurcation parameter $\epsilon^* = 5.59749821$. See (iv) of Theorem 2.

a) a limit cycle with a period about 31.25 of Hopf type for $\epsilon = 6$.
b) a periodic motion with two peaks in one period for $\epsilon = 8$.
c) Vance's spiral chaos for $\epsilon = 10$.

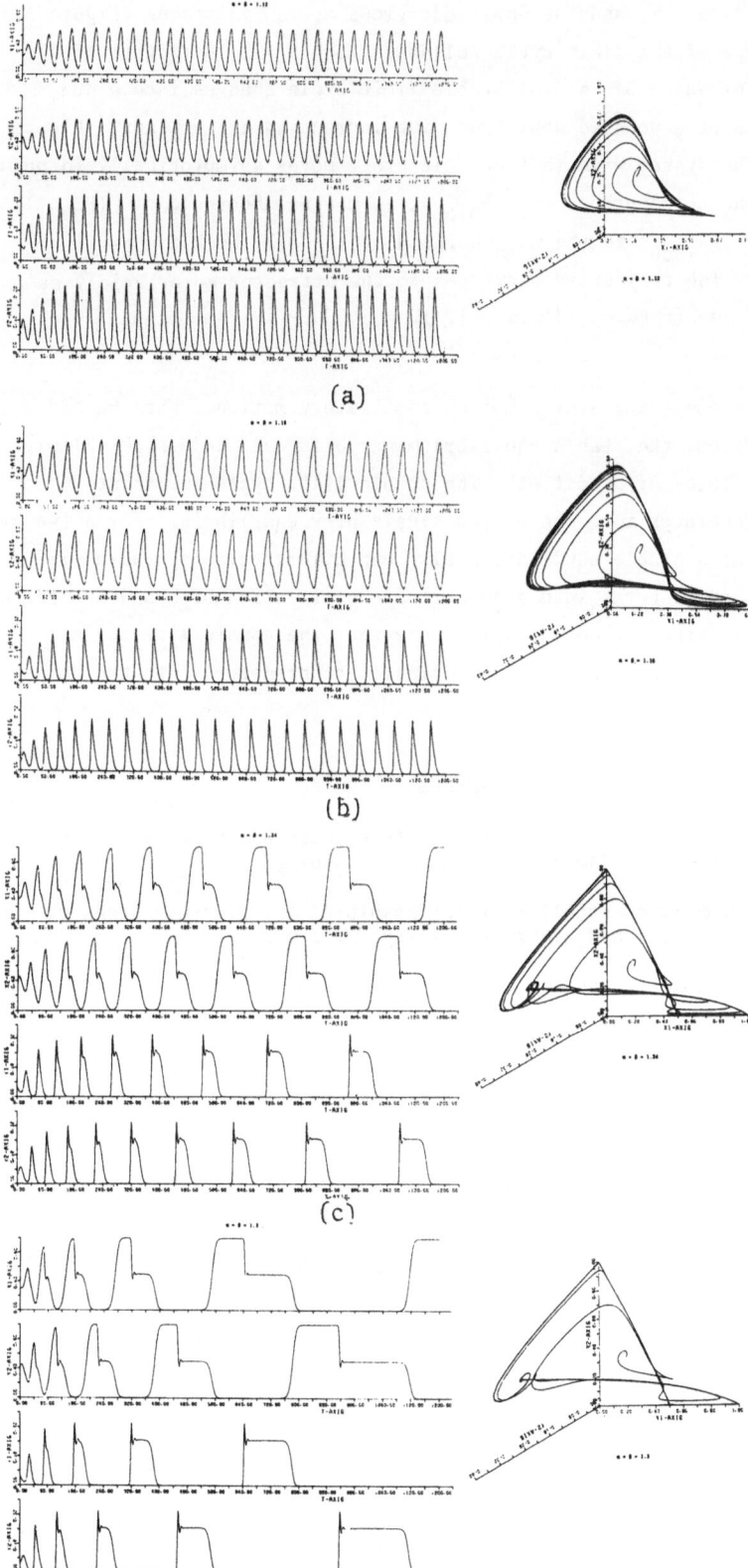

(a)

(b)

(c)

(d)

Figure 3 Each figure depicts one-dimensional x_1, x_2, y_1, y_2 population changes with respect to time t and three-dimensional x_1, x_2, y_2 population space for system (3) with $\varepsilon = 2.0$, $\mu = 0.6$, $d = 1.0$. The Hopf bifurcation value $\alpha = 1.0$. See Theorem 3. (a) a limit cycle with a period about 25.0 of Hopf type for $\alpha = 1.12$. (b) collapse of the shape of the limit cycle for $\alpha = 1.18$. (c) a nonperiodic oscillation of a bounded amplitude with increasing cycle times for $\alpha = 1.24$. (d) the cycle time increases comparing with (c) for $\alpha = 1.3$.

Hopf type appears at $\alpha > 1$, and the amplitude grows as α increases (Figure 3(a), $\alpha = 1.12$). The shape of the limit cycle collapses for $\alpha = 1.18$ (Figure 3(b)). For further increasing of value of α, the stable limit cycle changes into a non-periodic oscillation of a bounded amplitude with increasing cycle times (Figure 3(c), $\alpha = 1.24$). The system with this chaotic orbit moves asymptotically in population space from the neighborhood of (E_{0+00}), to the neighborhood of (E_{0+0+}), to the neighborhood of (E_{+000}) to the neighborhood of (E_{+0+0}), back to the first (E_{0+00}), and so on. The trajectory converges to the attractor more rapidly as an increasing value of α (Figure 3(d), $\alpha = 1.3$).

4. CONCLUSION

A two-species competing system has no oscillatory motion. When we add one predator to this system, the stable equilibrium may bifurcate to a stable limit cycle as increasing rates of predation. The stable limit cycle turns into a spiral chaos or degenerates into one of two single-prey equilibria. For a two-prey, two-predators system; a stable equilibrium also can bifurcate to a stable limit cycle and the limit cycle turns into a nonperiodic oscillation of bounded amplitude with increasing cycle times. These results show that the possibility for all species to coexist can be enlarged by the inclusion of predators into two competing species.

REFERENCES

Coste, J., J. Peyraud, and P. Coullet (1977): Asymptotic behaviors in the dynamics of competing species, *SIAM J. Appl. Math.* 39:516-543.

Fujii, K. (1977): Complexity-stability relationship of two-prey-one-predator species model: local and global stability, *J. Theor. Biol.* 69:613-623.

Gilpin, M.E. (1979): Spiral chaos in a prey-predator model, *Am. Nat.* 113:306-308.

Hsu, S.B. (1981): Predator-mediated coexistence and extinction, *Math. Biosci.* 54:231-248.

Marsden, J.E., and M. McCracken (1976): *The Hopf Bifurcation and its Application*, Springer-Verlag, New York.

Takeuchi, Y., and N. Adachi (manuscript): Existence and bifurcation of stable equilibrium in two-prey, one-predator communities.

Vance, R.R. (1978): Predation and resource partitioning in one predator-two prey model communities, *Am. Nat.* 112:797-813.

PART VI:

COMMUNITY ECOLOGY

COEVOLUTION

Simon A. Levin

1. INTRODUCTION

Mathematical models have long played a central role in evolutionary theory, primarily at its interface with population genetics. However, the approach most developed mathematically, that which builds upon one- or two-locus models in constant or random environments, is of limited value for the consideration of the evolution of most ecological characters and relationships.

In most cases of ecological interest, we do not know the detailed genetic basis of traits; but we do know that those traits are controlled through the action of many interacting loci. To deal with such situations, breeders developed a comprehensive theory of *quantitative* inheritance which, although it complements the discrete locus theory, is quite different in structure (Falconer 1960, Lewontin 1980). A central concept in the quantitative theory, the *heritability* h^2, is nothing more than a descriptive statistic which characterizes the relationship between the genotypic and phenotypic variances of a character, and thereby provides a bridge between the quantitative description of change and the detailed population genetic basis. Recent theoretical investigations of evolutionary change have reemphasized the value of the quantitative approach, and have had considerable impact on contemporary coevolutionary thought (Lande 1976, Slatkin and Lande 1976).

A second limitation upon much mathematical theory has been its attention to constant or randomly varying environments, without linking environmental change to the state of the evolving population. Populations exist within ecological communities, however loosely defined, and evolutionary changes affect and are affected by demographic changes within the population of interest (density- and frequency-dependent selection) and within other populations. The latter include more than what is usually understood as *coevolution*, a term often restricted to evolutionary changes in two or more species; but within this paper, we shall broaden the point of view to include all cases of evolution involving feedbacks mediated through other populations or even through the abiotic environment.

Levin (1983) and Slatkin and Maynard Smith (1979), from different perspectives, review the mathematical theory of coevolution, and the present note is simply a summary of Levin (1983). Levin (1983), in his concluding remarks, makes the gratuitous statement, "Coevolution presents a fascinating class of evolutionary problems"; but Lewontin (pers. comm.) argues that this is much too modest a claim. Coevolutionary problems in fact are not restricted to a special class, but are the most central ones in understanding natural communities and ecosystems. As Lewontin points out, the species defines its environment and changes its effective environment as it evolves and its pattern of distribution changes.

2. TIGHT COEVOLUTION: POPULATION GENETIC AND EPIDEMIOLOGICAL EVOLUTIONARY MODELS OF PARASITE-HOST COEVOLUTION

Despite the comments made earlier, there are some instances where the genetic bases of inheritance are well understood, and where these can be related to specific and independent genes. The most striking case involves the gene-for-gene systems underlying the relationship between cereal plants and the fungal pathogens -- rusts -- which attach them. The special nature of these *tightly coevolved* (e.g. Feeny 1975, Janzen 1980) systems, in which specific genes for host resistance may be matched to specific genes for parasite virulence, is the result of strong selection for particular characters. We may also encounter such simplified genetic relation-ships in other situations in which there has been similarly strong selection, e.g., as a result of intense anthropogenic stresses.

The literature on cereal-rust interactions has been extensive, and is re-viewed by Levin (1983); it builds upon the experimental work by Flor (1955, 1956) and the early theoretical work by Mode (1958, 1960, 1961). Characteristic of the models considered is that the epidemiological details are suppressed and replaced by a mass-action formulation for the probability of association of parasite and host.

One of the most recent treatments of this problem is due to Lewis (1981a,b). In Lewis (1981a), the host is treated as a diallelic diploid and the pathogen as a diallelic haploid. The pathogen fitness is specified in Table 1 for each possible host-parasite association; for the host in each pair, the fitness w is 1 minus the pathogen fitness.

Table 1 Pathogen Fitnesses in Combination with Given Hosts

Host Genotype

		AA	Aa	aa
Pathogen	B	α	β	γ
Genotype	b	γ	β	α

(1)

Letting x denote the (fixed) probability that a given host will be parasitized, and assuming that the frequencies of particular associations are proportional to the products of the frequencies of the associated types, Lewis obtains the model

$$p' = p \frac{pw_{AA} + (1-p)w_{Aa}}{p(pw_{AA}+(1-p)w_{Aa}) + (1-p)(pw_{Aa}+(1-p)w_{aa})} = p \frac{w_{A.}}{pw_{A.} + (1-p)w_{a.}}$$

(2)

for the host; here

$$w_{AA} = 1 - x + x[q(1-\alpha)+(1-q)(1-\gamma)] = 1 - x[q\alpha+(1-q)\gamma]$$

$$w_{Aa} = 1 - x + x[q(1-\beta)+(1-q)(1-\beta)] = 1 - x[\beta] \qquad (3)$$

$$w_{aa} = 1 - x + x[q(1-\gamma)+(1-q)(1-\alpha)] = 1 - x[q\gamma+(1-q)\alpha].$$

p and q are, respectively, the allelic frequencies of A (in the host) and B (in the pathogen) in the current generation, and p' is the frequency of A in the next generation. Similarly, for the pathogen, the relevant equation is

$$q' = q \frac{v_B}{qv_B + (1-q)v_b} , \qquad (4)$$

in which

$$v_B = p^2\alpha + 2p(1-p)\beta + (1-p)^2\gamma \qquad (5)$$

and

$$v_b = p^2\gamma + 2p(1-p)\beta + (1-p)^2\alpha. \qquad (6)$$

By symmetry, this system has an internal (polymorphic) equilibrium at $p = q = 0.5$, but this equilibrium is stable if and only if

$$\beta < \frac{\alpha+\gamma}{2} \sqrt{1 - 2 \frac{(\alpha-\gamma)^2}{(\alpha+\gamma)^2}} < \frac{\alpha+\gamma}{2} . \qquad (7)$$

This condition is stronger than is necessary to assure marginal overdominance for the host at equilibrium; thus, marginal overdominance is a necessary condition for stability (see also Levin and Udovic 1977). More generally, the oscillatory solutions which arise when β is increased beyond the threshold specified in (7) deserve further study in relation to the fluctuations found in many simplified agricultural systems.

In some circumstances, marginal overdominance is not possible; for example, in the consideration of cereal-rust systems one usually regards resistance as dominant and virulence as recessive. For this case, the model (2)-(6) is not applicable, and those models which are commonly used exhibit metastable behavior, with oscillations of nonstable type unless intraspecific frequency-dependent selection is assumed.

Intraspecific frequency dependence is actually a quite logical assumption for host-parasite systems, and probably contributes substantially to their stabilization. For example, as virulence increases in the pathogen population, host survival may be expected to decrease and this may mean reduced transmissiblity of the disease (however, for some diseases, increased virulence may also evoke behaviors, such as biting, which increase the spread of disease). Similarly, increased mean resistance to pathogens should decrease the incidence of disease and reduce the advantage of the resistant type. Such frequency-dependent feedbacks can introduce stabilizing effects on host-parasite systems (see for example Gillespie 1975). Gillespie demonstrates the existence of such feedbacks by taking an epidemiological approach that implicitly recognizes the demic structure of the parasite population.

Levin and Pimentel (1981), in considering the evolution of attenuation in the myxoma virus in relation to the European rabbit population in Australia, took an explicitly epidemiological approach to the problem of parasitic evolution. Levin (1983) generalized their treatment, and obtained the system

$$\frac{dS}{dt} = (r_0 S + r_1 I_1 + r_2 I_2 + r_3 I_3) - bS - \beta_1 S I_1 - \beta_2 S I_2 + v_1 I_1 + v_2 I_2$$

$$\frac{dI_1}{dt} = \beta_1 S I_1 - (b+\alpha_1) I_1 - v_1 I_1 + w_2 I_3 - \gamma_2 \beta_2 I_1 I_2$$

$$\frac{dI_2}{dt} = \beta_2 S I_2 - (b+\alpha_2) I_2 - v_2 I_2 + w_1 I_3 - \gamma_1 \beta_1 I_1 I_2 \qquad (8)$$

$$\frac{dI_3}{dt} = (\gamma_1 \beta_1 + \gamma_2 \beta_2) I_1 I_2 - (w_1 + w_2) I_3 - (b+\alpha_3) I_3,$$

in which S denotes susceptible hosts; I_1, I_2 denote hosts infected with viral strains 1 and 2; and I_3 signifies hosts infected with both strains. The parameters r_i are birth rates; v_i, w_i are recovery rates; b and $b+\alpha_i$ are death rates; β_i are transmission rates; and γ_i are secondary infection rates.

There are several possible outcomes for this system, including the competitive exclusion of either viral type or their stable coexistence. Unbounded behavior is also possible unless the parameters are appropriately constrained.

Any polymorphic (positive) equilibrium in (8) must satisfy the relations

$$I_2 = \frac{\beta_1}{\gamma_2 \beta_2 - w_2 Q} \left(S - \frac{b+\alpha_1+v_1}{\beta_1} \right) \qquad (9)$$

$$I_1 = \frac{\beta_2}{w_1 Q - \gamma_1 \beta_1} \left(\frac{b+\alpha_2+v_2}{\beta_2} - S \right) \qquad (10)$$

$$I_3 = Q I_1 I_2 = \frac{\gamma_1 \beta_1 + \gamma_2 \beta_2}{w_1 + w_2 + b + \alpha_3} I_1 I_2 . \qquad (11)$$

If the two multipliers in (9) and (10) are both positive, then

$$\frac{b+\alpha_2+v_2}{\beta_2} > S > \frac{b+\alpha_1+v_1}{\beta_1} \, , \qquad (12)$$

and

$$\gamma_2\beta_2w_1 > \gamma_1\beta_1(w_2+b+\alpha_3) \, . \qquad (13)$$

Equation (13) is given in Levin (1983); however, unfortunately the subscripts 1 and 2 are mistakenly interchanged. Stable equilibria of this sort have been shown to exist for special cases; but not withstanding the claim in Levin (1983), it has not been established in general that these (and the analogous ones with subscripts interchanged) are the only possible solutions. Moreover, the question of stability remains open for the general case, as does the existence of easily-interpreted necessary and sufficient conditions for the existence of solutions satisfying (12).

A broad selection of open questions remain regarding the interaction between the evolution of virulence in the parasite and of resistance in the host, the importance of seasonality and of multiple modes of transmission, and the relative importance of parasitism and resource abundance in limiting the host population.

3. DIFFUSE COEVOLUTION

Host-parasite interactions represent ideal systems for tight coevolution, because the fates of host and parasite are intimately linked. However, many problems of interest in the evolution of ecological communities are much more *diffuse*, involving many species with varying degrees of relationship to one another. Further, communities and ecosystems are loosely defined assemblages without fixed boundaries or stable composition; and although it is obvious that ecosystem characteristics such as diversity, successional relationships, and the biotic control of nutrient processing emerge from the loose coevolution of species, it is less obvious how to approach these central ecological questions. An intermediate situation involves clusters of interacting species which interact strongly as groups, but for which specific relationships are diffuse. Problems of this sort arise in the consideration of the chemical defenses of plants in response to insects or other pests (Feeny 1982), for often these do not have the finely tuned species-for-species relationship already discussed for the cereals and their rusts. Similar problems occur in predator-prey systems, which are by nature less specific than host-parasite relationships; in competition theory; and regarding the evolution of vertebrate immune systems.

Levin and Segel (1982) approach problems of this sort by modifying the standard approaches of quantitative genetics to deal with multiple species interactions. For details, the reader is referred to Levin and Segel (1982) or Levin (1983).

Innovative approaches to dealing with the evolution of ecological communities and ecosystems are essential if we are to achieve any understanding of their structure and function. Evolution proceeds on multiple hierarchical scales, and these hierarchies are confounded even more when system-level patterns are of interest. Evolution within one species is at times tightly linked to evolution within another, but more generally, the relevant feedbacks are mediated by processes at different levels of hierarchical organization. Finding ways to approach such problems represents one of the greal challenges in evolutionary theory, and will do much to bring closer together the distinct fields of population biology and ecosystem science.

Acknowledgement

It is a pleasure to acknowledge the support of the National Science Foundation under grant MCS 80-01618.

REFERENCES

Falconer, D.S. (1960): *Introduction to Quantitative Genetics*, Ronald Press, New York.

Feeny, P. (1975): Biochemical coevolution between plants and their insect herbivores, Pages 3-19 *in* L.E. Gilbert and P.H. Raven (editors), *Coevolution of Animals and Plants*, University of Texas Press, Austin and London.

_____ (1982): Coevolution of plants and insects, Chapter 11 *in* T.R. Odhiambo (editor), *Current Themes in Tropical Sciences, 2: Natural Products for Innovative Pest Management*, Pergamon Press, Oxford.

Flor, H.H. (1955): Host-parasite interaction in flax rust -- its genetics and other implications, *Phytopathology* 45:680-685.

_____ (1956): The complementary genic systems in flax and flax rust, *Advances in Genetics*, 8:29-54.

Gillespie, J.N. (1975): Natural selection for resistance to epidemics, *Ecology* 56:493-495.

Janzen, D.H. (1980): When is it coevolution? *Evolution* 34:611-612.

Lande, R. (1976): The maintenance of genetic variability by mutation in a polygenic character with linked loci, *Genetical Research* 26:221-235.

Levin, S.A. (1983): Some approaches to the modelling of coevolutionary interactions, *in* M. Nitecki (editor), *Coevolution*, University of Chicago Press, Chicago.

Levin, S.A., and D. Pimentel (1981): Selection of intermediate rates of increase in parasite-host systems, *Amer. Nat.* 117:308-315.

Levin, S.A., and L.A. Segal (1982): Models of the influence of predation on aspect diversity in prey populations, *J. of Math. Biol.* 14:253-285.

Levin, S.A., and J.D. Udovic (1977): A mathematical model of coevolving populations, *Amer. Nat.* 111:657-675.

Lewis, J.W. (1981a): On the coevolution of pathogen and hosts: I, General theory of discrete time coevolution, *J. Theor. Biol.* 93:927-951.

Lewis, J.W. (1981b): On the coevolution of pathogen and hosts: II, Selfing hosts and haploid pathogens, *J. Theor. Biol.* 93:953-985.

Lewontin, R.C. (1980): Models of natural selection, *in* C. Barigozzi (editor), *Vito Volterra Symposium on Mathematical Models in Biology*, Lect. Notes in Biomath. 38, Springer-Verlag, Heidelberg.

Mode, C.J. (1958): A mathematical model for the coevolution of obligate parasites and their hosts, *Evolution* 12:158-165.

_____ (1960): A model of a host-pathogen system with particular reference to the rusts of cereal, Pages 84-96 *in Biometrical Genetics*, Pergamon Press, New York.

_____ (1961): A generalized model of a host-pathogen system, *Biometrics* 17: 386-404.

Slatkin, M., and R. Lande (1976): Niche width in a fluctuating environment--density independent model, *Amer. Nat.* 110:31-55.

Slatkin, M., and J. Maynard Smith (1979): Models of coevolution, *Quart. Rev. Biol.* 54:233-263.

A COEXISTENCE MODEL FOR N-SPECIES USING NEAREST NEIGHBOUR
INTERACTION AND THE ROLE OF DIFFUSION

S.C. Bhargava

A model for N-species with interaction among the nearest neighbours as a
possible explanation of their coexistence in the environment is proposed and
analysed.

1. INTRODUCTION AND THE MODEL

The interactions among various species coexisting in a given environment are
known to be extremely complex. Without going into individual interactions, which
are subjects in themselves, we attempt a mathematical model of the coexistence of
species considering only the overall effect of the interactions. We shall begin by
assuming that:

a) it is the cooperative effect of various interactions which is responsible for
    the coexistence of species in a given environment;
b) the cooperative effect may be represented by the interaction among the nearest
    neighbours, that is, the ith species interacts only with (i+1)th and
    (i-1)th;
c) as a result of interaction between the ith and (i+1)th species the popula-
    tion density (N_i) of the ith species increases whereas it decreases as a
    result of its interaction with the (i-1)th species.

The growth rate of ith species is thus taken as

$$\dot{N}_i = N_i\{\alpha_i + \beta_i f(N_{i+1}) - \gamma_i g(N_{i-1})\}, \tag{1}$$

where the β_i and γ_i's are real positive constants and α_i's are real constants
which can take either positive values (indicating environmental support), or negative values (indicating the absence of environmental support).

It should be mentioned that our interaction is neither pure prey-predator
nor the commonly known food chains (Maynard Smith 1974, Freedman 1980) as some of
α's could be positive, that is, the species could have environmental support.
Moreover, our chain is closed unlike the food chain.

2. THE LOGARITHMIC INTERACTION

We first consider interaction functions f and g in Equation (1) to be of
logarithmic type. This type of function was first introduced by Gompertz (1825)
while investigating mortality rates and used later by Gomatam (1974) and others

(Pande, 1978) in the study of prey-predator interactions. Our growth equation (1) can then be rewritten in the form

$$\frac{d}{dt} \{\log(N_i)\} = \alpha_i + \beta_i \log(N_{i+1}) - \gamma_i \log(N_{i-1}).$$ (2)

This is a linear differential equation (LDE) in the variable $\log N_i$ and if there are N-species in the environment, LDE's for all of them can be added to yield

$$\frac{d}{dt} \{\log(N_1 N_2 \cdots N_N)\} = \sum_{i=1}^{N} \alpha_i + \sum_{i=1}^{N} \log(N_{i+1})(\beta_i - \gamma_{i+2}).$$ (3)

The interaction is such that species form an endless chain, therefore, in the above equation $N+1 \equiv 1$ and $N+2 \equiv 2$.

We notice that if

$$\left. \begin{array}{c} \sum\limits_{i=1}^{N} \alpha_i = 0, \\ \\ \text{and} \\ \\ \beta_i = \gamma_{i+2} \quad (i=1,2,\ldots,N), \end{array} \right\}$$ (4)

then equation (2) has a solution

$$N_1 N_2 \cdots N_N = \text{constant}.$$ (5)

This is quite a useful result because it allows variation in time of individual populations in such a manner that the product of the population densities remains constant. Further, none of the population densities can either vanish or rise indefinitely. This may be interpreted to mean that N-interacting species can co-exist in the environment if the interaction is defined through (2) and the parameters satisfy equation (4).

It should be mentioned that the conditions of coexistence mentioned in this section are in addition to the requirement of positivity of all populations at the equilibrium point. These are discussed in detail elsewhere (Goel, Maitra, and Montroll 1971, Strobeck 1973, Maynard Smith 1974, Freedman 1980).

3. THE QUADRATIC INTERACTION

We consider in this section the more common form of interaction functions $f(N_{i+1}) = N_{i+1}$ and $g(N_{i-1}) = N_{i-1}$. In particular, we consider the case of three interacting species with interaction amongst the nearest neighbours and study the role of diffusion in such a model. The growth equations are thus written as

$$\frac{\partial N_1}{\partial t} = \alpha_1 N_1 + \beta_1 N_1 N_2 - \gamma_1 N_1 N_3 + D_1 \frac{\partial^2 N_1}{\partial z^2}$$

$$\frac{\partial N_2}{\partial t} = \alpha_2 N_2 + \beta_2 N_2 N_3 - \gamma_2 N_2 N_1 + D_2 \frac{\partial^2 N_2}{\partial z^2} \qquad (6)$$

$$\frac{\partial N_3}{\partial t} = \alpha_3 N_3 + \beta_3 N_3 N_1 - \gamma_3 N_3 N_2 + D_3 \frac{\partial^2 N_3}{\partial z^2} ,$$

where D_1, D_2 and D_3 are the diffusion coefficients.

In the absence of diffusion, $D_1 = D_2 = D_3 = 0$. If in addition the other parameters satisfy equation (4), viz,

$$\alpha_1 + \alpha_2 + \alpha_3 = 0,$$

and

$$\beta_1 = \gamma_3; \qquad \beta_2 = \gamma_1; \qquad \beta_3 = \gamma_2 . \qquad (7)$$

then equation (6) has a solution given by

$$N_1 N_2 N_3 = \text{constant.} \qquad (8)$$

The above mentioned solution is neutrally stable, that is, any perturbation or different set of initial conditions will take the system to an entirely new trajectory. This makes the solution (8) uninteresting. We have, however, noticed a very interesting feature of the system (6) when we allow diffusion of species and work with parameter values not as given in (7) but close to them so that whole system is linearly unstable. This will require analysis of complete system described by the partial differential equations (6), commonly known as a reaction-diffusion system of equations.

The analysis we attempt (an approximate one) is based on a conjecture due to Landau (1959) which was later developed by Kogelman and di Prima (1970), according to which a linearly unstable system can in the presence of nonlinearities and diffusion saturate and give rise to travelling wave solutions at large times. Because of lack of space only a brief outline of the method is given here. For more details we refer to our earlier papers (Bhargava and Saxena 1977, Bhargava 1980).

First, we linearize our system (6) around the equilibrium point (N_{10}, N_{20}, N_{30}) of the system and write the growth equation (6) in the following matrix form

$$
\begin{pmatrix}
D_1 \partial^2/\partial z^2 & \beta_1 N_{10} & -\gamma_1 N_{10} \\
-\gamma_2 N_{20} & D_2 \partial^2/\partial z^2 & \beta_2 N_{20} \\
\beta_3 N_{30} & -\gamma_3 N_{30} & D_3 \partial^2/\partial z^2
\end{pmatrix}
\begin{pmatrix} n_1 \\ n_2 \\ n_3 \end{pmatrix}
- \frac{\partial}{\partial t}
\begin{pmatrix} n_1 \\ n_2 \\ n_3 \end{pmatrix}
=
\begin{pmatrix}
\gamma_1 n_1 n_3 - \beta_1 n_1 n_2 \\
\gamma_2 n_2 n_1 - \beta_2 n_2 n_3 \\
\gamma_3 n_3 n_2 - \beta_3 n_3 n_1
\end{pmatrix}
\tag{9}
$$

with $n_i = N_i - N_{i0}$.

Assuming that our disturbance is periodic (period 2L) we make a Fourier expansion of

$$
\Phi(z,t) = \begin{pmatrix} n_1(z,t) \\ n_2(z,t) \\ n_3(z,t) \end{pmatrix}.
\tag{10}
$$

The differential equation satisfied by the Fourier components ϕ_m's can be written as

$$
M_m(\phi_m) - \frac{\partial}{\partial t}(\phi_m) = F_m,
\tag{11}
$$

where

$$
\phi_m(t) = 2/L \int_{-L}^{L} \exp(-im\pi z/L)\Phi(z,t)dz,
\tag{12}
$$

$$
M_m(\phi_m) = 2/L \int_{-L}^{L} \exp(-im\pi z/L)M(\Phi)dz,
\tag{13}
$$

with the linear operator M as given in (9), i.e.,

$$
M_m =
\begin{pmatrix}
-D_1(m\pi/L)^2 & \beta_1 N_{10} & -\gamma_1 N_{10} \\
-\gamma_2 N_{20} & -D_2(m\pi/L)^2 & \beta_2 N_{20} \\
\beta_3 N_{30} & -\gamma_3 N_{30} & -D_3(m\pi/L)^2
\end{pmatrix}
\tag{14}
$$

F_m in (11) is the mth Fourier component of the nonlinear term appearing on the right hand side of equation (9). Analysis of the linear problem is fairly standard and we can easily find the point at which the instability occurs. The linear solution is written as

$$
\phi_m = \exp(-\sigma_m t) \cdot \phi_m(0).
\tag{15}
$$

Three eigenvalues σ_{m1}, σ_{m2}, σ_{m3} and the corresponding eigenvectors ϕ_m's can be easily found. The manner in which the instability will grow will depend on the form of the nonlinearity and its interaction with diffusion. The complete nonlinear differential equation (11) should, therefore, be analysed. The essence of the method we use is contained in following two steps.

1. We construct asymptotic solutions of (11) in terms of a small parameter ϵ, which will turn out to be $\left|\text{Re } \sigma_{11}\right|^{\frac{1}{2}}$. Since ϵ is a measure of the amplitude of the dominant mode ϕ_1, we write,

$$\phi_m(t) = \epsilon^{|m-1|+1} \psi_m(t). \tag{16}$$

That ϕ_1 is the dominant mode can be easily seen from the fact that $\exp(-\sigma_1 t)$ has the positive exponent which is maximum (Re $\sigma_1 < 0$).

2. The normalized Fourier components, ψ_m, are now expanded in terms of a complete set of eigenfunctions of the corresponding linear stability problems with unknown coefficients which depend only on time. Thus

$$\psi_m(t) = \sum_{\ell=1,2,3} A_{m\ell}(t) \phi_{m\ell}(0). \tag{17}$$

The index 1 labels three independent eigenvectors. Since $m = 1$ is the dominant mode, if A_{11} stabilizes the other will automatically do so. ψ_m's are also solutions of (11) but with a measure $\epsilon^{|m-1|+1}$, i.e., from (16) and (11) we obtain the following equations for ψ_0, ψ_1, and ψ_2.

$$[M_0 - \frac{\partial}{\partial t}]\psi_0 = F_0(\psi_1 \bar{\psi}_1) + O(\epsilon^2) \tag{18}$$

$$[M_2 - \frac{\partial}{\partial t}]\psi_2 = F_2(\psi_1 \psi_1) + O(\epsilon^2) \tag{19}$$

$$[M_1 - \frac{\partial}{\partial t}]\psi_1 = \epsilon^2 F_1(\psi_0, \psi_1, \psi_2) + O(\epsilon^4). \tag{20}$$

We solve (18) and (19) by using ψ_1 as a solution of the linear equation (right hand of (20) equal to zero). Once we have ψ_0 and ψ_2, we can solve for ψ_1 and thus can obtain the equation satisfied by $A(t) \equiv A_{11}(t)$ from (20) and (17) and completeness and normalizability of the eigenfunctions. This reads

$$\frac{d}{dt} A(t) + \sigma_{11} A(t) = \epsilon^2 B A(t) |A(t)|^2 + O(\epsilon^4). \tag{21}$$

where $\epsilon^2 = \text{Re } \sigma_{11}$. The expression for B will be different for different problems and can be obtained for the system described by equations (6). Equation (21) can admit 'permanent type' solutions of the form

$$A(t) = A_0 \exp(iwt), \tag{22}$$

where A_0 is the amplitude, which should be real and positive, and w is the frequency, such that

$$A_0^2 = \left(\frac{1}{\epsilon^2} \frac{Re\ \sigma_{11}}{Re\ B} \right) + O(\epsilon^4) \qquad (23)$$

$$w = Im\ \sigma_{11} + \epsilon^2 A_0^2\ Im\ B + O(\epsilon^4) \qquad (24)$$

Clearly, such solutions exist only if $Re\ \sigma_{11}$ and $Re\ B$ are of the same sign.

4. RESULTS AND COMMENTS

Our calculations of B for the system (6) indicates that for parameter values (expressed in arbitrary units)

$$\alpha_1 = -0.071; \qquad \alpha_2 = -0.151; \qquad \alpha_3 = 0.228; \qquad \beta_1 = 0.03;$$

$$\beta_2 = \gamma_1 = 0.02; \qquad \beta_3 = \gamma_2 = 0.012; \qquad \gamma_3 = 0.0306.$$

$Re\ B$ and $Re\ \sigma_{11}$ will have the same sign if diffusion coefficients are small enough, i.e., $D_1 = D_2 = D_3 \approx 5 \times 10^{-5}$. Thus a travelling wave solution will exist for these parameter values.

The model presented here has interesting features even for unrealistic but very simple form of functions f and g discussed in the text. The emphasis here is on the suggestion that inter-dependence of species in the environment could be considered by arranging them on an endless chain and taking into account interaction among the nearest neighbours.

REFERENCES

Bhargava, S.C. (1980): On the Higgins model of glycolysis. *Bull. Math. Biol.* 42: 829-836.

Bhargava, S.C., and R.P. Saxena (1977): Stable periodic solution of the reactive-diffusive Volterra system of equations. *J. Theor. Biol.* 67: 399-407.

Freedman, H.I. (1980): *Deterministic Mathematical Model in Population Ecology.* Marcel Dekker, New York.

Goel, N.S., S.C. Maitra and E.W. Montroll (1971): On the Volterra and other non-linear models of interacting populations. *Rev. of Mod. Phys.* 43: 231-276.

Gomatan, J. (1974): A new model for interacting populations. *Bull. Math. Biol.* 36: 347-364.

Gompertz, B. (1825): On the nature of function expressive of law of human mortality, and on a new mode of determining the value of life contingencies. *Phil. Tras. Roy. Soc.* 115: 513-585.

Kogelman, S., and R.C. di Prima (1970): Stability of spatially periodic super-critical flows in hydrodynamics. *Phys. Fluids* 13: 1-11.

Landau, L.D., and E.M. Lifshitz (1959): *Fluids Dynamics*. Pergamon Press, p. 103.

Maynard Smith, J. (1974): *Models in Ecology*. Cambridge, Cambridge University Press.

Pande, L.K. (1978): Ecosystems with three species one-prey and two-predators system in an exactly solvable model. *J. Theor. Biol.* 74: 591-595.

Strobeck, C. (1973): N species competition. *Ecology* 54: 650-654.

PERIODIC LOTKA-VOLTERRA SYSTEMS AND TIME
SHARING OF ECOLOGICAL NICHES

J.M. Cushing

1. PERIODIC LOTKA-VOLTERRA SYSTEMS

The Lotka-Volterra system

$$P_1' = P_1(b_1 - a_{11}P_1 - a_{12}P_2), \quad P_2' = P_2(b_2 - a_{21}P_1 - a_{22}P_2) \tag{1}$$

with positive, periodic coefficients $b_i = b_i(t) \geq 0 \ (\not\equiv 0)$, $a_{ij} = a_{ij}(t) \geq 0 \ (\not\equiv 0)$ was studied in Cushing (1980) where it was shown that to a large extent the dynamics of (1) with such periodic coefficients mimics that of the familiar classical case of (1) with positive constant coefficients. With constant coefficients (1) has a (unique) positive equilibrium if and only if b_2 lies in a certain interval determined by b_1 and the a_{ij} (namely, the interval with endpoints $b_1 a_{21}/a_{11}$ and $a_{22}b_1/a_{12}$) and this positive equilibrium is stable if and only if $\Delta = a_{11}a_{22} - a_{12}a_{21} > 0$. In Cushing (1980) it is shown more generally that with positive, periodic coefficients (1) has a positive periodic solution if the average of $b_2(t)$ lies in a certain interval (whose endpoints are averaged quantities which reduce to those above for the constant coefficients case) and that this periodic solution is stable if a certain averaged quantity (which reduces to Δ/a_{11} for constant coefficients) is positive.

The purpose of this note is to describe some more specific results for (1) when the periodic coefficients are derived from the MacArthur-Levins theory and to relate them to the idea of time sharing an ecological niche.

In the MacArthur-Levins theory of competition for a one dimensional resource niche the coefficients in (1) are given by

$$b_i = w_i \int_{-\infty}^{+\infty} R(\rho) f_i(\rho) d\rho, \quad a_{ij} = w_i w_j \int_{-\infty}^{+\infty} f_i(\rho) f_j(\rho) d\rho$$

where $w_i f_i(\rho) \geq 0$, $\int_{-\infty}^{+\infty} f(\rho) d\rho = 1$, is a *resource utilization function* for species i and $R(\rho)$ is the availability rate of the resource ρ (Christiansen and Fenchel 1977, May 1974). The simplest case is when the utilization functions are Gaussian:

$$f_i(\rho) = (2\pi W^2)^{-\frac{1}{2}} \exp(\rho - D_i)^2 / 2W^2$$

and $R(\rho) = R > 0$ is constant. Here W is the "niche width" and $d = |D_1 - D_2|$ the "niche separation". Then

$$b_i = w_i R, \quad a_{ii} = w_i^2/(4\pi W^2)^{\frac{1}{2}}, \quad a_{ij} = \delta w_i w_j/(4\pi W^2)^{\frac{1}{2}},$$

$$\delta = \exp(-(d/2W)^2) \leq 1 \tag{2}$$

and (1) has a positive, stable equilibrium as long as $d/W > 0$. Stability is weakened as $d \to 0$, however, in the sense that the smallest real part of the eigenvalues of the linearized system is a monotonically increasing function of d/W which vanishes at $d/W = 0$ when the niches coincide, (see the final graph in Section 3 ($\lambda = 0$) below).

A biological case can be made for time fluctuations and periodicities in any of the quantities R,W,d and w_i. Such periodicities in (2) lead to a system (1) which falls within the purview of the general theory in Cushing (1980). In this note, attention will be restricted to the case when all parameters are constant except the w_i which will be assumed periodic in time. Thus, as in classical theory, the resource availability, niche positions and niche separation are constant in time. Only resource utilization will vary periodically in time. Specifically, it will be assumed that

$$w_1 = w(1 + \lambda\cos \omega t), \quad w_2 = w(1 + \lambda a \cos(\omega t+\gamma))$$

$$0 \leq \lambda \leq 1, \quad 0 \leq a\lambda \leq 1, \quad 0 \leq \gamma \leq \pi$$

so that the resource utilization functions have the same averages $w > 0$, but vary cosinusoidally in time with period $2\pi/\omega$ and with relative amplitudes a and phase difference γ. A rescaling of time (from t to wRt) and of P_i to $wP_i/R(4\pi W^2)^{\frac{1}{2}}$ leads to (1) with the coefficients

$$b_1 = 1 + \lambda\cos \omega t, \quad a_{11} = (1 + \lambda\cos \omega t)^2, \quad a_{22} = (1 + \lambda a \cos(\omega t+\gamma))^2$$

$$b_2 = 1 + \lambda a \cos(\omega t+\gamma), \quad a_{12} = a_{21} = \delta(1 + \lambda\cos \omega t)(1 + \lambda a \cos(\omega t+\gamma)). \tag{3}$$

The goal is to describe various properties of the positive periodic solution of (1) and (3) and its stability as they depend on the parameters λ,ω,a,γ and δ. It is hoped that this will lead to some insights into the phenomenon of time sharing a resource niche as well as the effects that these periodicities have on the fundamental concepts of competitive coexistence and exclusion and of limiting similarity.

2. SOME ANALYTICAL RESULTS

When the amplitude λ is small, regular perturbation techniques can be used to derive lower order approximations to the positive periodic solution of (1) and (3). Tedious, but straightforward calculations show that $P_i(t) = (1+\delta)^{-1} + \lambda y_i(t) + O(\lambda^2)$ where

$$2y_i(t) = (A_1 + (-1)^{i+1} A_2) \cos \omega t + (B_1 + (-1)^{i+1} B_2) \sin \omega t$$

$$A_1 = -[1 + a \cos \gamma + \omega a \sin \gamma]/(1+\delta)(1+\omega^2)$$

$$B_1 = [-\omega(1 + a \cos \gamma) + a \sin \gamma]/(1+\delta)(1+\omega^2)$$

$$A_2 = -(\delta-1)[(\frac{\delta-1}{\delta+1})(1 - a \cos \gamma) + \omega a \sin \gamma]/[(\delta-1)^2 + (\delta+1)^2 \omega^2]$$

$$B_2 = (\delta-1)[-(\frac{\delta-1}{\delta+1})a \sin \gamma + \omega(1 - a \cos \gamma)]/[(\delta-1)^2 + (\delta+1)^2 \omega^2].$$

The Floquet exponents of the system linearized at this periodic solution are

$$e_1 = -1 + 0(\lambda), \quad e_2 = (\delta-1)/(\delta+1) + \theta\lambda^2 + 0(\lambda^3)$$

$$\theta = \theta(\delta) = \omega^2(a^2 - 2a \cos \gamma + 1)/8(1 + \omega^2) + 0(|\delta-1|).$$

Of particular interest is the case of very similar niches $\delta \sim 1$ (i.e. $d/W \sim 0$). For $\delta \sim 1$

$$P_i(t) \sim \frac{1}{2} + \lambda(A_1 \cos \omega t + B_1 \sin \omega t) + 0(\lambda^2)$$

$$e_2 \sim -\theta(1)\lambda^2 + 0(\lambda^3) \sim -\frac{\omega^2}{8}(1 + \omega^2)^{-1}(a^2 - 2a \cos \gamma + 1)\lambda^2 + 0(\lambda^3).$$

Note that e_2 is the smallest Floquet exponent and hence determines the strength of the stability of the positive periodic solution.

   For small amplitude oscillations $\lambda \sim 0$ and similar niches $\delta \sim 1$ some conclusions which can be deduced from these lower order terms are the following.

1. Since $\theta(1) \geq 0$ and since $\theta(1) = 0$ if and only if $\gamma = 0$, $a = 1$, it is seen that the presence of periodicities in the resource utilization functions $(\lambda \neq 0)$ decreases e_2 and hence promotes the stability of the competitive interaction, except possibly in or near the case of in-phase oscillations $(\gamma = 0)$ of equal amplitudes $(a = 1)$.

2. Since the maximum of $\theta(1)$ occurs for $a = 1$, $\gamma = \pi$, stability is maximized when the oscillations of the utilization functions are out-of-phase $(\gamma = \pi)$ and of equal amplitudes $(a = 1)$.

3. $\theta(1)$ and hence stability is increased by an increase of amplitude λ or frequence ω or relative amplitude $a > \cos \gamma$. Stability decreases with increasing $a \in [0, \cos \gamma]$.

4. The population mean values are, to lowest order, equal to $(1 + \delta)^{-1}$ and are independent of a, λ, ω and γ.

5. The amplitude $(A_1^2 + B_1^2)^{\frac{1}{2}} = [(a^2 + 2a \cos \gamma + 1)/16(1 + \omega^2)]^{\frac{1}{2}}$ of the oscillations in the population sizes $P_i(t)$ is minimized when $\gamma = \pi$. Thus, in this sense too, out-of-phase utilization of resources leads to maximal stability. This

amplitude is also decreased with increased frequence ω , but increased with increased relative amplitude a.

6. When $\gamma \sim 0$, the amplitude of P_1 increases with the phase difference γ while that of P_2 decreases with increases in γ .

7. The population sizes $P_i(t)$ oscillate nearly in-phase, except near $\gamma = \pi$.

Any one of these conclusions can be drastically altered if either $\delta \gg 1$ or $\lambda \gg 0$ or $a \sim 1$ and $\gamma \sim 0$.

3. NUMERICALLY FOUND RESULTS

I have carried out a great many numerical integrations of (1) and (3) for various parameter values. While corroborating the conclusions above these have also revealed some other interesting phenomena. The following graphs show a typical plot of the computed magnitude of the smallest Floquet exponent and the population maxima and minima as they change with the phase difference γ in resource utilization.

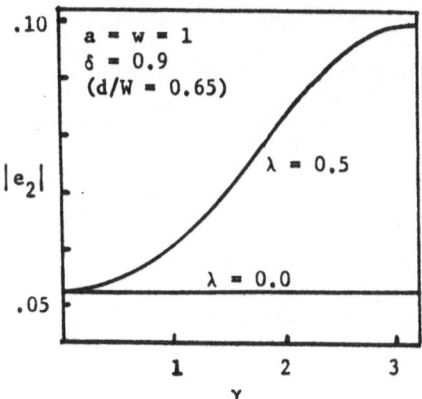 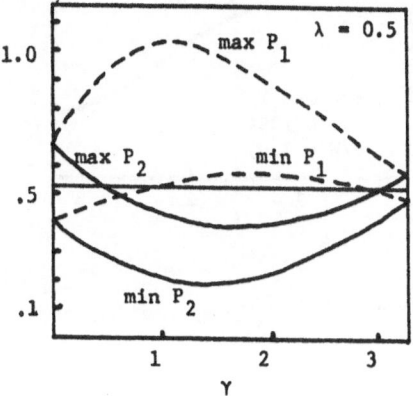

Figure 1

The monotonic increase in stability with increasing γ is however lost as the niches are brought closer together (δ is increased) as is shown in the next pair of graphs. Note not only the drop in system stability, but also the threatened extinction of P_2 (because of low population levels) for interactions only slightly out-of-phase ($\gamma \sim 1$), see Figure 2.

For widely separated niches ($\delta \sim 0$) the stability dependence on γ is reversed as is shown in the next graph (Figure 3). Thus, being *in*-phase is most advantageous for species with widely separated niches. The next pair of graphs below show the effects on stability of the amplitude λ and how they also qualitatively change for close versus widely separated niches (Figure 4). All of these computations were done with frequency $\omega = 1$.

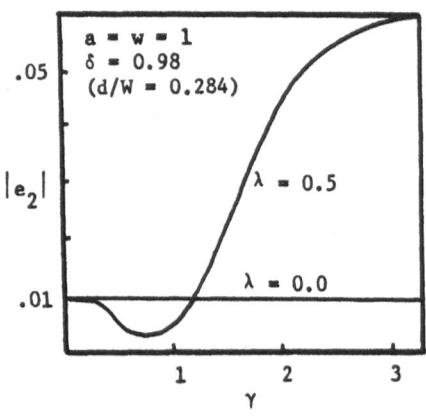 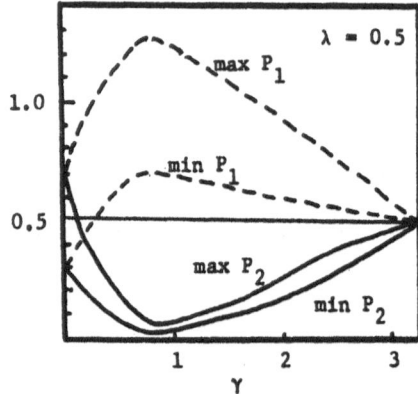

Figure 2

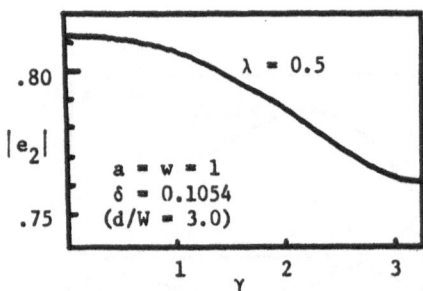

Figure 3

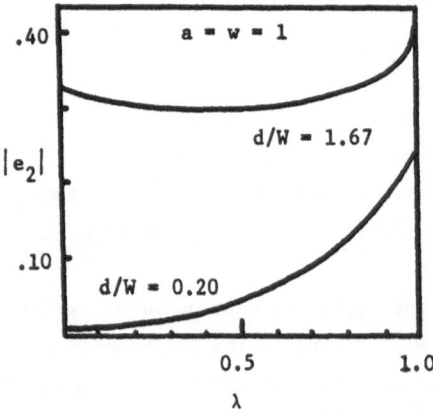 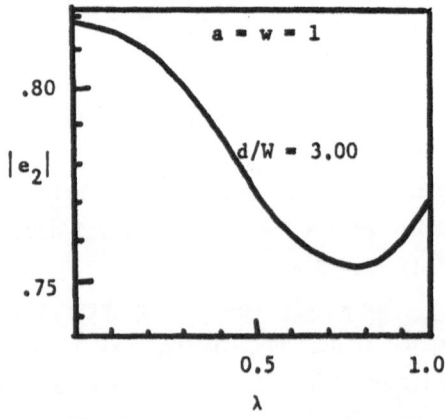

Figure 4

The final graph (Figure 5) shows the effect on stability of changes in the niches separation to width ratio d/W (for out-of-phase oscillations of maximum amplitude λ = 1) and allows a comparison with similar graphs of the classical case of constant coefficients λ = 0 (see Christiansen and Fenchel 1977, May 1974).

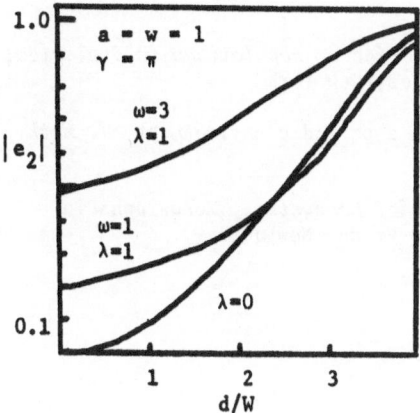

Figure 5

Thus, for frequency ω = 1 it is advantageous to periodically, out-of-phase utilize the resource niches (λ = 1, γ = π) only for niches sufficiently close together (d/W $\lesssim$ 2). For frequency ω = 3 the graphs for λ = 1 and λ = 0 were not found numerically to cross and hence it appears always advantageous to be out-of-phase. The "cross-over" of the graphs for λ = 1 and λ = 0 seems to be a complicated function of the frequence ω .

4. CONCLUSIONS

Holding other parameters constant, we considered periodicities in the resource utilization functions of two competing species modeled by the classical Lotka-Volterra-MacArthur-Levins theory. It was found for small niche separations (δ $\sim$ 1) that stable coexistence is enhanced when the periodic oscillations in the utilization functions are out-of-phase (γ = π) and are of maximum amplitude (λ = 1), i.e. the similar niches are "time shared". Short periods and large relative amplitudes also increased stability. These conclusions, however, may drastically alter and indeed be reversed for widely separated niches (δ $\sim$ 0).

It was also found that for close niches, small phase differences (γ $\sim$ 0) are disastrous for coexistence (actually worse than in-phase oscillations γ = 0), the species P_2 whose resource utilization function peaks earliest being threatened with extinction because of very low population levels. This rather unexpected result could have importance with regard to the possible evolution of time sharing in similar niches by two competing species who begin in a state of in-phase resource utilization. The above result would not allow a continuous evolution of the phase difference γ to an out-of-phase state without threatened extinction of one of the species.

REFERENCES

Christiansen, F.B., and T.M. Fenchel (1977): *Theories of Populations in Biological Communities*, Ecological Studies 20, Springer, New York.

Cushing, J.M. (1980): Two species competition in a periodic environment, *J. Math. Biology* 10:385-400.

May, R.M. (1974): *Stability and Complexity in Model Ecosystems*, Monographs in Population Biology 6, Princeton University Press, New Jersey.

NONOBLIGATE AND OBLIGATE MODELS OF MUTUALISM*

H.I. Freedman, John F. Addicott, and Bindhyachal Rai

1. INTRODUCTION

An interaction among organisms of different species is mutualistic, if the presence of each species enhances the per capita growth rate of the other. The dynamics of mutualistic systems are still poorly understood. Most studies of mutualism are directed at understanding the existence or evolution of mutualism rather than its dynamics, and theoretical ecologists have only begun to study mutualism intensively during the last eight years. However, there are many qualitatively different kinds of mutualistic systems (see Addicott 1983), and this will make understanding the dynamics of mutualism more difficult. Two factors contributing to the diversity of mutualistic systems are the number of species that must interact in order for there to be mutualism between two of them, and the degree of obligateness of the interaction. In this paper we address problems raised by incorporating these two factors into models of mutualism.

The number of species involved in a mutualistic interaction depends upon how each species benefits the other. Mutualistic benefits based upon modification of the abiotic environment or the direct transfer of nutrients from one organism to another require the interaction of just two species. Examples of 2-species mutualism include mycorrhizal associations, endozoic algae, lichens, symbiotic nitrogen fixation, gut faunas of termites and ruminants, and pollination (Henry 1966, 1967).

Mutualistic benefits arising from modification of predator-prey or competitive interactions involve interactions among at least three species. A mutualist may affect a predator-prey interaction to the benefit of either the predator or the prey (Addicott 1983), with the most common pattern being a mutualist deterring predation on a prey. For example, ants deter herbivores from feeding or plants (Bentley 1977) and deter predators from feeding on aphids (Addicott 1979), endozoic algae deter predators from feeding on protozoans (Berger 1980), and crustacea deter starfish from feeding on corals (Glynn 1976).

Mutualism based upon altering competitive interactions is less well known, probably reflecting the difficulty of detecting competition. Nevertheless, there are a number of examples of this type of mutualism. Springett (1968) describes a system involving the interaction of burying beetles, mites, and flies. The mites interfere with flies that compete with the larvae of the beetles for the resources in the burried carcass. Messina (1981) describes a case in which ants interfere with beetles that may compete with the membracids that the ants are tending. Wright

*Research for this paper was partially supported by the Natural Sciences and Engineering Research Council of Canada.

(1973) shows that, when a hydroid is present on the shells occupied by one species of hermit crab, it decreases the probability that another species of hermit crab will successfully compete for these shells. Quinlan and Cherret (1978) discuss mutualism between leaf cutting ants and fungi that the ants cultivate. The ants effectively remove competing fungi from the fungus gardens.

Mutualism also varies with respect to the degree to which one species is necessarily dependent upon the presence of the other for its existence. For example, the interaction between *Acacia* and *Pseudomyrmex* is obligate for both species (Janzen 1967), and many plants and their pollinators are obligate mutualists (e.g. Dodson et al. 1969). At the other extreme are those species that benefit from the presence of another species, but can persist in the absence of that species. Many temperate-zone mutualisms are facultative such as some ant-plant interactions (e.g. Inouye and Taylor 1978) and ant-aphid interactions (Addicott 1979). Whether mutualism is obligate or facultative for either or both species should have a significant effect upon the dynamics of a mutualistic system.

2. TWO SPECIES MODELS

2.1 *Nonoscillation.*

The most general model of 2-species mutualism modelled by autonomous ordinary differential equations, the so-called Kolmogorov model, and first given by Kolmogorov (1936) and discussed by Rescigno and Richardson (1967), Albrecht et al. (1974), Freedman (1980) is in fact a non-obligate model.

In May (1976, Chapter 4), there is some discussion of an obligate mutualism model. In either case, the model is given by

$$x_1' = x_1 f_1(x_1, x_2), \quad x_2' = x_2 f_2(x_1, x_2), \quad x_i(0) = x_{i0} > 0,$$

$$i = 1,2, \quad (' = \frac{d}{dt}) \tag{2.1}$$

If the assumption $(\partial f_i(x_1, x_2))/\partial x_i < 0, \quad x_1 > 0, \quad x_2 > 0, \quad i = 1,2$ is made, then it can be shown using Dulac's theorem (see Andronov, et al. 1973, Chapter 6) with Dulac function $x_1^{-1} x_2^{-1}$ that there are no nontrivial periodic solutions lying in the positive quadrant (see Goh, 1980). We will discuss implications of this in the rest of this section.

2.2 *Nonobligate model*

The hypotheses of the model as proposed by Kolmogorov (1936) and revised by Albrecht, et al. (1974) are not satisfied by the Lotka-Volterra model. We propose the following hypotheses for system (2.1).

(H1) $\qquad \dfrac{\partial f_i(x_1, x_2)}{\partial x_i} < 0, \quad \dfrac{\partial f_i(x_1, x_2)}{\partial x_j} > 0, \quad i,j = 1,2, \quad i \neq j.$

(H2)
$$\exists K_1 > 0, \ K_2 > 0 \ \ni \ (x_1-K_1)f_1(x_1,0) < 0$$

$$(x_2-K_2)f_2(0,x_2) < 0, \ x_1 > 0, \ x_2 > 0, \ x_i \neq K_i.$$

Note that if (H1) and (H2) are satisfied, by the implicit function theorem, $f_i(x_1,x_2) = 0$ can be solved for $x_2 = \phi_i(x_1)$, $i = 1,2$, such that $\phi_1(K_1) = 0$ and $\phi_2(0) = K_2$, giving the existence of the isoclines. Further, since $\phi_i'(x_1) =$

$- \dfrac{\partial f_i/\partial x_1}{\partial f_i/\partial x_2} > 0$, the isoclines are monotonically increasing functions which may or may not intersect. If they do not intersect, then all solutions enter the region between them and become unbounded. Hence it is reasonable to determine criteria for the intersection of $x_2 = \phi_i(x_1)$, $i = 1,2$.

Define $\phi_1(x_1) \equiv 0$ on $0 \leq x_1 \leq K_1$, define $\psi(x_1) = \phi_2(x_1) - \phi_1(x_1)$, and define

$$F(x_1) = \frac{\partial f_1(x_1,\phi_1(x_1))/\partial x_1}{\partial f_1(x_1,\phi_1(x_1))/\partial x_2} - \frac{\partial f_2(x_1,\phi_2(x_1))/\partial x_1}{\partial f_2(x_1,\phi_2(x_1))/\partial x_2} . \tag{2.2}$$

Then clearly $\psi'(x_1) = F(x_1)$ and

$$\psi(x_1) = K_2 + \int_0^{x_1} F(u)\,du.$$

Now since $\psi(0) > 0$, if

$$\lim_{x_1 \to \infty} \int_0^{x_1} F(u)\,du < -K_2 \tag{2.3}$$

then there is an equilibrium. If, in addition to (2.3) $F(u)$ is negative, then the equilibrium is unique. By computing the variational matrix, it is readily seen that this equilibrium is asymptotically stable. It is also easy to see that all solutions are bounded in the case of a unique equilibrium, which in the absence of nontrivial periodic solutions implies that it is globally asymptotically stable.

2.3 *Obligate model.*

So as to make system (2.1) a model of obligate mutualism, we propose the following hypotheses in addition to (H1).

(H3)
$$f_1(x_1,0) < 0, \qquad f_2(0,x_2) < 0$$

(H4)
$$\exists L_1(x_1) > 0, \ L_2(x_2) > 0, \ L_i(0) \geq 0, \quad i = 1,2,$$

$$f_1(x_1,L_1(x_1)) = f_2(L_2(x_1),x_2) = 0,$$

Hypothesis (H3) models the obligateness of the mutualism. Either population will go extinct in the absence of the other.

(H4) states that for each positive value of either population, there is a positive value of the other population which reverses the growth rate from negative to positive, i.e. the mutualism "works."

$x_2 = L_1(x_1)$ and $x_1 = L_2(x_2)$ are isoclines. If they do not intersect, then the mutualism is not effective, and all solutions go to zero. If there is only one intersection of the isoclines, then the corresponding equilibrium will be a saddle point, and solutions (except those initiating on the stable manifold) will either tend to zero or to infinity. Hence for a realistic model, we will require a second intersection of the isoclines giving a second equilibrium, $\hat{E}(\hat{x}_1, \hat{x}_2)$. $\hat{E}$ will be asymptotically stable. If there are no more intersections of the isoclines, then $\hat{E}$ is asymptotically stable in the large. See May (1980, Chapter 4, Figure 4.5) for typical such isoclines.

3. THREE SPECIES MODELS

In Addicott and Freedman (1983), we considered the interactions of predator-prey populations and competitive populations with a slow-growing mutualist. Here we consider the case where the dynamics of all three species must be considered. We take the point of view of obligate versus nonobligate mutualism. Details and proofs of the results cited here are found in Rai, et al. (1983).

3.1 *Predator-prey-mutualist.*

We propose as a model of predator and prey populations interacting with a mutualist (with respect to the prey) population the system

$$
\begin{aligned}
u' &= uh(u,x) \\
x' &= \alpha x g(u,x) - y p(u,x) \\
y' &= y(-x + c p(u,x)),
\end{aligned}
\tag{3.1}
$$

where u, x, y are the mutualist, prey and predator population numbers respectively. $h(u,x)$ is the specific growth rate of the mutualist and is a decreasing function of u, and an increasing function of x. $g(u,x)$ is the specific growth rate of the prey population and is a decreasing function of x. $p(u,x)$ is the predator functional response and is a decreasing function of u and an increasing function of x.

Since, in the absence of mutualism, under reasonable hypotheses the predator will not drive the prey (its food) extinct (see Freedman, 1980, Chapter 4), system (3.1) is viewed as a nonobligate model.

It is shown in Rai, et al. (1983), that under certain circumstances, the presence of the mutualist can substantially lower the predator population numbers and in some cases even drive the predator extinct.

3.2 *Competitor-competitor-mutualist.*

We propose as a model of two competing populations and a mutualist (to one of them) population the following system

$$u' = uh(u,x_1)$$
$$x_1' = \alpha x_1 [g_1(u,x_1) - q_1(u,x_1,x_2)] \qquad (3.2)$$
$$x_2' = x_2 [g_2(x_2) - q_2(x_1,x_2)],$$

where h and g_i are as in system (2.1), and where the competition functions q_i are increasing functions of x_1 and x_2, but q_1 is a decreasing function of u. Here u is a mutualist of x_1.

Since, in the absence of mutualism, x_1 may or may not be driven extinct by x_2, the mutualism may be obligate or nonobligate.

It is shown in Rai, et al. (1983), that the mutualism can reverse competitive outcome, i.e. from x_1 always going extinct to x_1 sometimes going extinct to x_1 never going extinct, and in these cases the mutualism is truly obligate.

REFERENCES

Addicott, J.F. (1979): A multispecies aphid-ant association: density dependence and species-specific effect, *Can. J. Zool.* 57:558-569.

_____ (1981): Stability properties of 2-species models of mutualism: Simulation studies, *Oeciologia* 49:42-49.

_____ (1983): Mutualistic interactions in population and community processes. *In* Price, P.W., C.N. Slobodchikoff and B.S. Gaud (Eds.). *A New Ecology: Novel Approaches to Interactive Systems.* John Wiley and Sons, New York, (in press).

Addicott, J.F., and H.I. Freedman (1983): On the structure and stability of mutualistic systems: Analysis of predator-prey and competition models as modified by the action of a slow-growing mutualist, (manuscript).

Albrecht, F., H. Gatzke, A. Haddad, and N. Wax (1974): The dynamics of two interacting populations, *J. Math. Anal. Appl.* 46:658-670.

Andronov, A.A., E.A. Leontovich, I.I. Gordon, and A.G. Maier (1973): *Qualitative Theory of Second Order Dynamic Systems*, Wiley, New York.

Bentley, B.L. (1977): Extrafloral nectaries and protection by pugnacious bodyguards, *Annu. Rev. Ecol. Syst.* 8:407-427.

Berger, J. (1980): Feeding behaviour of *Didinium nasutum* on *Paramecium bursaria* with normal or apochlorotic zoochlorellae, *J. Gen. Microbiol.* 118:397-404.

Dodson, C.H., R.L. Dressler, G.H. Hills, R.M. Adams, and N.H. Williams (1969): Biologically active compounds in orchid fragrances, *Science* 164:1243-1249.

Freedman, H.I. (1980): *Deterministic Mathematical Models in Population Ecology*, Marcel Dekker Inc., New York.

Glynn, P.W. (1976): Some physical and biological determinants of coral community structure in the eastern Pacific. *Ecol. Monogr.* 46:431-456.

Goh, B.S. (1976): Global stability in two species interactions, *J. Math. Biol.* 3:313-318.

_____ (1979): Stability in models of mutualism, *Amer. Natur.* 113:261-275.

_____ (1980): *Management and Analysis of Biological Populations*, Elsevier, New York.

Henry, S.M. (ed.), (1966): *Symbiosis, Vol 1. Associations of Microorganisms, plants, and marine organisms*, Academic Press, New York.

_____ (ed.), (1967): *Symbiosis, Vol. 2. Associations of Invertebrates, Birds, Ruminants, and other Biota*, Academic Press, New York.

Inouye, D.W., and O.R. Taylor, Jr. (1979): A temperate region plant-ant-seed predator-system: Consequences of extra floral nectar secretion by *Helianthella quinquenervis*, *Ecology* 60:1-7.

Janzen, D.H. (1967): Fire, vegetation structure, and the ant X acacia interaction in Central America, *Ecology* 48:26-35.

Kolmogorov, A.N. (1936): Sulla teoria di Volterra della lotta per l'esisttenza, *Gior. Instituto Ital. Attuari* 7:74-80.

May, R.M. (1976): *Theoretical Ecology Principles and Applications*, W.B. Saunders, Philadelphia.

Messina, F.J. (1981): Plant protection as a consequence of an ant-membracid mutualism: Interactions on goldenrod (*Solidago* sp.), *Ecology* 62:1433-1440.

Quinlan, R.J., and J.M. Cherret (1978): Aspects of the symbiosis of the leaf-cutting ant *Acromyrmex octospinosus* (Reich) and its fungus food, *Ecol. Entomol.* 3:221-230.

Rai, B., H.I. Freedman, and J.F. Addicott (1983): Analysis of three species models of mutualism in predator-prey and competitive systems, *Math. Biosci.* (to appear).

Rescigno, A., and I.W. Richardson (1967): The struggle for life, I: Two species, *Bull. Math. Biosphys.* 29:377-388.

Springett, B.P. (1968): Aspects of the relationship between burying beetles, *Necrophorus* spp. and the mite, *Poecilochirus necrophori* Vitz. *J. Anim. Ecol.* 37:417-424.

Tilman, D. (1978): Cherries, ants and tent caterpillars: Timing of nectar production in relation to susceptibility of caterpillars to ant predation, *Ecology* 59:686-692.

Vance, R.R. (1978): A mutualistic interaction between a sessile marine clam and its epibionts, *Ecology* 59:679-685.

Vandermeer, J.H., and D.H. Boucher (1978): Varieties of mutualistic interaction in population models, *J. Theor. Biol.* 74:549-558.

Way, M.J. (1963): Mutualism between ants and honeydew-producing Homoptera, *Ann. Rev. Entomol.* 8:307-344.

Wright, H.O. (1973): Effect of commensal hydroids on hermit crab competition in the littoral zone of Texas, *Nature (Lond.)* 241:139-140.

STABILITY OF COMMUNITY INTERACTION MATRICES

Harold M. Hastings

ABSTRACT

We sketch a conceptual, self-contained proof (Hastings, 1982a) of R.M. May's (1972) stability theorem for randomly assembled linear systems. Several ecological consequences follow. First, ecological constraints on interaction matrices (May, 1974; L.R. Lawlor, 1978) as well as organization into loosely coupled subsystems (cf. May, 1972) tend to enhance stability. Secondly, scaling results on interaction strength reconcile May's stability criterion with R.H. MacArthur's (1955) thesis that the existence of multiple energy pathways enhances stability. Finally, we analyze the effect of noise in these models.

1. INTRODUCTION

M.R. Gardner and W.R. Asbhy (1970) observed that large, randomly assembled linear (cybernetic) systems are less likely to be stable than smaller systems. R.M. May (1972) supplied a precise statement and proof, using E.P. Wigner's (1959) statistics of random matrices. These results appeared to challenge the classical wisdom that more complex ecosystems tend to be more stable than simpler systems, and MacArthur's (1955) explanation that stability depends upon multiple energy pathways in complex systems. In addition, Lawlor (1978) and May (1974) observed that ecological constraints such as shortness of food chains made real ecological interaction matrices far from random.

We (Hastings, 1982a) provided a simpler proof of the May-Wigner stability theorem (see Section 2, below) which allows analysis (cf. Hastings, 1982b) of the role of ecological constraints. These constraints tend to make real ecosystems more stable than random systems; see Section 4, below.

We also argue that replacing one interaction (say, predator eating prey) by a family of similar interactions, with consequent adjustments in interaction strength tends to enhance stability. This reconciles MacArthur's (1955) thesis that multiple energy pathways enhance stability with May's stability theorem.

Finally, our proof of the May-Wigner theorem yields a straight-forward analysis of the role of Gaussian noise upon dynamics of these model ecosystems near equilibrium.

We acknowledge helpful discussions with Drs. D. Cohen, R.M. May and G. Sugihara. These results were also announced at the Oak Ridge Food Web Workshop (Hastings, 1983).

2. THE MAY-WIGNER STABILITY THEOREM

We state and sketch a proof of the May-Wigner Stability Theorem for difference equations. Consider a community of n interacting species. Let x; denote the difference between the population level of the 1st species and its equilibrium value. Let $\underline{x} = (x_1, x_2, \ldots, x_n)$, or its transpose when appropriate. Let $M = (M_{ij})$ be the community interaction matrix; M_{ij} represents the effect of species j upon species i. The dynamics of such a community, near equilibrium, may be represented by the difference equation

$$\underline{x}(t+1) = M\underline{x}(t) \tag{1}$$

Random interaction matrices are parameterized by the *mean square interaction strength* α^2, the *size* n, and the *connectance* C. C denotes the fraction of the entries in M that are not zero; these entries are located independently so that the number of non-zero entries in each column represents a sample from a binomial distribution with variance α^2 and fourth moment of order α^4.

MAY-WIGNER STABILITY THEOREM *Let M be an $n \times n$ matrix of connectance C and mean square interaction strength α^2. Let $P(\alpha, n, C)$ be the probability that the corresponding differential system (1) has a stable equilibrium at 0. Let $\varepsilon > 0$. Then $P(\alpha, n, C) \to 1$ as $n \to \infty$ provided $\alpha^2 nC < 1 - \varepsilon$; conversely, $P(\alpha, n, C) \to 0$ as $n \to \infty$ for $\alpha^2 nC > 1 + \varepsilon$.*

We announce here a direct proof for matrices with connected underlying graphs. More precisely, the underlying graph of M (with one edge joining i and j if M_{ij} or M_{ji} is non-zero) is asymptotically almost surely connected if $C \geq (1+\varepsilon)\log n/n$, and asymptotically almost surely not connected if $C \leq (1-\varepsilon)\log n/n$ for any fixed positive ε (Bollobas, 1979, p. 143). We assume the former condition holds; in particular the theorem holds for any constant C.

Outline of proof (Hastings, 1982a). For motivation recall the Gerschgorin circle theorem: if every column sum (sum of absolute values of entries in a column) of M is less than 1, then all eigenvalues of M are less than 1 is size, and the system of difference equations $\underline{x}_{t+m} = M\underline{x}_t$ is stable.

In our case the expected *Euclidean* norm of each column is $\alpha\sqrt{nc}$, so we compute $\|M\underline{v}\|^2$ for all *unit* vectors $\underline{v}$:

$$\|M\underline{v}\|^2 = \sum_{j=1}^{n} \|\underline{M}_j\|^2 v j^2 + \sum_{\substack{j,k=1 \\ j \neq k}}^{n} (\underline{M}_j \cdot \underline{M}_k) v_j v_k \tag{2}$$

Here $\underline{M}_j$ denotes the jth column of M, and $\|\cdot\|$ denotes the Euclidean norm.

Assume $\alpha^2 nC < 1 - \varepsilon$. Then asymptotically almost surely each column of M has at most $(1+\varepsilon/2)nC$ non-zero entries, whose mean square is bounded by $(1+\varepsilon/2)\alpha^2$. Thus the first sum in (2) is asymptotically almost surely bounded by

$1 - 3\varepsilon^2/4$. In the second sum in (2), each term has mean 0 and variance $\alpha^4 nc^2 < 1/n$. Since $\sum(v_j v_k)^2 < (\sum v_j^2)^2 = 1$, the second sum is of order $1/n$, and thus asymptotically almost surely, for all unit vectors $\underline{v}$, $\|M\underline{v}\|^2 < 1$. Stability follows.

Instability in the case $\alpha^2 nC > 1 + \varepsilon$ is shown similarly.

3. THE EFFECT OF NOISE

We extend the basic model (1) to include noise (random fluctuations) as follows:

$$\underline{x}(t+1) = m\underline{x}(t) + \Delta\underline{w}(t), \qquad (3)$$

where the increments $\Delta\underline{w}(t)$ are selected independently from a symmetric distribution of fixed variance σ^2. We now compute the effect of these fluctuations upon the distribution of $\underline{x}(t)$.

Let M be a random matrix as defined above. Our proof of the May-Wigner stability theorem shows that multiplication by M is expected to multiply Euclidean lengths by $\alpha\sqrt{nC}$. If $\alpha^2 nC < 1$, for large t,

$$\underline{x}(t+1) = \Delta\underline{w}(t) + M\Delta\underline{w}(t-1) + M^2\Delta\underline{w}(t-2) + \ldots + M^{t-1}\Delta\underline{w}(1) + M^t\underline{x}(0)$$

$$\approx \sum_{r=0}^{t-1} M^r\Delta\underline{w}(t-r). \qquad (4)$$

Thus, since the increments $\Delta\underline{w}(t)$ are independent, for large t, $\underline{x}(t+1)$ has expected value 0 and variance $\sigma^2/(1-\alpha^2 nC)$. (If, in addition, the increments are bounded, say by b, then $\|\underline{x}(t+1)\|$ is bounded by $b/(1-nC)$, for large t.)

Thus, if $\alpha^2 nC$ is small, this variance has order σ^2 and is unlikely to cause the system to crash in ecological time; however, if $\alpha^2 nC$ is near 1, $\underline{x}(t+1)$ has variance $\gg \sigma^2$ and crashes become likely.

4. DISCUSSION

We have outlined a simple, conceptual proof of the May-Wigner stability theorem, and used the proof to analyze the effect of noise upon fluctuations about the equilibria. We conclude by briefly summarizing two additional ecological consequences of a detailed analysis of the May-Wigner Theorem.

(i) *Multiple energy pathways.* Increases in connectivity can increase stability provided that they yield sufficient decreases in interaction strength. Suppose, following MacArthur's (1955) discussion of multiple energy pathways, that the mean interaction strength, α, is inversely proportional to the mean number of interactions per species, nC. Then $\alpha^2 nC \propto 1/(nC)$, and increasing complexity increases stability! Cf. Hastings (1979).

(ii) *Ecological constraints*. Lawlor (1978) and May (1972, 1974) observed that community interaction matrices are far from random because of ecological constraints involving the length of food chains, no loops conditions, organization into loosely compiled subsystems, etc. These constraints tend to enhance stability for two reasons. They reduce the variance in the number of non-zero elements per column, and also reduce the size of the "covariance" or "interaction" terms $\underline{M}_j \cdot \underline{M}_k$ in (2). Thus the asymptotic stability results hold for relatively small values of n. Cf. Hastings (1982b) for further ecological discussion.

REFERENCES

Bollobás, B. (1979): *Graph Theory*, Graduate Texts in Math., Vol. 63, Springer, New York.

Gardner, M.R., and W.R. Ashby (1970): Connectance of large dynamic (cybernetic) systems: critical values for stability, *Nature* 228, 784.

Hastings, H.M. (1979): Stability considerations in community organization, *J. Theoret. Biol.* 78: 121-127.

_____ (1982a): The May-Wigner stability theorem for connected matrices, *Bull. Amer. Math. Soc.* (new series) 7:387-388.

_____ (1982b): The May-Wigner stability theorem, *J. Theoret. Biol.* 97: 155-168.

_____ (1983): Stability of community interaction matrices, Food Web Workshop, ed. De Angelis, D.L., Post, M., and Sugihara, G., Oak Ridge National Laboratory, *Technical Report* (in press).

MacArthur, R.H. (1955): Fluctuations of animal populations and a measure of community stability, *Ecology* 36:533-536.

May, R.M. (1972): Will a large complex system ever be stable? *Nature* 238:413-414.

_____ (1974): *Stability and Complexity in Model Ecosystems*, Princeton U. Press, Princeton.

Wigner, B. (1959): Statistical properties of real symmetric matrices with many dimensions, in *Proc. Fourth Canad. Math. Cong.*, ed. MacPhail, M.S., U. Toronto Press, Toronto, pp. 174-184.

BIOMASS FLOW, STRUCTURE AND STABILITY OF MODEL ECOLOGICAL SYSTEMS

A. Porati, M.I. Granero-Porati and R. Kron-Morelli

1. INTRODUCTION

The problem of the stability of mathematical models describing interacting populations has attracted, in the last years, the attention of many authors.

In particular the stability properties of model ecosystems have been investigated (both from the analytical and the numerical point of view), as a function of various parameters: number of species, connectedness, interaction strength (let us only recall the works of Gardner and Ashby (1970), May (1972), Roberts (1974), Siljak (1975), Tregonning and Roberts (1978)).

More recently it has been recognized that completely random constructed food webs can lead to biological absurdities. We recall, for example, the 3-species loop, in which species 1 feed on species 2 which feed on 3 which feed on 1 (Gilpin, 1975).

In this sense some attempts have been made (Kirkwood and Lawton (1981), Granero-Porati et al. (1982)) to construct "quasi random" models, that is models in which the interactions among species are not completely random, but subject to particular selection rules, in order to avoid biological absurdities.

This work is concerned with the numerical analysis of model ecosystems with structure. We focus, in particular, upon the relation: "biomass conversion efficiency - feasibility-stability".

2. BIOMASS CONVERSION EFFICIENCY

Our starting point is the well-known Generalized Lotka-Volterra (G.L.V.) equations regarding the temporal behaviour of a prey-predator community, i.e.:

$$\dot{N}_i = \beta_i N_i + \sum_{j=1}^{n} \alpha_{ij} N_i N_j \quad (i = 1,\ldots,n) \tag{1}$$

where N_i is the number of individuals, β_i is the autoincrease rate, α_{ij} $(i \neq j)$ the interaction terms and α_{ii} the selfregulation term of the i-th species.

If we consider the equations relating the biomasses M_i, we have:

$$\dot{M}_i = b_i M_i + \sum_{j=1}^{n} a_{ij} M_i M_j \quad (i = 1,\ldots,n) \tag{2}$$

Equations (2) are connected with (1) by the following relations:

$$\beta_i = b_i \qquad \alpha_{ij} = a_{ij} m_j \quad (\forall i,j) \tag{3}$$

where m_j is the mean mass of individuals belonging to the j-th species.

Equation (2) permits the introduction of another parameter, very interesting from the pure biological point of view, i.e. the so-called "Biomass Conversion Efficiency" γ (as introduced by De Angelis (1975)), that is the efficiency of biomass transfer between predator and prey. The parameter γ is defined as follows:

$$\gamma = \frac{|a_{ij}|}{|a_{ji}|} = \frac{|\alpha_{ij}m_i|}{|\alpha_{ji}m_j|} \tag{4}$$

where the i-th species preys on the j-th one.

We want now to examine the effect of γ on the stability of ecosystems, the structure of which we exposed in a recent work (Granero-Porati et al., 1982). Let us briefly recall the main features of our model. We consider n interacting species (the interactions are assumed to be of the prey-predator type), divided in two separated groups: p prey and n-p predators (p is an integer ranging from 1 to n-1). The "selection rules" we impose are the following:

(a) Prey are self-regulated and do not interact among themselves.

(b) Predators are without self-regulation and interactions among predators are random; the only obvious condition is

$$\text{sgn } a_{ij} = -\text{sgn } a_{ji} \quad (i,j = p+1,\ldots,n; \ i \neq j).$$

In this way, the graph (or, more precisely, the digraph) representing food webs is *always* connected, and the connectedness C (that is the number of existing branches over the total number of possible branches) is:

$$C = [1 - \frac{p(p-1)}{n(n-1)}] \tag{5}$$

The coefficients b_i and a_{ij} are chosen in this way:

$b_i = 0.1 \quad$ for $i = 1,\ldots,p \quad$ (prey)

$b_i = -0.01 \quad$ for $i = p+1,\ldots,n \quad$ (predators)

$a_{ii} = -0.01$.

The a_{ij} are drawn from a rectangular distribution $[-1,0]$ for $i < j; |a_{ji}| = |\gamma a_{ij}|$.

We consider now the influence of γ on the stability of the model ecosystem.

3. THE GENERAL CASE

We make some attempts on "quasi-random" systems (in the sense previously described), mainly to compare our results with the ones of previous authors (Kirkwood and Lawton, 1981).

4. THE CASE OF p = n - 1

In this case, as one can see from (5) the connectedness C assumes its minimum value, $C = 2/n$, and the number of branches of interaction is $n-1$.

We have already demonstrated (Granero-Porati et al., 1981) that the equilibrium point of a G.L.V. system of type (1) is, if feasible, *asymptotically stable* when the number of branches equals $n-1$ (i.e., when the representative graph is a *tree*).

We have examined the behaviour of the feasibility of systems of this kind (and, we recall, the number of feasible systems *must* correspond to the number of *stable* systems), for various values of γ, and for $n = 2,3,4,5$. The results are summarized in Figure 2. (The values of γ are the following:

$$\gamma = 0.0025, \ 0.0050, \ 0.0075, \ 0.010, \ 0.025$$
$$0.05, \ 0.075, \ 0.1, \ 0.25, \ 0.50, \ 0.75, \ 1).$$

It is easily seen that, for $n = 3,4,5$ the number of feasible (and stable as well) samples decreases monotonically when γ increases. When $n = 2$, the behaviour is completely different, and is in good agreement with a similar result reported by Kirkwood and Lawton (1981).

It must be noted that the case of $n = 2$ is a particular case of *simple alimentary chain*. This result led us to examine longer alimentary chains, with the aim of investigating if we are faced with a general behaviour or with an exception.

5. THE CASE OF ALIMENTARY CHAINS

First of all we note that in the graph representing alimentary chains are present n points and $n-1$ branches: the theorem of stability of feasible samples is consequently valid. We studied alimentary chains with $n = 2,3,4$. The range of γ is the same as reported. The results are summarized in Figure 3. One can see that (for three cases at least), the behaviour is very general: the percentage of stable samples increase monotonically when γ increases.

This behaviour can also be justified from the analytical point of view. In fact for $n = 3$ it follows that the equilibrium point is feasible (and, consequently, stable) when these two functions are positive:

$$x_1(\gamma) = A\gamma - B$$
$$x_2(\gamma) = C\gamma - D$$

For $n = 4$ we have feasibility when these following two functions are positive:

$$y_1(\gamma) = E\gamma^2 - F\gamma - G$$
$$y_2(\gamma) = H\gamma^2 - K\gamma - J$$

In all cases, we founded a behaviour like the one reported in Figure 1, that is an "optimal value" of γ for the homeostaticity (i.e. feasibility and stability) of the system.

It must too be noted that, in general, the feasibility of the system decreases when γ increases, but the percentage of stable samples over the feasible ones have a sharp peak.

One important exception occurs when the ecosystem is composed of n-1 prey and only one predator.

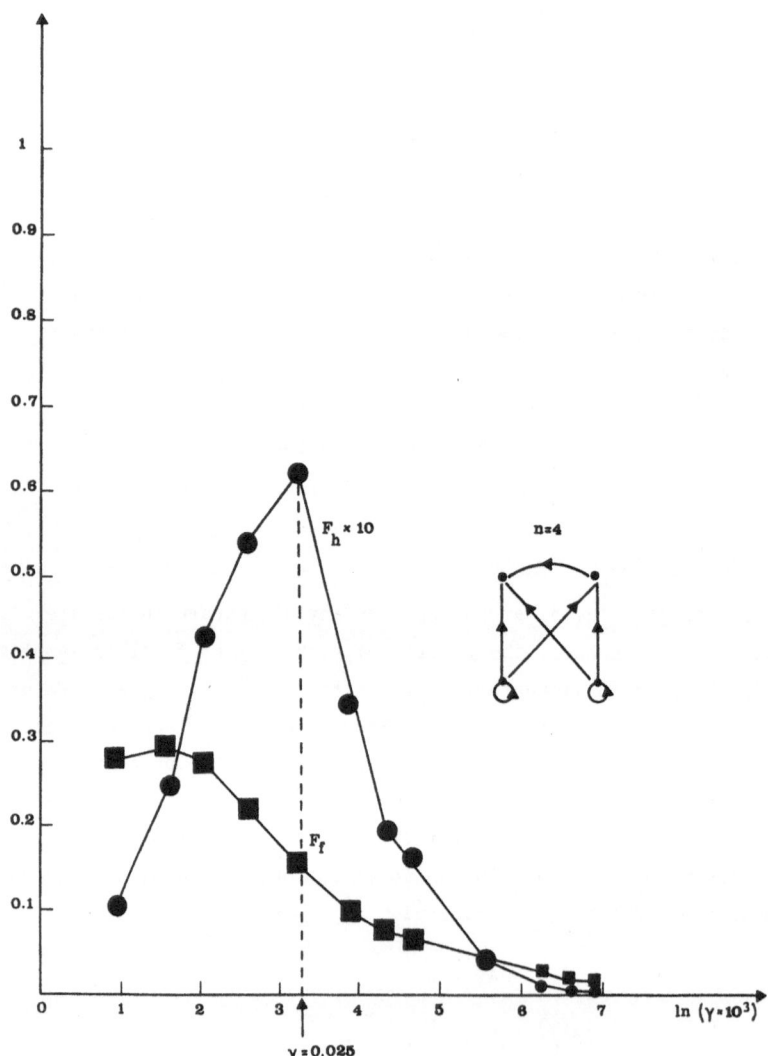

Figure 1 Frequency of feasible samples (F_f) and of homeostatic samples (F_h) versus biomass conversion efficiency (γ) in the case of 4 species (2 prey and 2 predator).

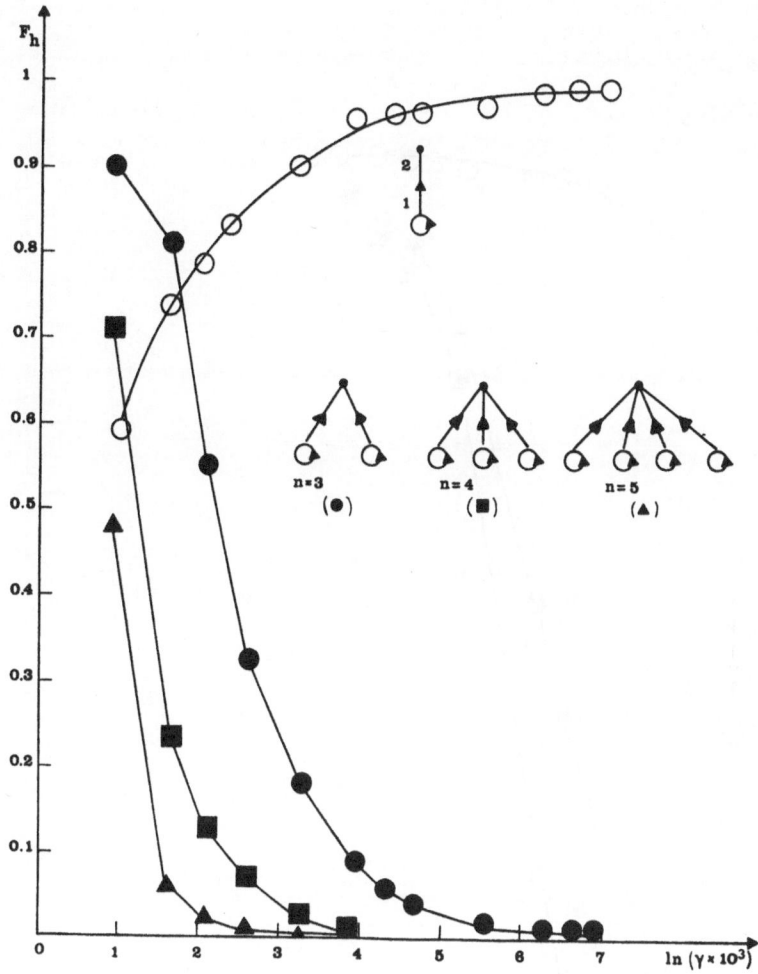

Figure 2 Frequency of feasible (and also homeostatic) samples versus γ in the case of p = n - 1 (n = 2,...,5).

(All the coefficients A,...,J are positive functions of random variables).

We also began to study the influence, on the same alimentary chain, of different values of γ for different trophic levels. In the case of n = 3 we selected $\gamma_1 = a_{21}/a_{12} = 0.1$ and $\gamma_2 = a_{32}/a_{23} = 0.4$.

The values of γ_i was then inverted ($\gamma_1 = 0.4$, $\gamma_2 = 0.1$), to investigate how the homeostaticity of the system is affected by different biomass conversion efficiency from level to level.

The preliminary results show that the percentage of homeostatic systems, F_h, is, in the first case, $F_h = 0.86$, and, in the second case, $F_h = 0.49$. This result (very preliminary, we recall) indicates that an alimentary chain is more stable when the biomass conversion efficiency is higher for higher trophic levels.

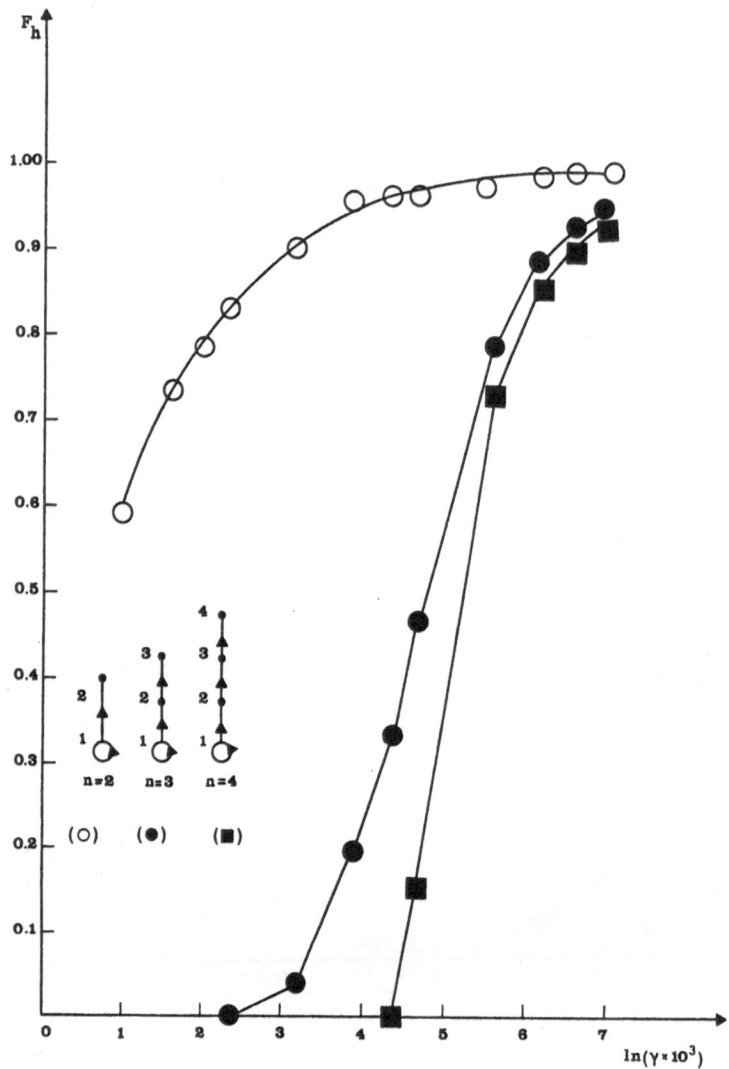

Figure 3   Frequency of feasible (and also homeostatic) samples in the case of
simple alimentary chains.

REFERENCES

De Angelis, D.L. (1975): Stability and connectance in food web models, *Ecology*
56:238-243.

Gardner, M.R., and W.R. Ashby (1970): Connectedness of large dynamical
(cybernetic) systems: critical values for stability, *Nature* 228:784.

Gilpin, M.E. (1975): Stability of feasible predator-prey systems, *Nature* 254:
137-139.

Granero-Porati, M.I., R. Kron-Morelli, and A. Porati (1981): Stability of model systems describing prey-predator communities, in *Quantitative Population Dynamics*, Statistical Ecology Series, Vol. 13:139-150, International Cooperative Publishing House.

_____ (1982): Random ecological systems with structure: stability-complexity relationship, *Bulletin of Mathematical Biology* 44:103-117.

Kirkwood, R.S.M., and J.H. Lawton (1981): Efficiency of biomass transfer and stability of model food-webs, *Journal of Theoretical Biology* 93:225-237.

Roberts, A. (1974): The stability of feasible random ecosystem, *Nature* 251:607-608.

Siljak, D.D. (1975): When is a complex ecosystem stable? *Mathematical Biosciences* 25:25-50.

Tregonning, K., and A. Roberts (1978): Ecosystem-like behaviour of random interaction model - I, *Bulletin of Mathematical Biology* 40:513-524.

IS DYNAMICAL SYSTEMS THEORY THE BEST WAY
TO UNDERSTAND ECOSYSTEM STABILITY?

William Silvert

ABSTRACT

Ecosystem models are usually based on the methodology of dynamical systems theory. This approach, although powerful, requires a high level of sophistication on the part of the modeller and can require an astonishing degree of complexity to represent effects which are biologically straightforward. Other ways of modelling ecosystems exist which are simpler and appear to correspond more closely to meaningful interpretations of ecological concepts such as stability. It may be more efficient to use these methods for ecological modelling in some circumstances as an alternative to dynamical systems theory.

1. INTRODUCTION

Models of ecosystems are often based on systems of first order differential equations, which involves an implicit assumption that ecosystems are dynamical systems and are thus describable by this methodology (Rosen 1970). It follows that many of the conceptual abstractions which play a basic role in dynamical systems theory, such as the various definitions of stability (Goh 1980), should correspond in some transparent way to similar concepts of concern to ecologists.

In many cases this approach is inefficient and produces models which are at best more complicated than they need to be, and at worst leads to incorrect conclusions. For example, systems in which threshold behaviour occurs are often very difficult to represent in terms of differential equations; even when it is technically possible to do so, it is usually simpler to break the trajectory into separate segments of finite length and to ignore the dynamics of the switching process at the points where these segments join.

The first part of the paper deals with the dynamics of systems regulated by on-off controllers and the problem of defining concepts like stability for such systems. The second part deals with the stability of complex food webs, and suggests that the uncritical use of the dynamical systems theory approach has led to results which may be completely at variance with reality.

2. DIFFICULTIES WITH DYNAMICAL SYSTEMS THEORY

The definition of dynamical systems theory used in this paper is based on Rosen's (1970) relatively strict definition of a dynamical system as one described by a set of first-order differential equations. The traditional way to analyze such systems is by studying their trajectories in phase space, and concepts like stability play a major role in the theory. Although this categorization is a narrow one (it excludes discrete systems, for example), it seems to encompass much of the theoretical work carried out in ecology.

It is by no means clear that the best way to understand an ecosystem is by following the complete trajectory of its development through time; for example, there is no convincing evidence that following community dynamics over the winter season contributes to our ability to predict the magnitude of spring algal blooms.

Biologically reasonable models are not always easy to express as systems of differential equations, especially if they involve discontinuous feedback or other types of nonlinear interaction. As a result, important mechanisms may have to be omitted to fit the Procrustean bed of dynamical systems theory. We may develop better models if we try to develop methods which fit realistic models, rather than modifying the models to fit our methods.

3. MODELS WITH SWITCHING

Many ecosystems contain components which change behaviour abruptly in response to changes in the physical or biological environment. For example, many predators either switch to another prey or even stop feeding altogether when the abundance of their preferred food falls below some threshold level. Other organisms change their behaviour in a discontinuous way in response to changes in light intensity or temperature. Such effects can be treated as limiting cases of continuous change, but to do so often leads to models which are unnecessarily complicated.

The modelling of an ecosystem in which one or more of the important components exhibits switching behaviour is analogous to the modelling of a heating system controlled by a thermostat. This system is not difficult to model if we ignore the dynamics of the switching process, and it is instructive to see some of the problems which can arise if it is treated as a continuous dynamical system.

If T is the temperature of a house, then its change with time can be described by a differential equation such as

$$dT/dt = H - k(T-T_{ext})$$

where H is the heating rate and T_{ext} is the outside temperature. This is a typical dynamical systems equation, but the second variable, H, varies discontinuously and takes only the two values 0 (furnace off) and H_{max} (furnace on). This cannot be represented by a differential equation (unless we use generalized functions, which seems like overkill for such a simple system); we can however write a set of equations which say that the furnace is on if it is cold,

$$H = H_{max} \quad if \quad T < T_1$$

and off if it is warm,

$$H = 0 \quad \text{if} \quad T > T_2$$

while in the intermediate range $T_1 \leq T \leq T_2$ a differential equation is required,

$$dH/dt = 0.$$

These equations are easy to analyze even when T_{ext} changes with time, but it is not easy to carry out the analysis in the language of dynamical systems theory. The trajectory of the system is only piecewise continuous, and the state of the system at the instant of switching is not defined. Since the furnace is either on or off, the system has two equilibrium states; when the furnace is off the temperature approaches $T = T_{ext}$, and when the furnace is on the temperature approaches $T = T_{ext} + H_{max}/k$. However, during a typical winter the system may rarely approach either one of these two equilibria, even though both are stable. It is far more relevant to consider the problem of maintaining the temperature at some intermediate value T_{opt} by turning the furnace on and off. At T_{opt} the temperature is always either rising or falling, so this is definitely not a stable point (it could be considered the centre of a stable limit cycle, but even this more sophisticated approach is tenable only when the external temperature is held constant, and it obscures the essential fact that a thermostatically controlled heating system maintains the same temperature range even if the external temperature changes, so long as it remains between T_{opt} and $T_{opt} - H_{max}/k$). We can make the range $T_2 - T_1$ as small as the maximum switching rate of the furnace allows without ever achieving a stable state. But what does this instability really mean? Suppose that $T_{opt} = 20°$ and we set the thermostat to go on at $T_1 = 19°$ and off at $T_2 = 21°$. Most homeowners probably cannot feel the difference between $19°$ and $21°$, so in any practical sense the system is stable, despite mathematical arguments to the contrary. This seems analogous to the existence of communities in which species coexist on a continuing basis even though the corresponding mathematical models are unstable. What is the significance of a mathematical concept of stability that cannot be tested because of practical limitations on the precision of measurements? We seem to have a distinction which is of little significance to either homeowners or ecologists, but which is important from a mathematical point of view.

From this example we see that it is possible for a system to be self-regulated without being stable (Ashby 1956). At the same time we see that unless we understand how the regulator works, it is very easy to confuse well-regulated behaviour with stability and thus to identify stability empirically in a system which is far from any true equilibrium point.

4. THE ROLE OF OMNIVORY

The stability of complex ecosystems has received a great deal of attention from mathematical ecologists. One aspect of this work which is interesting from the viewpoint of this paper is the role of omnivory in food web structure. It has generally been concluded that, as Pimm (1980) succinctly puts it, "Species that feed on more than one trophic level (omnivores) will not be frequent within a [stable] web." The conclusions of Pimm and others in the field are controversial and have generated considerable discussion, but most of this has focussed on specific features of the models used (randomly connected systems of Lotka-Volterra equations) rather than on the issue of whether stability is the appropriate concept to use. Since omnivores are frequently found in food webs which have at least the appearance of stability, it is worth looking at the problem of omnivory from both the mathematical and biological points of view in sufficient detail to identify possible inconsistencies between the two formulations.

It is easy to see why omnivory, or polyphagy in general for that matter, tends to destabilize Lotka-Volterra models. Stability arises because when a specialized predator grazes its prey down to low levels, its own feeding rate decreases and thus the predator population declines as well. This generates the neutrally stable trajectory associated with Lotka-Volterra predator-prey systems. If the predator has an alternate food source, however, then it will not decline as rapidly when only one of its prey species is over-grazed, and thus it is able to maintain enough grazing pressure on a declining prey to drive it towards extinction. This reduction in the response of the predator to changes in the prey population destabilizes the system. From a biological point of view it is equally easy to see why this sort of instability is unlikely to occur; when an omnivore discovers that one of its prey has become very scarce, it can switch to a more abundant source of food, which reduces the pressure on the endangered prey (Lawton et al. 1974). This enhanced negative feedback "stabilizes" the system, even though it may not become stable in a mathematical sense. It is rare for the feeding process to be so totally unselective that this kind of switching cannot occur. It is therefore reasonable to model the feeding behaviour of the omnivore as an on-off regulator like the thermostat discussed earlier; arguing that the omnivore will wipe out one of its prey species is like saying that central heating will cook the inhabitants of a house, since in both cases the fact that the forcing function will turn itself off is ignored.

As regulators, omnivores can be more effective than other predators because they affect several trophic levels simultaneously. Consider for example an omnivore like the northern anchovy (*Engraulis mordax*), which generally feeds on copepods but is able to filter phytoplankton when copepods are scarce (Leong and O'Connell 1969). This strategy has a stabilizing effect on the phytoplankton, because they are grazed by anchovy only when they are not being grazed by herbivorous copepods; it has a stabilizing effect on the copepods, since they are grazed only when they are abundant; and it tends to stabilize the anchovy population by enabling the fish to stave off starvation when zooplankton stocks are low. The reverse behaviour has

been found in the omnivorous copepod *Calanus pacificus* which is normally phytophag-
ous but can switch to carnivory in response to changes in food density (Landry
1981). However, these stabilizing effects do not drive the system towards a stable
equilibrium point; in both cases phytoplankton feeding involves different behaviour
from that required for feeding on zooplankton (filtering rather than individual
capture) and thus the predators can alternate between the two feeding modes without
ever reaching an equilibrium state, in the same way that a thermostat perpetually
switches the furnace between on and off states.

5. CONCLUSIONS

All ecosystems exhibit nonlinear cause-effect relationships, many of them
quite pronounced. The usual approach to mathematical analysis, which is based on
the methodology of dynamical systems theory, is to treat the nonlinearities as
higher-order perturbations and to assume a high degree of continuity in the system
properties. Stability analysis is a natural consequence of this approach, and when
systems which are biologically regulated through switching behaviour are modelled
it is necessary to invoke sophisticated concepts of stable trajectories, limit
cycles, and the like. I do not think that it is necessary or even very helpful to
use such a high degree of mathematical abstraction to analyze a simple system like a
thermostatically-controlled heating system or a selective predator. In such cases
one is better off focussing on the role of nonlinearities in the system and treating
the corresponding effects as discontinuous, or at least as spread-out step functions.

The important practical difference between the two approaches is that stable
behaviour is most likely to be found in smoothly varying models with linear feedback, while models containing discontinuous switched regulators are very unlikely to
have stable equilibrium states in the vicinity of the region in which these regulators maintain them. Thus it is entirely possible for a well-regulated system to
appear stable on the basis of ecological observations but to be unstable in a
mathematical sense.

Acknowledgements

I have enjoyed the opportunity to discuss the ideas contained in this paper
with a number of colleagues at the Gordon Conference in Theoretical Biology and
Biomathematics, Tilton, N.H., and at the International Conference on Population
Biology, Edmonton, Alberta, both held in June 1982. I wish to express particular
thanks to P.L. Donaghay, B. Hargrave, C.M. Hawkins, A.R. Longhurst, T. Platt,
R. Rosen, W.R. Smith, and G.N. White, III, for especially useful suggestions and
observations.

REFERENCES

Ashby, W.R. (1956): *An Introduction to Cybernetics*, Chapman and Hall, London

Goh, B.-S. (1980): *Management and Analysis of Biological Populations*, Elsevier Scientific, Amsterdam.

Landry, M.R. (1981): Switching between berbivory and carnivory by the planktonic marine copepod *Calanus pacificus*, *Mar. Biol.* 65:77-82.

Lawton, J.H., J.R. Beddington, and R. Bonser (1974): Switching in invertebrate predators, in *Ecological Stability* (M.B. Usher and M.H. Williamson, Eds.), Chapman and Hall, London, pp. 141-158.

Leong, R.J.H., and C.P. O'Connell (1969): A laboratory study of particulate and filter feeding of the northern anchovy (*Engraulis mordax*), *J. Fish. Res. Bd. Can.* 26:557-582.

Pimm, S.L. (1980): Properties of food webs, *Ecology* 61:219-225.

Rosen, R. (1970): *Dynamical System Theory in Biology*, V. 1, John Wiley and Sons, New York.

STABILITY IN COMPARTMENTAL MODELS

G.G. Walter

ABSTRACT

The relative stability of a closed strongly connected compartmental model of
an ecosystem is studied. The mean first passage time of the associated regular
Markov chain is used as an index of stability. An expression for this index in
terms of the eigenvalues and flow rates is derived. Several special cases,
including mammillary systems and rosettes are studied. Two models involving
specialist predators vs. generalists are compared and the former found to be more
stable.

1. INTRODUCTION

Compartmental models, by which the flow of energy or nutrients through an
ecosystem can be studied, have dual aspect: one structural and the other dynamic.
The structural aspect leads to a directed graph (digraph) in which the compartments
are the vertices and the flows are the arcs. The dynamic aspect on the other hand
leads to a system of differential equations, which in the case of linear donor controlled flow rates, has the form

$$\frac{dX}{dt} = AX. \tag{1.1}$$

Here $X^T = [x_1 \ x_2 \ ... \ x_n]$ is the vector of levels in each compartment and $A = [a_{ij}]$
is the matrix whose elements a_{ij} are the relative flow rates from compartment j
to compartment i when $i \neq j$, and $a_{ii} = -(a_{i1} + a_{i2} + ... + a_{i,i-1} + a_{i,i+1} +$
$... + a_{in})$. We have assumed, and will throughout this work, that the system is
closed, i.e. has no input from or output to the environment.

Rather that using a differential equation, one can use a difference equation
to study the dynamics by merely replacing the derivative by a difference quotient,
say $\Delta X/h$. If the time step h is sufficiently small this can in turn be con-
verted to a Markov chain, i.e. (1.1) is approximately

$$X(t+h) = X(t) + hAX(t) = (I+hA)X(t). \tag{1.2}$$

The matrix $I + hA = P$ is the transition matrix of a Markov chain since its columns add up to 1 and its elements are non-negative and ≤ 1 for small h. The
solutions to the equation (1.1) have a point of stable equilibrium since by
Hearon's Law (Hearon, 1963), the real parts of all eigenvalues are non-positive.
However the relative magnitude of the eigenvalues may differ considerably, and
hence the time needed to return to equilibrium after a perturbation may vary in

different systems.

In order to measure the return time we use a concept borrowed from Markov chains, the mean first passage time. (See Kemeny and Snell, 1960, p. 79.) It is the expected time needed for an average molecule of nutrient to pass through the system and return to the starting point. This concept is valid only for regular Markov chains which the equation (1.1) leads to when the associated directed graph is strongly connected. (See Walter, 1979.)

Another measure of return time is the resilience index introduced in Walter (1980). The two are compared in Walter (manuscript) at least for compartmental models whose relative flow rates are all the same. In this work we first derive a general expression for the mean first passage time and then apply it to particular cases. The case of specialist systems are then compared to generalists. We shall assume throughout that the system is closed and strongly connected , i.e. there are no sinks or sources in the system.

2. A GENERAL THEOREM

In most closed ecosystem models in which the flow of nutrients is studied, there is a central compartment through which all food chains pass. It usually corresponds to a decomposer compartment. If the direct graph is strongly connected, i.e. if there is a path between any two vertices (compartments), then it has the form of an advanced rosette (see Roberts, 1976, p. 222). The mean first passage time will be measured from this central vertex back to itself.

Since the directed graph is strongly connected and since the diagonal element of $P = I + hA$ are non-zero, the Markov chain will be regular. (See Roberts, 1976, p. 289.) Such chains have a unique probability vector $\underset{\sim}{w}$ all of whose components are positive such that

$$P\underset{\sim}{w} = \underset{\sim}{w}. \tag{2.1}$$

The mean first passage time m from compartment 1 back to itself is given by

$$m = w_1^{-1} \tag{2.2}$$

where w_1 is the first component of the stationary vector $\underset{\sim}{w}$ of (2.1).

In terms of the matrix A associated with the compartment model we have

$$A\underset{\sim}{w} = \underset{\sim}{0}. \tag{2.3}$$

Hence $\underset{\sim}{w}$ is the equilibrium solution to the differential equation (1.1). Its components are all positive and add up to 1, and except for magnitude, it is unique.

Another way of interpreting $\underset{\sim}{w}$ is as the eigenvector corresponding to the eigenvalue 0. Since all the other eigenvalues have a negative real part, the solution $X(t)$ to (1.1) must approach a multiple $\alpha\underset{\sim}{w}$ at $t \to \infty$. That is, $\underset{\sim}{w}$ is a feasible stable equilibrium solution to (1.1).

We wish to establish a relation between m and certain structural properties of the system. We first relate $\underline{w}$ to the eigenvalues of $A, 0, \lambda_2, \ldots, \lambda_n$.

LEMMA 2.1 *Let* A *be the matrix of a strongly connected compartmental model,* $A*$ *its adjoint matrix,* $\underline{w}$ *its normalized equilibrium solution. Then*

$$A* = \alpha \begin{bmatrix} w_1 & w_1 & \cdots & w_1 \\ w_2 & w_2 & \cdots & w_2 \\ \cdots & \cdots & & \cdots \\ w_n & w_n & \cdots & w_n \end{bmatrix} = \hat{\alpha} \begin{bmatrix} \underline{w} & \underline{w} & \cdots & \underline{w} \end{bmatrix}$$

where $\alpha = \lambda_2 \lambda_3 \cdots \lambda_n$.

This lemma whose proof is omitted, allows us to find an expression for m rather easily.

THEOREM 2.2 *Let* $A, A*, \underline{w}$ *be as in Lemma 2.1, and suppose that furthermore the directed graph of the model is an advanced rosette; let* α_i *denote the total weights of arcs leaving the* ith *vertex. Then* m, *the mean first passage time, is given by*

$$m = (-1)^{n-1} \prod_{i=2}^{n} \lambda_i / \alpha_i . \tag{2.4}$$

Since the mean first passage time m is just w_1^{-1} and since by Lemma 2.1 $A_{11} = \alpha w_1$ we have

$$m = \frac{\alpha}{A_{11}} .$$

But now we are assuming that the directed graph of the model is an advanced rosette with the central vertex 1. If we remove this central vertex and all arcs associated with it, the resulting directed graph will no longer be strongly connected. Indeed, its strongly connected components will consist of individual vertices. Its matrix will therefore be triangular or at least can be put into triangular form by numbering the vertices appropriately. Moreover, this matrix will be identical to the matrix whose determinant is A_{11} except possibly the diagonal elements. Hence A_{11} is the determinant of a triangular matrix and therefore is the product of the elements on the diagonal of this matrix. But each diagonal element is the negative of the sum of the flow rates of all flows leaving that compartment, i.e.

$$A_{11} = \prod_{i=2}^{n} (-a_{i1}-a_{i2}-\cdots-a_{in}) = (-1)^{n-1} \prod_{i=2}^{n} \alpha_i \qquad (2.5)$$

This combined with Lemma 2.1 give us our conclusion.

REMARK 2.3 From the same considerations we can obtain other expressions for α and hence for m. From (2.5) we see that

$$\alpha = \sum_{i=1}^{n} A_{ii}$$

and hence that

$$m = 1 + \sum_{i=2}^{n} A_{ii}/A_{11}. \qquad (2.6)$$

3. SOME SPECIAL CASES

In this section we explore the consequences of Theorem 2.2 applied to special cases. We consider the mammillary system, its generalization to a rosette, and compare systems whose compartments have a specialized feeding strategy to those with a generalist strategy.

3.1. *Mammillary system*

This system (see Jacquez, 1972, p. 55) has a central vertex joined to every other vertex by a cycle of length 2. For example Figure 1 is such a system with 5 vertices.

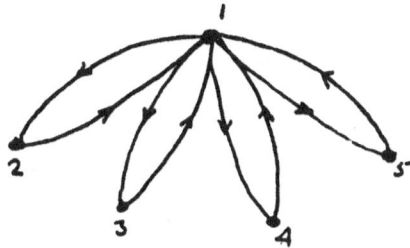

Figure 1 A mammillary system

By Theorem 2.2

$$m = (-1)^{n-1} \prod_{i=2}^{n} \frac{\lambda_i}{a_{i1}} .$$

However, the equilibrium solution may also be found directly since it satisfies

$$a_{1i}w_1 = a_{i1}w_i \quad i = 2,3,\ldots,n.$$

Hence

$$m = 1 + \frac{a_{12}}{a_{21}} + \frac{a_{13}}{a_{31}} + \ldots + \frac{a_{1n}}{a_{n1}} . \tag{3.1}$$

This is not as contradictory as it may first seem since, if there are more compartments and the flow rates are the same, the probability of staying in compartment 1 is reduced. Two systems are comparable only if the total flow rate leaving compartment 1, i.e. $\sum_{j=2}^{n} a_{1j}$, is the same in both.

3.2. *Rosettes*

If the mammillary system is extended to a system in which the cycles are of arbitrary length, then the directed graph is still a rosette. Each cycle in the rosette considered as a model itself has a mean first passage time which we denote by m_i for the ith cycle.

The equilibrium solution for a single cycle is very easy to calculate,

$$a_{12}w_1 = a_{13}w_2 = a_{34}w_3 = \ldots = a_{n1}w_n \tag{3.2}$$

and therefore

$$m = 1 + \frac{a_{12}}{a_{23}} + \frac{a_{12}}{a_{34}} + \ldots + \frac{a_{12}}{a_{n1}} . \tag{3.3}$$

If the rosette has 2 cycles (bicycle) then there are 2 equations each of which is similar to (3.2). Hence m satisfies

$$m = 1 + \frac{a_{12}(1)}{a_{23}(1)} + \ldots + \frac{a_{12}(1)}{a_{n_1 1}(1)} + \frac{a_{12}(2)}{a_{23}(2)} + \ldots + \frac{a_{12}(2)}{a_{n_2 1}(2)}$$

$$= m_1 + m_2 - 1 .$$

This is easily extended to k cycles and

$$m = \sum_{i=1}^{k} m_i - k + 1$$

3.3. *Generalist vs. Specialist*

In order to compare the relative stability of a system in which the apex predators are generalists vs. one in which they are specialists, i.e. each specializing on a single prey, we construct the two models shown in Figure 2. Their

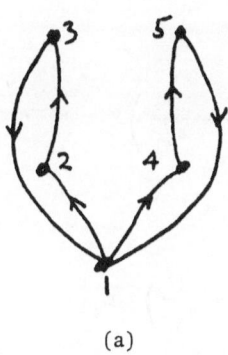

(a)

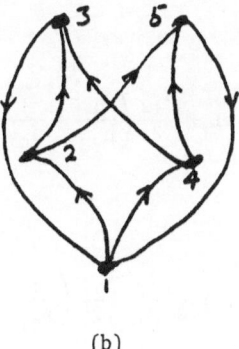

(b)

Figure 2 Specialist model (a) and generalist (b) in which 3 and 5 are apex predators

matrices are respectively

$$A_s = \begin{bmatrix} -a_{11} & 0 & a_{31} & 0 & a_{51} \\ a_{12} & -a_{23} & 0 & 0 & 0 \\ 0 & a_{23} & -a_{31} & 0 & 0 \\ a_{14} & 0 & 0 & -a_{45} & 0 \\ 0 & 0 & 0 & a_{45} & -a_{51} \end{bmatrix}$$

and

$$A_g = \begin{bmatrix} -a_{11} & 0 & a_{31} & 0 & a_{51} \\ a_{12} & -b_{22} & 0 & 0 & 0 \\ 0 & b_{23} & -a_{31} & b_{43} & 0 \\ a_{14} & 0 & 0 & -b_{44} & 0 \\ 0 & b_{25} & 0 & b_{45} & -a_{51} \end{bmatrix} .$$

We have implicity assumed that the equivalent arcs have the same flow rates in the two models. For simplicity we assume that b_{23} and b_{25} are approximately the same and equal to $b_{22}/2$. The same is true for b_{43} and b_{45}. Moreover since specialists are supposed to be more efficient than generalists in their food uptake we assume that $a_{23} > b_{23} + b_{43}$ and $a_{45} > b_{25} + b_{45}$.

The mean first passage time for the specialist case is

$$m_s = 1 + \frac{a_{12}}{a_{23}} + \frac{a_{12}}{a_{31}} + \frac{a_{14}}{a_{45}} + \frac{a_{14}}{a_{51}}$$

while

$$m_g = 1 + \frac{a_{12}}{2}\left(\frac{1}{b_{23}} + \frac{1}{a_{51}} + \frac{1}{a_{31}}\right) + \frac{a_{14}}{2}\left(\frac{1}{b_{43}} + \frac{1}{a_{51}} + \frac{1}{a_{31}}\right).$$

Hence we have

$$m_g > 1 + \frac{a_{12}}{a_{23}} + \frac{a_{12}}{2}\left(\frac{1}{a_{51}} + \frac{1}{a_{31}}\right) + \frac{a_{14}}{a_{45}} + \frac{a_{14}}{2}\left(\frac{1}{a_{51}} + \frac{1}{a_{31}}\right) = m_s$$

provided a_{51} and a_{31} are about the same.

Thus the specialists have a smaller mean first passage time and therefore a system containing them is more stable. This is not altogether unexpected, but it does contradict the hypothesis that greater complexity (associated with the generalists) leads to greater stability.

REFERENCES

Hearon, John Z. (1963): Theorems on linear systems, *Ann. N.Y. Acad. Sci.* 108: 368-68.

Jacquez, J.A. (1972): *Compartmental Analysis in Biology and Medicine*, Elsevier, Amsterdam.

Kemeny, J.J., and J.L. Snell (1960): *Finite Markov Chains*, Van Nostrand, Princeton, N.J.

Roberts, Fred S. (1976): *Discrete Mathematical Models*, Prentice Hall, Englewood Cliffs, N.J.

Walter, G.G. (1979): Compartmental Models, Digraphs, and Markov Chains, in *Compartmental Analysis of Ecosystem Models*, (Matis, Patten & White eds.) Intern. Co-op Pub., Fairland, Md. pp. 295-310.

_____(1980): Stability and structure of compartmental models of ecosystems, *Math. Biosci.* 51:1-10.

_____(manuscript): Passage time, resilience, and structure of compartmental models.

PART VII:

RESOURCE MANAGEMENT

SEARCH THEORY IN ECOLOGY AND RESOURCE MANAGEMENT

Colin W. Clark and Marc Mangel

Can there be a theory of "rational search"? After all, the word "search" already conveys the impression that the searcher does not know where to look. But it is seldom the case that the searcher has *no* information on the likely location of his object, or objects -- indeed, *the very process of searching itself provides such information*. We might then say that search is "rational" if it maximizes the accumulation of information.

Here, however, we adopt a somewhat more explicit notion: a search strategy will be considered optimal if it maximizes the expectation of some specified payoff, depending of course on the particular application.

The prototypical search situation in ecology is *predation* (including foraging). Until recently, however, the literature on optimal foraging theory has had little contact with search theory. For example, the "marginal value theorem" of optimal foraging (Charnov 1976) asserts that a forager should remain in a given "patch" until its feeding rate falls below the average feeding rate for the entire environment. But in fact the forager cannot *know* with certainty either of these values, since both are the result of a sampling process, namely the search for food within the patch, and over the entire feeding area. Consequently, the optimal foraging strategy must be more complicated than suggested by the deterministic marginal value theorem (Oaten 1979, Green 1980, Iwasa et al 1981).

Man also is an inveterate searcher -- after truth, jobs, enemy submarines ... and fish! We begin with a model of the fisherman's search problem (Mangel 1981, 1982; Mangel and Clark 1982).

1. SEARCHING FOR FISH

Consider a fishing fleet of N vessels, which exploits distinct fish stocks on several fishing grounds $A_1, \ldots, A_s$. The seasonal abundance of fish on each ground A_i is a random variable λ_i. Explicitly, λ_i equals the average rate of encounter of fish schools on ground A_i, by a single vessel, during a particular fishing season. Searching for fish on A_i is modelled by a Poisson process with parameter λ_i:

$$\text{Pr (one encounter in } t, t+dt) = \lambda_i dt \tag{1}$$

Thus λ_i varies from ground to ground, and on each ground, also varies from season to season. The fishermen do not know the current values of λ_i, but we assume that each λ_i has a known prior p.d.f. $f_i(\lambda_i)$, obtained from the historical record of catch and effort for each ground. For simplicity the λ_i will be assumed independent, although the case of dependence is easily covered by our theory.

It is both convenient and appropriate to specify f_i as gamma distribution:

$$f(\lambda) = f(\lambda; \nu, \alpha) = \frac{\alpha^\nu}{\Gamma(\nu)} \lambda^{\nu-1} e^{-\alpha\lambda} \tag{2}$$

(suppressing subscripts i for typographical simplicity). The mean and variance of this distribution are given by

$$\mu = \nu/\alpha, \quad \sigma^2 = \nu/\alpha^2 \tag{3}$$

so that the coefficient of variation is $CV = 1/\sqrt{\nu}$.

Assume that the vessels make M trips per season, each to a specified fishing ground A_i. We pose the question, what is the optimal allocation of vessels to the various grounds during each trip? It is assumed that the objective of the fleet is to maximize total seasonal catch. (More complicated objective functions, involving differential costs or prices, are easily accommodated.)

Possible strategies would include the following:

1. *Nonadaptive strategy:* maximize $\bar{\lambda}_i = E^{f_i}\{\lambda_i\}$. In this strategy, all M vessels are sent to the ground where historical catches have been greatest.

2. *Nonhistorical strategy:* maximize $\hat{\lambda}_i$. The vessels are distributed evenly over the n grounds during the first trip, and $\hat{\lambda}_i$ is the observed catch rate on A_i. Subsequently, all M vessels are sent to the (apparently) most productive ground.

(In this preliminary model we shall assume that the fishing fleet is small, so that no noticeable depletion occurs as the result of fishing down the stock on a given ground.)

3. *Passive adaptive strategy:* first maximize $\bar{\lambda}_i$; then "update" λ_i; switch to another ground (max $\bar{\lambda}_i'$) if appropriate; etc. Here, all vessels are first sent to the best fishing ground. If fishing turns out to be poor there, the vessels then switch to the next most promising ground, etc.

What distinguishes these strategies, clearly, is the way in which information is used. The first two strategies ignore, respectively, current and historical information. The third uses both, but in general does not generate the *optimal* amount of information. This should be clear: strategy 3 will completely miss any ground where current abundance is high but average historical abundance is less than the maximum. It seems obvious that the best strategy would involve some "probing" to determine whether such cases exist. Such a strategy is called an *active adaptive strategy.*

In order to proceed, we first discuss the process of "updating" of prior information. We have, for any ground,

$$\text{Pr (n encounters, by k vessels, in time t, given } \lambda)$$

$$= p(n,kt,\lambda) = \frac{(\lambda kt)^n}{n!} e^{-\lambda kt} \tag{4}$$

(the vessels search independently). Let $f(\lambda:n,k)$ denote the posterior distribution of λ, given that k vessels encounter n schools in time t. The Bayes formula gives

$$f(\lambda:n,k) = \frac{p(n,kt,\lambda)f(\lambda)}{\int p(n,kt,\mu)f(\mu)d\mu} \tag{5}$$

Carrying out the indicated integration, using the gamma prior (2), we easily find that

$$f(\lambda:n,k) = f(\lambda:\nu+n,\alpha+kt) \tag{6}$$

The posterior distribution is thus again a gamma distribution (the gamma and Poisson distributions are "conjugate"), but with *updated parameters*:

$$\nu' = \nu + n, \quad \alpha' = \alpha + kt \tag{7}$$

This immediately gives an updated estimate of abundance:

$$\bar{\lambda}' = \frac{\nu'}{\alpha'} = \frac{\nu+n}{\alpha+kt} \tag{8}$$

which may be contrasted with the "nonhistorical" estimate

$$\hat{\lambda} = \frac{n}{kt} \tag{9}$$

The sampling (fishing) process also decreases the uncertainty in the estimate of λ, since the updated CV' is $1/\sqrt{\nu'} = 1/\sqrt{\nu+n}$. An example is given in the following table; here $\nu = 4$, $\alpha = 0.4$, so $\bar{\lambda} = 10$ and CV = 0.5; also kt = 1.

Table 1 Nonhistorical vs. Updated Estimates of λ for Various Value of n = Number of Schools Encountered

n	$\hat{\lambda}$	$\bar{\lambda}'$	CV'
5	5	6.3	0.33
10	10	10.0	0.27
20	20	17.1	0.20

The computational simplicity of this approach should be noted. Unfortunately this simplicity breaks down if priors other than the gamma are used; however we do not consider this an important limitation. More significant, perhaps, is the assumption that depletion can be ignored. The formulas become considerably more complex (but still easily computed on a micro) if depletion is included (Mangel and Clark 1982); in practice even this may not be important, since most of the updating information will come from the early stages of probing, in which depletion is minor.

Let us press on to the question of the optimal active search strategy. Here the computation, in full generality, becomes formidable -- as always in stochastic dynamic programming. We shall therefore adopt some further simplification as we proceed. Let $J_m(\vec{\nu},\vec{\alpha})$ denote the maximum expected total catch, given that m fishing trips remain in the season, and given *current* parameter estimates $\vec{\nu} = (\nu_1,\ldots,\nu_n)$ and $\vec{\alpha} = (\alpha_1,\ldots,\alpha_n)$. These parameters have been obtained from the prior values by updating according to catches taken so far in the season.

For $n = 1$ we have simply

$$J_1(\vec{\nu},\vec{\alpha}) = N \max_i \frac{\nu_i}{\alpha_i} \tag{10}$$

where the length of one trip is taken as the basic unit of time. For $n = 2$ we obtain

$$J_2(\vec{\nu},\vec{\alpha}) = \max_{\Sigma k_i = N} \left(\sum k_i \nu_i/\alpha_i + E\{J_1(\vec{\nu'},\vec{\alpha'})\} \right) \tag{11}$$

where $k_i \geq 0$ represents the number of vessels allocated to ground A_i in the current period, and where $\vec{\nu'}$, $\vec{\alpha'}$ are the updated estimates. The expectation in (11) is given by

$$E\{\ldots\} = \sum_{n_1=0}^{\infty} \cdots \sum_{n_s=0}^{\infty} J_1(\nu_i+n_i,\alpha_i+k_i)Pr(n_1) \cdots Pr(n_s) \tag{12}$$

with

$$Pr(n_i) = \int_0^{\infty} Pr(n_i:\lambda)f(\lambda:\nu_i,\alpha_i)d\lambda$$

$$= \frac{k_i^{n_i}}{n_i!} \frac{\alpha_i^{\nu_i}}{(\alpha_i+k_i)^{n_i+\nu_i}} \frac{\Gamma(n_i+\nu_i)}{\Gamma(\nu_i)} \tag{13}$$

for $k_i \neq 0$, and $Pr(n_i) = 1$ or 0 for $n_i = 0, n_i > 0$ respectively, if $k_i = 0$.

Numerical computation of (12) is quite feasible, although in general the sum contains very many small terms, so that some form of heirarchical summation is needed. The process (11) may be iterated to obtain $J_3, J_4, \ldots$, but of course the numerical calculations rapidly become formidable. In the following example only J_2 has been computed.

We consider the case of two grounds A_1, A_2; the prior parameters for each ground are listed in Table 2. The average catch rate on A_2 is

Table 2 Prior Parameters for A_1, A_2 ($\bar{\lambda}$ Is the Expected Number of Schools Encountered Per Day Fishing)

	ν_i	α_i	$\bar{\lambda}_i$	cv_i
A_1	4.0	0.4	10	0.5
A_2	0.08	0.01	8	3.54

slightly less than on A_1, but the fluctuations are much more pronounced on A_2.

Table 3 shows the expected catches, for $T = 2$ and $T = 10$ trips of 5 days' duration each, by a

Table 3 Expected Catches, for a Single Vessel Making T Trips Per Session, Under Passive and Adaptive Strategies

	Expected Total Catch		
T	Passive (A_1)	Active (A_2)	% Improvement
2	104.6	119.8	14.5
10	541.5	758.0	40.0

single vessel. Recall that in the passive case, the vessel first fishes A_1, and switches to A_2 only if the updated estimate $\bar{\lambda}'$ is less than $\lambda'_2 = 8$. For $T = 2$ trips this strategy is 4.6% bettern than the nonadaptive strategy of simply fishing A_1 for both trips (expected catch = $2 \times 10 \times 5 = 100$ schools).

The optimal strategy is for the vessel to first fish A_2; this is 19.8% better than the nonadaptive strategy, and 14.5% better than the passive adaptive. The reason for this is clear: since abundance fluctuates wildly on A_2, that ground should be fished first in order to discover whether abundance is unusually high there.

For the case of $T = 10$ trips, the active strategy is far superior to the passive. In fact the figures given in Table 3 apply only to a "partially active" strategy, in which the information obtained on trip 1 is used to determine which ground will be fished for the remaining 9 trips. But in actuality the vessel might switch grounds more than once -- if $\bar{\lambda}_2' < \bar{\lambda}_1'$ trip 2 will be to A_1, but then $\bar{\lambda}_1$ will be updated to λ_1' and another switch will occur if $\lambda_1' < \lambda_2'$ etc. Thus Table 3 somewhat underestimates the comparative advantage of the active strategy.

In the case of a fleet of $N > 1$ vessels, one can determine the optimal allocation of vessels between different fishing grounds. In the above example, with $N = 10$ it turns out that *one* vessel should be sent to A_2 on the first trip (Mangel and Clark 1982).

The lessons to be learned from the above exercise are first, that *probing* for information is important, especially where uncertainty is high, and second that the value of information clearly depends on the degree to which it is used. In Table 3, for example, the expected "value" of search information is 19.8 schools of fish for a two-trip season, but at least 258 schools for a 10-trip season.

It is also worth noting that if the N vessels fish *competitively* an optimal search strategy will probably not be used, and catch rates may be less than optimal. For further discussion of this point, as well as for extensions of the above model (depletion effects, nonuniform school size, correlated priors, etc.) we refer to our original paper (Mangel and Clark 1982).

2. FORAGING AND FLOCKING STRATEGY

The ideas expressed above can also be applied to the question of foraging strategy of natural predators (Clark and Mangel 1982). As pointed out by Oaten (1979), foraging strategy may be strongly influenced by considerations of uncertainty and information. For example, consider the question of when a forager should switch from feeding in a given "patch" to searching for a new patch. The stochastic version of the marginal value theorem would assert that the switch should be made when the expected capture rate in the patch falls below the overall expected rate. Both of these rates can be estimated by a Bayesian updating procedure based on the sequence of samples obtained by the forager (Green 1980, Isawa et al 1981).

The study of foraging strategy takes on a new dimension if the possibility of social foraging ("flocking") is taken into consideration. Roughly speaking, a flock of n foragers will generate information on local abundance at roughly n times the rate of a single forager, although effects of crowding and interference will tend to reduce this as n grows large. On the other hand, the flock will deplete the local supply of food at n times the individual rate, unless forage is so abundant that individual foragers become satiated. In Clark and Mangel (1982) we describe a simple model incorporating these three effects; the resulting relationship between flock size and average individual feeding rate is shown in Figure 1. The upper curve arises when forage is scarce and patchy, the lower when it is either abundant or uniformly distributed. Search information is valuable clearly only in the former case, and it is in this case that flocking yields an advantage to individual foragers. There is some field observation in support of this prediction (Cody 1971).

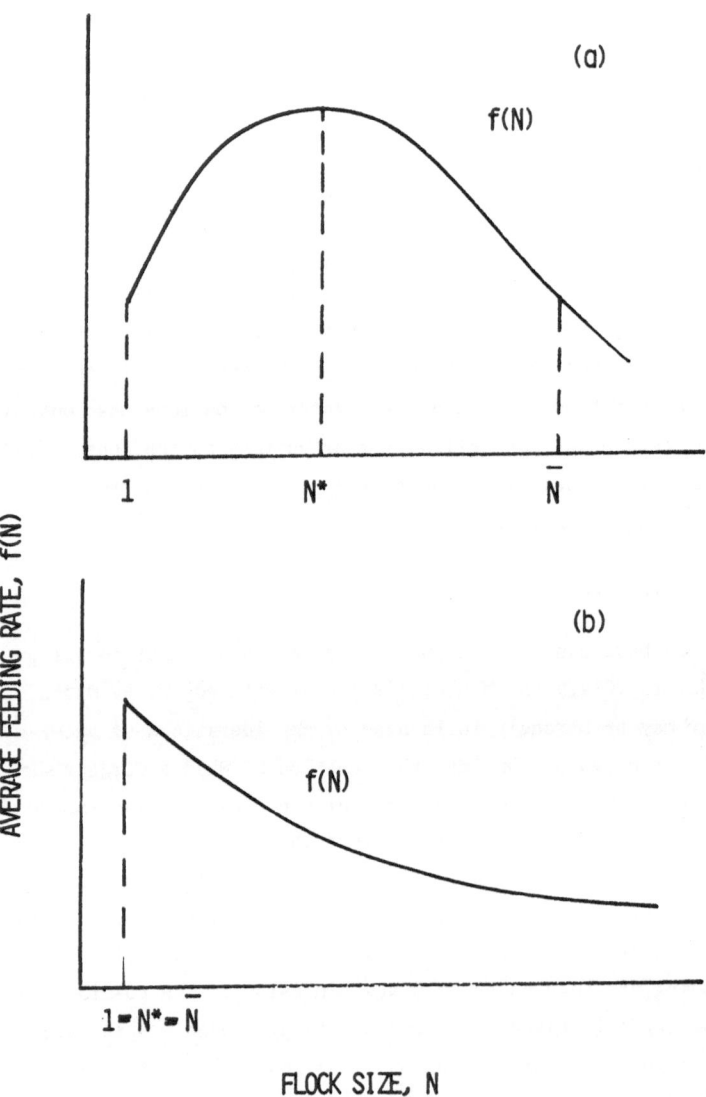

Figure 1   Average individual feeding rate for flocks of size N:
      (a) highly patchy distribution;
      (b) less patchy distribution.

What size flocks will be observed? In the case shown in Figure 1(b), the individual obviously gains no advantage by flocking. In case (a), however, there should be a tendency to form flocks, and indeed the flocks observed should be considerably *larger* than the optimal size N^*. For imagine that a flock of size N^* has formed, and that an additional forager arrives. If he feeds alone, his expected feeding rate will be $f(1)$, which is less than $f(N^*+1)$; this holds for any newcomer as long as $N < \bar{N}$, where $f(\bar{N}) = f(1)$. Thus, unless some mechanism whereby "breakaway" flocks can form, one would expect to observe flocks of size $\bar{N}$, with at most a slight advantage over the individual feeding rate.

A field test (albeit unintentional) of this prediction has been reported by Caraco and Wolf (1975), who analyzed data on the sizes of foraging groups of lions in the Serengeti plain. Lions formed foraging groups only when prey was scarce (and large enough to feed several lions), but the observed sizes of groups were well in excess of the optimum sizes calculated by Caraco and Wolf. This apparently irrational behavior was attributed by the authors to other unknown causes, such as increased breeding success resulting from foraging in large groups. Our theory, however, explains the observations without invoking other mechanisms.

3. SUMMARY

In this paper we have suggested that questions of uncertainty and information may be important determinants of behavior of both natural and human searchers after prey. We feel that the theory could be greatly expanded, and that both field observations and controlled experiments could be conducted in the light of the theory.

REFERENCES

Caraco, T. and L.L. Wolf (1975): Ecological determinants of group sizes of foraging lions, *Amer. Nat.* 109:343-352.

Charnov, E.L. (1976): Optimal foraging: the marginal value theorem, *Theor. Pop. Biol.* 9:129-136.

Clark, C.W. and M. Mangel (1982): Foraging and flocking strategies: information in an uncertain environment, Manuscript.

Cody, M.L. (1971): Finch flocks in the Mohave desert, *Theor. Pop. Biol.* 2:142-158.

Green, R.F. (1980): Bayesian birds: a simple example of Oaten's stochastic model of optimal foraging, *Theor. Pop. Biol.* 18:244-256.

Iwasa, Y., M. Higashi and N. Yamamura (1981): Prey distribution as a factor determining the choice of optimal foraging strategy, *Amer. Nat.* 117:710-723.

Mangel, M. (1981): Search for a randomly moving object, *SIAM J. Appl. Math.* 40:327-338.

_____ (1982): Search effort and catch rates in fisheries, *Eur. J. Oper. Res.* 11:361-366.

Mangel, M., and C.W. Clark (1982): Uncertainty, search, and information in fisheries, *J. du Conseil Intern. Expl. Mer.*, in press.

Oaten, A. (1979): Optimal foraging in patches: a case for stochasticity, *Theor. Pop. Biol.* 12:263-285.

MODELS OF PHEROMONE RELEASE FOR PEST CONTROL

IN WHICH MALES ATTRACT FEMALES

Hugh J. Barclay

ABSTRACT

In some insect species aggregation for mating is mediated via pheromones released by the males which attract females, such as in the Boll Weevil (*Anthonomus grandis*). Discrete density independent models are developed for the capture and annihilation of females of a pest species by means of male-baited pheromone traps. Four models are developed involving both male and female monogamy and polygamy and also delayed male mating. Control is little affected by male mating frequency but is made substantially easier with delayed mating or female polygamy. The dynamics of pest control are very similar regardless of which sex produces the attractant unless females remate regularly, in which case species in which the males release the sex pheromone are easier to control.

1. INTRODUCTION

In most insects in which the use of sex attractant pheromones has been observed it is the females that release pheromones that attract males for mating (Shorey and Gaston, 1967). Many examples of males releasing pheromones attractive to females are also known although many of these pheromones are the male aphrodisiacs which operate only at very close range (Jacobson, 1972). This paper focusses on the cases in which females seek out males for mating in response to male-produced pheromones and is similar to another paper dealing with males attracted to female-produced sex pheromones (Barclay and van den Driessche, in preparation). The models in this paper are all extensions of a model developed by Knipling and McGuire (1966, model VI).

2. THE MODELS

Model I: the basic model.

This model forms the basis of all the other models presented here. The order of events assumed is: (a) overnight mortality, (b) morning emergence of both sexes, (c) tally of numbers, (d) mating. Letting f_i be mated (fertile) females, v_i be virgin females, m_i be males (all on day i), a be the birth rate, s be the daily survivorship and m_0 be the number of caged males or pheromone equivalents, we get the equations:

$$f_{i+1} = sv_i \cdot \left(\frac{m_i}{m_i + m_0} \right) + sf_i \tag{1}$$

$$v_{i+1} = af_{i-k} \tag{2}$$

$$m_{i+1} = af_{i-k} + sm_i \tag{3}$$

The proportion of virgin females attracted to caged males is $m_0/(m_i+m_0)$ and these are killed upon contact with the traps. In this model females only search for males once per day. The constant k is the incubation time between egg laying and adult emergence; any mortality incurred during this period is absorbed into the parameter a. In each of Equations (1) to (3) the first term on the right represents recruitment and the second term is survivorship. Virgins will all either mate or be attracted to caged males and killed daily since males are assumed to mate daily, given the chance, and males outnumber virgins.

Equations (1) to (3) have two equilibrium points (EP), a stable EP at $f = m = v = 0$ and an unstable positive EP given in Table 1. A linearized stability analysis (not shown) confirms the stability of the two EP. Referring to Table 1, high values for both the birth rate, a, and survivorship, s, result in greater difficulty of control than do lower values. These results exactly parallel those for systems involving male attraction to female produced pheromones (Barclay and van den Driessche, in preparation) and the same behaviour holds true for all the models presented here.

Model II: attraction with no killing.

Here virgins attracted to caged males are not killed but are simply deprived of mates for that day. Only Equation (2) of the basic model is affected by this, becoming

$$v_{i+1} = af_{i-k} + sv_i \left(\frac{m_0}{m_i+m_0} \right) \tag{4}$$

which also has two EP, one at (0,0,0) and one given in Table 1 and in which both $\hat{m}$ and $\hat{f}$ here are exactly (1-s) times those values in model I, while $\hat{v}$ incurs a similar, though not identical, reduction from $\hat{v}$ of model I. Thus if daily survivorship is low, depriving virgins of mates for one day is almost as effective as killing them. For large s this method is very ineffective.

Model III: male monogamy and delayed male mating.

Here males produce sex pheromones only when ready to mate, d days after eclosion, and cease pheromone production upon mating so that the proportion of males surviving is s^d of those emerging. Only those males not yet having mated are tallied (m_i). This model becomes Equations (1) and (2) unchanged plus

$$m_{i+1} = af_{i-k-d}/p + sm_i[1-v_i/(m_i+m_0)] \tag{5}$$

Table 1. The Equilibrium Values for Each of the Four Models in Which
(a) Males Release Sex Pheromones, (b) Females Release Sex Pheromones

Model	Variable	Male Pheromoes	Female Pheromones	♀/♂*
	v	$\dfrac{m_0(1-s)^2}{as-(1-s)}$	$\dfrac{v_0(1-s)}{as-(1-s)^2}$	$\sim 1/(1-s)$
I	m	$\dfrac{m_0(1-s)}{as-(1-s)}$	$\dfrac{v_0(1-s)^2(a+1-s)}{(a-(1-s))(as-(1-s)^2)}$	$\sim 1-s$
	f	$\dfrac{m_0(1-s)^2}{a(as-(1-s))}$	$\dfrac{v_0(1-s)^2}{(a-(1-s))(as-(1-s)^2)}$	~ 1
	v	$\dfrac{m_0(1-s)^2(a-(1-s))}{a(as-(1-s))}$	$\dfrac{v_0(1-s)^2}{as-(1-s)^2}$	~ 1
II	m	$\dfrac{m_0(1-s)^2}{as-(1-s)}$	$\dfrac{v_0 a(1-s)^2}{(a-(1-s))(as-(1-s)^2)}$	~ 1
	f	$\dfrac{m_0(1-s)^3}{a(as-(1-s)}$	$\dfrac{v_0(1-s)^3}{(a-(1-s))(as-(1-s)^2)}$	~ 1
	v	$\dfrac{apm_0(1-s)^2}{(as-(1-s))(a-p(1-s))}$	$\dfrac{pv_0(1-s)}{as-p(1-s)}$	$\sim 1/(1-s)$
III	m	$\dfrac{m_0(1-s)}{as-(1-s)}$	$\dfrac{av_0(1-s)^2}{(as-p(1-s))(a-(1-s))}$	$\sim 1-s$
	f	$\dfrac{pm_0(1-s)^2}{(as-(1-s))(a-p(1-s))}$	$\dfrac{pv_0(1-s)^2}{(as-p(1-s))(a-(1-s))}$	~ 1
	v	$\dfrac{m_0(1-s)}{as-(1-s)}$	$\dfrac{v_0(1-s)(a-(1-s))}{a(as-(1-s))}$	~ 1
IV	m	$\dfrac{m_0}{as-(1-s)}$	$\dfrac{av_0(1-s)^2}{(as-(1-s))(a-(1-s))}$	$\sim (1-s)^2$
	f	$\dfrac{m_0(1-s)}{a(as-(1-s))}$	$\dfrac{v_0(1-s)^2}{a(as-(1-s))}$	$\sim 1-s$

*These quotients are approximate and serve to give an indication of the relative size of the two equilibrium values. For these approximations, m_0 and v_0 can be taken as being the same.

where $1/p = s^d$ and the positive EP is given in Table 1. If $d = 0$, then $p = 1$ and males mate immediately; in that case the EP is only slightly larger than that from model I and the ease of control is hardly affected. If $p > 1$ and mating is delayed, then the EP is substantially further from the origin than that from model I and control is much easier than with immediate mating.

Model IV: daily female remating.

In this model females remate every day. Since males are necessarily in excess with a 1:1 sex ratio and since both virgins and mated females are attracted to the male-baited traps and killed, all females are either killed at traps or mated on any given day. Equations (2) and (3) remain the same but Equation (1) becomes

$$f_{i+1} = s(v_i + f_i)[m_i/(m_i + m_0)] \tag{6}$$

and they have an unstable positive EP given in Table 1, in which $\hat{v}$, $\hat{m}$, and $\hat{f}$ are all exactly $1/(1-s)$ times the corresponding values for model I. If s is large (i.e. high survivorship), then this system is much easier to control than is model I since both mated females and virgins are killed daily at the pheromone traps. Clearly if females mated several times per day, the number killed at traps would be greater and the effectiveness would be even greater than with daily mating.

3. DISCUSSION

The use of sex pheromones for pest control is potentially very attractive since they are clean and used only in minute quantities compared with conventional pesticides. There are many problems associated with the method but most of them are technical difficulties of implementation rather than problems with undesirable side effects (Minks, 1977).

It is evident from Table 1 that, relative to the basic model, (a) attraction of females to baited traps without killing them is much less efficient than trapping for annihilation unless daily survivorship is very low; (b) there is little difference between the ease of control with male monogamy and daily male remating; however, the time taken to the first mating is very important, control being much easier if mating is delayed; (c) regular female remating makes control much easier. It appears that any species in which males release sex attractants and in which males have an extended maturation period following emergence while females remate regularly is an ideal candidate for control by means of male-baited annihilation traps.

Comparing the equilibria obtained from the four models here with those from the corresponding four models with female pheromone release (Barclay and van den Driessche, in preparation) the differences between the EP values for the two sets of models (Table 1) tend to cancel out in all except the last model where the EP with male pheromone release is much higher than that with female pheromone release unless daily survivorship is extremely low. Thus if females remate regularly, then species in which males release the sex attractants are much easier to control than species in which the females release the pheromone, other things being equal.

Acknowledgements

I thank Pauline van den Driessche for discussions on the topic of this paper and for providing the stability analysis for model I.

REFERENCES

Jacobson, M. (1972): *Insect Sex Pheromones*. Academic Press, New York, 382 pp.

Knipling, E.F. and J.U. McGuire (1966): Population models to test theoretical effects of sex attractants used for insect control. *Agric. Info. Bull.* 308, *USDA*, 20 pp.

Minks, A.K. (1977): Trapping with behavior modifying chemicals: feasibility and limitations. In: H.H. Shorey and J.J. McKelvey (Editors), *Chemical Control of Insect Behavior*, John Wiley, New York, 385-394.

Shorey, H.H. and L.K. Gaston (1967): Pheromones. In: *Pest Control - Biological, Physical and Selected Chemical Methods*. W.W. Kilgore and R.L. Doutt (Editors), Academic Press, New York, 241-265.

AGE- AND SIZE-SPECIFIC MODELS IN THE DUNGENESS CRAB FISHERY

Louis W. Botsford

ABSTRACT

Over the past thirty years, development of age-specific models in which the
number of older individuals affects survival of the young (e.g., through cannibalism)
has yielded several practical results. Briefly, (1) periodic solutions arise
under the condition that the slope of the recruitment survival function is less
than an age-structure dependent parameter and (2) the period of these solutions is
roughly twice the average age of influence on production and survival of offspring.
These results have been applied to determination of the cause of cyclic fluctuations
in the northern California Dungeness crab fishery. Recent analysis of the influence
of size-structure and a delayed response of fishing effort to abundance have changed
our view of this fishery. Since the expected period of oscillations is greater, the
previously observed disparity between model period and observed period is much less
and the previous rejection of cannibalism as a mechanism appears to have been
premature.

1. INTRODUCTION

The ultimate goal in the construction and analysis of mathematical models of
populations is presumably to provide a better understanding of real populations.
Achievement of this goal depends critically on the manner in which the results of
analysis are related to observations of real populations. Unfortunately, this
aspect of population biology is usually either ignored or dismissed by alluding to
a vague correspondence between the model and a whimsical view of how the biological
world might operate. The main points of this paper are (a) to briefly review useful results from age-specific population models with density-dependent recruitment,
and (b) to show how these results and several recent further developments have
been applied to a practical problem: identification of the cause of observed cycles
in the northern California Dungeness crab fishery.

The catch record from this crab fishery has varied in a cyclic fashion for
the past thirty years (Figure 1). The same cycles appear in recent estimates of
legal abundance (i.e., of males larger than 159 mm carapace width). This behavior
has been attributed to (a) a cyclic environmental variable (e.g., upwelling),
(b) a predator-prey relationship (e.g., with man as the predator) and (c) a
density-dependent recruitment relationship (e.g., cannibalism). At this time there
is little positive, reasonable evidence for the first category and the second
category has been dismissed. However, the cause of cyclic behavior of the crab

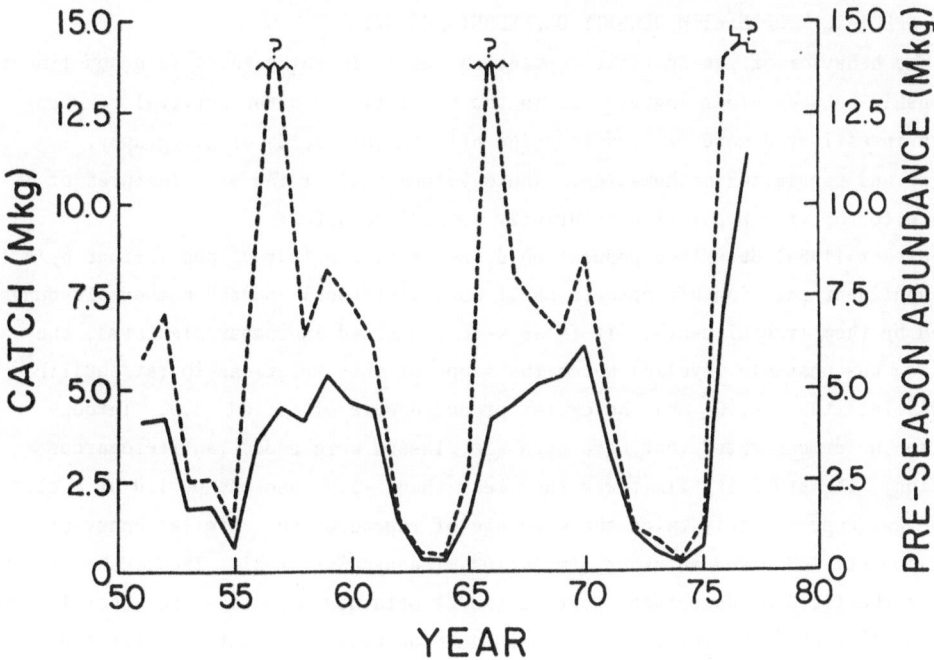

Figure 1. Catch (——) and estimated abundance (-‑-) for northern California
Dungeness crab fishery. (? indicate uncertain, high values)

population remains unknown. The studies summarized here involve the application of
mathematical modelling results to determination of the specific mechanism(s) in the
third category which can be responsible for the observed cycles.

Mathematical models can be used to determine the specific mechanism causing
cyclic behavior by comparing the behavior of models to behavior of the population.
Conceptually, a population model consists of a set of assumptions from which
conclusions are drawn through mathematical analyses (i.e., if A_1, A_2, A_3 ... then
C). One of the assumptions could be a proposed causal mechanism. The conclusion
could be a characteristic of total population numbers (e.g., the period of cycles)
or a characteristic of individuals (e.g., the rate of cannibalism at different
densities). Conclusions from analysis are compared to field observation. If the
outcome is positive (i.e., projected behavior matches observed behavior), then
little is learned and that mechanism is merely one of many possible causes. If
however the outcome is negative, that mechanism can be rejected *provided* all other
assumptions are known to be true. Logically, the negative outcome (not C) implies
that one of the assumptions is not true (i.e., not A_1 or not A_2 or ...).
Further work may be required to determine which assumption is invalid.

2. AGE-SPECIFIC MODELS WITH DENSITY-DEPENDENT RECRUITMENT

The behavior of age-specific population models in which there is a non-linear relationship between older individuals in the population and the survival of young has been investigated more or less independently in the fields of demography, fisheries and population mathematics. The developmental of the main features of relevance to the crab population is briefly summarized here.

Ricker (1954) described population dynamics of semelparous populations by a single nonlinear relationship between adult numbers (stock) and the number of young produced by them (recruitment). If these were expressed in comparable terms, the population was unstable (cyclic) which the slope of this relationship (at equilibrium) was less than -1.0 and the cycles produced were of period 2.0. Through simulation he demonstrated that when more age classes were added (an iteroparous population), the stability limit was then less than -1.0 and the period of cycles produced was approximately twice the mean age of reproduction. Similar behavior was subsequently observed in other simulations (Pennycuick et al. 1968, Usher 1972).

In the field of demography Keyfitz (1972) obtained conditions for oscillation from a model in which reproductive rate depended on past reproduction. Lee (1972) analyzed two discrete versions of the renewal equation: one in which fertility depended on cohort size and one in which it depended on the size of the labor force. Frauenthal (1975), in his investigation of the Easterlin effect, analyzed a continuous time renewal equation in which reproductive rate depended in a non-linear way on total fertility. Rorres (1976) investigated global and local stability of a continuous time model in which maternity depended on population size.

In a fisheries context, Allen and Basasibwaki (1974) analyzed a Leslie matrix model with a stock-recruitment relationship. They analytically determined conditions under which oscillations would occur for one case of the model Ricker (1954) had used.

Botsford and Wickham (1978) analyzed a renewal equation with non-linear recruitment: $R(t) = B(t)f[C(t)]$, where $R(t)$ = recruitment rate, $B(t)$ = reproductive rate, $F[\cdot]$ = recruitment survival and $C(t)$ = total population size as it affects recruitment (a weighted integral of the age density over all age in which the weighting function reflects the relative effect on survival of an individual at age a; c.f. Rorres (1976)). From a linear analysis about the single equilibrium, they showed that as the normalized slope of f at equilibrium became increasingly negative, the real part of the pair of dominant eigenvalues became more positive. They obtained numerical solutions for the value of normalized slope at which the real part was zero, the point at which oscillations could occur. This point depended on age structure. It was typically more negative (i.e., a more stable population) when the age structure was broad and flat (many age classes) and became less negative (a less stable population) when the age structure was peaked and narrow (few age classes). This result was shown analytically for a case with some simplifying assumptions (c.f. Frauenthal 1975). The period of oscillations was typically twice the mean age of influence of older individuals or production of the young. This was shown analytically for only one simplified case.

Rorres (1979) performed a similar analysis to Botsford and Wickham's (1978) and obtained essentially the same results regarding stability conditions and period. Levin (1981) has analyzed a discrete time, age model with a stock recruitment relationship to obtain similar results regarding the dependence of stability on the shape of the age structure. Cushing (1980), using a continuous time model with a complex recruitment function also found that narrowing the ages of reproduction promotes instability. Swick (1981) has analyzed behavior of a continuous model as the slope of the recruitment survival function is tuned beyond the point at which simple oscillations arise. Gurtin and Levine (1982) have also shown that periodic behavior can arise in continuous time models in which recruitment survival depends on total fecundity and total population numbers (see also Levine, this volume).

3. CYCLES IN THE DUNGENESS CRAB FISHERY

The conditions under which cycles could occur and the period of the cycles expected from modelled mechanisms have been used to illuminate the cause of cycles in the crab fishery. The condition under which oscillatory behavior could occur was proposed by Botsford and Wickham (1978) as a field test of potential causal mechanisms. In order to cause the observed cycles a specific mechanism must produce a change in recruitment survival (f) with effective population size that is greater than a certain minimum that depends on age structure. To test each mechanism recruitment mortality must be determined over several years, and the relative effect of each individual on recruitment mortality must be determined. Two mechanisms were identified in the crab population: cannibalism and an egg predator worm whose population size could depend on the size of the female crab population. Field data are being collected on both cannibalism and egg predation.

Botsford and Wickham (1978) also constructed a preliminary age-specific model of the Dungeness crab population. They used cannibalism as the recruitment survival mechanism, in addition to assumed growth rates, mortality rates, constant harvest rates, other life history parameters and an absence of environmental influences. The period of cycles from this model was much less than the period of the observed cycles (six versus nine or ten years). However, since the model was based on many untested assumptions and any variation in these that altered the mean age of influence could change the period, they did not reject cannibalism on the basis of this disparity in periods.

McKelvey et al. (1980) developed a discrete time and age model of this same population with two different ages at which density-dependent recruitment mortality could act. One was at the age of settlement (reflecting cannibalism) and the other was one year earlier (reflecting density-dependent fecundity). Since the latter led to a longer lag between the age at which recruitment mortality acts and the mean age of adult influence on recruitment mortality, the period produced by that mechanism was greater and matched the observed period. The period produced when recruitment mortality acted at the cannibalism point did not match the observed period. On that basis they rejected the cannibalism hypothesis as a cause of the cycles in crab catch.

The preliminary model of Botsford and Wickham (1978) and the model used by McKelvey et al. (1981) both contain significant departures from reality. One is the assumption that harvest rate is a high positive constant. Methot and Botsford (in press) developed an estimate of abundance (Figure 1) that indicated a lagged response of effort to changes in abundance. Botsford et al. (in press) formulated a model in which harvest rate depended on catch in the previous year. As the degree of past dependence was increased two characteristics changed: (a) the period of cycles increased and (b) stability decreased up to a point, then increased. The latter was explained as an initial decrease because a feedback system was being allowed greater overshoot, then increasing stability because that overshoot allowed more individuals to reach older ages, thus producing a more stable age structure.

Use of an unrealistic growth pattern represents a second possible departure from reality. In the age-specific model adopted, all individuals in an age class enter the fishery at age 3.5, whereas in the fishery individuals may enter (become greater than legal size) at ages 3.5, 4.5, and 5.5 years because of size dispersion in the growth process. To reflect this possibility a size-specific model was formulated. Behavior of size-specific models is in general different from age-specific models. However, Botsford (in preparation) has developed a means of collapsing a size-specific model so that it can be interpreted in age-specific terms. Using this approach he showed that if after size structure is introduced the age of largest contribution to the fishery is the same as the age of entrance in the age-specific model, the period does not change appreciably. However, introduction of the likely crab growth pattern alters the age of largest contribution to 4.5 rather than 3.5 years. As a result, both the period of the cycles and stability of the population are increased.

In summary, analyses of the two more realistic features added to the model thus far, a delayed effort response and a slower growth pattern, have increased the period expected from a model with cannibalism from approximately six years in the initial model to more than eight years. Thus, as pointed out by Botsford (1981) in response to McKelvey et al. (1980), their rejection of cannibalism on the basis of disparity between model period and observed period was premature.

4. DISCUSSION

The cause of cyclic behavior of the crab population is not yet known.

Further work on the influence of environmental variables and more detailed modeling of the egg-predator worm are in progress. There are several recruitment mechanisms that mimic observed behavior fairly closely, but we have no guarantee that one of them is the correct one. Formulation and testing of further hypotheses and further testing of current hypotheses through increasingly realistic models are required.

This work points out the minimal value of finding that a model matches observed behavior. (I haven't even mentioned the sunspot model, which produces quite realistic cycles.) More is learned when model behavior does not match population behavior. However, in that case a hypothesis can be rejected only if the remaining structure and parameter values of the model are well known.

REFERENCES

Allen, R.L., and P. Basasibwaki (1974): Properties of age structure models for fish populations. *J. Fish. Res. Board Can.* 31: 1119-1125.

Botsford, L.W. (1981): Comment on cycles in the northern California Dungeness crab population. *Can. J. Fish.Aquat. Sci.* 38(10): 1295-1296.

Botsford, L.W. (1983): Effect of realistic individual growth rates of expected behavior of the northern California Dungeness crab fishery (in preparation).

Botsford, L.W., and D.E. Wickham (1978): Behavior or age-specific, density-dependent models and the northern California Dungeness crab *(Cancer magister)* fishery. *J. Fish. Res. Board Can.* 35: 833-843.

Botsford, L.W., R.D. Methot and W.E. Johnston (in press): Effort dynamics of the northern California Dungeness crab *(Cancer magister)* fishery. *Can. J. Fish. Aquat. Sci.*

Cushing, J.M (1980): Model stability and instability in age-structured populations. *J. Theor. Biol.* 86: 709-730.

Frauenthal, J.C. (1975): A dynamic model for human population growth. *Theor. Popul. Biol.* 8: 64-73.

Gurtin, M.E., and D.S. Levine (1982): On populations that cannibalize their young. *SIAM J. Appl. Math.* 42: 94-107.

Keyfitz, N. (1972): Population waves, *in* T.N.E. Greville (Editor), *Population Dynamics*. Academic Press, Inc., New York.

Lee, R. (1974): The formal dynamics of controlled populations and the echo, the boom and the bust. *Demography* 11: 563-585.

Levin, S. (1981): Age structure and stability in multiple-age spawning populations, *in* T. Vincent and J. Skowronski (Editors), *Renewable Resource Management*. Lecture Notes in Biomathematics, Vol. 40. Springer Verlag, New York, N.Y. 236 p.

McKelvey, R., D. Hankin, K. Yanosko and C. Snygg (1980): Stable cycles in multi-stage recruitment models: an application to the northern California Dungeness crab *(Cancer magister)* fishery. *Can. J. Fish. Aquat. Sci.* 37: 2323-2345.

Methot, R.D., and L.W. Botsford (1982): Estimated pre-season abundance in the California Dungeness crab *(Cancer magister)* fisheries. *Can. J. Fish. Aquat. Sci.* 39: 1077-1083.

Ricker, W.E. (1954): Stock and recruitment. *J. Fish. Res. Board Can.* 11: 559-623.

Rorres, C. (1976): Stability of an age-specific population with density-dependent fertility. *Theor. Pop. Biol.* 10: 26-46.

_____ (1979): Local stability of a population with density-dependent fertility. *Theor. Pop. Biol.* 16: 283-300.

Swick, K.E. (1976): Stability and bifurcation in age-dependent population dynamics. *Theor. Pop. Biol.* 20: 80-100.

Usher, M.B. (1972): Developments in the Leslie Matrix model, pp. 29-60 *in* J.N.R. Jeffers (Editor), *Mathematical Models in Ecology*. British Ecol. Soc. Symp. 12.

HARVESTING UNDER SMALL DEMOGRAPHIC UNCERTAINTY

F.B. Hanson and C. Tier*

The constant effort harvesting of a renewable biological resource is considered. The single year class population grows logistically in the absence of harvesting and is perturbed by small demographic stochasticity. The asymptotic behavior of the expected extinction times and expected harvest yields, as well as its coefficient of variation, are computed as functions of harvesting effort.

Clark (1976) discusses deterministic fishery models including the Schaefer model which combines logistic growth with constant effort harvesting. However, the Schaefer model, being a single species model without delays, can not model the fluctuations often found in fisheries (May, et al. 1978, or Murphy 1977). May, et al. (1978) have examined the effects of a randomly varying environment, with dynamics for the stock size x expressed by,

$$dx(t) = [rx(1-x/k)-Ex]dt + sx(t)dw(t), \qquad (1)$$

where r is the intrinsic growth rate, k is the carrying capacity, E is the effort, w(t) is normalized Gaussian white noise and s is a coefficient. Equation (1) is a stocahstic perturbation of r in the Schaefer model. This is called environmental stochasticity by May because it affects the population as a whole through r and is discussed in detail by Turelli (1977). Model (1) does not permit extinction of the stock, but does have a stationary distribution which was used by May et al. (1978) to find the expected yield and the coefficient of variation for several recruitment models in addition to the logistic. They also document the frequent use of the logistic in fisheries. See also Beddington and May (1977), and see Ludwig (1980) for the economic aspects of this model.

On the other hand, several fisheries have collapsed to very low levels short of extinction. Murphy (1977) cites seven examples of sardine, herring and anchovy fisheries that have suffered widely fluctuating yields and have collapsed, some permanently. This and the work of May et al. (1977) suggest the study of models in which extinction is possible as with random fluctuations due to demographic stochasticity. Demographic stochasticity can be motivated by the diffusion approximation of a branching process (cf. Ludwig 1976, or Tier and Hanson 1981, for ecological examples). The essential biological dynamics can be expressed as

$$dx(t) = [rx(1-x/k)-Ex]dt + (x/N)^{\frac{1}{2}}dw(t), \qquad (2)$$

*Research supported by NSF Grants MCS 79-01718 and MCS 81-01698 and DOE Grant AC02-78ERO-4650.

where N is some large parameter making the noise small. The obvious difference between (1) and (2) is that the noise coefficient in the environmental model (1) is at least linear while in the demographic model (2) it behaves like $x^{\frac{1}{2}}$ as $x \to 0$. The infinitesimal parameters for (2), in the Ito sense, are

$$M(x) = Rx(1-x/K) \quad \text{and} \quad V(x) = x/N \tag{3}$$

where $R = r - E$ and $K = k(1-E/r)$ are the harvesting reduced logistic parameters. We note that Goh (this Proceedings) has indicated that the demographic approach may be more applicable to whales due to their low fecundity.

For this diffusion, extinction at $x = 0$ is now possible in finite time and there is no nontrivial stationary distribution. Two problems are considered. The first is the expected extinction time $T(x)$ which satisfies

$$\frac{1}{2} VT''(x) + MT'(x) = -1; \quad T(+0) = 0; \quad T'(+\infty) = 0. \tag{4}$$

The boundary at $x = 0$ is absorbing and the boundary at infinity is asymptotically reflecting. The second concerns the expected yield,

$$\langle Y \rangle (x^0,t) = \langle E \cdot x \rangle = \int_0^\infty (Ex)p(x,x^0,t)dx \tag{5}$$

where the probability density $p(x,x^0,t)$ satisfies the forward equation

$$P_t = \frac{1}{2} (Vp)_{xx} - (Mp)_x, \quad p(x,x^0,0^+) = \delta(x-x^0), \tag{6}$$

$Vp \to 0$ as $x \to +0$ (absorbing), $Mp - \frac{1}{2}(Vp)_x \to 0$ as $x \to +\infty$ (asmyp. reflecting),

where x^0 is the initial population size (called x in (4)).

The exact solution of (4) using two integrations is

$$T(x) = 2N \int_0^x \int_z^\infty \{\exp[2NH(y,z)]/y\}dydz \tag{7}$$

$$H(y,z) = F(z) - F(y) \quad \text{and} \quad F(x) = R(K-x)^2/2K. \tag{8}$$

The double integral cannot be evaluated in simple closed form, but can be asymptotically approximated for small noise, $N \gg 1$ (more precisely, $RKN \gg 1$ so that N does not have to be large when K is large) using a two dimensional extension of Laplace's method that depends on the critical points and gradient of $H(y,z)$ (cf. Hanson and Tier 1981, 1982). When $x < K$ and $0 < E < r$, we expand the integral about the isolated maximum $(K,0)$ of H to get

$$T(x) \sim (\pi/KR^3N)^{\frac{1}{2}}\exp(KRN)[1 - \exp(-2RNx)]. \tag{9}$$

Outside an "extinction layer", $x > 0(1/N)$, $T(x)$ is exponentially large if $KRN \gg 1$ as exhibited in Figure 1.

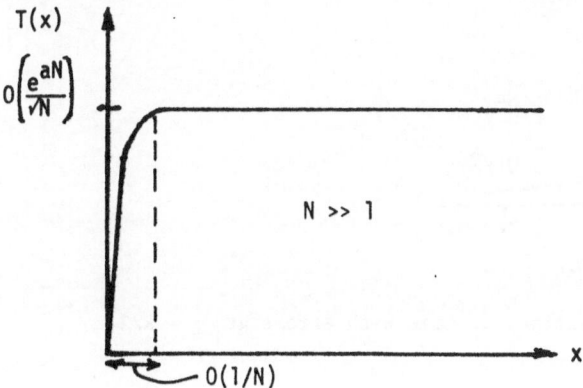

Figure 1   Qualitative variation of extinction time with stock size

We may examine the qualitative variation of $T(x)$ with E by focusing on the MSY (maximum sustainable yield) stock size $x = k/2$ when $E < r/2$. In this case,

$$T(x) \sim (\pi/kr^3 N)^{\frac{1}{2}} \begin{cases} \exp(krN) - 1, & E \to +0 \\ 4[\exp(krN/2)-\exp(-krN/4)], & E \to r^-/2. \end{cases} \tag{10}$$

Note that the exponential order of T changes as the effort E increases as is shown in Figure 2. This is consistent with the real examples of collapses of Clupeoid fisheries described by Murphy (1977).

In the special case when $E > r/2$, or $x > K$ in the general case, the leading order contribution given in (9) must be supplemented by contributions from the nonisolated critical points of H. For example, $T(x)$ is reduced to order $N^{\frac{1}{2}}$ as $E \to r^-$ and this is also indicated in Figure 2.

The second problem is to compute the expected yield, $\langle Y \rangle(x^o,t)$ by finding the transition density, $p(x,x^o,t)$, satisfying (6). The solution of (6) can be expressed in an eigenvalue expansion,

$$p(x,x^o,t) = \sum_{n=1}^{\infty} \exp(-\lambda_n t)P_n(x,x^o) + P_0(x,x^o)\delta(x), \tag{11}$$

where the $P_n(x,x^o)$ satisfy the eigenvalue problem for $n \geq 1$,

$$\frac{1}{2}(VP_n)_{xx} - (MP_n)_x = -\lambda_n P_n, \tag{12}$$

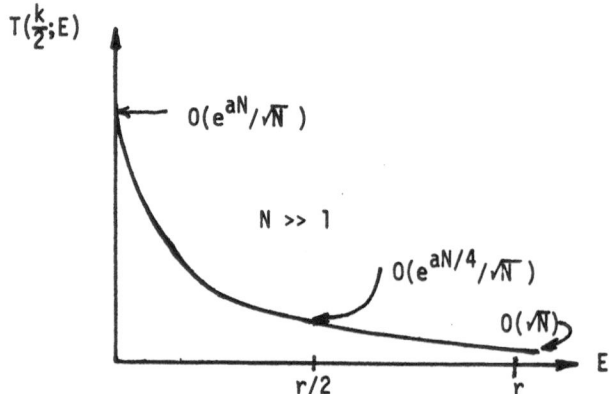

Figure 2 Qualitative variation of extinction time with effort at x = k/2

with the same absorbing and asymptotically reflecting boundary conditions as in (6).
The last term in (11) represents the discrete, extinction absorption probability.
For large time and for nonsmall x, the density is well approximated by the term in
(11) associated with the principal (or minimum) eigenvalue, λ_1, i.e. the quasi-
stationary distribution

$$p(x,x^o,t) \sim P_1(x,x^o)\exp(-\lambda_1 t). \tag{13}$$

The principal eigenvalue is shown in Ludwig (1975), Matkowsky and Schuss (1977),
Tier and Hanson (1981), to be exponentially small and represented by the reciprocal
of the extinction time in (9) for x outside the extinction layer. We have

$$\lambda_1 \sim 1/T_1(E) \sim (kr^3N/\pi)^{\frac{1}{2}}(1-E/r)^2\exp(-krN(1-E/r)^2), \quad N \gg 1, \tag{14}$$

while the other eigenvalues, λ_n, are order n for n > 1 (cf. Tier and Hanson,
1981). Since λ_1 is exponentially small, P_1 may be found by approximating (12)
by

$$\frac{1}{2}(VP_1)_{xx} - (MP_1)_x \simeq 0, \tag{15}$$

with the solution

$$P_1(x,x^o) \sim 2NC_1(x^o)\exp(-NR(x-K)^2/K)/x. \tag{16}$$

This is referred to as Wright's formula and in this case is not integrable at x = 0.
Thus (16) may be thought of as an "outer solution" of singular perturbation theory
and requires a boundary layer correction near x = 0, to give an integrable density,

$$P_1(x,x^o) \sim [C_{11}(x^o)W_1(b,u)+C_{12}(x^o)W_2(b,u)]\exp(RNx)/x, \qquad (17)$$

where W_1 and W_2 are the Whittaker functions used in Tier and Hanson (1981) with parameter $b = \lambda_1/R$ and $u = 2RNx$. The constants C_{11} and C_{12} are determined by requiring integrability at $x = 0$ and matching (17) as $u \to +\infty$ with (16) as $x \to 0$. We find that

$$C_{11} \sim -C_{12} \quad \text{and} \quad C_{12} \sim 2N \exp(-KRN)C_1. \qquad (18)$$

The normalization constant $C_1 = C_1(x^o)$ must be determined from the initial condition (i.e. setting $t = 0$ in (11)). The asymptotics developed for the adjoint problem to (12) by Tier and Hanson (1981) and the biorthogonality of the adjoint eigenfunctions can be used to show that

$$C_1(K) \sim \frac{1}{2} (RK/N)^{\frac{1}{2}}, \qquad N \gg 1, \qquad (19)$$

independent of other normalization constants, where we have taken $x^o = K$. Since $P_1(x,K)$ is now determined to leading order, we compute the expected yeild as

$$<Y> \sim E \exp(-\lambda_1 t) \int_0^{\infty} xP_1(x)dx. \qquad (20)$$

Due to the form of (20), only the outer solution and Laplace's method need be used to obtain,

$$<Y> \sim kE(1-E/r)\exp(-t/T_1(E)), \qquad (21)$$

where $KRN \gg 1$ and $1 \ll t < T_1(E)$. Here $T_1(E)$ is the extinction time associated with λ_1 in (14). In Figure 3, we graph the expected yield $<Y>$ versus effort at a fixed but large value of t using (21). We note that $<Y>$ and its maximum go to zero as $t \to \infty$ because extinction is certain. This is contrasted with the fixed deterministic MSY and the results in May et al. (1978) where a genuine stationary distribution exists.

Using a similar analysis, the coefficient of variation of the yield is given by

$$CV[Y] = \{Var[Y]\}^{\frac{1}{2}}/<Y> \sim \{\exp[t/T_1(E)]-1\}^{\frac{1}{2}}. \qquad (22)$$

The CV becomes large as t gets as large as $T_1(E)$ or larger.

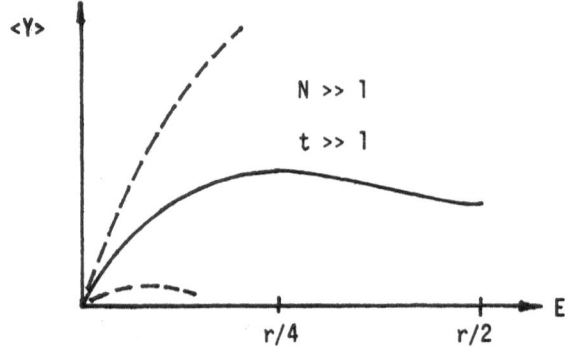

Figure 3 Summary of expected yield versus effort.
The dashed lines represent one standard deviation about the mean
{$\langle Y \rangle (1 \pm CV[Y])$}.

REFERENCES

Beddington, J.R., and R.M. May (1977): Harvesting natural population in a randomly fluctuating environment, *Science* 197:463-465.

Clark, C.W. (1976): *Mathematical Bioeconomics,* Wiley, New York.

Goh, B.S. (this Proceedings): Management of fish and whale populations.

Hanson, F.B., and C. Tier (1982): A stochastic model of tumor growth, *Math. Biosci.* 61:73-100.

Ludwig, D. (1975): Persistence of dynamical systems under random perturbations, *SIAM Rev.* 17:605-640.

_____ (1976): A singular perturbation problem in the theory of population extinction, *SIAM-AMS Proc.* 10:87-104.

_____ (1980): Harvesting strategies for randomly fluctuating populations, *J. Cons. Int. Explor. Mer.* 39(2):168-174.

Matkowsky, B.J., and Z. Schuss (1977): The exit problem for randomly perturbed dynamical systems, *SIAM J. Appl. Math.* 33:365-382.

May, R.M., J.R. Beddington, J.W. Horwood and J.G. Shepherd (1978): Exploiting natural populations in an uncertain world, *Math. Biosci.* 42:219-252.

Murphy, G.I. (1977): Clupeoids, in *Fish Population Dynamics,* J.A. Gulland, ed., Wiley, New York.

Tier, C., and F.B. Hanson (1981): Persistence in density dependent stochastic populations, *Math. Biosci.* 53:89-117.

Turelli, M. (1977): Random environments and stocahstic calculus, *Theor. Pop. Biol.* 12:140-178.

RESOURCE RECOVERY TIME: JUST HOW DESTABILIZING IS IT?

Len Nunney

1. INTRODUCTION

Time delays are an inevitable part of all ecological systems. Energy transfer between trophic levels is more-or-less instantaneous but it takes some time for
energy input to be reflected in reproductive output and it generally takes even
longer for the offspring produced to themselves reach reproductive maturity. In
dynamic terms this effect prevents a population from responding immediately to favorable conditions, which in turn has the potential of destroying stabilizing negative feedback loops. A general rule which has been suggested is that instability
results if a system is subject to a time lag which is long compared to the natural
period of the system (May, 1973a; Maynard Smith, 1974). More simply, if a time
delay is made long enough then eventually a locally stable equilibrium will become
unstable (see for example Adams et al., 1980 and Cushing, 1980 and the references
therein). If such a conclusion is correct then we must be very cautious in the application of results derived from continuous differential models. It also suggests
that population cycles can conveniently be explained by invoking a single long time
lag.

Is the general rule regarding the effect of time lags too simplistic? I will
show that one source of delay, resource recovery time, does not inevitably lead to
instability and in fact can result in greater stability.

2. THE "DELAYED LOGISTIC" EQUATION

May (1973a) suggested that the delayed form of the logistic model used by
Hutchinson (1948) may be interpreted as describing the dynamics of herbivores grazing
upon vegetation which takes a time T to recover:

$$\frac{dN(t)}{dt} = rN(t)[1 - \frac{N(t-T)}{K}]$$
(1)

Before discussing whether or not the delayed logistic equation (1) reflects
the effects of a resource recovery time delay, consider the local stability of an
equilibrium of (1) as T is increased. The largest real part of the eigenvalues
of the system is shown in Figure (1a), from which it can be seen that the equilibrium is locally stable if T < π/2r (see May, 1973a). Also of interest is that for
a small but non-zero value of T the system is at its most stable. A similar
pattern was shown by a related model considered by Beddington and May (1975). The
cause of the effect can be seen by considering a growing population approaching K
(the carrying capacity and equilibrium of the model (1)). The delayed regulation

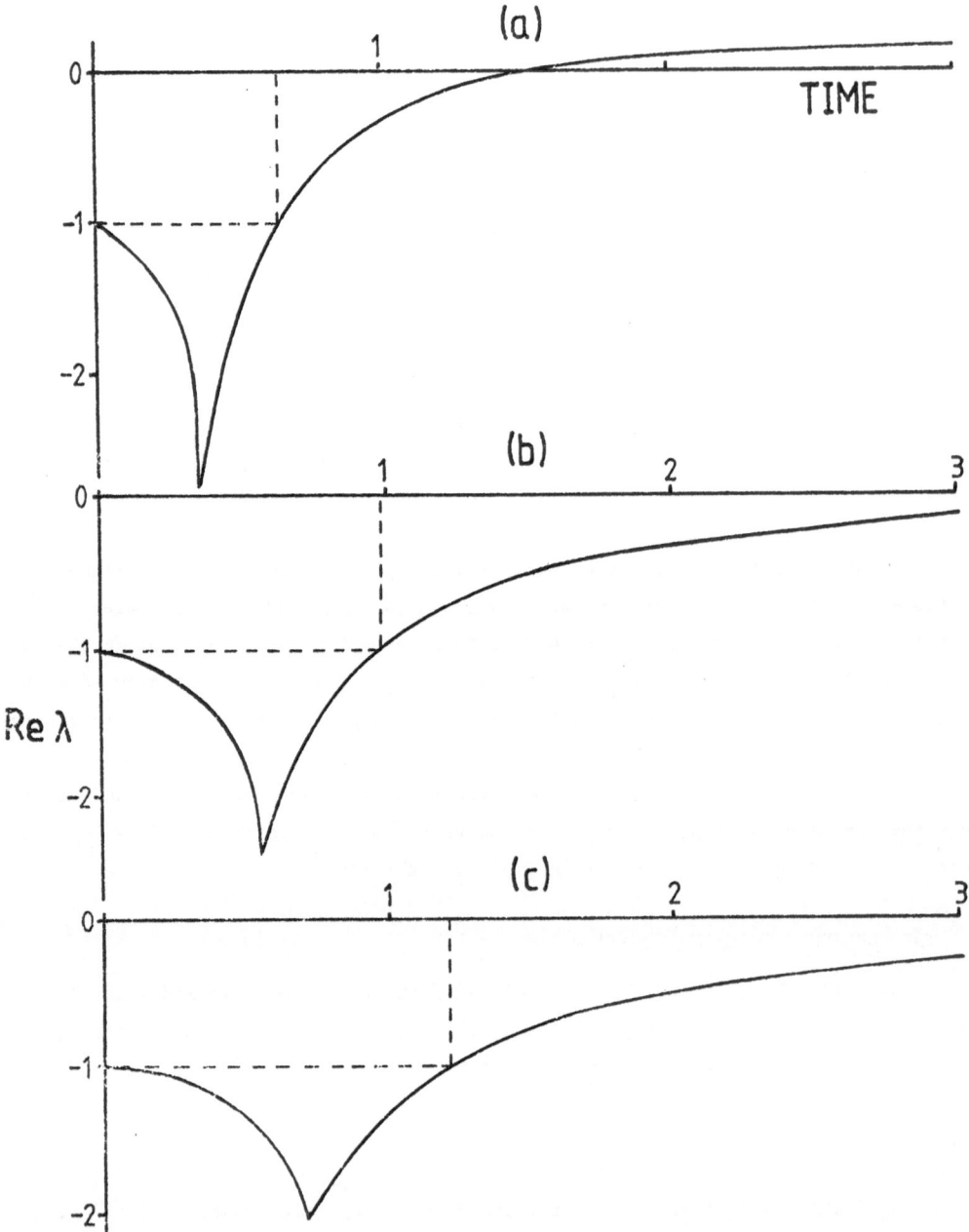

Figure 1   Stability in time-lagged forms of the logistic equation.   (a) the "delayed
logistic" equation   (1);   (b) the equation (5) with   (dK/dN) =-1;   (c) the
equation (5) with   (dK/dN) = -0.5.   Each curve is drawn such that the
largest real value of the eigenvalues   (Re λ) is scaled to its value when
the time lag is zero.   The reciprocal of this value is used to scale the
time lag (TIME) axis.   The dashed lines define the region of enhanced
stability.

term determines that, close to K, the population grows faster than it would if undelayed because the feedback reflects an historical population farther from K. Initially, as T is increased from zero, this results in a faster approach to the equilibrium until the approach becomes so rapid that it causes a growing population to overshoot K. At this point the system begins to be governed by a complex root and exhibits damped oscillation. As T is increased further the overshoot becomes more marked and, eventually, undamped.

Note that the delayed logistic equation (1), despite being the paradigm for the rule relating the natural period (taken as the reciprocal of the largest real part of the eigenvalues; in this case $1/r$) to the maximum time delay compatible with stability, also exhibits increased stability for small T which suggests a careful examination of some alternative models.

3. RESOURCE RECOVERY

Gurney et al. (1980) have criticized the use of the delayed logistic equation (1) when the time delay arises from generation time. However this leaves the possibility that the delay T reflects a resource recovery time (May, 1973a). It is a plausible suggestion, but it remains unjustified and, in the light of the following argument, appears unjustifiable.

The ratio N/K in the logistic equation is generally considered to reflect the ratio of demand to the supply of resources. In other words, it is K, the carrying capacity of the environment, which measures resource availability. It the resources take a time T to recover then this will be reflected in the supply term K and not in the demand. It follows that historical effects upon the resource level are most appropriately included in the logistic model by making K a function of N(t-T):

$$\frac{dN(t)}{dt} = rN(t)\left[1 - \frac{N(t)}{K(N(t-T))}\right] \tag{2}$$

Linearizing (2) for small displacements x(t) we obtain:

$$\dot{x}(t) + rx(t) - r(dK/dN)x(t-T) = 0 \tag{3}$$

where dK/dN is evaluated at equilibrium. Given that K is a decreasing function of N, an equilibrium of (3) will exhibit absolute asymptotic stability provided:

$$(dK/dN)_{\hat{N}} > -1 \tag{4}$$

(El'sgol'ts and Norkin, 1973). In other words, given condition (4), the equilibrium of a system defined by (2) will be locally stable regardless of the length of the time delay. The limit of condition (4) occurs if:

$$K(N(t-T)) = \hat{N}^2/N(t-T) \tag{5}$$

and if K is any less responsive than this then the system will be absolutely stable. Since K is now a function, $\hat{N}$ defines the equilibrium.

The stability of a system obeying the relationship (5) and of a system which is half as responsive ($dK/dN = 0.5$) is shown in figure (1b) and (1c) respectively. Note that the figures (1) are scaled in terms of their return times (i.e. natural period) and that in both figures (1b) and (1c) the systems do no worse in terms of stability measured by the largest real part of their eigenvalues when subject to a delay equal to their return time than when there is no delay. Even in the case of the delayed logistic equation (1) (see figure (1a)) stability is enhanced by a time delay provided that it is less than about 65% of the return time. Bearing in mind that equation (2) is almost certainly more realistic than the delayed logistic we can propose a new general rule: the presence of a time delay due to resource recovery will almost certainly enhance stability provided that it is less than the return time of the undelayed system.

4. TWO-LEVEL MODELS

From the preceding section we can draw two conclusions: realistic time-delayed systems may exhibit absolute stability and they may exhibit enhanced stability compared to an equivalent non-delayed system. However, in the modelling of ecological systems realism is a somewhat relative term since it can always be improved upon and it is important to examine whether these conclusions are robust. One important modification which can be made is to consider explicitly the dynamics of the resources. In such a two-level model, a delay due to resource recovery becomes a maturational time lag in the lowest trophic level. Newly recruited resources are unavailable as a food supply for a period T, and during this maturation period it is also assumed that the resources are reproductively immature. An appropriate model is given by:

$$\frac{dN(t)}{dt} = N(t)F(R(t))$$

$$\frac{dR(t)}{dt} = B(R(t-T)) - D(R(t)) - P(N(t),R(t)) \tag{6}$$

where F is an increasing function converting resource intake by the "predators" into dynamic changes at the higher trophic level. The functions B, D and P reflect birth, death and predation of the resource population, R. It is assumed that P is an increasing function of the nubmer of predators, N.

The detailed analysis of the system (6) is presented elsewhere (Nunney, in preparation). Of importance for the present discussion is that an equilibrium of (6) exhibits absolute stability provided that:

$$\frac{\partial (D+P)}{\partial R} > \left| \frac{dB}{dR} \right| \tag{7}$$

at the equilibrium. Notice that it is the absolute magnitude of the time-lagged birth process which is of importance in promoting absolute stability. Compare (7) to the equivalent condition for local stability when $T = 0$:

$$\frac{\partial (D+P)}{\partial R} > \frac{dB}{dR} \tag{8}$$

One system which is at the limit of condition (7) is the Lotka-Volterra predator-prey model. Adding the time lag we have:

$$\frac{dN(t)}{dt} = N(t)[caR(t)-e]$$
$$\frac{dR(t)}{dt} = rR(t-T) - aN(t)R(t) \tag{9}$$

The linearized system can be expressed in terms of r, the reproductive rate of the resource R, and e, the death rate for the predator population N. Measuring small displacements of R by $x(t)$ we have:

$$\ddot{x}(t) + r\dot{x}(t) + erx(t) - r\dot{x}(t-T) = 0 \tag{10}$$

Using (10) to calculate eigenvalues we find that the local stability of the Lotka-Volterra model is always enhanced by the addition of a resource recovery time. The unlagged Lotka-Volterra equations have an infinite return time, but the period of their oscillations is $2\pi/(er)^{\frac{1}{2}}$. From figure (2) it can be seen that the enhancement of stability is only of consequence for time lags that are less than the period of the system, a result in complete accord with the findings from the single-level models.

5. DISCUSSION

The question posed in the title can now be answered. Time delays due to resource recovery are not the potent source of instability that they might be expected to be; in fact they are not very destabilizing at all. Very often even infinite time delays cannot destabilize an otherwise stable equilibrium; both the single-level model (2) and the two-level model (6) exhibit zones of absolute asymptotic stability. In both types of model the criterion for absolute stability relies on a ratio of effects acting in the present to effects acting in the past. A similar result arises in the analysis of predator-prey difference equations (Nunney, 1980).

Modifications to the simple relationship between the length of a time delay and return time have been made by May (1973b) in the context of trophic interactions and by Beddington and May (1975) in the context of unstable equilibria. In addition,

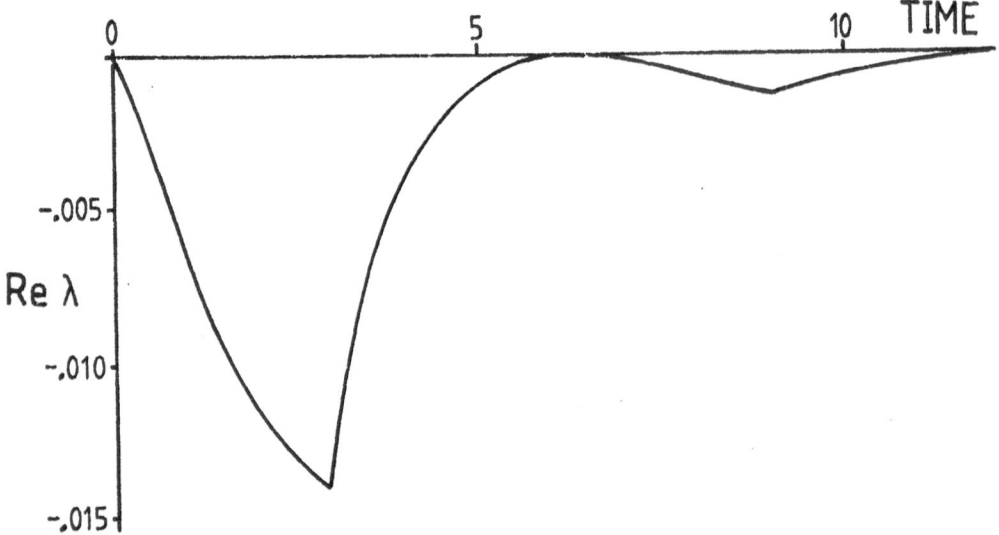

Figure 2   Stability and the Lotka-Volterra predator-prey model. The axes are as
in figure (1) except that the time lag axis is scaled to $(er)^{-\frac{1}{2}}$ and
the Re λ axis is scaled by the choice of r = e = 0.1.

Cushing (1980) noted a complex interaction between maturation time and stability.
However, a general reappraisal of the effect of time delays has not appeared to be
necessary. The present work suggests that, at least when time delays arise from
resource recovery, time lags which are shorter than the return time of the system
tend to increase local stability. This result, taken in conjunction with the finding
that even long time lags may be muted in their effects upon stability, suggests that
the citing of time delays as a source of natural instability requires careful
justification.

REFERENCES

Adams, V.D., D.L. DeAngelis, and R.A. Goldstein (1980): Stability analysis of the
time delay in a host-parasitoid model, *J. Theoret. Biol.* 83:43-62.

Beddington, J.R., and R.M. May (1975): Time delays are not necessarily destabilizing, *Math. Biosci.* 27:109-117.

Cushing, J.M. (1980): Model stability and instability in age structured populations,
J. Theoret. Biol. 86:709-730.

El'sgol'ts, L.E., and S.B. Norkin (1973): *Introduction to the Theory and
Application of Differential Equations with Deviating Arguments,* Academic
Press, New York.

Gurney, W.S.C., S.P. Blythe, and R.M. Nisbet (1980): Nicholson's blowflies revis-
ited, *Nature,* 287:17-21.

Hutchinson, G.E. (1948): Circular causal systems in ecology, *Ann. N.Y. Acad. Sci.* 50:221-246.

May, R.M. (1973a): *Stability and Complexity in Model Ecosystems*, Princeton Univ. Press, Princeton, New Jersey.

_____(1973b): Time-delay versus stability in population models with two and three trophic levels, *Ecology* 54:315-325.

Maynard Smith, J. (1974): *Models in Ecology*, Cambridge Univ. Press.

Nunney, L. (1980): The influence of the type 3 (sigmoid) functional response upon the stability of predator-prey difference models, *Theoret. Pop. Biol.* 18: 257-278.

PART VIII:

EPIDEMOLOGY AND IMMUNOLOGY

INFLUENCE OF INFECTIOUS DISEASE ON THE GROWTH OF A
POPULATION WITH THREE GENOTYPES

K. Beck*, J.P. Keener and P. Ricciardi

1. INTRODUCTION

Studies of population dynamics must ultimately take into account the effects of population genetics. This paper deals with the evolutionary pressure exerted on the host population from the presence of a vector-borne infectious disease. Two historical examples of such influence are the maintenance of the sickle cell anemia gene in the regions of Africa where malaria is endemic, and evolution of genetic resistance to the disease myxomatosis in the wild rabbit population of Australia, (Allison 1954, Caldwell and Rucknagel 1973, Fenner and Ratcliffe 1965). In the latter case an important factor in the history of myxomatosis was the evolution of the virus itself which this paper will not deal with. The model presented here is a one locus - two allele situation in which the three genotypes differ slightly in their susceptibility to, recovery from, and death from the infectious disease. Intrinsic death rates are also allowed to vary slightly. The model is expressed in terms of differential equations. Similar use of differential equations in problems of population genetics may be found in Anderson and May (1979), Beck (1982), Butler et al (1982), Freedman and Waltman (1978), Hoppensteadt (1975), Kemper (ms.), May and Anderson (1979).

2. DEVELOPMENT OF THE MODEL

The model is derived from a simple epidemic model where the host population consists of individuals who are susceptible to the disease, infective, and recovered from (and immune to) the disease. Let x,y,z be the number of individuals who are susceptible, infective and recovered, respectively. Then an individual moves from class x to class y at a rate aw, from y to z at rate r or else dies at rate d. This gives us

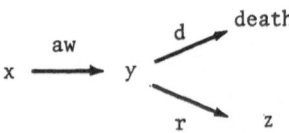

where w is the number of infective vectors.

In addition the host population is assumed to consist of three genotypes (the AA, Aa, aa genotypes of a one locus - two allele problem). This gives us

---

*Research from this paper was partially supported by NSF Grant MCS-80-15359.

$$\dot{x}_1 = \frac{u^2}{N} b - (\Delta(N) + \varepsilon\mu_1 N) \frac{x_1}{N} - (a + \varepsilon\alpha_1) x_1 w$$

$$\dot{x}_2 = \frac{2uv}{N} b - (\Delta(N) + \varepsilon\mu_2 N) \frac{x_2}{N} - (a + \varepsilon\alpha_2) x_2 w$$

$$\dot{x}_3 = \frac{v^2}{N} b - (\Delta(N) + \varepsilon\mu_3 N) \frac{x_3}{N} - (a + \varepsilon\alpha_3) x_3 w$$

$$\dot{y}_1 = (a + \varepsilon\alpha_1) x_1 w - (\Delta(N) + \varepsilon\mu_1 N) \frac{y_1}{N} - (r + \varepsilon\eta_1 + d + \varepsilon\delta_1) y_1$$

$$\dot{y}_2 = (a + \varepsilon\alpha_2) x_2 w - (\Delta(N) + \varepsilon\mu_2 N) \frac{y_2}{N} - (r + \varepsilon\eta_2 + d + \varepsilon\delta_2) y_2 \qquad (1)$$

$$\dot{y}_3 = (a + \varepsilon\alpha_3) x_3 w - (\Delta(N) + \varepsilon\mu_3 N) \frac{y_3}{N} - (r + \varepsilon\eta_3 + d + \varepsilon\delta_3) y_3$$

$$\dot{z}_1 = (r + \varepsilon\eta_1) y_1 - (\Delta(N)) + \varepsilon\mu_1 N) \frac{z_1}{N}$$

$$\dot{z}_2 = (r + \varepsilon\eta_2) y_2 - (\Delta(N)) + \varepsilon\mu_2 N) \frac{z_2}{N}$$

$$\dot{z}_3 = (r + \varepsilon\eta_3) y_3 - (\Delta(N)) + \varepsilon\mu_3 N) \frac{z_3}{N}$$

where x_1, x_2, x_3 are the number of susceptibles of type AA, Aa, aa respectively, y_1, y_2, y_3 are the number of infectives of types AA, Aa, aa respectively, and z_1, z_2, z_3 are the number of recovereds of type AA, Aa, aa respectively, $u = x_1 + y_1 + z_1 + \frac{1}{2}(x_2 + y_2 + z_2)$, $v = x_3 + y_3 + z_3 + \frac{1}{2}(x_2 + y_2 + z_2)$ and $N = u + v$. In (1) b and $\Delta(N)/N$ represent the natural intrinsic birth and death rates of the entire population, and w is the number of infective vectors which for simplicity is assumed to be constant. Also, $a + \varepsilon\alpha_i$, $i = 1,2,3$, represent the rate at which the genotypes contract the disease; $d + \varepsilon\delta_i$, $i = 1,2,3$, represent the death rates from the disease; $r + \varepsilon\eta_i$, $i = 1,2,3$, represent the recovery rates from the disease. Note that $(u^2)/(N^2)$, $(2uv)/(N^2)$, $(v^2)/(N^2)$ are the probabilities of being born of type AA, Aa, aa, respectively. We will assume throughout that a, r, d, b > 0, $\Delta(N) > 0$ for $N > 0$ and also that $\Delta(N)/N$ remains bounded as $N \to 0$.

Note that u is one half of the number of A genes and v is one half of the number of a genes. The size of u and v is our ultimate concern and hence, after dividing each component of (1) by N and renaming $(x_1)/N$ as x_1 etc., we will work with the system

$$\dot{x}_1 = b(u^2 - x_1) - (a + \varepsilon\alpha_1) x_1 w - \varepsilon\mu_1 x_1 + x_1 \beta$$

$$\dot{x}_2 = b(2uv - x_2) - (a + \varepsilon\alpha_2) x_2 w - \varepsilon\mu_2 x_2 + x_2 \beta$$

$$\dot{x}_3 = b(v^2 - x_3) - (a + \varepsilon\alpha_3) x_3 w - \varepsilon\mu_3 x_3 + x_3 \beta$$

$$\dot{y}_1 = (a + \varepsilon\alpha_1) x_1 w - by_1 - (r + \varepsilon\eta_1 + d + \varepsilon\delta_1) y_1 - \varepsilon\mu_1 y_1 + y_1 \beta \qquad (2)$$

$$\dot{y}_2 = (a+\varepsilon\alpha_2)x_2 w - by_2 - (r+\varepsilon\eta_2+d+\varepsilon\delta_2)y_2 - \varepsilon\mu_2 y_2 + y_2\beta$$

$$\dot{y}_3 = (a+\varepsilon\alpha_3)x_3 w - by_3 - (r+\varepsilon\eta_3+d+\varepsilon\delta_3)y_3 - \varepsilon\mu_3 y_3 + y_3\beta$$

$$\dot{z}_1 = (r+\varepsilon\eta_1)y_1 - bz_1 - \varepsilon\mu_1 z_1 + z_1\beta$$

$$\dot{u} = -(d+\varepsilon\delta_1)y_1 - \tfrac{1}{2}(d+\varepsilon\delta_2)y_2 - \varepsilon\mu_1(x_1+y_1+z_1) - \tfrac{1}{2}\varepsilon\mu_2(x_2+y_2+z_2) + u\beta$$

where $\beta = \sum\limits_{i=1}^{3} [\varepsilon\mu_i(x_i+y_i+z_i) + dy_i + \varepsilon\delta_i y_i]$. Note that $\sum\limits_{i=1}^{3} (x_i+y_i+z_i) = 1$.

3. BEHAVIOR OF THE MODEL FOR $\varepsilon = 0$.

When $\varepsilon = 0$ each of the genotypes is identical with respect to contracting, recovering from, and dying from the disease. In this case we will show that Hardy-Weinberg proportions are approached.

Let $T_x = x_1 + x_2 + x_3$, $T_y = y_1 + y_2 + y_3$. From (2) with $\varepsilon = 0$ we obtain the system

$$\dot{T}_x = b(1-T_x) - awT_x + dT_x T_y$$

$$\dot{T}_y = awT_x - bT_y - (r+d)T_y + dT_y^2. \tag{3}$$

THEOREM 1. *The region* $U = \{(T_x,T_y) \mid T_x, T_y > 0 \text{ and } T_x + T_y < 1\}$ *is a positively invariant region of* (3). U *contains precisely one critical point of* (3) *which is the ω-limit set of all trajectories of* (3) *with initial conditions in* U.

Sketch of Proof. The Poincaré index of the boundary of U is 1 and hence U contains at least one critical point. One may easily show that another critical point of (3) lies outside of U. It can be shown that if two critical points of (3) lie in U then one of them is a saddle point, contradicting Poincaré index 1. The absence of limit cycles follows from an application of Dulac's criterion with density function $B(T_x,T_y) = 1/(T_x T_y)$.

LEMMA 2. *Let* $(\hat{T}_x,\hat{T}_y)$ *be the critical point of* (3) *in the region* U. *Then*

$$\lim_{t\to\infty} \frac{x_1 + \tfrac{1}{2}x_2}{u} = \lim_{t\to\infty} \frac{x_3 + \tfrac{1}{2}x_2}{v} = \lim_{t\to\infty} \frac{x_2}{2uv} = \hat{T}_x \ ;$$

$$\lim_{t\to\infty} \frac{y_1 + \tfrac{1}{2}y_2}{u} = \lim_{t\to\infty} \frac{y_3 + \tfrac{1}{2}y_2}{v} = \lim_{t\to\infty} \frac{y_2}{2uv} = \hat{T}_y \ .$$

PROOF. Let $P = (x_1 + \tfrac{1}{2}x_2)/u$, $Q = (y_1 + \tfrac{1}{2}y_2)/u$. Then

$$P' = b(1-P) - awP + dpQ$$

$$Q' = awP - bQ - (r+d)Q + dQ^2$$

which is precisely system (3). The other limits are similar.

LEMMA 3. $\lim_{t \to \infty} u = p$ *exists and*

$$\lim_{t \to \infty} \frac{x_1 + y_1}{x_2 + y_2} = \lim_{t \to \infty} \frac{x_1}{x_2} = \lim_{t \to \infty} \frac{y_1}{y_2} = \frac{p^2}{2p(1-p)} \ ,$$

$$\lim_{t \to \infty} \frac{x_2 + y_2}{x_3 + y_3} = \lim_{t \to \infty} \frac{x_2}{x_3} = \lim_{t \to \infty} \frac{y_2}{y_3} = \frac{2p(1-p)}{(1-p)^2} \ .$$

Lemma 3 tells us that Hardy-Weinberg proportions are approached. Combining Lemmas 2 and 3 one may show that solutions of (2) for $\varepsilon = 0$ approach $(x_1, x_2, x_3, y_1, y_2, y_3, z_1, u) = (p^2 \hat{T}_x, 2p(1-p) \hat{T}_x, (1-p)^2 \hat{T}_x, p^2 \hat{T}_y, 2p(1-p) \hat{T}_y, (1-p)^2 \hat{T}_y, p^2(1 - \hat{T}_x - \hat{T}_y), p)$.

4. THE GENERAL CASE

Returning to system (2) we see that this system may be written in the form

$$\frac{ds}{dt} = F(s) + \varepsilon G(s) \tag{4}$$

where $s = (x_1, x_2, x_3, y_1, y_2, y_3, z_1, u)$, $F(s)$ and $G(s)$ are 8-dimensional vectors. We have already seen in the last section that $F(s) = 0$ has a one parameter family of solutions $s_0(p) = (p^2 \hat{T}_x, 2p(1-p) \hat{T}_x, (1-p)^2 \hat{T}_x, p^2 \hat{T}_y, 2p(1-p) \hat{T}_y, (1-p)^2 \hat{T}_y, p^2(1 - \hat{T}_x - \hat{T}_y), p)$ where $(\hat{T}_x, \hat{T}_y)$ is the critical point of (3) in the region U.

THEOREM 4. *For* ε *sufficiently small, solutions of* (4) *are of the form* $s(t) = s_0(p(t)) + \varepsilon s_1(t, \varepsilon)$ *where* s_1 *is uniformly bounded for all* $t > 0$ *provided that* $p(t)$ *is a solution of*

$$\frac{dp}{d\tau} = \frac{p(1-p)\hat{T}_y}{(d\hat{T}_x^2 \hat{T}_y aw + db \hat{T}_y^2 - \hat{T}_x baw)} [-p\gamma_1 + (2p-1)\gamma_2 + \gamma_3(1-p)]$$

$$+ p(1-p)[-p\mu_1 + (2p-1)\mu_2 + (1-p)\mu_3] \tag{5}$$

where $\gamma_i = b[\delta_i(d\hat{T}_y - aw\hat{T}_x) + \eta_i d\hat{T}_y] + d\hat{T}_x \alpha_i (\hat{T}_x aw - b)]$, $i = 1, 2, 3$, $\tau = \varepsilon t$, *and that* (5) *has a uniform-asymptotically stable critical point.*

Sketch of Proof. The null space of $[F'(s_0(p))]^T$ is generated by

$$\hat{Q}(p) = \left(\frac{\hat{T}_x aw}{b} (p-1), \ \frac{\hat{T}_x aw}{b} (p-\tfrac{1}{2}), \ \frac{\hat{T}_x aw}{b} p, \ p-1, \ p-\tfrac{1}{2}, \ p, \ 0, \ \frac{\hat{T}_x aw}{\hat{T}_y d} \right).$$ (This calculation

and several which follow were done with the aid of a computer and the symbolic

manipulator REDUCE.) Let $Q(p) = \dfrac{\hat{Q}(p)}{\langle \hat{Q}(p), s_0'(p) \rangle}$ where $\langle \cdot, \cdot \rangle$ denotes the usual

inner product for vectors. We look for solutions of (4) of the form $s(t) = s_0(p(t)) + \varepsilon s_1(t, \varepsilon)$. This will be a unique representation of s if in addition we require that $\langle Q(p), s_1'(t,\varepsilon) \rangle = 0$. Differentiating (4) yields

$$s_0' p' + \varepsilon \frac{\partial s_1}{\partial t} = F(s_0) + \varepsilon F'(s_0)s_1 + R_1(\varepsilon^2) + \varepsilon G(s_0) + \varepsilon^2 G'(s_0)s_1 + R_2(\varepsilon^3).$$

Multiplying by Q (and noting that $F(s_0) = 0$) yields

$$p' = \varepsilon \langle Q, G(s_0) \rangle + O(\varepsilon^2).$$

Hence
$$\frac{dp}{d\tau} = \langle Q, G(s_0) \rangle + O(\varepsilon)$$

where $\tau = \varepsilon t$. This is Equation (5). The validity of this approximation follows from Hoppensteadt (1966).

The following lemma determines the sign of the coefficients in (5).

LEMMA 5. $d\hat{T}_y - aw\hat{T}_x < 0$ *and* $d\hat{T}_x^2\hat{T}_y aw + bd\hat{T}_y^2 - \hat{T}_x baw < 0$.

COROLLARY 6. *Let* $\mu_1 = \mu_2 = \mu_3$. *For* ε *sufficiently small,* $u(0) > 0$, $v(0) > 0$, *if* $\gamma_3 > \gamma_2 \ge \gamma_1$ *then* $\lim_{t \to \infty} u(t) = 0$.

COROLLARY 7. *Let* $\mu_1 < \mu_2 < \mu_3$, $\gamma_2 > \gamma_1$, $\gamma_2 > \gamma_3$, $u(0) > 0$, $v(0) > 0$. *For* ε *sufficiently small solutions of* (4) *are of the form*

$$s(t) = s_0 \left(\frac{1}{1 + \dfrac{c\gamma_1 - c\gamma_2 - \mu_1 + \mu_2}{c\gamma_3 - c\gamma_2 - \mu_3 + \mu_2}} \right) + \varepsilon s_1(t,\varepsilon).$$

whenever $c(\gamma_1 - \gamma_2) > \mu_2 - \mu_1$ *and* $c(\gamma_3 - \gamma_2) > \mu_2 - \mu_3$ *where*

$$c = \frac{\hat{T}_y}{bd\hat{T}_y^2 - b\hat{T}_x^2 aw + d\hat{T}_y \hat{T}_x^2 aw}.$$

5. DISCUSSION

In the case of increased genetic resistance to myxomatosis in the rabbit population of Australia, it is assumed that one gene conveys an advantage in terms of a reduced transmission rate, a decreased death rate or an increased recovery rate. At the same time it is assumed that the gene does not convey any advantage or disadvantage in the absence of myxomatosis. This is the case covered by Corollary 6. The condition $\mu_1 = \mu_2 = \mu_3$ says that the gene does not change the intrinsic death rate, while $\gamma_3 > \gamma_2 \geq \gamma_1$ means that genotype a has an advantage with respect to myxomatosis. Corollary 6 then tells us that genotype aa is the only survivor, i.e. gene A dies out of the population.

The case involving malaria and the sickle cell anemia gene is more compli-cated. There is evidence that an individual with sickle cell trait, i.e. the heterozygote, has an increased resistance to malaria over either of the homozygotes. So in this case $\gamma_2 > \gamma_1$ and $\gamma_2 > \gamma_3$. (Let A be the normal gene and a the sickling gene). But the individual with sickle cell anemia, i.e. the aa homo-zygote, has serious disadvantages in other respects. Such individuals rarely live to reproduce. The heterozygote also has some disadvantage in terms of a decreased life expectancy. Hence $\mu_1 > \mu_2 > \mu_3$. This is the special case covered by Corollary 7. The constant c is a negative number and so the conditions $c(\gamma_1-\gamma_2) > \mu_2 - \mu_1$ and $c(\gamma_2-\gamma_2) > \mu_2 - \mu_3$ are measures of the advantage conveyed by the gene versus the disadvantages conveyed by the gene. If the advantage conveyed to the heterozygote is larger than the accompanying disadvantage in terms of decreased life expectancy than the sickling gene will persist.

The above examples are of course only two out of many possible cases. In general, the possible outcomes are $p \to 0$, $p \to 1$ or $p \to p_0$, $0 < p_0 < 1$, depending on parameters. As long as this convergence is exponential, Theorem 4 is valid. Theorem 4 is a very general result which the reader may use to work out many spe-cial cases.

REFERENCES

Allison, A.C. (1954): Protection afforded by sickle-cell trait against subtertian malaria infection, *British Med. Jour.*: 290-294.

Anderson, R.M. and R.M. May (1979): Population biology of infectious diseases: Part I, *Nature*, 280: 361-367.

Beck, Karen (1982): A model of the population genetics of cystic fibrosis in the United States. *Math. Biosci.* 58: 243-257.

Butler, G.J., H.I. Freedman and Paul Waltman (1982): Global dynamics of a selec-tion model for the growth of a population with genotypic fertility differ-ences. *J. Math. Biol.* 14: 25-35.

Caldwell, E.S. and D.L. Rucknagel (1973): The genetics of sickle cell syndrome, in *Sickle Cell Disease*. E.F. Mammen, G.F. Anderson and M.I. Barnhart (Editors), F.K. Schattaner Verlag.

Fenner, F. and R. Ratcliffe (1965): *Myxomatosis*, Cambridge University Press.

Freedman, H.I. and Paul Waltman (1978): Predator influence in the growth of a population with three genotypes. *J. Math. Biol.* 6: 367-374.

Hoppensteadt, Frank (1966): Singular perturbations on the infinite interval. *Trans. Amer. Math. Soc.* 123: 521-535.

_____ (1975): *Mathematical Theories of Populations: Demographics, Genetics and Epidemics.* SIAM Regional Conference Series in Applied Mathematics.

Kemper, John T. (manuscript): The evolutionary effect of endemic infectious disease: continuous models.

May, Robert M. and R.M. Anderson (1979): Population biology of infectious diseases: Part II. *Nature* 280: 455-461.

INTERACTION OF ANTITUMOR CELLS: COMPETITION AND INTERFERENCE

Stephen Merrill

ABSTRACT

The competition for tumor cells by several antitumor populations and the interference resulting from various cell products is examined. The setting investigated here involves natural killer (NK) cells interacting with a small tumor and macrophages being activated and attacking the same tumor. Activated macrophages are potent producers of prostaglandins which suppress natural killer activity while interferon produced by NK cells enhances macrophage killing.

A mathematical model of the interactions of the two cell-types is examined with respect to the system's ability to eliminate a tumor. The question addressed is a problem of biologic control. A pest (tumor) is controlled by a predator (NK's) until a second predator (or pesticide) is introduced, possibly allowing the pest to escape the natural control.

1. INTRODUCTION

Several cell-types, natural killer (NK) cells, macrophages and T lymphocytes, have the ability to destroy tumor cells. Along with their cytotoxic abilities, each of these antitumor cells also produces substances which affect their functioning as well as the functioning of the other cell populations. These substances include interferons, prostaglandins and lymphokines.

In situations where several cell-types may be actively present in a single tumor, not only is competition for the tumor present but also the interference and enhancement resulting from the products of these cells. As the goal of the immune system is to drive the tumor to extinction, one wonders if this competition and interference might aid in tumor survival?

NK cells are lymphoid cells of the spleen and peripheral blood with the ability to lyse tumor cells (primarily of carcinogen or spontaneous origin) without prior exposure to the tumor ("priming"). Macrophages also have some natural cytotoxicity but the greatest antitumor effects occur after nonspecific macrophage activation. This activation is usually accomplished by administration of BCG, C. Parvum, LPS or other materials of bacterial origin. Herberman (1980) contains extensive descriptions of the properties of NK cells and macrophages with regards to tumor cytotoxicity.

It has long been recognized (Evans, 1972) that macrophages are plentiful at a tumor site, although a mechanism to explain this phenomen and the physiological role of these macrophages is not known. Recent evidence suggests however that this macrophage infiltration is unrelated to an immune response of the host or the presence of tumor-associated antigens on the tumors (Evans and Eidlen, 1981). Tumor products have been shown to attract macrophages (Metzer, et al., 1975) whose presence may be promoting tumor growth (Salmon and Hamberger, 1978). The tumor products are generally effective only on activated macrophages which show an increase both in cytotoxic activity and prostaglandin (PG) production (Russell, et al., 1977).

PG is suggested as a tumor promoting substance by the following:

1) PG suppresses NK cell activity;
2) PG is present in many tumors and elevated levels are found in blood and urine of patients bearing tumors;
3) mice have an increase in carcinogen induced tumors when PG's are administered while PG inhibitors have reduced the growth of experimental tumors while restoring NK activity; (Herberman, 1981).

A T-lymphocyte response against a tumor has also been suggested as an avenue of escape for a tumor (Prehn, 1976) and suppressor cells (both macrophages and suppressor T cells) have been recognized as branches of the immune response which favor tumor growth (Naor, 1979).

In this paper, the effect of activated macrophages on NK-tumor interaction is examined using a model developed in Merrill (1981, manuscript).

2. THE MODEL

The model of NK-tumor interaction developed in Merrill (1981, manuscript) involved the populations of "pre-NK" cells (NK_p), mature NK cells (NK_m), tumor (T) and the concentration of interferon (IF). The binding of NK cells to tumors causes the production of interferon which helps recruit pre-NK's to become mature. General functional forms are given for rates involving tumor, necessary due to variations in tumors with respect to growth, NK susceptibility and interferon response. The model is

$$\frac{dNK_p}{dt} = S_1 - k_1 NK_p IF - k_2 NK_p$$

$$\frac{dNK_m}{dt} = k_1 NK_p IF - k_3 NK_m + k_2' NK_p$$

$$\frac{dIF}{dt} = S_2 - k_4 IF + k_5(NK_m,T) - k_1' NK_p IF - k_7(IF,T)$$

$$\frac{dT}{dt} = \gamma(IF,T) - k_6(NK_m,IF,T)$$

(1)

with constants $S_1, S_2, k_1', k_1, k_2', k_2, k_3, k_4 \geq 0$, $k_2 - k_2' > 0$.

The non-negative functions k_5, k_6, k_7 and γ satisfy

$$k_5(0,T) = k_7(0,T) = k_5(\dot{NK}_m,0) = k_7(IF,0) = 0,$$

$$\frac{\partial k_5}{\partial NK_m} > 0, \quad \frac{\partial k_5}{\partial T} > 0, \quad \frac{\partial k_6}{\partial NK_m} > 0, \quad \frac{\partial k_6}{\partial T} > 0, \quad \frac{\partial \gamma}{\partial IF} \le 0,$$

with $\gamma(IF,0) = k_6(0,IF,T) = k_6(NK_m,IF,0) = 0$ and initial condition $(NK_p(0), NK_m(0), IF(0), T(0)) \ge 0$.

Assuming PG reduces the fraction of functioning NK cells, let

$$NK_m = \widetilde{NK}_m + \hat{NK}_m$$

where $\widetilde{NK}_m$ and $\hat{NK}_m$ are the populations of normal and (temporarily) suppressed mature NK cells. Also, as PG is cleared rapidly in the lungs and not stored in tissue, the concentration of PG is nearly proportional to the number of activated macrophages, M.

The model to be examined here is

$$\frac{dNK_p}{dt} = S_1 - k_1 NK_p IF - k_2 NK_p$$

$$\frac{d\widetilde{NK}_m}{dt} = k_1 NK_p IF + k_2' NK_\rho - k_3 \widetilde{NK}_m - k_8 M \widetilde{NK}_m + k_9 \hat{NK}_m$$

$$\frac{d\hat{NK}_m}{dt} = k_8 M \widetilde{NK}_m - k_9 \hat{NK}_m - k_3 \hat{NK}_m$$

$$\frac{dIF}{dt} = S_2 - k_4 IF + k_5(\widetilde{NK}_m,T) - k_1' NK_p IF - k_7(IF,T) \tag{2}$$

$$\frac{dM}{dt} = k_{10}(IF,M,T)$$

$$\frac{dT}{dt} = \gamma(IF,T) - k_6(\widetilde{NK}_m,IF,T) - k_{11}(IF,M,T)$$

where parameter and function description is as (1) with

$$k_8, k_9 \ge 0, \quad k_{11}(IF,0,T) = k_{11}(IF,M,0) = 0,$$

$$\frac{\partial k_{11}}{\partial M} > 0, \quad \frac{\partial k_{11}}{\partial T} > 0 \quad \text{and} \quad \frac{\partial k_{10}}{\partial M}(IF,0,T) \neq 0.$$

We are assuming that the difference between macrophage production and consumption of interferon is negligible. In Merrill (1981, manuscript) (1) was shown to have a unique equilibrium $(x_0,y_0,z_0,0)$ with T coordinate 0. This point was asymptotically stable if

$$\lambda_0 = \frac{\partial \gamma}{\partial T}(z_0,0) - \frac{\partial k_6}{\partial T}(y_0,z_0,0) < 0$$

and unstable if $\lambda_0 > 0$.

In either case, the $T = 0$ hyperplane was contained in the stable manifold. For (2), an equilibrium point, $E = (NK_{P_0}, \widetilde{NK}_{m_0}, \widehat{NK}_{m_0}, IF_0, M_0, 0)$, must satisfy

$$\frac{dNK_{P}}{dt} = 0, \quad \frac{d(\widetilde{NK}_m + \widehat{NK}_m)}{dt} = 0 \quad \text{and} \quad \frac{dIF}{dt} = 0$$

so that $NK_{P_0} = x_0$, $\widetilde{NK}_{m_0} + \widehat{NK}_{m_0} = y_0$ and $IF_0 = z_0$.

Assuming $k_{10}(z_0, M, 0) = 0$ determines M as a function M_0 of z_0, one finds

$$\widehat{NK}_{m_0} = \frac{k_8}{k_9 + k_3} M_0 \widetilde{NK}_{m_0}.$$

This implies $\widetilde{NK}_{m_0} = (\frac{1}{1 + (k_8/k_9 + k_3)M_0}) y_0$.

THEOREM 1 *If* $(\partial k_{10}/\partial M)(E) < 0$ *then the equilibrium point* E *of* (2) *is asymptotically stable if*

$$\chi_0 = (\frac{\partial \gamma}{\partial T} - \frac{\partial k_6}{\partial T} - \frac{\partial k_{11}}{\partial T})\Big|_E < 0$$

and unstable if $\chi_0 > 0$. *Moreover the stable manifold of* E *always has dimension* ≥ 5. *If* $(\partial k_{10}/\partial M)(E) > 0$, E *is unstable.*

Proof. Linearize (2) about E and use the Routh Hurwitz criteria.

Viewing M_0 as a parameter, setting

$$f(M_0, y_0) = \frac{\partial k_6}{\partial T}(E) + \frac{\partial k_{11}}{\partial T}(E),$$

we now look for conditions under which M_0 is small and

$$f(M_0, y_0) < \frac{\partial \gamma}{\partial T}(z_0, 0) \quad (\chi_0 > 0)$$

yet $\lambda_0 < 0$, a case in which NK action alone would eliminate a small tumor but macrophage interference prevents it. To locate these regions, examine where $\lambda_0 = 0$ and $(\partial f/\partial M_0)(0, y_0) < 0$ (addition of any macrophages makes it worse). To illustrate (see example in Merrill, manuscript) let

$$k_6(NK_m, IF, T) = \frac{k_6' NK_m T}{K_1 + K_2 NK_m + T}$$

and
$$k_{11}(IF,M,T) = \frac{k'_{11}MT}{K'_1+K'_2M+T}$$

Then
$$\frac{\partial f}{\partial M_0}(0,y_0) = \frac{-k'_6K_1(k_8/k_3+k_9)y_0}{(K_1+K_2y_0)^2} + \frac{k'_{11}}{K'_1}.$$

Along $\lambda_0 = 0$ curve,

$$\frac{\partial \gamma}{\partial T}(z_0,0) = \frac{k'_6y_0}{K_1+K_2y_0}.$$

Thus $\frac{\partial f}{\partial M_0}(0,y_0) < 0$ along this curve if

$$k'_{11}(K_1+K_2y_0) < K_1K'_1\frac{k_8}{k_9+k_3}\frac{\partial \gamma}{\partial T}(z_0,0).$$

From this example, activated macrophages can be harmful even if they are quite effective in tumor destruction. Macrophage activating agents (BCG, C. Parvum) are used in immunotherapy. Care must be taken to prevent *promotion* of tumor growth.

REFERENCES

Evans, R. (1972): Macrophages in syngeneic animal tumors, *Transplantation* 14: 468-473.

Evans, R., and D.M. Eidlen (1981): Macrophage accumulation in transplanted tumors is not dependent on host immune responsiveness or presence of tumor-associated rejection antigens, *J. Reticuloendothel. Soc.* 30:425-437.

Herberman, R.B. (ed.) (1980): *Natural Cell-Mediated Immunity Against Tumors*, Academic Press, New York.

_____(1981): Natural killer (NK) cells and their possible roles in resistance against disease, *Clin. Immunol. Rev.* 1:1-65.

Meltzer, M.S., R.W. Tucker, and A.C. Breuer (1975): Interaction of BCG-activated macrophages with neoplastic and non-neoplastic cells lines in vitro: Cinemicrographic analysis, *Cell Immunol.* 17:30-42.

Merrill, S. (1981): A model of the role of natural killer cells in immune surveillance, Part I, *J. Math. Biol.* 12:363-373.

_____(1982): A model of the role of natural killer cells in immune surveillance, Part II, *J. Math. Biol.* (to appear).

Naor, D. (1979): Suppressor cells: permitters and promoters of malignancy?, *Adv. Cancer Res.* 29:45-125.

Prehn, R.T. (1976): Do tumors grow because of the immune response of the host?, *Transplant. Rev.* 28:34-42.

Russell, S.W., G.Y. Gillespie, and A.T. McIntosh (1977): Inflammatory cells in solid murine neoplasms III. Cytotoxicity mediated in vitro by macrophages recovered from disaggregated regressing Moloney sarcomas, *J. Immunol.* 118: 1574-1579.

Salmon, S.E., and A.W. Hamberger (1978): Immunoproliferation and cancer: A common macrophage-derived promoter substance, *Lancet* 1:1289-1290.

A MODEL FOR INFECTION

R. Shonkwiler and Maynard Thompson

1. INTRODUCTION

There have been numerous mathematical studies on the spread of epidemics. Many useful and interesting results have been discovered thereby. However, to our knowledge all have treated the process beginning with an individual's exposure to the causative agent and leading to incidence of disease in this individual on a purely empirical basis. Typically these models assume the aggregate number of new cases of illness to be proportional to the present number of cases. This may be completely appropriate in relation to the conclusion of the study but sheds no information on the processes of infection itself. It is known that illness among laboratory animals exposed to disease agents often result in only a fraction of the animals actually incurring the disease. Further it is known that mammals possess at least two defense mechanisms against disease organisms. Thus it is possible that some degree of exposure to the causative agent for a disease need not result in onset of illness. Indeed all organisms are constantly exposed to disease agents with frequently no ill effect.

In order to study in more detail the process of infection leading to illness for an exposed individual, we envision three distinct divisions of the phenomenon. First the causative agent must gain access to its target tissue. Second, the invader attempts to carry out its life processes in spite of the host's defenses. And third, some criteria must be met to warrant a judgement of illness in the host. In this paper we discuss each of these in more detail below and present some results of computer simulation. We observe that the model actually pertains to a wider range of afflictions than invasion by micro-organisms. For example chemical or radio-logical maladies are also covered.

2. PENETRATION

The pathogen must make its way from the external environment to its own particular target tissue. Micro-organisms utilize a variety of techniques for this purpose such as burrowing, ingestion, inhalation, vectoring, etc. Often one of these is followed by drifting within the circulatory system. Chemical agents can diffuse across physiological barriers.

A simple model for this process is to assume that the number of pathogenic agents, called immigrants, which penetrate in a given time Δt is proportional to the number of these agents in the immediate external environment of the host. This penetration rate or transmissivity τ is a parameter assumed to be characteristic of a given type of pathogen and susceptible and possibly the environment.

Of course it is assumed that the required local density of pathogenic agents, which we refer to as the intensity I, is available as input to the model. It may vary as a function of time. The governing equation is

$$\Delta N = \tau I(t)(\Delta t).$$

This scheme would also apply to chemical and radiological agents.

3. SURVIVAL AND GROWTH

In general the host possesses biological defenses to an invasion. It can happen that immigrating pathogens are destroyed immediately. In any case, biological invaders also possess mechanisms for survival and reproduction and so a kind of stochastic battle ensues.

The problem from the point of view of the pathogen is that of colonizing a remote hostile habitat completely similar to the problem of biological colonization of an island. The latter has been extensively studied by Goel and Richter-Dyn (1974).

If the life style of the invader is sufficiently simple in that it can be approximated as binary fission, then the theory of birth and death processes may be used here. This introduces two parameters, the birth rate λ and the death rate μ. One could determine the birth rate in independent experiments. However the death rate is not so easily determined and so is an experimental parameter. It includes natural mortality along with the effects of the hosts immunological response.

If, in addition, one makes the Markov assumption that only the present state affects the future, then the waiting times for the birth and death process are exponential and the equations governing the process can be explicitly written. Let $P_t(X = x)$ be the probability that the random variable X for colony size is x at time t for a process starting from a single individual at time zero. Also let $E(t)$ be the expected colony size at time t.

Then for $\lambda \neq \mu$,

$$E(t) = e^{(\lambda-\mu)t}, \quad P_t(X = 0) = \alpha, \quad P_t(X = x) = (1-\alpha)(1-\beta)\beta^{x-1}$$

where

$$\alpha = \frac{\mu(e^{(\lambda-\mu)t}-1)}{\lambda e^{(\lambda-\mu)t}-\mu}, \quad \beta = \frac{\lambda(e^{(\lambda-\mu)t}-1)}{\lambda e^{(\lambda-\mu)t}-\mu};$$

and if $\lambda = \mu$, $E(t) = 1$,

$$P_t(X = 0) = \alpha = \beta = \frac{1}{1+(1/\lambda t)},$$

$$P_t(X = x) - (1-\alpha)^2\alpha^{x-1}.$$

From these equations it is possible to make some observations about the consequences of our assumptions. Especially from the equation for the expected value, we see that the fate of the colony N is relatively insensitive to either λ or μ separately but is highly sensitive to their difference $\lambda - \mu$. Even starting with small colony sizes, where the expected value is not so important as the extinction probability, this effect is still in evidence.

Interpreting this biologically we can say that an immune system need only be slightly better than the pathogens ability to reproduce in order to be effective. Simulation studies show that parameter values $\lambda = 3.00$ and $\mu = 2.95$ lead to 100% attack rates nearly every time.

From another point of view, a strain of high fecundity need only to be slightly better at reproducing than the ability of an immune system to destroy them in order to inflict illness.

Still another conjecture is that for evolutionary established diseases the two parameters λ and μ are very nearly the same.

From the modelling point of view this means that effectively we are only concerned with one experimental parameter here, the difference $\lambda - \mu$.

Another consequence of these equations is the possibility of classifying diseases as to the importance of transmissivity versus fecundity. Thus a pathogen of high fecundity need only penetrate once in small numbers to result in disease while a non-prolific or even non-reproducing pathogen can rely on long or repeated exposures to cause illness.

All non-biological diseases are included in the latter case.

4. CRITERIA FOR ILLNESS

When should a victim be judged sick?

The mere presence of invading organisms is not a reliable condition. Clearly an exposure to a pathogen for which one has immunity results in immediate extinction of the immigrants and the victim should not be judged ill.

We consider two schemes, one based on colony size and the other based on colony duration. In the former a judgement of illness is made when the colony size of the invading pathogens reaches a predetermined level. For the case of a non-biological affliction, some measure of the dose is used in place of colony size. This criterion has two virtues. If the birth rate is not extremely close to the death rate, then either exponential growth or decay ensues so that a judgement of illness here will be insensitive to the chosen critical level. Biologically this criterion is consistent with a saturation effect that may apply to the susceptibles immune system. Such a phenomenon is well-known in enzyme mediated chemistry.

A second scheme for judging illness consults the time integral of colony size. Should this accumulated time-amount reach a pre-set level illness is assessed. This scheme is consistent with, for example, a toxin producing infection wherein any given colonizer can have an increasingly deleterious effect over time.

5. CONCLUSIONS

It is hoped that the model will explain the process beginning with exposure and culminating with illness. Currently there is a great debate over the possibility of a threshold effect in exposure to chemical or radiological agents and the validity of extrapolating experimental data as to their toxicity. Presumably a valid model for infection could shed light on this subject.

Overall the model involves only two (sensitive) experimental parameters, transmissivity and the difference between birth rate and death rate. Nevertheless, through computer simulations the model appears consistent with attacks ranging from highly virulent bacteria agents to chronic infections to chemical poisoning.

REFERENCES

Goel, N., and Richter-Dyn, N. (1974): *Stochastic Models in Biology*, Academic Press, New York.

Shonkwiler, R., and M. Thompson (1982): Common source epidemics I: A stochastic model, *Bul. Math. Biol.* 44:259-269.

_____ (1982): Common source epidemics II: Toxoplasmosis in Atlanta, *Bul. Math. Bio.* 44:377-398.

A CYCLIC EPIDEMIC MODEL WITH TEMPORARY
IMMUNITY AND VITAL DYNAMICS

P. van den Driessche

ABSTRACT

A cyclic epidemic model with constant length of immunity is constructed for a constant population. Below a certain threshold value the disease dies out, while above this value there is an endemic equilibrium; vital dynamics are shown to have a stabilizing effect on this equilibrium. This threshold condition is modified when the assumption of a constant population is dropped, and death due to the disease is incorporated.

1. INTRODUCTION

Most deterministic models of infectious diseases are formulated by dividing the constant populations into classes, for example into those individuals which are susceptible (S), infectious (I) and removed by immunity (R). Movement from the susceptible to infectious class is assumed to be governed by the mass action law, and individuals recover (leave the infectious class) at a rate proportional to their number. Some diseases confer a period of temporary immunity before a removed individual is again susceptible. In Hethcote et. al. (1981) a cyclic SIRS model with constant temporary immunity is shown to exhibit periodic solutions for certain parameter values. It is conjectured there that vital dynamics (births and deaths) have a stabilizing effect on the endemic equilibrium, and this conjecture is shown here in Section 2 to be true.

In the last few years Anderson and May have developed and studied many models in which the classical assumption of a constant population is dropped. They investigate the way parasitic infections influence their host population; see Anderson (1979) for a review. Such models give interplay betwen epidemiology and population ecology, as well as evolution (see Levin and Pimentel, 1981). In particular, Anderson and May (1979) consider an SIRS model of a disease in mice which incorporates immigration and mortality due to disease; recovered mice lose immunity at a rate proportional to their number. In Section 3 a veriable population SIRS model similar to this but assuming a constant temporary immunity is considered.

Finally, in Section 4, a fairly general transcendental equation which occurs in these and other delay problems is investigated. Attention is focused on the possibility of stability switching which has recently been investigated by Cooke and Grossman (1982).

2. CLOSED POPULATION SIRS MODEL

The notation is the same as Hethcote et. al. (1981) with the susceptible, infective, removed fraction of the population denoted respectively by S, I, R. There is a constant temporary immunity ω, contact rate β in the mass action term, and recovery rate γ; and a constant birth and (equal) death rate b is introduced. Thus the cyclic SIRS model under the above assumptions can be represented schematically by:

$$
\begin{array}{ccccc}
 & \beta SI & \gamma I & & \\
b \rightarrow S & \rightarrow & I \rightarrow & R^{\omega} \rightarrow & S \\
 & \downarrow & \downarrow & \downarrow & \\
 & bS & bI & bR &
\end{array}
$$

The integral equation for $I(t)$ for this model is

$$I(t) = I_0 e^{-(\gamma+b)t} + \int_0^t \beta S(x) I(x) e^{-(\gamma+b)(t-x)} dx; \qquad (1)$$

or equivalently the rate of change of I with time for $t \geq \omega$ is

$$I'(t) = -(\gamma+b)I + \beta SI$$

with $I(0) = I_0$, the initial infective fraction, which is assumed positive. The fraction of the initial removed population that is still removed at time t, denoted by $R_0(t)$, is assumed to be continuous and nonincreasing on $[0,\omega]$ and zero for larger t. The removed fraction is given by

$$R(t) = R_0(t) e^{-bt} + \gamma \int_{-\omega}^0 I(t+u) e^{bu} du. \qquad (2)$$

The total population does not change, so the susceptible fraction is known from

$$S(t) + I(t) + R(t) = 1. \qquad (3)$$

These equations can be combined into an integrodifferential equation for I:

$$I'(t) = -(\gamma+b)I(t) + \beta I(t) \left[1 - I(t) - R_0(t)e^{-bt} - \gamma \int_{-\omega}^0 I(t+u)e^{bu} du \right]. \qquad (4)$$

As discussed in Hethcote et. al. (1981), S, I, R remain between 0 and 1, and this model is epidemiologically and mathematically well posed. The (S,I,R) coordinates of the equilibrium points are $(1,0,0)$ and (S_e, I_e, R_e) where $S_e = 1/\sigma$, $I_e = (1-S_e)(1 + \gamma(1-e^{-\omega b})/b)^{-1}$, and R_e is given by (3). Here the contact number σ is given by $\sigma = \beta/(\gamma+b)$, and for $\sigma \leq 1$ all solutions approach the only equilibrium point $(1,0,0)$ as t approaches infinity; thus the disease dies out. This gives a threshold condition in terms of the epidemiological parameters.

Considering $\sigma > 1$, periodic solutions of (4) for $t \geq \omega$ are equivalent to those of the equation with $R_0(t)$ ignored. Translating the origin using $I = I_e(1 + X)$ and letting $t = \omega\tau$, gives

$$X'(\tau) = -\omega\beta I_e(X(\tau)+1)\left[X(\tau) + \omega\gamma\int_{-1}^{0} X(\tau+v)e^{\omega bv}dv\right]. \tag{5}$$

The linearization of this about $X = 0$ has characteristic equation

$$\Delta = \lambda + \omega\beta I_e\left[1 + \omega\gamma\int_{-1}^{0} e^{(\lambda+\omega b)v}dv\right] = 0, \tag{6}$$

and roots of this govern behaviour about the endemic equilibrium.

For equation (6), $\lambda = 0$ is not a root, and any complex roots occur as conjugate pairs. For purely imaginary roots $\lambda = i\mu$ $(\mu > 0)$

$$-\frac{1}{\gamma\omega} = \frac{\omega b(1-e^{-\omega b}\cos\mu) + \mu e^{-\omega b}\sin\mu}{\omega^2 b^2 + \mu^2}, \tag{7}$$

$$\mu = \frac{\omega\beta\omega\gamma I_e(\mu(1-e^{-\omega b}\cos\mu) - \omega be^{-\omega b}\sin\mu)}{\omega^2 b^2 + \mu^2}. \tag{8}$$

Alternatively writing the characteristic equation as

$$\lambda^2 + (\omega\beta I_e+\omega b)\lambda + \omega\beta\omega bI_e + \omega\beta\omega\gamma I_e(1 - e^{-(\lambda+\omega b)}) = 0, \tag{9}$$

the real and imaginary parts give the relations

$$-\mu^2 + \omega\beta\omega bI_e + \omega\beta\omega\gamma I_e(1 - e^{-\omega b}\cos\mu) = 0, \quad \mu(\omega\beta I_e + \omega b) + \omega\beta\omega\gamma I_e e^{-\omega b}\sin\mu = 0.$$

For $\mu \in (\mu_\ell, \mu_u)$, equation (7) defines $\omega\gamma > 0$, and then (8) given $(\sigma-1) > 0$ from I_e formula. Here $\mu_\ell(\mu_u)$ lies in the 3rd (4th) quadrant and $\to \pi^+(2\pi^-)$ as $b \to 0^+$. A family of imaginary root curves can be found by increasing μ by $2k\pi$. The lowest imaginary root curve is shown in Figure 1 for $\omega b = 0.05, 0.1, 0.3$. As $\mu \to \mu_\ell(\mu_u)$, $\omega\gamma \to \infty$ and $(\sigma-1) \to 0$. This is in contrast to the case with $b = 0$ in which $(\sigma-1) \sim 2\omega\gamma$ as $\mu \to \mu_u = 2\pi^-$. In fact as ωb increases the range of $\sigma - 1$ values inside the imaginary root curve loop becomes small, and the minimum value moves in the direction of increasing $\omega\gamma$ and decreasing $\sigma - 1$. From (6), $Re(\lambda) < 0$ for $\omega\gamma < 1$, and any root with $Re\,\lambda \geq 0$ is bounded by $|\lambda| \leq \omega\beta I_e(1 + \omega\gamma)$, so the equilibrium point (S_e, I_e, R_e) is locally asymptotically stable for parameter values

below the lowest imaginary root curve. Thus the presence of vital dynamics (b > 0)
does have a stabilizing effect on the model, in that the parameter set supporting
periodic solutions is decreased. An estimate of the region of stability in the
$(\sigma-1, \omega\gamma)$ parameter plane can be obtained by applying Rouché's theorem.

THEOREM 1. *If* $\beta^2 I_e^2 - 2\beta\gamma I_e + b^2 > 0$ *then the endemic steady state* (S_e, I_e, R_e) *is
locally asymptotically stable.*

PROOF. Consider (9) with

$$f(\lambda) = \lambda^2 + (\omega\beta I_e + \omega b)\lambda + \omega\beta I_e(\omega\gamma + \omega b), \qquad g(\lambda) = -\omega\beta\omega\gamma I_e e^{-(\lambda + \omega b)}.$$

The inequality above assures that $|f| > |g|$ on $C_1 \cup C_2$ where $C_1 = \{iy : y \in [\eta, -\eta]\}$
where $\eta = \omega\beta I_e(1 + \omega\gamma) + \varepsilon$; $\varepsilon > 0$ and $C_2 = \{z : |z| = \eta, \arg z \in [-\pi/2, \pi/2]$. Note
that η is the upper bound of $|\lambda|$ as stated previously. As f has no zeros in-
side $C_1 \cup C_2$, the function $f + g$ has no zeros with non-negative real parts.
Thus Rouché's theorem and a continuity argument complete the proof.□

The curve $\omega^2\beta^2 I_e^2 - 2\omega\beta\omega\gamma I_e + \omega^2 b^2 = 0$ is found numerically for fixed ωb,
and the region below this curve is the stability region. This analytical estimate,
which approximates a line for large $\omega\gamma$, does not give the extent of the stability
region found numerically (where large $(\sigma-1)$ values give stability for all $\omega\gamma$),
see Figure 1 where the curve for $\omega b = 0.3$ is shown as a broken line.

Numerical solution, by a fourth order method, of the equation (5) confirms
the influence of including vital dynamics in this model. For example with $\sigma - 1 = 4$,
$\omega\gamma = 6$ and $\omega b = 0.3$, this model tends to the constant endemic level $(S_e, I_e, R_e) =$
$(.2, .13, .67)$, but with $b = 0$ periodic oscillations are sustained, see Hethcote
et. al. (1981). A (non-cyclic) SIR model with constant infectious period and
vital dynamics is considered by Grossman (1980), and the non-trivial equilibrium is
shown to be locally stable. A similar result is obtained in Hethcote and Tudor
(1980) for the model when immunization is included. Thus the presence of a period
of temporary immunity and the cyclic nature of the model are again important for
periodicity. In the case with no vital dynamics Hethcote et. al. (1981, Theorem
3.1) prove analytically that for fixed $\sigma > 1$, as $\omega\gamma$ increases and passes the
lowest imaginary root there is a Hopf bifurcation to a locally asymptotically stable
periodic solution. Numerical results give similar periodic solutions in this case
when vital dynamics is included, but the doubling back of the imaginary root curves
could indicate more complicated behaviour.

3. SIRS MODEL WITH VARIABLE TOTAL POPULATION

The model constructed previously is modified to allow for variation in the
total population $N = S + I + R$, where now variables are numbers of individuals in

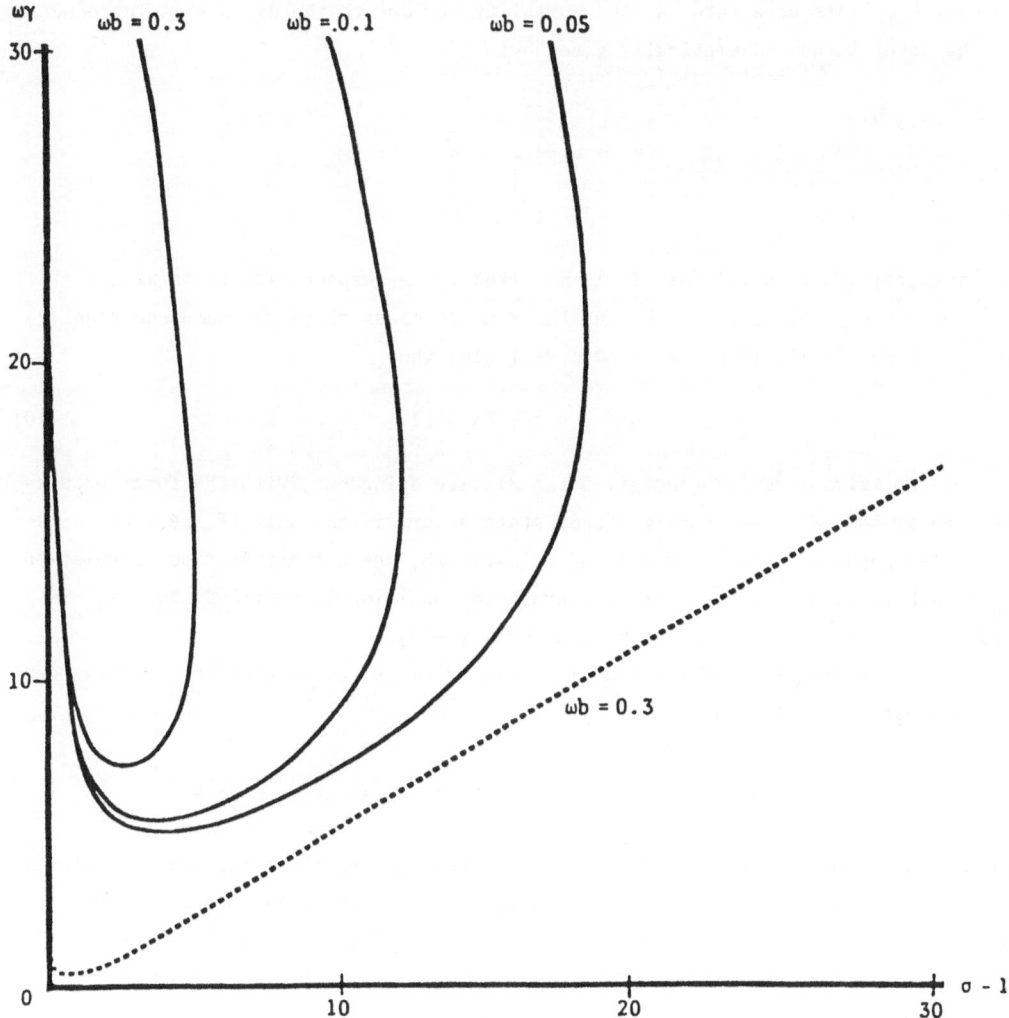

Figure 1 The lowest imaginary root curves are shown as solid curves. For

$\omega b = 0.05$, $\mu \in (1.0103\pi, 1.9999\pi)$

$\omega b = 0.1$, $\mu \in (1.021\pi , 1.9994\pi)$

$\omega b = 0.3$, $\mu \in (1.069\pi , 1.993\pi)$

The broken curve is $\omega^2\beta^2 I_e^2 - 2\omega\beta\omega\gamma I_e + \omega^2 b^2 = 0$ for $\omega b = 0.3$.

each class (rather than fractions as in equation (3)). Anderson and May (1979)
show that infectious diseases can regulate their host populations, and can change
the threshold condition. In their model it is assumed that removed individuals
lose immunity at a rate proportional to their number, thus the formulation is a set
of four ordinary differential equations any three of which are independent, see
Anderson and May (1979, equations (1)-(4)). This assumption is modified here to
allow a constant temporary immunity as included above. Immigration into the

susceptible class at a rate   a   and mortality   c   due to the disease is introduced.
So the model is now schematically given by:

$$a \rightarrow S \xrightarrow{\beta SI} I \xrightarrow{\gamma I} R^{\omega} \rightarrow S$$

$$\downarrow \quad \downarrow \quad \downarrow$$

$$bS \quad (b+c)I \quad bR$$

The integral equation (1) for   I   is modified in the exponential terms with
$\gamma + b + c \equiv \alpha$ replacing $\gamma + b$, and the removed class obeys the same equation (2).
The total population is now a dynamic variable, thus

$$N'(t) = a - bN - cI. \tag{10}$$

The trivial equilibrium gives the disease dying out, with the total population equal to a/b. An endemic steady state occurs if and only if $a\beta > b\alpha$ with
the total population having value $N_e = (a - cI_e)/b$, where the infectious population
$I_e = (a - b\alpha/\beta)(\alpha - \gamma e^{-b\omega})^{-1}$. Thus the threshold condition is modified, and N_e is
reduced by infection, as in Anderson and May (1979).

Linear stability of the endemic equilibrium is now governed by the transcendental equation (c.f. (9)):

$$\lambda^2 + (\omega\beta I_e + \omega b)\lambda + \omega\beta\omega(b+c)I_e + \omega\beta\omega\gamma I_e(1 - e^{-(\lambda + \omega b)}) = 0, \tag{11}$$

with an additional negative root -ωb. This equation exhibits the same qualitative
features as shown in Figure 1 (for $a \equiv b$, $c \equiv 0$), with minima occurring for
larger values of $\omega\gamma$, and the same doubling back behaviour. Thus limit cycle solutions for the model with delay are again possible, but they were not found in the
corresponding ordinary differential equation model, Anderson and May (1979).

4.   TRANSCENDENTAL EQUATION INVESTIGATION

The inclusion of a constant lag in the removed class of the   SIRS   models
leads to a transcendental equation (rather than a polynomial) associated with linearization about the endemic equilibrium. Equations (9) and (11) are of second
order, a review of stability results for such equations is given by Chuma and van
den Driessche (1980); see also the standard text by Bellman and Cooke (1963). Most
results concentrate on regions of stability, but recently Cooke and Grossman have
focused on the possibility of multiple stability - instability switches for second
order transcendental equations. The inclusion of more classes in an epidemic model,
or investigation of a model in which all roots remain coupled (c.f. the model in
Section 3 where the root -ωb uncouples) would increase the order of the transcendental equation.

So consider the general transcendental equation:

$$F(z) \equiv P(z) + Q(z)e^{-\omega z} = 0, \tag{12}$$

where $P, Q \neq 0$ are relatively prime polynomials with real, constant coefficients, degree P > degree Q, and $P(0) + Q(0) \neq 0$. The lag $\omega > 0$ is explicitly in only the exponential term. Consider purely imaginary roots $z = iy$ with $y > 0$ (as complex roots occur as conjugate pairs). Then writing $P_r = \text{Re}(P(iy))$, $P_i = \text{Im}(P(iy))$ and similarly for Q; equating real and imaginary parts gives: $\tan(\omega y) = (-P_iQ_r+P_rQ_i)/(P_rQ_r+P_iQ_i)$ with an obvious interpretation when the denominator is zero. Note also that $|P(iy)| = |Q(iy)|$.

Assuming no roots at infinity, transition from stability to instability or the reverse, corresponds to a purely imaginary root. Such a root is assumed to be simple, and so the direction of crossing the imaginary axis is governed by the sign of the derivative of $\text{Re}(z)$ with respect to the lag ω where $z = iy$ and $F(iy) = 0$. Denote this by:

$$s = \text{sign} \left\{ \frac{\partial(\text{Re } z)}{\partial \omega} \text{ at } z = iy \right\} = \text{sign} \left\{ \text{Im} \left(-\bar{P} \frac{\partial P}{\partial z} + \bar{Q} \frac{\partial Q}{\partial z} \right) \text{ at } z = iy \right\},$$

where $\bar{P}, \bar{Q}$ denotes the conjugate of P, Q, respectively.

For a second degree equation take $P(z) = z^2 + Az + B$, $Q(z) = Cz + D$; then $s = \text{sign } \{A^2-C^2-2B+2y^2\}$. This result is given by Cooke and Grossman (1982, Section 5), who also show that the assumptions above are true for this equation. They give conditions on the parameters so that multiple switching from stability-instability-stability occur.

The equations (9) and (11) have coefficients involving I_e, which depends on the lag ω. To incorporate this, P and Q in (12) depend on ω, but this leads to a complicated expression for s. So the view is taken here that I_e and ωb are fixed, and (11) is written (with $z = \lambda/\omega$) as:

$$z^2 + (\beta I_e + b)z + \beta I_e \alpha - \beta \gamma I_e e^{-\omega b} e^{-\omega z} = 0. \tag{13}$$

In this form results above (see also Cooke and Grossman (1982, Section 4)) are directly applicable, with quadratic P and constant Q, and give $s = \text{sign}\{2y^2 + (\beta I_e + b)^2 - 2\beta I_e \alpha\}$. Purely imaginary roots are given by

$$y^4 + y^2((\beta I_e + b)^2 - 2\beta I_e \alpha) + \beta^2 I_e^2 (\alpha^2 - \gamma^2 e^{-2\omega b}) = 0. \tag{14}$$

When vital dynamics and mortality due to disease are ignored (i.e. $b \equiv 0$, $c \equiv 0$), see Hethcote et. al. (1981), the constant term in (14) is zero, only one positive y^2 is possible, and s is positive. Thus every root which crosses the imaginary axis does so from left to right and there is instability for all ω after the first crossing. Retaining vital dynamics and disease mortality, the model fits into Case 2, Section 4 in Cooke and Grossman (1982). There can be a finite number of stability switches, at the first switch for values on the lowest imaginary root curve, the endemic equilibrium loses stability to a periodic solution. Subsequent switches are not completely understood, but it seems possible that stability could be regained for increased lag in the removed class.

Acknowledgements

It is a pleasure to thank H.W. Hethcote and H.W. Stech for their help, and Joe Chuma for doing the numerical calculations. Financial support was provided by NSERC A-4645.

REFERENCES

Anderson, R.M. (1979): The persistence of direct life cycle infectious disease within populations of hosts, in *Some Mathematical Questions in Biology*, (S.A. Levin, Ed.), Amer. Math. Soc. , pp. 1-68.

Anderson, R.M., and R.M. May (1979): Population biology of infectious diseases, I, *Nature* 280:361-367.

Bellman, R., and R.L. Cooke (1963): *Differential-Difference Equations*, Academic Press, New York.

Chuma, J., and P. van den Driessche (1980): A general second-order transcendental equation, *Appl. Math. Notes* 5:85-96.

Cooke, K.L., and Z. Grossman (1982): Discrete delay, distributed delay and stability switching, *J. Math. Anal. Appl.* 86:592-627.

Grossman, Z. (1980): Oscillatory phenomena in a model of infectious diseases, *Theor. Pop. Biol.* 18:204-243.

Hethcote, H.W., H.W. Stech, and P. van den Driessche (1981): Nonlinear oscillations in epidemic models, *SIAM J. App. Math.* 40:1-9.

Hethcote, H.W., and D.W. Tudor (1980): Integral equation models for endemic infectious diseases, *J. Math. Biology* 9:37-47.

Levin, S.A., and D. Pimentel (1981): Selection of intermediate rates of increase in parasite-host systems, *Amer. Nat.* 11:308-315.

Bio-mathematics

Managing Editor: S. A. Levin

Volume 8
A. T. Winfree

The Geometry of Biological Time

1979. 290 figures. XIV, 530 pages
ISBN 3-540-09373-7

The widespread appearance of periodic patterns in nature reveals that many living organisms are communities of biological clocks. This landmark text investigates, and explains in mathematical terms, periodic processes in living systems and in their non-living analogues. Its lively presentation (including many drawings), timely perspective and unique bibliography will make it rewarding reading for students and researchers in many disciplines.

Volume 9
W. J. Ewens

Mathematical Population Genetics

1979. 4 figures, 17 tables. XII, 325 pages
ISBN 3-540-09577-2

This graduate level monograph considers the mathematical theory of population genetics, emphasizing aspects relevant to evolutionary studies. It contains a definitive and comprehensive discussion of relevant areas with references to the essential literature. The sound presentation and excellent exposition make this book a standard for population geneticists interested in the mathematical foundations of their subject as well as for mathematicians involved with genetic evolutionary processes.

Volume 10
A. Okubo

Diffusion and Ecological Problems: Mathematical Models

1980. 114 figures, 6 tables. XIII, 254 pages
ISBN 3-540-09620-5

This is the first comprehensive book on mathematical models of diffusion in an ecological context. Directed towards applied mathematicians, physicists and biologists, it gives a sound, biologically oriented treatment of the mathematics and physics of diffusion.

Springer-Verlag
Berlin
Heidelberg
New York

Journal of Mathematical Biology

ISSN 0303-6812   Title No. 285

Editorial Board:
H.T.Banks, Providence, RI; **H.J.Bremermann,** Berkeley,
CA; **J.D.Cowan,** Chicago, IL; **J.Gani,** Lexington, KY;
K.P.Hadeler (Managing Editor), Tübingen;
F.C.Hoppensteadt, Salt Lake City, UT; **S.A.Levin**
(Managing Editor), Ithaca, NY; **D.Ludwig,** Vancouver;
L.A.Segel, Rehovot; **D.Varjú,** Tübingen in cooperation
with a distinguished advisory board.

The **Journal of Mathematical Biology** publishes papers in
which mathematics leads to a better understanding of biological phenomena, mathematical papers inspired by biological research and papers which yield new experimental data
bearing on mathematical models. The scope is broad, both
mathematically and biologically and extends to relevant
interfaces with medicine, chemistry, physics, and sociology.
The editors aim to reach an audience of both mathematicians
and biologists.

Contents:

Springer-Verlag
Berlin
Heidelberg
New York

Subscription information and sample copy upon request

Lecture Notes in Biomathematics

QH 352 .P578 1983

Population biology